Qualitative Research Methods in Human Geography

Fourth Edition

Qualitative Research Methods in Human Geography

Edited by
Iain Hay

OXFORD UNIVERSITY PRESS

Oxford University Press is a department of the University of Oxford.
It furthers the University's objective of excellence in research, scholarship,
and education by publishing worldwide. Oxford is a registered trade mark of
Oxford University Press in the UK and in certain other countries.

Published in Canada by
Oxford University Press
8 Sampson Mews, Suite 204,
Don Mills, Ontario M3C 0H5 Canada

www.oupcanada.com

First and second editions published by Oxford University Press, 253 Normanby Road,
South Melbourne, Victoria, 3205 Australia. Copyright © 2000, 2005 Iain Hay.
Third Edition published by Oxford University Press Canada in 2010.

Library and Archives Canada Cataloguing in Publication

Qualitative research methods in human geography / edited by Iain
Hay.—Fourth edition.

Includes bibliographical references and index.
ISBN 978-0-19-901090-5 (paperback)

1. Human geography—Methodology—Textbooks. 2. Qualitative
research—Textbooks. I. Hay, Iain, 1960–, editor

GF21.Q83 2016 304.2072 C2015-907565-3

Cover image: Jasper James/Getty Images

Oxford University Press is committed to our environment.
This book is printed on Forest Stewardship Council® certified paper.

Printed and bound in Canada

5 6 7 — 22 21 20

Contents

Boxes and Figures

Boxes

Figures

Notes on Contributors

Jamie Baxter, BA (Hons) (Queen's) 1989, MA (McMaster) 1992, PhD (McMaster) 1997, is an Associate Professor of Geography at the University of Calgary and Graduate Chair at the University of Western Ontario. He has also worked as an associate professor of geography at the University of Calgary. His research interests involve environmental risks from hazards, community responses to technological hazards, geography of health, noxious facility siting, and methodology. Most of his projects involve mixed-method case studies of risk perception in communities living with a technological hazard (e.g., waste, urban pesticides). His current projects involve renewable energy facilities—specifically community perceptions of waste-to-energy and wind turbine developments. Jamie has published this work in a wide array of journals, including *Transactions of the Institute of British Geographers*, *Journal of Risk Research*, *Risk Analysis*, *Environmental Planning and Management*, *Environment and Planning A & C*, and *Social Science and Medicine*.

Richard Bedford, BA (Auckland) 1965, MA (Auckland) 1967, PhD (Australian National University) 1972, is Professor of Migration Studies at Auckland University of Technology, Emeritus Professor at the University of Waikato, and President of the Royal Society of New Zealand. Prior to taking up the chair of geography at the University of Waikato in 1989, he was on the staff of the University of Canterbury. His research interests are mainly in the field of population dynamics, especially migration, in the Asia–Pacific region. He is a Fellow of the Royal Society of New Zealand and in 2008 was made a Companion of the Queen's Service Order in recognition of his services to geography. He has published widely on population movement and development in the Pacific Islands, including a book co-authored with Harold Brookfield, Tim Bayliss-Smith, and Marc Latham in the Cambridge Human Geography series on the colonial and post-colonial experience of eastern Fiji (1989). His articles have appeared in journals such as *Asia-Pacific Viewpoint*, *Asian and Pacific Migration Journal*, *Contemporary Pacific*, *Espaces Populations Sociétés*, *GeoJournal*, *Journal de la Société des Oceanistes*, *Journal of International Migration and Integration*, *International Migration*, *New Zealand Geographer*, *New Zealand Population Review*, and *Oceania*.

Lawrence D. Berg, BA (Dist.) (Victoria) 1988, MA (Victoria) 1991, DPhil (Waikato) 1996, began his career as a lecturer in the School of Global Studies at Massey University, Aotearoa/New Zealand. Lawrence now works in Canada at the University of British Columbia where he has been a Canada Research Chair and is now Professor of Human Geography and Co-director of the UBC Centre for Social, Spatial and Economic Justice. He has a diverse range of academic interests in radical geography, focusing especially on issues relating to analyses of identity politics and place and the emplaced cultural politics of knowledge production. Lawrence has been editor of both *SITES* and *The Canadian Geographer*, and he is one of the founding editors of *ACME: An International E-Journal for Critical Geographies*. He remains on the Editorial Collective of ACME, and he is active in the open-access movement. Lawrence's current research involves a wide range of participatory action research projects aimed at contesting and disrupting various taken-for-granted forms of everyday social relations and practices that contribute to oppression in a range of spaces.

Matt Bradshaw, BA (Hons) (Tasmania) 1989, MEnvStudies (Tasmania) 1992, PhD (Tasmania) 2000, is Principal Fisheries Management Officer, Commercial Fisheries Branch, Wild Fisheries Management Section, Tasmanian Government. His current academic interests include economic geography, the social impact assessment of fisheries, and community involvement in local government planning. He has published in *Antipode*, *Applied Geography*, *Area*, *Australian Geographical Studies*, *Environment and Planning A*, *Fisheries Research*, and *New Zealand Geographer*, among others.

Jenny Cameron, DipTeach (Brisbane College of Advanced Education) 1983, BAppSc (Queensland University of Technology) 1990, MA (Sydney) 1992, PhD (Monash) 1998, is Associate Professor in the Discipline of Geography and Environmental Studies at the University of Newcastle. Jenny has conducted focus groups and facilitated workshops as part of her participatory action research working with economically marginalized groups to develop community economic projects. Her most recent work has been with grassroots community enterprises, particularly documenting the issues faced by these enterprises. This research features in her recent co-authored book with J.K. Gibson-Graham and Stephen Healy, *Take Back the Economy: An Ethical Guide for Transforming Our Communities* (University of Minnesota Press, 2013). She has published on this research in journals that include *Antipode* and *Local Environment*. She is also committed to communicating research outcomes to a general audience and has produced a documentary on asset-based community economic development and several resource kits.

Meghan Cope, AB (Vassar) 1989, MA (Colorado) 1992, PhD (Colorado) 1995, is Professor and Chair of the Department of Geography at the University of Vermont. Meghan's research focuses on young people's mobility and access to public space in Vermont towns in the context of a larger interest in young people, place, and identity across Europe and North America. Her work has always blended qualitative research with a sustained interest in "the everyday." Meghan's experiences have also led her to develop work on blending qualitative research with geographic information systems (GIS), and she co-edited a book with Sarah Elwood on this topic in 2009 (*Qualitative GIS*, Sage). She has published more than 20 articles and chapters on methods, children's geographies, urban post-industrial decline, feminist geography, and participatory research and pedagogy.

James Craine, BA (Ohio University) 1975, MA (California State University, Northridge) 2000, PhD (San Diego State University/UC Santa Barbara) 2006, is Professor of Geography at California State University, Northridge, where he teaches cultural and human geography and cartographic design. His most recent books are *A Companion to Media Geography: Place/Space/Media* (Ashgate, with Paul Adams and Jason Dittmer, 2014) and *The Fight To Stay Put: Social Lessons through Media Imaginings of Urban Transformation and Change* (Franz Steiner Verlag, with Giorgio Hadi Curti and Stuart Aitken, 2013). He is also a co-editor of *Aether: The Journal of Media Geography* and has published extensively on various topics related to media geography, affectivity, and geovisualization.

Dydia DeLyser, BA (California, Los Angeles) 1992, MA (Syracuse) 1996, PhD (Syracuse) 1998, is Associate Professor of Geography at Louisiana State University. Her qualitative research—largely on tourism, landscape, and social memory in the American West—has been both ethnographic and archival. She teaches cultural geography, qualitative methods, and writing at the graduate and undergraduate levels. Dydia has published more than a dozen articles and chapters on these topics in such journals as the *Annals of the Association of American Geographers*, *Area*, *Journal of Geography in Higher Education*, *Journal of Historical Geography*, *cultural geographies*, and *Social and Cultural Geography* as well as a book, *Ramona Memories: Tourism and the Shaping of Southern California* (University of Minnesota Press, 2005). She serves as North American editor of *cultural geographies*.

Robyn Dowling, BEc (Hons) (Sydney) 1988, MA (British Columbia) 1991, PhD (British Columbia) 1995, is Professor in the Department of Geography and Planning at Macquarie University, NSW, Australia. Her broad research interests lie at the intersection of cultural and urban geography. Past research explored the material and cultural geographies of home in Australian suburbia and gendered aspects of transport and mobility, while her current research focuses on the changing governance and social patterns of suburban Australia. Robyn's publications include papers in *Antipode*, *Environment and Planning D*, *Australian Geographical Studies*, *Housing Studies*, *Social and Cultural Geography*, and *Urban Geography* as well as a book with Alison Blunt entitled *Home* (Routledge, 2006).

Kevin Dunn, BA (Hons) (Wollongong) 1990, PhD (Newcastle) 1999, is Professor in Human Geography and Urban Studies at the University of Western Sydney. Like other contributors to this volume, Kevin has a broad range of academic interests, including geographies of racism, transnationalism and migrant settlement, identity and place, media representations of place and people, and the politics of heritage and memorial landscapes. His PhD focused on opposition to mosques in Sydney. He co-authored *Introducing Human Geography* (Longman-Pearson Education Australia, 2000), which won the Australian Award for Excellence in Publishing, and *Landscapes* (Pearson, 2003) and has published more than 80 chapters and articles in various books and in journals including *Australian Geographer, Australian Geographical Studies, Environment and Planning A* and *D, Social and Cultural Geography*, and *Urban Studies.*

Colin Gardner, BA (Hons) (Cambridge) 1975, MA (UCLA) 1977, MA (Cambridge) 1979, PhD (UCLA) 1998, is Professor of Critical Theory and Integrative Studies at the University of California, Santa Barbara, where he teaches in the departments of Art, Film & Media Studies, Comparative Literature and the History of Art and Architecture. Since 2011 adjunct faculty in Geography at San Diego State University. His most recent book is *Beckett, Deleuze and the Televisual Event: Peephole Art* (Palgrave Macmillan, 2012), a critical study of Samuel Beckett's experimental work for film and television and its connection to Gilles Deleuze's ontology of the image in *Cinema 1* and *Cinema 2.* Gardner is also the author of critical essays on Tomas Gutierrez Alea's seminal Cuban film, *Memories of Underdevelopment* in James Craine, Giorgio Curti, and Stuart Aitken's collection *The Fight To Stay Put: Social Lessons Through Media Imaginings of Urban Transformation and Change* (Franz Steiner Verlag, 2013), as well as an essay on Stan Douglas's video work for the 2014 Ashgate collection, *A Companion to Media Geography: Place/Space/ Media*, edited by Paul Adams, James Craine, and Jason Dittmer.

Karen George, BA (Hons) (Adelaide) 1984, MA (Australian National University) 1988, PhD (Adelaide) 1994, is a historian and writer. She has operated *Historically Speaking* since 1993. She was oral historian for the City of Adelaide (1993–2001). She teaches oral history workshops for Oral History Australia SA/NT. She is author of *A Place of Their Own: The Men and Women of War Service Land Settlement at Loxton after World War II* (Wakefield Press, Adelaide, 1999) and *Finding Your Own Way: A Guide to Records of Children's Homes in SA* (Nunkuwarrin Yunti of SA, 2005) and contributor to Doreen Mellor and Ann Haebich, eds, *Many Voices: Reflections on Experiences of Indigenous Child Separation* (National Library of Australia, 2002). She was Research Historian for the Children in State Care Commission of Inquiry in South Australia (2005–8) and State-based Historian for the Find & Connect web resource project, www.findandconnect.gov.au (2011–2014). Karen conducts interviews for the National and State Libraries and other organizations. She uses oral history in exhibitions, multimedia, websites, and books.

Iain Hay, BSc (Hons) (Canterbury) 1982, MA (Massey) 1985, PhD (Washington) 1989, GradCertTertEd (Flinders) 1995, MEdMgmt (Flinders) 2004, LittD (Canterbury) 2009, is Matthew Flinders Distinguished Professor of Human Geography in the School of Environment at Flinders University, South Australia. His principal research interests revolve around geographies of domination and oppression. He is the author or editor of ten books including *Geographies of the Super-Rich* (Elgar 2013). He is currently Editor-in-Chief of *Geographical Research* and Vice-President of the International Geographical Union. He is a former President of the Institute of Australian Geographers. In 2006, Iain received the Prime Minister's Award for Australian University Teacher of the Year and in 2014 was admitted as a Fellow of the UK's Academy of Social Sciences.

Richard Howitt, BA (Hons) Dip Ed (Newcastle) 1978, PhD (New South Wales) 1986, is Professor of Human Geography and Director of the Macquarie-Ryde Futures Partnership at Macquarie University, NSW, Australia. He is author of *Rethinking Resource Management* (Routledge, 2001) and co-editor of *Resources, Nations and Indigenous Peoples* (Oxford University Press, 1996, with J. Connell and P. Hirsch). Richie works on issues of Indigenous rights, local governance, resource management, social impact assessment, social justice, and geographical education. He received the Australian Award for University Teaching (Social Science) in 1999 and is a Distinguished Fellow of the Institute of Australian Geographers.

Jay T. Johnson, BA (Kansas) 1986, MSW (Kansas) 1991, PhD (Hawai'i) 2003, is Associate Professor in Geography and Indigenous Studies as well as Director of the Indigenous Geographies Research Center at the University of Kansas, USA. He is the co-editor of *A Deeper Sense of Place: Stories and Journeys of Collaboration in Indigenous Research* (Oregon State University Press, 2013) with Soren Larsen. His research interests concern the broad area of Indigenous peoples' cultural survival with specific regard to the areas of resource management, political activism at the national and international levels, and the philosophies and politics of place, which underpin the drive for cultural survival. Much of his work is comparative in nature but focuses predominantly on New Zealand, the Pacific, and North America. Jay received the Association of American Geographers Enhancing Diversity Award in 2014 and an American Council of Learned Societies Collaborative Fellowship along with Soren Larsen for their co-authored book project, *being-together-in-place*.

Robin Kearns, BA (Auckland) 1981, MA (Hons) (Auckland) 1983, PhD (McMaster) 1988, is Professor in the School of Environment at the University of Auckland. He has published *Putting Health into Place: Landscape, Identity and Well-Being* (Syracuse University Press, 1998) and *Culture/Place/Health* (Routledge, 2002) with

Wilbert Gesler, as well as many refereed articles and book chapters. His latest book is *The Alterlives of the Psychiatry Asylum* (Ashgate, 2015), with Graham Moon and Alun Joseph. His current research interests include neighbourhood design and physical activity, geographies of mental health, aging, and place, and the social dynamics of coastal settlements. Robin is an editor of *Health and Place* (Elsevier). He maintains a range of involvements within the wider research community.

Sara Kindon, BA (Hons) (Durham) 1989, MA (Waterloo) 1993, DPhil (Waikato) 2012, is Associate Professor in Human Geography and Development Studies at Victoria University of Wellington, New Zealand. Sara engages feminist praxis, participatory research, and community development, particularly within Indigenous and refugee-background communities. Her thesis work thought through aspects of complicity, power, and desire in a participatory video research project with a North Island *iwi* and received the New Zealand Geographical Society Award for Best Doctoral Thesis in 2013. She also co-ordinates participatory research with refugee-background youth in Wellington and leads a network to support refugee-background students at Victoria. She published *Participatory Action Research Approaches and Methods: Connecting People, Participation and Place* (Routledge, 2007 with Rachel Pain and Mike Kesby) and has published in a number of journals and books. In 2010, she was awarded a national Ako Aotearoa Tertiary Teaching Excellence Award.

Clare Madge, Bsc (Hons) (Birmingham) 1986, PhD (Birmingham) 1991, is a Reader in the Department of Geography at the University of Leicester, UK. Her research interests are eclectic, ranging through postcolonial and feminist sensibilities, social media and Internet-mediated research, and creative forms of world-writing, particularly poetry. Clare has published widely in international journals, within and beyond geography. One day she will get time to write the book that has been fermenting inside for some time.

Pauline M. McGuirk, BA (Hons) (Dublin) 1986, Dip Ed (Dublin) 1987, PhD (Dublin) 1992, is Professor of Human Geography at the University of Newcastle, NSW, Australia, where she is Director of the University's Centre for Urban and Regional Studies. Pauline's research focuses on the interaction of structure, politics, and practice in reconfiguring urban governance, urban policy, and urban space. She has published widely in leading international journals, with more than 50 refereed publications. Pauline is co-editor of *Progress in Human Geography* (2015–) and on the editorial boards of *Territory, Politics, Governance*; *Irish Geography*, *Geographical Research*; and *Geographical Compass*. She was awarded Fellowship of the Geographical Society of New South Wales in 2008 and was elected Fellow of the Institute of Australian Geographers in 2009.

Juliana Mansvelt, BA (Hons) (Massey) 1989, PhD (Sheffield) 1994, is an Associate Professor in Geography in the School of People, Environment and Planning, Massey University, New Zealand. A social geographer, her teaching and research interests lie in the geographies of everyday life with emphasis on the socialities, spatialities, and subjectivities surrounding ageing and consumption. She is author of *Geographies of Consumption* (Sage, 2005) and editor of *Green Consumerism: An A-Z Guide* (Sage, 2011) and has contributed to a number of books and journals on topics related to learning and teaching, qualitative research, ageing and consumption. In 2006, Juliana was awarded a New Zealand National Tertiary Teaching Award for sustained excellence and in 2013 she received the New Zealand Geographical Society's Distinguished Service Award for distinguished service to geography.

Janice Monk, BA (Hons) (Sydney) 1958, MA (Illinois at Urbana-Champaign) 1963, PhD (Illinois at Urbana-Champaign) 1972, is Professor in the School of Geography and Development, at the University of Arizona. For more than two decades, she served as executive director of the Southwest Institute for Research on Women at the University of Arizona where she was responsible for funded research and educational projects supported by an array of governmental agencies and private foundations. She was president of the Association of American Geographers (2001–2) and has been honoured by the Association of American Geographers, the Institute of Australian Geographers, the National Council for Geographical Education, and the Royal Geographical Society (with the Institute of British Geographers). Janice's interests include gender studies, the history of women in American geography, and geography in higher education. Among her numerous publications are *The Desert Is No Lady* (Yale University Press, 1987, 1989; University of Arizona Press, 1997, with V. Norwood), *Full Circles* (Routledge, 1993, with C. Katz), *Women of the European Union* (Routledge, 1996, with M.D. Garcia-Ramon), *Encompassing Gender* (Feminist Press, 2002, with M.M. Lay and D.S. Rosenfelt), and *Aspiring Academics* (Pearson Education, 2009, with Michael Solem and Kenneth Foote). Her articles appear in such journals as *Annals of the Association of American Geographers*, *Journal of Geography in Higher Education*, *Journal of Geography*, and *Professional Geographer*.

Phillip O'Neill, BA (Hons) Dip Ed (Macquarie) 1975, MA (Hons) (Macquarie) 1986, PhD (Macquarie) 1995, is a Professorial Research Fellow at the University of Western Sydney. Phillip is a member of the international editorial boards of the *Journal of Economic Geography*, *Environment and Planning A*, *Human Geography*, and *Geography Compass*. He is a regular newspaper columnist and a frequent national media commentator and public speaker on economic change and regional development issues. Phillip is widely published in the fields of industrial, urban, and regional economic change and has a keen interest in the financing and operation of infrastructure in urban areas.

Eric Pawson, MA, DPhil (Oxford) 1975, is Professor of Geography at the University of Canterbury, New Zealand. He teaches at all levels in the curriculum, from large first-year classes to PhD students, and has a particular interest in community-based learning, developing this as one form of social response to the Christchurch earthquakes. His research focuses on environmental history, rural landscapes, and rural futures. He is the co-author of *Seeds of Empire: The Environmental Transformation of New Zealand* (IB Tauris, 2011) and co-editor of *Making A New Land: Environmental Histories of New Zealand* (Otago University Press, 2013) and *Active Learning and Student Engagement* (Routledge 2010). He is a holder of the Distinguished New Zealand Geographer Medal, received a National Tertiary Teaching Excellence Award in 2009, and the University of Canterbury Teaching Medal in 2013.

Michael Roche, MA, PhD (Canterbury) 1983, is Professor of Geography in the School of People, Environment and Planning at Massey University at Palmerston North in New Zealand. His publications based on archival research largely on historical and contemporary aspects of forestry and agriculture have appeared in *Historical Geography*, *Journal of Historical Geography*, *New Zealand Forestry*, *New Zealand Geographer*, *South Australian Geographical Journal*, the *Historical Atlas of New Zealand*, and "Te Ara the online encyclopedia of New Zealand." He is a Life Fellow of the New Zealand Geographical Society, currently a member of the New Zealand Geographic Board Ngā Pou Taunaha o Aotearoa, and has previously served as an editor of the *New Zealand Geographer*. He is a holder of the Distinguished New Zealand Geographer Medal (2010).

Matthew W. Rofe, BEd (Avondale College) 1993, BA (Hons) (Newcastle) 1996, PhD (Newcastle) 2002, is Senior Lecturer at the School of Natural and Built Environments (Urban and Regional Planning) at the University of South Australia. Matthew has worked at a number of Australian universities, including the University of Newcastle, Sydney, and Adelaide. The central theme of his research agenda involves unravelling the complexity of human landscapes and the often conflicting meanings that are attached to place. Currently, Matthew is researching city marketing and media constructions of place in Australia and the pursuit of world heritage status and tourism in Asia.

Stan Stevens, BA (Hons) (California, Berkeley) 1983, MA (California, Berkeley) 1986, PhD (California, Berkeley) 1989, is Senior Lecturer in Geography at the University of Massachusetts, Amherst. His research in cultural and political ecology has involved more than three decades of collaborative ethnographic fieldwork with Sharwa (Sherpa) communities, organizations, and co-researchers in the Khumbu (Mount Everest) region of Nepal. He is also active in promoting rights-based conservation through the International Union for Conservation of Nature (IUCN) and the ICCA Consortium. Stan is author of *Claiming the High Ground: Sherpas, Subsistence, and Environmental Change in the Highest Himalaya* (University of California Press, 1993) and editor and contributor to *Indigenous Peoples, National Parks, and Protected Areas: a New Paradigm Linking Conservation, Culture, and Rights* (University of Arizona Press, 2014) and *Conservation through Cultural Survival: Indigenous Peoples and Protected Areas* (Island Press, 1997).

Elaine Stratford, BA (Flinders) 1984, BA (Hons) (Flinders) 1986, PhD (Adelaide) 1996, is Associate Professor of Geography, University of Tasmania. Her research centres on various theoretical and empirical place-based, political, and ecological puzzles related to environmental planning, community, cultural expression, and governance in island places. Elaine teaches undergraduate human geography and graduate-level environmental planning. She has published widely in leading journals such as *Transactions of the Institute of British Geographers*, *Geoforum*, *Environment and Planning A* and *B*, *Health and Place*, and *Political Geography* as well as in numerous edited collections.

Gordon Waitt, MA (Hons Soc Sci) (Edinburgh) 1985, PhD (Edinburgh) 1988, is Professor of Human Geography, Department of Geography and Sustainable Communities, University of Wollongong, NSW, Australia. He has published widely on the geographies of tourism, sexuality, gender, sports, festivals, and sustainability. He co-authored *Introducing Human Geography* (Longman-Pearson Education Australia, 2000), *Gay Tourism: Culture and Context* (Howarth Publications, 2006) and *Household Sustainability: Challenges and Dilemmas in Everyday Life* (Edward Arnold, 2013). He has also written numerous articles in various journals, including the *Annals of the Association of American Geographers*; *Annals of Tourism Research*; *Australian Geographer*; *Environment and Planning D*; *Gender, Place and Culture*; *Social & Cultural Geography*; *Transactions of the Institute of British Geographers*; and *Urban Studies*. Gordon is actively involved in the professional activities of the Geographical Society of New South Wales.

Hilary P.M. Winchester, BA (Hons) (Oxford) 1974, MA (Oxford) 1977, DPhil (Oxford) 1980, is Provost at Central Queensland University, Australia. She has previously held roles as Pro Vice Chancellor and Vice President: Organisational Strategy and Planning at the University of South Australia and Pro Vice Chancellor (Academic) at Flinders University. Before that she worked for a decade at the University of Newcastle, NSW, during which time she held various positions including head of the Department of Geography and Environmental Science and president of the Academic Senate. Hilary has a wide range of interests in population, social, and cultural geography, specifically including marginal social groups (especially one-parent families), the social construction of place, the geography of gender, and urban social and landscape planning. She is co-author of *Landscapes: Ways of Imagining the World* (Pearson, 2003) and author of more than 100 articles and book chapters. She is a Fellow of the Institute of Australian Geographers and a former editor of Australian Geographical Studies.

Jamie Winders, BA (Kentucky) 1998, MA (British Columbia) 2000, PhD (Kentucky) 2004, is Associate Professor of Geography at Syracuse University. Her research examines immigration, racial politics, and social belonging. She is the author of *Nashville in the New Millennium: Immigrant Settlement, Urban Transformation, and Social Belonging* (Russell Sage, 2013) and co-editor of *The Wiley-Blackwell Companion to Cultural Geography* (Wiley-Blackwell, 2013). Jamie is completing a new introduction to cultural geography, in which new/social media play a significant role. She teaches human, cultural, and urban geography, as well as courses on race, migration, qualitative methods, postcolonial and feminist theories, and public policy. She is associate editor of *International Migration Review* and editorial board member for several geography journals. She has published in outlets including *Annals of the Association of American Geographers*, *Transactions of the Institute of British Geographers*, and *Antipode* and has an ongoing collaboration with Barbara Ellen Smith on social reproduction.

Preface

Qualitative Research Methods in Human Geography provides a succinctly written, comprehensive, and accessible guide to thinking about and practising qualitative research methods. On the basis of reviewer suggestions and changes in the field, this fourth edition has been revised and expanded to include three new chapters. The first of these examines feminist and Indigenous approaches; the second, visual methods; and the third, new media in qualitative research. The book is aimed primarily at an audience of upper-level undergraduate students, but feedback on ways in which the previous editions have been used indicates that the volume is of considerable interest to people commencing honours theses and post-graduate study, and indeed, some professional geographers with many decades of research experience have let me know—quietly in some instances—that they have learned a great deal from the previous editions. Chapters have been written with the dual intent of providing novice researchers with clear ideas on how they might go about conducting their own qualitative research thoroughly and successfully and offering university academics a teaching-and-learning framework around which additional materials and exercises on research methods can be developed. The text maintains its dedication to the provision of practical guidance on methods of qualitative research in geography.

Without realizing it at the time, I started work on this book in 1992 when I was asked to teach Research Methods in Geography at Flinders University in South Australia. I developed lectures and extensive sets of notes for my classes. I also referred students to helpful texts of the time, such as Bernard (1988), Kellehear (1993), Patton (1990), Sarantakos (1993), and Sayer (1992). All of these books came from disciplines other than geography or referred to the social sciences in general. It disturbed me that despite the renewed emphasis on qualitative research and the teaching of qualitative methods in the discipline, no geographer had produced an accessible text on the day-to-day practice of qualitative research.

In the meantime, during their regular visits to my office, publishers' representatives asked what sorts of books might be useful in my teaching. I repeatedly mentioned the need for a good text that dealt with qualitative research methods in geography. That message came back to haunt me when I was asked whether I would like to write a book on qualitative methods for Oxford University Press. I declined but offered instead to try to draw together the expertise and energies of a group of active and exciting geographers from Australia and Aotearoa/New Zealand to produce an edited collection.

Of course, during the period when I planned the first edition of this book and fulfilled the terms of the publishing contract, a new group of research methods texts emerged! This group included Flowerdew and Martin (1997), Kitchin and Tate (2000), Lindsay (1997), and Robinson (1998). Despite their various merits, none of these books deals exclusively, or as comprehensively, with qualitative methods as the first edition of this volume. While a small number of other useful volumes on research methods in geography have now emerged (e.g., Clifford and Valentine 2009; DeLyser et al. 2009; Limb and Dwyer 2001; Moss 2002), *Qualitative Research Methods in Human Geography* continues to be received very well, not only in Australia and Aotearoa/New Zealand but also much farther afield. Indeed, the book's growing popularity in North America led to its shift from Oxford University Press, Australia, to Oxford University Press, Canada, when the third edition was produced.

This fourth edition is revised and expanded to respond to constructive comments made about its predecessors. For instance, across the book the well-received emphasis on reflexivity and ethical practice has been maintained, and several chapters have been modified substantially in recognition of the ways in which the Internet and other electronic technologies are affecting the conduct of qualitative research. The book also includes review exercises associated with each chapter to help readers develop key skills and knowledge as well as a brief new appendix providing an example of fieldnote practice.

In overview, *Qualitative Research Methods in Human Geography* is subdivided loosely into three main sections: "introducing" qualitative research in human geography, "doing" qualitative research, and "interpreting and communicating" qualitative research. Issues considered important within these broad sections are also raised, where appropriate, within chapters: matters of ethical practice, data reduction, and communicating research results are among the more prominent. Continuing a shift that began with the second edition, this edition includes more authors from beyond Australia and Aotearoa/New Zealand, welcoming additional contributions now from the United States and Canada as well as from the United Kingdom. With this shift has come a subtle change in the use of examples. Where earlier editions of the book drew most heavily on examples drawn from the Asia–Pacific region, this edition expands the spatial scope a little. I hope this extension will make the book more informative and interesting to the book's growing international audience.

As I note above, I have subdivided the book into three sections: "introducing," "doing," and "interpreting and communicating." To some extent, it may be imprudent to impose such separations. I acknowledge, for instance, the ways in which the important act of writing—which might be seen superficially as a part of communication—is embedded in the process of "doing" qualitative research in human geography. Similarly, "doing" qualitative research typically implies ongoing interpretation and reinterpretation. The ordering of the book's contents is the product of the medium in which the contents are communicated (a book of linear structure). It is a device to help make more quickly comprehensible a book of 20 chapters. And it follows the organizing sequence used in many research methods classes—such as the one that gave rise to this book—for which this might serve as a textbook.

The first section of the book deals with "introducing" qualitative research methods in human geography. In Chapter 1, Hilary Winchester and Matthew Rofe situate qualitative research methods within the context of geographic inquiry. They outline the range of qualitative techniques commonly used in human geography and explore the relationship between those methods and the recent history of geographic thought. On this foundation, Robyn Dowling in Chapter 2 builds a review of some critical issues associated with qualitative research. These issues include power relationships between researchers and their co-researchers, questions of subjectivity and intersubjectivity, and points of ethical regulation and concern. Similar issues are taken up in further detail by Richie Howitt and Stan Stevens who—drawing from their considerable experience working with Indigenous communities in Australia and in Nepal respectively—present in Chapter 3 an illuminating and engaging conversation on ethics, methods, and relationships in cross-cultural qualitative research. Chapter 4 by Jay T. Johnson and Clare Madge is a wonderful new addition to the book. Reviewers of the third edition sought new chapters on feminist and Indigenous methodologies (as well as many other new chapters I might add!), but publishing budgets limited severely the space available. So I approached Jay and Clare asking if they might consider a new collaboration and write a chapter together. To my delight they agreed and the outcome is a tremendous contribution exploring "empowering methodologies" through an examination of feminist and Indigenous research approaches. In Chapter 5, Jan Monk and Dick Bedford draw from their vast range of research and research management experience to discuss compelling ways of introducing or proposing qualitative research projects to others. Chapter 6 by Elaine Stratford and Matt Bradshaw examines the difficult but vitally important matters of design and rigour in qualitative research. The final chapter in this section examines in more detail some of the methodological issues raised in Chapter 6 and explores research design issues associated with case studies in detail. In this chapter, Jamie Baxter sets out the value of case study methodology for understanding specific situations as well as for developing theory.

The second section of the book focuses on the "conduct" of qualitative research methods in human geography. Each of the chapters offers concise yet comprehensive, similarly structured outlines of good practice in some of the main forms of qualitative research practice employed in geography.

First, Kevin Dunn draws on his extensive experience of interviewing in cross-cultural settings to provide a complete and valuable outline of interviewing practice. Acknowledging some of the effects of the Internet on qualitative research practice, this long but rich chapter includes substantial material on computer-mediated communications (CMC) interviewing. Karen George and Elaine Stratford then set out the basic scope of oral history, discussing the ways in which its practice differs from interviewing and describing how oral history can enhance our understandings of space, place, region, landscape, and environment. In Chapter 10, Jenny Cameron considers the research potential of focus groups in geography, outlines key issues to take into account when planning and conducting successful focus groups, and provides an overview of strategies for analyzing and presenting the results. This chapter includes material on online focus groups.

The book then turns to two chapters on "texts." They deal respectively with archival research and the use of questionnaires in qualitative human geography. Historical geographer Mike Roche provides an insightful and substantially revised chapter on the use of archival resources, a sometimes neglected yet enormously productive and rewarding activity for human geographers, and an area being transformed by digital technologies. In Chapter 12, Pauline McGuirk and Phillip O'Neill explore the ways in which questionnaire surveys—including those offered online—can be used in qualitative research in human geography. It may be helpful to read this chapter in conjunction with the chapter on research design and rigour by Stratford and Bradshaw. The next chapter is the second major new addition to this edition. In it, Jim Craine and Colin Gardner offer a challenging and faar-reaching introduction to visual methodologies, blurring the lines you will see between matters "textual" and "visual." As part of their discussion they introduce discourse analysis, a topic given additional detailed attention by Gordon Waitt in his chapter describing how a Foucauldian approach to understanding "texts" might be applied. In Chapter 15, Robin Kearns reviews the purposes and practice of observational techniques in human geography, giving attention to photographic and video observation.

In a fine chapter new to this fourth edition made vital by the ways in which technologies are changing, Jamie Winders examines new media and their impacts on the qualitative study of human geographies. In Chapter 16, Jamie explores two questions in particular: first, what new research topics do new media raise for geographers? and, second, how can geographers use new media themselves as research tools? Then, taking up some of the issues about participant status for researchers that Robin Kearns introduced in Chapter 15 as well as matters of empowering research raised by Jay Johnson and Clare Madge, Sara Kindon offers an excellent chapter on participatory action research. There is also a strong complementarity between Sara's chapter and that by Richie Howitt and Stan Stevens: it is helpful to consider them in conjunction with each other.

Three chapters in a section on "interpreting and communicating" follow. In the first of these, Meghan Cope presents a substantially revised chapter on coding in qualitative research. It is worth reading Meghan's chapter together with that by Gordon Waitt on discourse analysis. Meghan's chapter has been modified substantially for this edition to embrace discussion of relationships between computers, qualitative data, and geographic research. The book then turns from explicit discussions of interpretation to matters of communication, but with Juliana Mansvelt and Lawrence Berg, who explore this matter in detail in Chapter 19, I would again like to emphasize the dialogic, relational nature of the connection between communicating and research. Mansvelt and Berg discuss some of the practical and conceptual issues surrounding representation of qualitative research findings. As I note above, they make the point that writing not only reflects our findings but constitutes how and what we know about that work. Finally, for this fourth edition, Eric Pawson and Dydia DeLyser have made significant amendments to their vital chapter, helping to fill a lacuna in research methods training. Very often, education in research methods provides advice on how to conceptualize or "do" research, but offers less counsel on interpretation and communication. This book, with its chapters on coding and communicating, goes a long way to rectifying this oversight and Pawson and DeLyser's chapter is central to that project. They focus on the effective communication of research to wide and diverse audiences, taking careful account of dramatic contemporary changes in communication possibilities. Importantly too, they give serious consideration to the oft-overlooked credibility and intensely persuasive power of qualitative research.

This book has chapters of varying lengths, "densities," and complexity. While some of the chapters are challenging to read, I would like to think that the challenges lie in comprehending the ideas presented, not in wading through obfuscatory text. In every case, authors have striven to present material that is as clear and well-illustrated as possible. As an editor with a long-standing interest in effective communication, I have devoted considerable attention to the matter of clarity.

While the chapters of this volume offer a comprehensive overview of qualitative research methods in human geography, the book is not intended as a "one-stop" resource or prescriptive outline (indeed, how could it be?) for qualitative researchers and students. Instead, it is meant as a starting point and framework. Accordingly, each chapter directs readers to a number of useful resources that may be consulted to follow up the material introduced here. There is, of course, a consolidated list of all references cited for readers who might wish to inquire even further. Chapters also include review questions and review exercises intended to support individual reflection as well as classroom engagement. These questions might also serve as prompts for ideas for quite different exercises. Indeed, if you have discovered or created any useful exercises to illustrate some of the matters covered within these chapters, I would be delighted to hear about them.

The book features an extensive glossary. While the individual authors have tried to ensure that their chapters are written in a language that is accessible to undergraduate readers, there are—without doubt—terms that may be somewhat alien to many readers. The glossary should help resolve that sort of difficulty.

Many of the terms in the glossary are drawn from the lists of key words associated with each chapter.

I owe thanks to many people who have been involved in the production of this book. Students in research methods classes at Flinders University encouraged me to put the book together and offered honest and helpful comments on many of its chapters. Each chapter was reviewed by at least two experts in the field who provided timely, critical, and comprehensive opinion. In alphabetical order, with their affiliation at the time of their review, these reviewers were Stuart Aitken (San Diego State), Kay Anderson (Durham), Nicola Ansell (Brunel), Alison Barnes (Western Sydney), Andrew Beer (Flinders), Alison Blunt (Queen Mary, London), Mark Brayshay (Plymouth), Michael Brown (Washington), Jacquie Burgess (University College, London), Jenny Cameron (Griffith), Garth Cant (Canterbury), Bev Clarke (Flinders), Mike Crang (Durham), Leah Gibbs (Wollongong), Jon Goss (Hawai'i), Penny Gurstein (British Columbia), Ellen Hansen (Emporia State), Andy Herod (Georgia), Richie Howitt (Macquarie), Mark Israel (Flinders), Jane Jacobs (Melbourne), Lucy Jarosz (Washington), Ron Johnston (Bristol), Agnieszka Leszcynski (Queen's, Ontario), Hannah Macpherspon (Brighton), Minelle Mahtani (New School), Murray McCaskill (Flinders), Pauline McGuirk (Newcastle), Jessica McLean (Macquarie), Eric Pawson (Canterbury), Meryl Pearce (Flinders), Chris Philo (Glasgow), Joe Powell (Monash), Lydia Mihelic Pulsipher (Tennessee), Lyn Richards (La Trobe), Noel Richards (Flinders), Chantelle Richmond (Western Ontario), Mark Riley (Exeter), Mark Rosenberg (Queen's, Ontario), Regina Scheyvens (Massey), Pamela Shurmer-Smith (Portsmouth), Robert Summerby-Murray (Mount Allison), Lynn Staeheli (Colorado—Boulder), Sarah Turner (McGill), Bryan Wee (Colorado—Denver) and Eileen Willis (Flinders).

I must also express a special note of thanks to the contributors to this volume. As with previous editions, authors have had to put up with repeated emailed requests and comments from me, often as long as the chapters themselves! I admire the authors' tolerance, persistence, and fortitude in the face of both my editing style and any personal and professional adversity they may have encountered while working on the project. I thank each of them for their efforts, their patience, and their continuing dedication to this very successful collection.

Iain Hay
Adelaide

PART

"Introducing" Qualitative Research in Human Geography

Qualitative Research and Its Place in Human Geography

Hilary P.M. Winchester and Matthew W. Rofe

Chapter Overview

This chapter has two main objectives. The first is to provide an overview of established and developing qualitative research methods in human geography. This chapter considers and categorizes the range of methods commonly used in human geography together with examples of the ways in which those methods are used to provide explanation. The techniques introduced here are expanded in later chapters. Second, the chapter examines briefly the changing theoretical context within which qualitative research has developed in human geography and links debates about method to wider theoretical perspectives in geography.

Introduction

Contemporary human geographers study not only tangible people and places, but also the intangible and fluid. These intangible objects of qualitative study include **discourses**, identities, and places of inscription. The complex and varied research questions arising in human geography require a multiplicity of conceptual approaches and methods of enquiry. Increasingly, the research methods used are qualitative ones intended to elucidate individual experiences, social processes, and human environments.

Qualitative research methods in human geography have reached a phase of maturity after a prolonged period of development since the 1980s. As such, they have come to rely on a well-established range of techniques. This chapter provides an overview of qualitative methods within human geography, grounded within the context of broader theoretical and methodological developments within the discipline. It outlines both the established and the developing methods that human geographers use in their explorations of the complex phenomena related to people and place. It sets the scene for the approaches and techniques which are explored in more detail in later chapters.

This chapter will inevitably establish categories of methods. However, these are not intended to be fixed or constraining but are designed as an organizing construct. Crang argued in 2005b (225) that "there is . . . a maturity about qualitative methods in geography" which brings with it "a certain conventionality of approaches." This categorization recognizes those conventional approaches as well as those that are more recent and exploratory. A number of the issues we raise in this introduction are more fully developed later in the volume, particularly those that relate to ethical practices (Chapters 2 and 3), the positioning of the author relative to the audience (Chapters 2 and 19), as well as the broad issues of *rigour*, **objectivity**, *trustworthiness*, **credibility**, **dependability**, and **confirmability** (Chapters 6 and 7).

This chapter also outlines the arguments about, and differences and similarities between, qualitative and **quantitative methods**. We discuss the ways in which human geographers have attempted to resolve the apparent dichotomy between the qualitative and quantitative, including through **triangulation** and **mixed methods**. We argue that the maturing of qualitative methods has led to an increased acceptance of their validity and hence a lesser emphasis on the need for the ostensible *objectivity of quantitative methods*. The apparent polarization between quantitative and qualitative, **objective** and **subjective** (see, for example, Johnston 2000) appears to be a false dichotomy. As Sui and DeLyser Sui (2012, 111) contend, "this divide has hindered cooperation, collaboration, and constructive engagement of diversity." Methodological differences among the various branches of geography should not be taken as indicators of opposition, but as a sign of healthy debate and intellectual vigour within the discipline. Finally, the chapter focuses particularly on links between theory and method and raises crucial issues of ethics, authorship, and power. Intense arguments about method are often as much to do with researchers' beliefs and feelings about the structure of the world as about their regard for a particular research method. So, for example, the choice to use **participant observation** or in-depth **interviews** may reflect the attitude of researchers about their research and their need for deep knowledge and understanding. This is commonly referred to as the researcher's position or more succinctly as **positionality**. For example, Billo and Hiemstra (2013) consider ways in which their personal and field lives informed each other and, as women working within a feminist paradigm, they also consider concepts of reflexivity and embodiment. Other geographers may hold equally strong, but perhaps less obvious, views about the order, structure, measurement, and knowability of phenomena. In complex ways, **ontology** (beliefs about the world) and **epistemology** (ways of knowing the world) are linked to the methods we choose to use for research (Grbich 2007; Hesse-Biber and Leavy 2004; Merriman 2012; Pile 2010).

What Is Qualitative Research?

What Questions Does Qualitative Research Answer?

Qualitative research is used in many areas of human geography. In a broad sense, qualitative research is concerned with elucidating human environments and human experiences within a variety of conceptual frameworks. The term "research" is used here to mean the whole process from defining a question to analysis and interpretation. The research methods involve several aspects or steps (as with quantitative research methods) including the definition of the research problem, development of hypotheses, research design, the gathering of data, and analysis to derive meaning. Qualitative methods employ techniques of data gathering and analysis that differ from quantitative techniques. A simple example is that gathering information about human behaviour and motivation through qualitative techniques is likely to include in-depth interviews, while quantitative techniques might use questionnaires. "Method" is used as a much more specific term for the investigative technique employed.

An extensive range of techniques is used in many different situations. Some of the variety of oral techniques that spring to mind range from interviews with imams about the construction of migrant Muslim identity and places (Fridolfsson and Elander 2013), through participant observation with homeless children (Winchester and Costello 1995), to questionnaires via telephone survey of rural South Australians to determine their attitudes to racism (Forrest and Dunn 2013). Other techniques involve the **deconstruction** of media events and **textual** material, for example in the place-making of former industrial cities such as Port Adelaide (Szili and Rofe 2007) or of current mega-events such as the *Tour de France* (Ferbrache 2013).

Inevitably, it is difficult to summarize the questions addressed by such a variety of research. However, it is instructive to recall the answer to a similar question posed three decades ago in relation to statistical analysis in geography. In that text, Ron Johnston (1978, 1–5) argued that the gamut of statistical techniques answered two fundamental questions. Those questions were about either the relationships between phenomena and places or the differences between them. The elegant simplicity of Johnston's questions can be paralleled by a different but similarly broad pair of questions that qualitative research is trying to answer.

The two fundamental questions tackled by qualitative researchers are concerned either with **social structures** or with individual experiences. This dualism may be hard to disentangle in practice, but it is of fundamental importance in explanation. The behaviour and experiences of an individual may be determined not so much by their personal characteristics but by their position in the social structure.

The first question may be phrased as:

Question One: *What are the shapes of societal structures, and by what processes are they constructed, maintained, legitimized, and resisted?*

The structures that geographers analyze may be social, cultural, economic, political, or environmental. Structures may be defined as internally related objects or practices. Andrew Sayer (2000, 13) gives the example of the landlord–tenant relationship in which structures exist in relation to private property and ownership, rent is paid between the two parties, and the structure may survive a continual turnover of individuals. Further, he emphasizes that tenants almost certainly exist within other structures; for example, they may be students affected by educational structures or migrants constrained by racist structures. The coexistence of rented housing, students, and minority groups produces a complex linkage and mutual reinforcement of structures within which individuals live out their lives. As Sayer (2000, 13) asserts, "[i]n the social world, people's roles and identities are often internally related, so that what one person or institution is or can do, depends on their relation to others." Qualitative geographers balance the fine line between the examination of structures and processes on the one hand and of individuals and their experiences on the other. Structures constrain individuals and enable certain behaviours, but in some circumstances individuals also have the capacity to break rather than reproduce the mould. However, it must be recognized that individuals do not have all-powerful free will and ability to challenge powerful structures embedded in our societies. Structures such as capitalism, patriarchy, or racism cast long shadows over human experience and opportunity. Qualitative geographers must be aware that an overemphasis on structures and processes rather than individuals could lead to a dehumanized human geography. Sayer (2000) considers that the key question for qualitative researchers about structures may be phrased as "What is it about the structures which produce the effects at issue?" Geographers have studied structures qualitatively in a number of ways. A significant focus has been on the ways in which they are built, reproduced, and reified: for example, Kay Anderson (1993) analyzed the documentary history that has led to stigmatization of the suburb of Redfern in Sydney as an Australian "ghetto." More recently, Wendy Shaw (2000; 2007) examined the physical and ideological context of Redfern's stigmatization as a manifestation of the disproportionate politics of race that disempowers Indigenous Australians, and by 2013, shows how Aboriginal Australians live ordinary lives in this highly contested site (Shaw 2013). In this example, the structures are essentially indistinguishable from the processes that build, reinforce, and contest them. However, at the level of analysis, both authors have chosen to emphasize one particular scale of social agency for the purpose of framing, understanding, and representing the processes evident in Redfern. Similarly, other authors have considered either the material or the symbolic **representations** of structures: in their 1994 analysis of British merchant banking, McDowell and Court emphasize media representations of banking patriarchs and the importance of dress and body image for younger female and male bankers. A further aspect of the investigation of structures may be concerned with their oppressive or exclusionary nature: Gill Valentine's (1993) research on lesbians' experiences of the workplace, young people's "coming out" experiences (Valentine, Skelton, and Butler 2003) and the gendered dimensions of drinking **landscapes** (Holloway, Valentine, and Jayne 2009) reveal the complex ways in which workplace and home structures naturalize sexual and gender norms, thereby contributing to the oppression and marginalization of individuals who do not conform to these norms. Most of the qualitative geographical work on structures in fact emphasizes the processes and relations that sustain, modify, or oppose those structures, rather than focusing specifically on their form and nature.

The second fundamental question considered by qualitative human geographers is concerned with individual experiences of structures and places:

Question Two: *What are individuals' experiences of places and events?*

Individuals experience and understand the same events and places differently. Giving voice to individuals allows viewpoints to be heard that otherwise might be silenced or excluded: Jenny Pickerill's (2009) examination of the complex and at times contested negotiations between Indigenous Australians and environmental groups gives voice to Aboriginal perspectives on naturalized notions of land and country (see also Johnson and Murton 2007). For Pickerill (2009, 66) these negotiations ". . . provide clues as to how commonalities across difference . . . [can be] built." The upsurge of academic research into Indigenous knowledge and politics is a global phenomenon creating what Johnson et al. (2007, 118) refer to as ". . . anti-colonial geographies." At the heart of this project is the opening up of geography to previously unheard voices typically positioned as curios on the margin. However, as Wilson and Peters (2005) reveal in their examination of the spatial identities of First Nations people in Canada, Indigenous populations are increasingly urban. Indigenous peoples' existence within and identification with the urban "core" of the nation-state disrupts naturalized understandings of Indigeneity, thereby blurring the boundaries between the core and the margin. For Native Hawaiian academic Renee Pualani Louis (2007, 137) decolonizing Indigenous voices has heralded the emergence of Indigenous-specific and/or -sensitive methodologies that can help ". . . invigorate and stimulate geographical theories and scholarship while strengthening Indigenous peoples' identities and supporting their efforts to achieve intellectual self-determination." Louis's (2007, 130) published challenge, "can you hear us now?," is a contemporary answer to earlier calls to give "voice" to silenced groups (McDowell 1992a), which in itself reflects a wider political project answering David Smith's (1977, 370) challenge to develop a ". . . new paradigm for geography . . . [that being] a . . . geography about real people."

Dissident or marginalized stories may be "given voice" through diaries, oral histories, recordings of interviews and conversations or the use of "alternative" rather than "mainstream" media. Participant observation by immersion in particular settings allows multiple viewpoints to be heard and acknowledged. Boyd's (2013) year-long participant observation, augmented with interviews, of Vancouver's dance music culture provides a nuanced study of both the freedom and constraints of dancing within gendered and sexualized frameworks. An earlier study into party culture by Winchester, McGuirk, and Everett (1999) gave voice not only to the partying young people (i.e., high school students celebrating the end of their high school years) but also to agents of control (for example, police). In a study of the daily geographies of caregivers Wiles (2003, 1307) found that "the social and physical aspects of the many interconnected scales and places which caregivers negotiate on an everyday basis both shape and are shaped by caregiving." In other words, the experience itself could not be analytically separated from the structures that form the context for that experience.

The experiences of individuals and the meanings of events and places cannot necessarily be generalized, but they do constitute part of a multifaceted and fluid reality. Qualitative geographical research tends to emphasize multiple meanings and interpretations rather than seeking to impose any one "dominant" or "correct" interpretation.

Types of Qualitative Research

It is clear, even from the brief preceding section, that qualitative research in geography has been used for more than thirty years to address a huge range of issues, events, and places and that these studies utilize a variety of methods. Indeed in the years since Crang (2005b, 225) asserted the "maturity of qualitative methods in geography" there has been significant further development and a growing acceptance of their legitimacy and power. At the same time, there has been a fuller recognition of the constraints imposed by the positionality of the researcher, the gatekeepers to knowledge, and the selection of participants. In recent years a range of new techniques has emerged particularly focused on **performativity**, popular and everyday culture and the mediation and representation of experience through Internet technology and social media.

While this "maturity" can be said to reflect a rigorous and well-laid epistemological foundation, Crang (2005b, 225) cautions that with it has also come ". . . a certain conventionality of approaches." While a number of new, so-called "experimental" techniques are emerging, (such as **discontinuous writing**,

photo-elicitation, and **go-along interviews**), the weight of qualitative analysis rests on a limited range of techniques that have become conventional rather than cutting-edge. Reflecting this view, Davies and Dwyer (2007, 257) assert that oral techniques (e.g., interviews, focus groups), **textual analysis**, and associated observationally based ethnographies embody a "suite" of methods that ". . . remains the backbone of qualitative research in human geography." For Thrift (2000, 3) the emergence of a "convention" of qualitative approaches within human geography is based on this narrow range of techniques that may unwittingly register a narrow range of life, brought back from the field, and represented in "nice," even predictable ways. In a similar vein, Silverman (2005) has cautioned that the widespread acceptance of interviewing as a specific method should not be accompanied with complacency in how we as researchers understand the complexity of this process. Silverman's caution urges us to never lose sight of the socially dynamic nature of interviewing and so fail to critically engage with these dynamics. These are valid critiques and may be likened to a mid-life crisis of the maturity of qualitative methods. In this vein this section discusses the three main "conventional" types of qualitative research approaches employed in human geography: the oral (primarily interview-based), the textual (creative, documentary, and landscape), and the observational. It also considers a number of emerging techniques that are presently gaining acceptance within human geography.

Oral Methods

Clearly, the most popular and widely used methods of data collection are oral. Talking with people as research participants encompasses a wide range of activities and techniques. The spoken testimony of people is used in ways that range from the highly individualistic (oral histories and autobiographies) to the highly generalized (the individual as one of a **random sample**). The latter type of survey technique borders on the quantitative in which responses can be counted, cross-tabulated, and analyzed statistically. The former approach, often achieved through **oral history** methods, lies at the more qualitative, individualistic end of the spectrum. Such methods are considered in more detail in Chapters 8 and 9. A middle ground is occupied by the popular technique of using **focus groups** (see Chapter 10). Valentine and Waite (2012) used focus groups differentiated by sexual orientation, age, gender, ethnicity, and religion to analyze the complex landscape of difference and understanding. The focus-group excerpts presented in Valentine and Waite's paper are noteworthy for the rich and multi-layered data they represent. Similarly, Pearce et al. (2010) used focus groups and face-to-face interviews to examine perceptions of and responses to drought in rural South Australia.

Another form of rich data can be derived from respondent storytelling. Wright et al. (2012) employed storytelling as a key method of engaging with Indigenous Australian communities. The use of such techniques also enables the researcher to engage with their local, academic, and surrounding communities in a process of knowledge exchange that can empower and give voice to the researched as well as the researcher. The range of ways in which **oral methods** are utilized in geography—allowing subjects to speak in their own voice—is outlined in Box 1.1. Oral approaches can be used across the range of the research questions outlined in the previous section, from answering questions about individual meanings and experiences at the biographical end of the spectrum, to answering research questions about societal structures at the survey end. Surveys are undertaken to obtain information from and about individuals that is not available from other sources. While an interview is undertaken with an individual, a survey involves a more standardized interaction with a number of people. Oral surveys of personal information, attitudes, and behaviour usually (but not inevitably) utilize questionnaires. Questionnaires tend to be more closely structured and ordered than interviews, and every **respondent** answers the same question in a standard format. These are discussed fully in Chapter 12. Oral methods will often merge into a textual analysis, as the interviews, oral histories, or focus groups may result in a recorded and transcribed text that can be analyzed in various ways. Fridolfsson and Elander (2013) combine interviews with textual analyses to help understand the construction of Muslim migrant identity in Sweden.

Textual Methods

The second major type of qualitative research methods used in human geography is textual. Such texts are wide-ranging but more diffuse in the human geography literature than the oral testimonies described above. Textual analyses are generally predicated on a constructionist epistemology (i.e., that the world

BOX 1.1 **Oral Qualitative Methods in Human Geography**

General Method	Specific Method	Research Questions
Biography	Autobiography Biography Oral history	Individual ↑
Interviews	Unstructured Semi-structured Structured Focus groups—open-ended	↕
Surveys	Surveys—structured Questionnaires—structured	↓ General/Structural

is socially constructed and mediated). In human geography, Meinig's (1979) *The Interpretation of Ordinary Landscapes: Geographical Essays* represented a watershed. In this collection of seminal essays, Meinig and his contributing authors charted an understanding of the socially constructed nature of the human world and an approach to its understanding and interpretation. Essentially, their approach was deconstructionist in nature, before the term was popularized in the late 1980s. Textual analysis is firmly rooted in deconstruction which, according to Pratt (1994, 468), represents ". . . an attempt to undo claims to truth and coherence by uncovering the incoherences within texts." Textual analysis, through deconstruction, actively engages with the spoken and unspoken meanings or discourses encoded within a text.

Important groups of textual methods utilize creative, documentary, and landscape sources as well as the texts created from oral methods as mentioned in the preceding section. Creative texts are likely to include poems, fiction, films, art, and music. Documentary sources may include maps, newspapers, planning documents, and even postage stamps. Waitt and Head (2002) examined the role of postcards in Australian frontier mythology, Rose (2003a) used family photographs in a study of domestic space, while Dittmer (2010) employed comic books to innovatively expose the complexity of non-mainstream storytelling. Landscape sources may be very specific, such as the individual gardens as a site of remembrance for the deceased (Ginn 2014) or the militarization of the landscape (Pearson 2012). Frequently, landscape sources are more general, such as the landscapes of suburbia as indicators of social status (Duncan 1992). The analysis of creative sources, including fictional literature, film, art, and music, has shown increasing complexity in recent years. (For a survey of this field, see Winchester and Dunn [1999] and for a brief Australian introduction, see Carroll and Connell [2000]. See also Chapters 13 and 15 of this volume.) Lukinbeal and Zimmerman (2008) use a crisp lens on the geography of cinema by suggesting a triple focus on author, text, and reader. Geographers have searched such sources for underlying structures, looking at paintings, for example, to understand changing perceptions of landscape (Heathcote 1975, 214–17; Lowenthal and Prince 1965) or using film to examine both the impact of city restructuring and the ways in which it is represented. For example, Law, Wee, and McMullan (2011) examine the city–state development of Singapore through Eric Khoo's (2005) film *Be With Me* and casts various parts of the city as spaces of parts of the life course—youth, middle- and old age—showing particularly how the landscapes are constituted through state policies and the intersections of class, gender, and age. Visual methodologies are also being applied to nature documentaries to create what Hayden Lorimer (2010) refers to as "more-than-human geographies." In the past decade, a plethora of methodological monographs has been produced, exploring and explaining the epistemologies and practices of textual analysis. Within human geography, Rose's (2001) *Visual Methodologies*,

Winchester, Kong, and Dunn's (2003) *Landscapes: Ways of Seeing the World*, and Wylie's (2007) *Landscape* are prime examples. Each of these texts reveals the fecund data available through deconstructive analysis.

Written texts have also been used as the source of underlying discourses that underpin and legitimate social structures. Analysis of media representations demonstrates their myth-making power, whether related to myths of the inner city (Burgess and Wood 1988) or to the imagery of national identity (Pickles 2002) and both urban and rural place-making (for urban examples, see Dunn, McGuirk, and Winchester 1995 and Schollmann, Perkins, and Moore 2000, and for rural examples, Winchester and Rofe 2005 and Rofe and Winchester 2007). Herman (1999) has argued that changing placenames in Hawai'i reveal a transformation from Hawai'ian political and cultural economy into Western capitalist forms, while Sparke (1998) shows convincingly some of the ways in which *The Historical Atlas of Canada* is enmeshed in the post-colonial politics of that country, and Hudson (2013) discusses the cultural politics of the naming of waterfalls, such as the naming of the Victoria Falls, as an aspect of colonial possession.

A significant source of textual analysis is the landscape itself; this view of landscape as text has shifted over the last decades from being "controversial" to "conventional." Landscape here is understood as the entwining of social meaning with, and its expression through, the physical environment. This entwining is crucial to the creation of *place* as a definable and knowable geographic location. Thus, landscape is a specific and highly contextual "way of seeing" (Winchester, Kong, and Dunn 2003). As a social construction, physical landscapes can be "read" to reveal their social meanings and intentions. Reading a landscape is a challenging process, but it offers important insights into the history of places and struggles over them. Lewis (1979, 12) astutely argued 30 years ago that the ". . . human landscape is our unwitting autobiography . . . [where] our cultural warts and blemishes are . . . exhibited for anybody who wants to find them and knows how to look for them." The use of landscape as text to help define the identity of place can be a powerful way of researchers engaging with their regions and with their communities. Landscapes are also co-opted into the creation of national identities (e.g., Ferbrache 2013). The argument that landscape may be read as text is epitomized in the work of Duncan and Duncan (1988) in which the residential landscape is decoded of its social nuances. A study of roadside memorials in the Australian state of New South Wales concluded that the roadside crosses and flowers were indicative not only of individuals' behaviour but also of a "problematic masculinity" characterized by aggression, fast driving, and recklessness (Hartig and Dunn 1998). Textual analyses of particular landscapes, such as model housing estates, use techniques derived from **semiotics** (the language of signs) to demonstrate literally the inbuilt naturalization of social roles according to gender and family status (Mee 1994). From a feminist perspective, Massey (1994) demonstrates how planning decision-making processes behind seemingly innocuous land-use decisions such as the provision of sporting facilities reveal ". . . systems of gender relations" (Massey 1994, 189) that are inequitable. Rofe (2007), after Massey (1994), has problematized the belief that the revitalization of derelict industrial areas heralds the emergence of a more progressive and equitable city, concluding that inequitable gender relations remain implicitly encoded within the post-industrial landscape. Schein (1997) interprets landscape architecture, insurance mapping, and other elements of a "discourse materialised" to explore the ways they symbolize and constitute particular cultural ideals. Waitt and McGuirk (1996) examine both documentary and landscape texts to explore the selective representation of the heritage site of Millers Point in Sydney; the choice of particular buildings as "heritage" both reflects and reproduces a white, male, colonizer's view of Sydney's history while silencing other views and voices. Bishop (2002) uses the Alice Springs-to-Darwin railway as a corridor of "difference, struggle and reconciliation" in the redefinition of national identity and its relationship with the land. While landscape is an important platform for textual analysis, more recent attention has been focused on other sites of inscription, in particular the human body, where social structures and processes play out in visible and personal form.

Participatory Research Methods

The third significant type of qualitative research method in human geography consists of forms of participation within the event or environment that is being researched (see Chapters 3, 4, 14, and 17). The most common form of qualitative geographical research involving participation is participant observation. Within participant observation, there may be a wide variation in the role of the observer from passive to proactive (Hammersley and Atkinson 1983, 93). All forms of observation involve problems of how the author should be positioned in relation to the subject of the research (for example, Smith 1988). In particular, very active participation may clearly influence the event that is being researched, while researchers who are personally involved (for example, researching the community in which they grew up) may find it hard to wear their "community" and "researcher" hats at the same time. We have first-hand experience of these issues from our research into the scripting and performance of a hyper-masculine motorcycling identity (Rofe and Winchester 2003). Rofe was a long-standing member of the subculture under study. While his background afforded detailed insight into and access to a community notoriously suspicious of "outsider" scrutiny, it had to be balanced against the potential for bias in the interpretation and reporting of results. To avoid these problems, we employed a creative tension within the research and writing process whereby understanding and explanation were not just determined between us and the research participants but also negotiated between ourselves as researchers. Clearly, issues surrounding the position of the researcher or his or her **reflexivity** are critical in participant observation. Despite such tensions, participant observation allows the researcher to be—at least in part—simultaneously "outsider" and "insider," although differences in social status and background are hard to overcome (Moss 1995). The positioning of the researcher in relation to the "researched" raises some significant ethical issues, especially if the research is covert (Evans 1988, 207–8). It can, however, have some important advantages. For example, a student doing research at a fast-food outlet might be given paid work there (see Cook 1997 for a highly readable account of this and other student projects). In-depth participant observation is essentially indistinguishable from ethnographic approaches, which often involve lengthy fieldwork. That fieldwork can enable meaningful relationships to develop with the research subjects and may facilitate deep understanding of the research context (Cooper 1994; 1995; Eyles 1988, 3). Cook's conclusion (1997, 127) that participant observation is the means or method by which ethnographic research is undertaken is useful in this context.

New Foci

Beyond the backbone of the methods outlined above, a range of additional techniques have begun to emerge. These techniques stem from the backbone methods but, reflecting the evolving nature of inquiry in human geography, constitute new tools to address emerging questions. Rather than representing radically new methods, these experimental techniques embody adaptations of existing ones aimed at making them "dance a little" (Latham 2003; 2000). Such techniques include discontinuous writing (Meth 2003; Morrison 2012), photo-elicitation, and **diary interviews** (Harper 2002; 2003; Latham 2003), and go-along or walking interviews (Kusenbach 2002; Evans and Jones 2011). Latham's (2003) use of **diary photographs** and associated interviews is most instructive in terms of the rich insights available into seemingly ordinary experiences and their geographies offered by experimentation with methods. Observing that "[e]veryday life and everyday culture are two of the great frontiers of contemporary human geography," Latham (2003, 1996) encouraged his respondents to keep personalized journals recording their social-geographic activities. The resulting journals represented deeply personalized and flexible "time–space collages" that according to Latham (2003, 2009) ". . . can . . . enliven our sense of what human geography should look like." Even more ephemeral aspects of human existence are pursued through other, newly emerging techniques such as observant listening and participant sensing (Wood and Smith 2004). Davies and Dwyer (2007) position the development of these techniques as stemming from human geographers' increasing interest in the geographies of performance and emotion. For Wood and Smith (2004, 534), the ". . . challenge of understanding emotions has . . . fuelled a search for new ways of knowing: ways that range beyond the visual and representational traditions which have for so long dominated social thought." In these pursuits, these more recent techniques offer new insights into the lived reality of the social world, pioneering what Wood, Duffy, and Smith (2007) refer to as "unspeakable geographies" (see also Davidson, Bondi, and Smith 2005). The unspeakable nature of such geographies relates to the inherent difficulties of capturing emotion, which by its nature is elusive, and the problems associated with its conversion into research data and representation as research findings. This echoes Law's (2004, 3) belief that human agency is ". . . emotive and embodied, rather than cognitive" and therefore fluid. Musing on the challenges facing qualitative researchers, Davies and Dwyer (2007, 258) wonder whether ". . . the world is so textured as to exceed our capacity to understand it," let alone represent it. In response to this, authors such as Duffy and Waitt (2011) encourage us to develop more finely tuned techniques that enable us to "listen to place." They use "sound diaries" undertaken at a music festival in New South Wales to help understand how music and sound helped constitute place (Duffy and Waitt 2011, 125). In the face of the question of representing emotive geographies, new experimental techniques such as those briefly identified in the preceding discussion are essential.

Recent developments in qualitative human geography place a strong focus on the individual in three increasingly important ways. First, qualitative studies have identified **bodily performance** and performativity as important aspects of the continued social construction of human behaviour. This performance has been strongly identified as an element of gendered behaviour. Most studies of embodied performance consider dress and scripting in the workplace and demonstrate how the body is socially constructed to enforce and reproduce those expectations. This is most obviously seen in the adoption of work dress, such as uniforms and suits, as in McDowell and Court's (1994) important study of bankers in the City of London. Bell et al. (1994) also showed how such expectations could be subverted. Recently Gorman-Murray (2012) unpacked aspects of such embodiment in masculine domesticity, which included the shedding of work clothes, and other tactile elements which contributed to their emotional well-being and the construction of gendered identities in the home. In this study, Gorman-Murray used interviews, time-diaries, and home tours with follow-up interviews.

The second aspect of the recent focus on the individual is the concept of the human body as a place of inscription. This shifts the focus of textual methods from the element of reading within the technique to the site on which the text is inscribed. The human body is a powerful site of inscription because of its personal and sacrificial nature. Examples include the anorexic body and the politics of weight loss (Longhurst 2012), the tattooed body (Botz-Borstein 2013), and the ways in which the body as weapon and the pain associated with full-contact martial arts operates as an indicator of individual strength and communal bonding (Green 2011). Murton (2012) explores the concept of the "geographical self" to illuminate Māori thought about the complex interactions between self, body, landscape, and place. Writing in a themed issue of *Geographical Research*, Sharpe and Gorman-Murray (2013) have categorized geographical approaches to the body as either "positioning" bodies (within the social structure) or "affective" bodies, the site of agency or transgression.

The third aspect of the increased focus on the study of the individual is the developing study of social media as individual behaviour and as text, and the processes of social networks. This is a growing field because of the explosion of mobile technology devices and the massive development of social networking. The uptake of social media networks such as Facebook, LinkedIn, YouTube, and Twitter offers new insights into the dynamic nature of geographic connections through cyberspace. As Warf (2011, 1) contends " . . . cyberspace offers profound real and potential effects of social relations, everyday life, culture, politics and other social activities." Salient examples of these potentials include Meek's (2012) research into the use of YouTube to bring wider attention to the plight of child-soldiers in Uganda and Woon Chih's (2011) work on digital protest in China responding to the Iraq war of 2003. These works exemplify what may be termed "new knowledge politics" (Elwood and Leszcynski 2012) and in particular to the apparent liberation of media sources. Fekete and Warf (2013) analyzed the role of information technology in the Arab Spring but cautioned against technological determinism in understanding these events. Control and censorship of the Internet is undertaken by many governments (see Warf 2011) as they seek to resist challenges to their authority.

Thatcher (2013) and Power et al. (2013) demonstrate clearly how new digital technologies, such as Google Street View and Microsoft's Pedestrian Route Production software, draw upon notions of spatial stigmatization, consequently reinforcing already established geographies of fear and moral landscapes. An important aspect of qualitative research in human geography is its ability to engage communities, understanding their lives and environment and striving to empower those communities and giving them voice as active participants in the research endeavour to understand ourselves and our environment. In these ways, qualitative research can contribute to community engagement, knowledge transfer, and the demystification of social science research.

The Contribution of Qualitative Techniques to Explanation in Geography

In this chapter we have discussed two fundamental questions of geographic inquiry: those concerned with individuals and those concerned with social structures. We have also indicated three main groups of methods: the oral, the textual, and the participatory. We have identified an increasing focus on the individual as the research subject, through performativity, through the body as a site of inscription, and through the textual analysis of social media and social networking sites. There is no simple relationship between the method used and the research questions posed. It is tempting to say that oral methods may be directed predominantly toward elucidating the experiences of individuals and their meanings; however, this is overly simplistic. People's own words do tell us a great deal about their experiences and attitudes, but they may also reveal key underlying social structures. In work on lone fathers, Winchester (1999) found that the in-depth interviews illuminated underlying structures of patriarchy and masculinity in ways that were much more profound than anticipated. Depths of individual anger and despair reflected mismatches between those individuals' romanticized expectations of marriage and gendered behaviour and their actual experiences of married life. In this sense, the oral method chosen elucidated both individual experiences and social structures in the holistic sense that would most frequently be associated with participant observation.

Similarly, it might appear that textual methods would most commonly be employed to throw light on the social processes that underpin, legitimate, and resist social structures. This generalization would probably be more widely accepted than any equation of oral methods with research questions that focus on the individual. Textual methods have indeed been used to analyze some of the many social processes studied by contemporary human geographers. Examples that spring to mind include the discursive construction of place (Dunn, McGuirk, and Winchester 1995; Mee 1994; Szili and Rofe 2007), processes of social exclusion (Duncan and Duncan 1988), marginalization (McFarlane and Hay 2003; Hay, Hughes, and Tutton 2004), and expressions of "problematic" masculinity (Rofe and Winchester 2003; Cowen and Siciliano 2011). In this latter example, even those individuals constructed as deviant, other, or subordinate can come to represent a "surplus" labor force for the military (Cowen and Siciliano 2011).

A fruitful area of study in human geography focuses on the body and on our embodied experiences. Longhurst's (1995) original study of the experiences of pregnant women in shopping malls reveals how the embodied experiences of individuals (of feeling marginalized, of needing more toilets, of being uncomfortable in public places) are indicative of the way certain bodies may be socially constructed as **"Other"** (i.e., oppositional to or outside the mainstream) to be confined to particular places and roles, medicalized, and marginalized. Resistance to such marginalization is manifested in a number of ways, documented notably in Longhurst's (2000) study of "bikini babes": pregnant women who participated in and thereby destabilized a beauty contest for bikini-clad women. The study of the body as text, as performance, or as social construction illuminates some of the richness of methods that cannot easily be pigeonholed into the types of qualitative method and types of geographical explanation identified for convenience earlier in this chapter.

The Relationship between Qualitative and Quantitative Geography

During the late twentieth century, the pendulum of geographical methods within human geography swung firmly from quantitative to qualitative methods. The two have traditionally been characterized as in opposition or as conflicting methods. Qualitative methods have been in the ascendant since the 1980s. The trend toward the resurgence of qualitative sources and methods in geography has been chronicled in and stimulated by recent books and collections on qualitative methods and mixed methods (Brannen 1992b; Denzin and Lincoln 2008; Eyles and Smith 1988; Flowerdew and Martin 1997; Grbich 2007; Hesse-Biber and Leavy 2004; Holland, Pawson, and Shatford 1991; Limb and Dwyer 2001; Lindsay 1997; Moss 2002). Typically, the perceived gulf between qualitative and quantitative methods has been presented as a series of dualisms. Hammersley (1992) listed seven "polar opposites" between qualitative and quantitative methods (Box 1.2). Similarly, Brannen (1992a) characterized qualitative approaches as viewing the world through a wide lens and quantitative approaches as viewing it through a narrow lens. A dualistic view of methods is highly problematic, as Hammersley (1992, 51) recognized: it represents quantitative methods as focused, objective, generalizable and, by implication, value-free. On the other hand, qualitative methods are often presented as soft and subjective, an anecdotal supplement, somehow inferior to "real" science.

Such a view misleadingly represents quantitative methods as objective and value-free; increasingly, this assumption about the nature of science has been questioned (see Chapter 19). Our choice of research questions and methods—what and how we study—reflects our values and beliefs. For example, much early feminist geography uncovered sexist assumptions in how geographers had typically studied and

Dualisms Identified between Qualitative and Quantitative Methods

Qualitative methods	Quantitative methods
Qualitative data	Quantitative data
Natural settings	Experimental settings
Search for meaning	Identification of behaviour
Rejection of natural science	Adoption of natural science
Inductive approaches	**Deductive approaches**
Identification of cultural patterns	Pursuance of scientific laws
Idealist perspective	Realist perspective

Source: After Mostyn (1985) and Hammersley (1992).

measured human behaviour (Monk and Hanson 1982). Measurements of migrants and shoppers that ignored "half of the human population" were clearly shown to be lacking objectivity, instead being value-laden and in many cases strongly coloured by naturalized assumptions about gendered roles and behaviour. If one acknowledges the subjectivity and value-laden nature of all research methods, then the apparent gap between the two groups of methods is reduced dramatically. Geographers using qualitative methods often declare their personal subjectivity and possible sources of **bias**, summarizing their own positionality, (i.e., their background as researchers and their relationship to the research and to its intended audience). Further, the cultural "turn" within human geography has made such openness more commonplace across the discipline, encompassing not only social and cultural geographers but those with more economic and political interests as well (for example, Coe, Kelly, and Yeung 2007; Thrift 1996). Indeed, researchers who define their own position in relation to their research may be more objective than their colleagues who point to the supposed objectivity of quantitative methods and fail to reveal the many subjective influences that shape both the research question and the explanations that they put forward. However, Bourdieu (2003, 282) takes aim at an excess of reflexivity within the social sciences generally. Specifically, he ridicules the self-indulgent "fashion" of ". . . observing oneself observing, observing the observer in his [sic] work of observing" (Bourdieu 2003, 282). Given the intellectual complexity and introspection for which Bourdieu is renowned, this is a most interesting comment. As Crang (2005b, 226) clarifies, "Bourdieu frets that . . . reflexivity recreates the myth of the exceptional researcher set apart from their respondents not now by the clarity of their knowledge, but by their level of introspection, doubt and anxiety." In short, equating "objectivity" with the quantitative and subjectivity with the qualitative is highly contested (Philip 1998). This contest is discussed further in Chapter 19.

As qualitative methods have become pre-eminent within the discipline, they have increasingly had to be justified in a scholarly environment that had come to value measurement and scientific observation more highly than individual experience or social process. Within a largely unfriendly hegemonic scientific framework, advocates of qualitative studies have generally drawn from three arguments. First, some studies of individual experiences, places, and events have been represented as essentially non-generalizable case studies that have meaning in their own right but are not necessarily either representative or replicable (for example, Donovan 1988). The second argument, appropriate to some large-scale studies, has been to suggest that these studies have generated sufficient data to allow general, and sometimes quantified, conclusions to be drawn from the research (for example, Rofe 2003; 2009). Third and more usually, however, qualitative methods have been justified as a complementary technique, as an adjunct or precursor to quantitative studies from which generalizations can be drawn and as explorations in greater depth as part of multiple methods or triangulation (Burgess 1982a). In making these arguments, which are all interrogated more fully throughout this volume, qualitative geographers have often been on the defensive, aiming to present their studies as legitimate in their own right and as research that produces not just case studies or anecdotal evidence but that has added immensely to the geographical literature through powerful forms of geographical explanation, including analysis, theory-building, and geographic histories. Mixed methods are increasingly being used in an effective and powerful way without the self-consciousness evident in some of the studies from the 1990s (Longhurst 2000; Nolan 2003). This lack of self-consciousness in drawing on a variety of methods is very clear in some of the newer contributions to qualitative geographical research already mentioned in this chapter, such as those by Fridolfsson and Elander (2013) and Pearce et al. (2010).

Classically, qualitative and quantitative methods, such as interviews combined with questionnaires, are seen as providing both the individual and the general perspective on an issue (for example, England 1993), while similar arguments have been raised for mixed methods more broadly (McKendrick 1996; Philip 1998). This triangulation of methods and use of multiple methods are sometimes deemed as offering a cross-checking of results in that they approach a problem from different angles and use different techniques. Brannen (1992a, 13), however, has argued that data generated by different methods cannot simply be aggregated, because they can only be understood in relation to the purposes for which they were created. This question of purpose is intimately related to the theoretical perspectives from which the techniques derive and is considered further in Chapter 6.

The History of Qualitative Research in Geography

As a discipline, geography enjoys a long history. For much of that time, geographical work has been dominated by qualitative research of a scholarly and informed but unquantified nature, drawing assessments of evidence from both physical and human environments (see, for example, Powell 1988). It should be recognized that qualitative methods of many sorts were used widely throughout the twentieth century, particularly in the development and writing of sensitive and nuanced regional geography, such as that of Oskar Spate on the Indian subcontinent (Spate and Learmonth 1967), in the landscape school of both human and physical geographers with "an eye for country" (see, for example, Sauer's 1925 and Zelinsky's 1973 work on American landscapes and culture), but also in interviewing and field observation (Davis 1954; Wooldridge 1955). The postwar era of the "quantitative revolution" may in hindsight be seen as an aberration rather than the revolutionary **paradigm** (mode of thought) that it was claimed to be at the time (Wrigley 1970).

The early history of geographical thought has been represented classically as a series of paradigm shifts, each triggered by dissatisfaction with the previous prevailing paradigm. A schematic representation of paradigm shifts in academic geography is presented in Box 1.3 (Holt-Jensen 1988; Johnston 1983; Wrigley 1970). For example, dissatisfaction with the crude environmental determinism of the early twentieth century prompted the study of unique places around the world. When this regional approach degenerated into stale layers of facts and geographers had completely marginalized themselves from the academy by their commitment to "the region"—both as object of study and research method—then an alternative, more credible to the academic community, was sought. The strategic alliances of the discipline shifted away from history and geology to newer and more innovative disciplines such as psychology and economics. By the 1960s, the quest for academic credibility, combined with a technological and data revolution, propelled more scientific ways of thinking into the discipline. This scientific approach combined the use of quantitative method, model-making, and hypothesis-testing. The regional idiosyncrasies were condemned as "old hat," and geographers turned themselves into spatial scientists. This very compressed "history" has some validity for the earlier years of the twentieth century, although the notion of paradigm shifts has been challenged (for a concise review, see Gregory 1994).

The notion of paradigm shifts essentially becomes inapplicable in the confusing and exciting world of post-quantitative human geography (Billinge, Gregory, and Martin 1984). It is recognized that in recent human geography, there are coexistent, contradictory, and competing communities of scholars adhering to different views of the world, different schools of thought, and different approaches to research questions. Box 1.3 shows that the recent period is occupied by a number of competing viewpoints jostling for space and credibility. The reactions against normalizing spatial science have spawned a huge diversity of approaches; by the early 1980s, radical, feminist, and environmental geographers were reasserting the importance of the social, the agency of the individual, and the particularity of place. Qualitative research requiring qualitative methods reasserted its respectability.

From the schema outlined in Box 1.3, a few major points can be drawn:

1. The period of spatial science is unique and aberrant in focusing on quantitative methods.
2. The paradigm shifts within geography have involved an increasing separation in the methods and philosophies of human geography from those of physical geography. The reactions against "scientific" geography established since the 1970s have drawn human geography more and more into the realms of critical social science, while physical geography has remained essentially within the scientific paradigm.
3. The questions that geographers have asked have oscillated between elucidating general trends and patterns in one period to examination of the individual and unique in subsequent phases of geographic inquiry.

BOX 1.3 Paradigm Shifts and Research Methods in Geography from the Twentieth Century*

Time	Paradigm	Research Questions	Research Methods	Characteristics	Trends
Early 20th century	Exploration/discovery	Discovery	Exploration	Colonial	
↕	Classical geography	General/theoretical/contextual	Qualitative/quantitative	Environmental determination	Decrease in spatial and time scale
	Regional geography	Unique/empirical	Qualitative/regional delineation and description	Region both method and object of study	
	Spatial science	General/theoretical and empirical	The "quantitative revolution"	Scientific method	Increasing separation of physical from human geography
1980s +	Critical social science	Theoretical/structural/individual	Qualitative	Pluralist	Increasing diversity in approaches
	Radical	Theoretical/structural			
	Feminist	Structural/individual			Rise of environmental studies
	Phenomenological	Unique/individual	Qualitative	Local	
	Postmodern	Theoretical			
	Postcolonial/subaltern	Theoretical/empirical	Qualitative	Global and local	Includes embodiment

* The shifts identified post-1980 are most relevant to human geography rather than to geography in general.

The multitude of post-quantitative approaches allows both individual and structural research questions to be tackled.

4. In general, the scale of geographic inquiry has shifted from the global, to the regional, to the local. However, recent writings are concerned not only with the specifics of individual experiences and places but have re-engaged with both the theoretical and the global.
5. The new foci outlined in this list on embodiment and social media provide new tools to address the two fundamental questions outlined earlier of societal structures and individual experiences.

Qualitative methods are currently used by all the major groups within the critical social science approaches utilized in human geography and identified in the lower half of Box 1.3. Much of the drive for qualitative research has come initially from humanistic geography of the late 1970s, which focused geographers sharply on values, emotions, and intentions in the search to understand the meaning of human experience and human environments (for example, Ley 1974). Another significant influence in the reassertion of qualitative methods has been the work of feminist geographers establishing links between the personal and the political. A clear example of this might be Cupples and Harrison's (2001) work on media representations of sexual assault by establishment figures or Mackenzie's earlier (1989) work on women and environments in a postwar British city. This approach also predominates in recent studies that reconfigure the "everyday" domestic spaces of women as entwined with very public circuits of community and even globalization (see, for example, Isabel Dyck 2005 on female caregivers). However, this may be becoming less evident now as feminist geography becomes mainstreamed (Johnson 2012). Similarly, and as discussed earlier, studies that might be grouped as **postcolonial research** increasingly give voice to people defined as "Other," enabling multiple interpretations of events to be heard. The power of this scholarship has been not just an opening up of geographic inquiry but equally an embracing of what Joanne Sharp (2009) would refer to as the transformative power of emotional geographies. Much qualitative work within contemporary human geography cannot be clearly categorized within any of the schools of thought listed in the final section of Box 1.3 (critical social science) but is concerned with the broad questions of elucidating human environments and human experiences within a variety of conceptual frameworks.

Contemporary Qualitative Geography: Theory–Method Links

Contemporary human geography adopts a broad range of research methods. Although not tied specifically to particular theoretical and philosophical viewpoints, the methods discussed in the preceding sections of this chapter are often more frequently associated with one standpoint than with another. For example, feminist geography is often associated with qualitative methods through a naturalized association of the feminine with the "softer" qualitative approaches. However, feminist questions can be stated within a variety of theoretical frameworks and have used a variety of methods (Lawson 1995; Johnson 2012).

Qualitative methods have been used more widely in human geography throughout the past century than is commonly believed. They have been used in conjunction with quantitative methods in a search for generality and have also been used to explain difficult cases or to add depth to statistical generalizations. Above all, they have traditionally been used as part of triangulation or multiple methods in a search for validity and corroborative evidence. However, qualitative methods have also been used in different conceptual frameworks to reveal what has previously been considered unknowable—feelings, emotions, attitudes, perceptions, and cognition. Overwhelmingly, qualitative methods have been used to verify, analyze, interpret, and understand human behaviour of all types.

Studies that utilize qualitative methods in their own right to express individual meanings are much more limited in number. Although humanistic geographers of the 1980s laid claim to this territory, the output of the humanistic school per se was both limited and short-lived. Even by 1981, Susan Smith was calling for "rigour" in humanistic method in a way akin to the current calls for **validity** and **replicability** and (by implication) respectability (Baxter and Eyles 1997; Philip 1998).

Similarly, the aims of critical realism expressed by Sayer and Morgan (1985) gave pre-eminence to individuals' actions and their meanings, yet contributions in this mould to geography literature have been slim. The schools of thought that may have made the greatest contribution to answering qualitative research questions have been the feminist and the post-structural (including the postmodern and the postcolonial). Both these frameworks recognize that multiple and conflicting realities coexist. They deliberately give voice to those silenced or ignored by hegemonic (modern, colonial) views of histories and geographies. They embody and acknowledge previously anonymous individuals. Paradoxically, however, the voice of the oppressed not only speaks for itself: it is part of a wider whole. Reality is like an orchestra: post-structural approaches differentiate the instruments and their sounds and bring the oboe occasionally to centre stage; usually dominated by the strings, the minor instruments too have a tune to play and a thread that forms a distinct but usually unheard part of the whole. It is the voices of the women and children, the colonized, the Indigenous, the minorities that, when released from their silencing, enable a more holistic understanding of society to be articulated (for example, Lane et al.'s 2003 examination of Aboriginal and mining company conflicts over the Coronation Hill mine project in Northern Australia and Hibbard, Land, and Rasmussen's 2008 examination of the potential role of planning processes in empowering Aboriginal people).

Sayer and Morgan (1985) make the point that exactly the same research technique can be used in different ways for different purposes according to the theoretical stance of the researcher. Interviews, for example, may be used to gain access to information from gatekeepers about structures or to give voice to silenced minorities. John McKendrick (1996, Table 1) considers the relationship between methods and their applications in different research traditions. He contrasts the use of interviews in the humanistic tradition to "explore the meaning of the migration of each individual migrant" with their use in a postmodernist framework, whereby in-depth interviews with women may be used "to 'unpack' their rationalizations of their migrations."

Qualitative methods raise an immense number of difficult issues that are considered in more detail in several of the chapters of this volume. Among these issues are concerns over authorship, audience, language, and power (discussed in many of the chapters of this book). Ethnographic research is often highly complex, within which the individual subject and the audience for the research are often intermingled and mutually dependent. The position of the author as observer in relation to the object of research raises issues of power relations and control. The engagement of the researcher does not necessarily allow the voices of the researched to speak, since they are mediated through the researcher's experience and values. The language of research reporting may also exclude those researched, although language varies according to the audience toward which the research is directed. The key issue of the outsider gazing, perhaps voyeuristically, at those defined as "Other" is an intractable problem that needs to be recognized even if it cannot be solved. Even participant observation cannot surmount inbred and naturalized class differences, as demonstrated by Canadian geographer Pamela Moss (1995) in her immersion in manual labour in hotels. Moss never managed to bridge the cultural, linguistic, and social gap between herself as middle-class researcher and the housemaids who spent their lives in manual labour. The engagement with human research subjects raises significant questions about ethical research practice, which are now being addressed seriously within the discipline.

Summary

This chapter outlines the maturing of qualitative methods in human geography and categorizes these into three major categories: oral, textual, and participatory methods. The range of methods is used to answer two broad research questions relating either to the experiences of individuals or to the social structures within which they operate. We also draw attention to some of the more recent trends in qualitative methods, including the focus on individual behaviours, performativity, the use and texts of social media, and the body as a site of inscription. Qualitative methods have often been categorized as oppositional to quantitative methods, yet in many respects this is a false dichotomy. The differences between qualitative and quantitative methods are related to the conceptual frameworks from which they have been derived. In elucidating human experiences, environments, and processes, qualitative methods attempt to gather, verify, interpret, and understand the general principles and structures that quantitative methods measure and record. Further, qualitative methods have very frequently been used in conjunction with other methods; however, increasingly, qualitative methods are being used alone to explore human values, meanings, and experiences. This reflects the maturation of qualitative approaches and their widespread acceptance within the scholarly community. It is clear that in human geography, qualitative methods have come of age, and that human geographers are bringing new and effective approaches to the fundamental questions of societal structures and individual experiences.

Key Terms

bias
bodily performance
confirmability
credibility
deconstruction
dependability
diary interviews
diary photographs
discontinuous writing
discourse
epistemology
focus group
go-along interviews
interviews/interviewing
landscape
mixed methods
objective
ontology
oral history
oral methods
"Other"
paradigm
participation/participant observation
performativity
photo-elicitation
postcolonial research
positionality
quantitative methods
random sample
reflexivity
replicability
representation
respondent
semiotics
structures/social structures
subjective
text/textual
textual analysis
triangulation
validity

Review Questions

1. What research questions may be answered by qualitative methods? Give examples.

2. What are the significant newly developing areas of qualitative research methods used in human geography?
3. How may different types of qualitative methods be linked to theoretical approaches within the discipline?
4. Outline two key issues in the quantitative/qualitative debate.

Review Exercise

Select one of the references listed in the suggested reading below. Read this reference and consider:

a. whether the research question is about individual experience or societal structures or processes, or a hybrid of the two;
b. whether the qualitative method used is predominantly oral, textual, or participatory;
c. what is the major insight this article has produced for you?

Useful Resources

Billo, E. and N. Hiemstra, 2013. "Mediating messiness: expanding ideas of flexibility, reflexivity, and embodiment in fieldwork." *Gender, Place & Culture: A Journal of Feminist Geography* 20 (3): 313–328. Outlines the tension between theory and practical fieldwork for two feminist geographers undertaking PhD study in Ecuador. It also considers embodiment from the researcher's perspective, including the emotional and physical toll taken from working in marginalized and isolated communities.

Botz-Bornstein, T. 2013. "From the stigmatized tattoo to the graffitied body: Femininity in the tattoo renaissance." *Gender, Place & Culture: A Journal of Feminist Geography*, 20 (2): 236–252. Examines the body as a blank canvas for the creation of text and describes the creation of a "tattoo space" as a shift from traditional tattoos to body graffiti.

Davies, G., and C. Dwyer. 2007. "Qualitative methods: Are you enchanted or are you alienated?" *Progress in Human Geography* 31 (2): 257–66. Provides an overview of methodological advances in three significant subfields of human geography: embodied and emotional geographies, geographies of nature, and performing places.

Duffy, M. and G. Waitt. 2011. "Sound diaries: a method for listening to place." *Aether: The Journal of Media Geography* 7: 119–136. Examines geographies of sound, using a readable example of a music festival, and examines the way sound creates place.

Ferbrache, F. 2013. "Le Tour de France: A cultural geography of a mega-event." *Geography* 98: 144–151. A short and relatively straightforward article which interprets the constructions of the Tour as a national icon, as a site of resistance, and as a festival, a spectacle and a consumer experience.

Fridolfsson, C. and I. Elander. 2013. "Faith and place: Constructing Muslim identity in a secular Lutheran society." *Cultural Geographies* 20 (3): 319–337. Uses interviews with imams and textual analyses from the print media and Internet to interpret the construction of migrant Muslim identity in Sweden.

Winchester, H.P.M., and M.W. Rofe. 2005. "Christmas in the 'Valley of Praise': Intersections of the rural idyll, heritage and community in Lobethal, South Australia." *Journal of Rural Studies* 21: 265–79. Provides significant insight into the development and use of a qualitatively based multi-methodology. Specific methods successfully employed in this research include interviews, participant observation, textual analysis, and landscape deconstruction.

Woon, C.Y. 2011. "'Protest is just a click away!' Responses to the 2003 Iraq War on a bulletin board system in China." *Environment and Planning D: Society and Space* 29 (1): 131–149. Examines a Chinese virtual protest against the Iraq War, which, while apparently a vehicle for freedom of expression, was allowed by Chinese authorities as a foreign event of little immediate impact and was thus socially constructed and bounded.

Power, Subjectivity, and Ethics in Qualitative Research

Robyn Dowling

Chapter Overview

This chapter introduces issues that arise because qualitative research typically involves interpersonal relationships, interpretations, and experiences. I discuss three issues of which qualitative researchers need to be aware: (1) the formal ethical issues raised by qualitative research projects; (2) the power relations of qualitative research; and (3) objectivity, subjectivity, and intersubjectivity. Rather than advocating simple prescriptions for dealing with these issues, the chapter proposes that researchers be "critically reflexive."

Introduction: On the Social Relations of Research

Chapter 1 outlined the types of research questions asked by qualitative researchers—namely, our concerns with the shape of societal structures and people's experiences of places and events. This chapter takes you one step closer to conducting qualitative research. It discusses some of the implications of research as a social process. Collecting and interpreting social information involves personal interactions. Interviewing, for example, is essentially a conversation, albeit one contrived for research purposes. Interactions between two or more individuals always occur in a societal context. Societal norms, expectations of individuals, and structures of power influence the nature of those interactions. For instance, when you are conducting a focus group, you may find men talking more than women and people telling you what they think you want to hear. Societal structures and behaviours are not separate from research interactions. This places all social researchers in an interesting position. We may use a variety of different methods to understand society, but those methods cannot be separated from the structures of society. The converse is also true. The conduct of social research necessarily has an influence on society and the people in it. By asking questions or participating in an activity, we alter people's day-to-day lives. And communicating the results of research can potentially change social situations.

Both qualitative and quantitative researchers recognize this lack of separation between research, researcher, and society. What typically distinguishes qualitative researchers' approach to this issue is the emphasis they give to it. For those who use the qualitative methods discussed in this volume, the interrelations between society, the researcher, and the research project are of critical and abiding significance. They permeate all methods and phases of research. These relationships cannot be ignored and raise key issues that must be considered when designing and conducting research. This chapter outlines three issues that arise because of the social nature of research and suggests an approach to dealing with them.

The chapter begins with a discussion of research ethics and the philosophies and practicalities of institutional review of research ethics within universities. It introduces key criticisms of institutional review and moves beyond these guidelines to introduce the concept of critical reflexivity. The chapter then focuses on the ways power traverses the conduct of qualitative methods. Finally, the chapter considers the significance that a researcher's subjectivity and intersubjective relations have for the collection of qualitative data.

A word of caution to begin. In general, most of the chapters that follow offer practical guides to different qualitative methods. They explain how to be a participant observer, how to conduct a focus group, and so forth. This chapter is different in two important respects. First, it is not about any specific method. It is about issues common to all of the methods discussed in this volume. This chapter should be used in combination with those on particular methods to help you think about the specific challenges you are likely to encounter in your research. Second, this chapter does not—and cannot—offer hard and fast rules on conducting ethical research that is responsive to matters of power and **intersubjectivity**. The conduct of good, sensitive, and ethical research depends in large part on the ways you deal with your unique relationships with research participants, peers, and other organizations at particular times in particular places.

Research Ethics and Institutional Review of Research Ethics

All research methods necessarily involve ethical considerations. Decisions about which research topics to pursue, appropriate and worthwhile methods of investigation, "right" ways to relate to sponsors of and participants in research, and appropriate modes of writing and communication of results involve ethical questions. These questions include how researchers ought to behave, the role of research in the pursuit of social change, and whether and how research methods are "just." **Research ethics**, broadly defined as being about "the conduct of researchers and their responsibilities and obligations to those involved in the research, including sponsors, the general public and most importantly, the subjects of the research" (O'Connell-Davidson and Layder 1994, 55), constitute an issue that must be dealt with in your research. Researchers' ethical practice is regulated in a number of ways. The Association of American Geographers (2005), for example, has published a statement on professional ethics for its members that is designed "to encourage consideration of the relationship between professional practice and the well-being of the people, places and environments that make up our world." The late twentieth century saw increasing formalization of ethical review (Israel and Hay 2006). Important here are statements on ethical research by government agencies charged with the oversight of research. In Canada, for example, the "Tri-Council policy statement: Ethical conduct for research involving humans" (Canadian Institutes of Health Research et al. 2010) provides both general guidelines and prescriptive rules for ethical conduct. These statements and their attendant rules are then implemented by ethics committees (known, for example, as institutional review boards [IRBs] in the United States, research ethics boards [REBs] in Canada, and human research ethics committees [HRECs] in Australia), typically associated with organizations where a considerable amount of research is conducted, such as schools, hospitals, and Indigenous organizations, as well as universities.

University ethics committees are of most relevance to student research, although in some research, the ethics committees of other institutions like schools may also need to be consulted. In most universities, post-graduate students or those in charge of courses are required to obtain the formal approval of a university ethics committee before beginning research that involves people. University ethics committees focus on the researcher's responsibilities to research subjects and apply guidelines about what researchers should and should not do. It is useful to consider such formal guidelines as a first step in thinking through the social context of your research. The committee may not evaluate your research design but will want to know the aims of the research and the methods you will use. It will be concerned primarily with your responsibilities to research participants with regard to matters of privacy, informed consent, and harm.

Privacy and Confidentiality

Qualitative methods often involve invading someone's privacy. You may be asking very personal questions or observing interactions in people's homes that are customarily considered private. Ethics committees are concerned that these private details about individuals not be released into the public domain. Accordingly, you may have to show that your original field notes, tapes, and transcripts will be stored in a safe place where access to them will be restricted. You may also need to ensure that your research does not enable others to identify your informants. There are various ways of ensuring the anonymity of informants, including using pseudonyms and masking other identifying characteristics (for example, occupation, location) in the written version of your research. You should note, however, that when dealing with significant public figures, it is sometimes not possible or desirable to ensure anonymity. For example, in O'Neill's (2001) research on the Australian-based transnational BHP (Broken Hill Proprietary Company Limited), he identified both the firm and the executives with whom he spoke.

Informed Consent

For most geographical research, participants must consent to being part of your research. In other words, they have to give you permission to involve them. However, this criterion is somewhat stricter than a simple "yes, you can interview me." It must be ***informed* consent** (see the helpful discussion of this in Chapter 3). Informants need to know exactly what it is that they are consenting to. You need to provide participants with a broad outline of what the research is about, the sorts of issues you will be exploring, and what you expect of them (for example, the amount of time required to complete an interview). Most ethics committees recognize that there are exceptions to informed consent. Simple observation of people in a place like a public shopping mall, for example, may not need the explicit consent of those individuals. Indeed, it may be physically impossible to secure the consent of everyone involved. Sometimes informed consent may be waived, although an ethics committee will typically ask you to justify that decision. There are some relatively rare instances when research may involve deception. Deception happens when research participants either do not know that you are a researcher (for example, Routledge's 2002 work in India) or do not know the true nature of your research. Set against the principle of informed consent, deception is clearly an ethically difficult issue, and you should think carefully and seek advice before contemplating a research project that involves deception. Moss's (1995) discussion of why and how she used deception in her study of domestic workers may be useful.

Harm

Your research should not expose yourself or your informants to harm—physical or social. As social scientists, it is highly unlikely that you will be subjecting people to physical harm. You may, however, be bringing them into contact with "psycho-social" harm. You may raise issues that may be upsetting or potentially psychologically damaging. This does not mean that your research cannot proceed. Rather, it means your research should cater to this possibility, such as being prepared to provide contact details of a counselling service if required (e.g., Hutcheson 2014). You should also avoid putting yourself at risk during the research. For example, a young woman planning a participant observation project on single women's safety at night on public transport could meet a very cautious response from an ethics committee or research supervisor because of the potential danger to her while conducting that work.

Ethics and Online Research

The rise of the Internet has transformed qualitative research, opening up a plethora of new sources of data such as social media (Curtis and Mee 2012) or YouTube (Longhurst 2009), new ways of collecting qualitative data like online surveys, and new ways of recruiting research participants. These methods and sources pose myriad challenges to research ethics, especially to informed consent, privacy, and confidentiality. Using online questionnaires brings confidentiality to the fore. Online surveys are often conducted using commercial software and web hosting services, which means that the researcher doesn't have complete control over protecting participants' identities. The web service, for example, may track IP [Internet Protocol] addresses and hence enable identification of participants. Researchers should hence be careful to use organizations with privacy and data security policies consistent with institutional privacy requirements.

Using the various elements of Web 2.0 as sources of qualitative data is even more contentious in terms of privacy and informed consent. Are blog postings or Facebook comments, to take just two common examples, public or private? If they are considered public, then most researchers would be comfortable using them. But what if it is a subscriber-only blog? Is it ethical for a researcher to subscribe to the blog and use the information for research purposes? Who should they inform and what does informed consent in this case look like? Some web resources provide their own answer to this question through statements that specify by whom and how the information in the blog postings can be used. But not all do. One suggestion is to gain permission from list owners to observe and record information and interactions online (Hesse-Biber and Griffing 2013). In other situations a direct approach to participants through the blog may be more appropriate, following Buchanan's (2011) claim that the onus is on researchers to identify themselves and gain informed consent. There is nonetheless agreement that gaining informed consent is a key issue that needs to be considered when setting the methodological parameters of a qualitative research project involving online sources.

Criticisms of, and Moving Beyond, Ethical Guidelines

Although important, ethical rules and ethics committees are not unproblematic for qualitative researchers in human geography. At a general level, the universalist ethical stance and biomedical model of research embedded in REBs involve rigid codes that cannot always deal with "the variability and unpredictability of geographic research" (Hay 1998, 65). More specific criticisms of how the issues of informed consent, privacy, confidentiality, and harm are framed within REBs, and their potential incompatibilities with qualitative research, now abound. They include the individualized nature of informed consent when a community's or collective consent may be more appropriate (Butz 2007) and the ways the unpredictable nature of qualitative research make gaining informed consent difficult. Whether guarantees to privacy and anonymity can and should be given if illegal behaviours are involved is also at issue. Finally, in relation to harm, it is not always possible to predict the impact of research on participants, especially over a longer term (Bailey 2001).

Critical Reflexivity

Despite these problems, ignoring institutional ethical review is neither desirable nor possible (Martin 2007). Instead, ways of practically negotiating with these codes are preferable. For Israel and Hay (2006, 142), this involves a commitment to theoretically informed, self-critical ethical conduct, revolving around awareness of how to identify and resolve ethical dilemmas when they arise. Implicit in practical engagements with ethical codes is the suggestion that, as geographers engaged in research, we must constantly consider the ethical implications of our activities. Because research is a dynamic and ongoing social process that constantly throws up new relations and issues that require constant attention, self-critical awareness of ethical research conduct must pervade our research. Our engagement with ethical behaviour does not end when we submit our research proposal to an ethics committee.

As the forgoing discussion of ethical research conduct might imply to you, human geographers have come to appreciate more fully than ever before the social nature and constitution of our research. Indeed, we now recognize and acknowledge this location through the concept of **critical reflexivity**. **Reflexivity**, as defined by Kim England (1994), is a process of constant, self-conscious scrutiny of the self as researcher and of the research process. In other words, being reflexive means analyzing your own situation as if it were something you were studying. What is happening? What social relations are being enacted? Are they influencing the data?

Critical reflexivity is difficult but rewarding. It is rewarding in that, as some of the examples used in this chapter indicate, it can initiate new research directions. Critical reflexivity is, however, difficult in two respects. First, many geographers do not write about the research process in their published work. Linda McDowell (1998) comments, for example, that many of the details of how her merchant banking research proceeded and, as a result, her reflections on the process do not

The Research Diary as a Tool for the Reflexive Researcher

Your efforts to be reflexive will be enhanced if you keep a research diary. The contents of a research diary are slightly different from those of a fieldwork diary. While a **fieldwork diary**, or field notes, contains your qualitative data—including observations, conversations, and maps—a research diary is a place for recording your reflexive observations. It contains your thoughts and ideas about the research process, its social context, and your role in it. You could start your research diary by writing answers to the questions posed in the checklist set out in Box 2.3 at the end of the chapter.

appear in the book based on that research. However, discussions of reflexivity are becoming more common, and a guide to some of them is provided in the Useful Resources section at the end of the chapter. Second, reflexivity is difficult because we are not accustomed to examining our engagement with our work with the same intensity as we regard our research subjects. You may be helped in this matter by keeping a **research diary**, as outlined in Box 2.1.

The rest of the chapter focuses on two of the important issues about which you need to be reflexive: power and subjectivity. For readers who are especially interested, Chapter 19 expands on the discussion of reflexivity.

Power Relations and Qualitative Research[1]

One important outcome of the social character of qualitative research is that research is also interwoven with relations of power. Power intersects research in a number of ways.

It can enter your research through the stories, or interpretations, you create from the information you gather. Power is involved here because knowledge is both directly and indirectly powerful. Knowledge is directly powerful through its input into policy. Some studies are specific analyses of policy issues, and their results have a direct impact on people's lives. Knowledge is also indirectly powerful. The stories you tell about your participants' actions, words, and understandings of the world have the potential to change the way those people are thought about. Power relations also exist beyond the relationship between researcher and participant. Qualitative research often investigates issues that are the subject of intense societal and political scrutiny, like Indigenous disadvantage or childhood poverty.

Power is also involved in earlier parts of the research process. In undertaking qualitative research, you are attempting to understand—participating in, and sometimes creating—situations in which people (yourself included) are differently situated in relation to social structures. Both you and your informants occupy different "speaking positions." Not only do you and your informants have different intentions and social roles, but you also have different capacities to change situations and other people (see Liamputtong 2007).

Social researchers typically enter one of three different sorts of power relations in parts of their work. **Reciprocal relationships** are those in which the researcher and the researched are in comparable social positions and have relatively equal benefits and costs from participating in the research. You may, for example, be conducting focus groups with your fellow students on how they are adjusting to university life. Although not absent, power differences in this relationship are minimal compared with two other sorts of relationships. The other two types of research relationships are forms of **asymmetrical relationship**, characterized by significant differences in the social positions of researcher and those being researched. On the one hand, those being studied may be in positions of influence in comparison to the researcher. Interviews with senior executives of large corporations or the "super-rich" can fall into this category because of such people's relative access to cultural and financial resources (see Hay and Muller 2012). This kind of asymmetrical research relationship was first termed ***studying up*** by anthropologist Laura Nader (1969). On the other hand, the researcher may be in a position of greater power than the research participant. This is termed a **potentially exploitative relationship**. Kate Swanson's research with young people who beg on the streets of Ecuador is an example of a potentially exploitative relationship with ethical and power dimensions that have been considered thoughtfully (Swanson 2007; 2008).

Power cannot be eliminated from your research, since it exists in all social relations. Human geographers typically have one of two responses to issues of power. The first, responding directly to potentially exploitative relationships, is to involve participants in the design and conduct of the research. In their research on young people's experience of crime and victimization, Pain and Francis (2003) explicitly sought young people's perceptions of what required investigation and then gave participants a number of opportunities to verify or refute the researchers' interpretations. Pain and Francis also conducted a number of meetings and workshops to disseminate their findings. Through these two strategies, they hoped to effect some social change and alter potentially exploitative relationships.

Such participatory forms of research are not always the most appropriate for every research project or for student researchers. A second response that recognizes and negotiates relations of power is critical reflexivity. For example, when collecting data, our responsibility to the research participants is such that we should not take advantage of someone's less powerful position to gather information. In the case of homeless youth, for example, you would not make the possibility of gaining access to shelter dependent on that person's participation in the study. But you cannot eliminate the power dimension from your research, since it exists in all situations. The best strategy is to be aware of, understand, and respond to it in a critically reflexive manner. Critical reflexivity does not necessarily mean altering your research design, but it does imply that you reflect constantly on the research process and modify it where appropriate. When you are formulating your topic, think about the various ethical and power relations that may be enacted during your research (see, for example, Box 2.2). Are you happy with the situation? Would you like to do anything differently? Could you justify your actions to others? You should also think about how you communicate the results. Have you reflected as faithfully as possible what you have been told or have observed without reproducing stereotypical representations? Are you presenting what you heard and saw or what you expected to hear and see? Remember, the stories you tell may change the worlds in which you and your research participants live (for more detail on this matter, see Chapter 19).

Increasingly, research in human geography is connected to organizations outside the university. Government, industry, or community groups may directly commission human geographers to carry out research on a particular topic, or student research projects may be carried out in consultation with such groups. These situations introduce new power dynamics to the research process. There may be attempts to control methodological protocols, research findings, and their dissemination, or you may find that your desire to please others involved in the research raises some moral dilemmas. If such issues arise, awareness of how other researchers have navigated such relationships is a necessary first step (see Hallowell, Lawton, and Gregory 2005). Israel and Hay (2006, 112) usefully employ the notion of research integrity as an additional prompt for all researchers. Research integrity encompasses intellectual honesty, accuracy, fairness, and collegiality. Its first two elements, honesty and accuracy, can guide researchers' interactions with other organizations involved in research projects. Sponsor dissatisfaction with research findings, for example, can be discussed by reiterating the rigour of the research methodology and the researcher's confidence in the robustness of the analysis.

Researchers also have relationships with peers. Undergraduate students may conduct a research project as a group activity, graduate students may be working on one element of a larger research project directed by their supervisors, or you may be the geographer on a multidisciplinary research team. The second two aspects of research integrity—fairness and collegiality—come into play here. Nurturing relationships through open and transparent communication is one method of successfully negotiating such situations, although again, the issue of power relations (e.g., between student and supervisor, contract researcher and employer) remains.

Sexism in Research

Sexism can present problems in many different sorts of research projects. Eichler (1988) identifies four primary problems of sexism in research:

- *androcentricity/gynocentricity*: A view of the world from male/female perspectives respectively. For instance, concepts of "group warfare" developed through reference to men's experiences only.
- *overgeneralization*: A study is only about one sex but presents itself as applicable to both sexes. For example, a study that is exclusively concerned with men's location decisions might be misleadingly entitled "Residential location decisions in Vancouver."
- *gender insensitivity*: Ignoring gender as an influential factor in either the research process or the interpretation. For instance, a study of the geographical effects of a free trade agreement that fails to consider any gender-specific effects.
- *double standards*: Identical behaviours or situations are evaluated, treated, or measured by different means or criteria (for example, drawing different conclusions about men and women on the basis of identical answers to a survey or aptitude test).

Eichler also identifies three "derived" forms of sexism:

- *sex appropriateness*: The notion that some characteristics and behaviours are accepted as being more appropriate for one sex than for the other (for example, designing a research project on parental perceptions of children's play space and interviewing women only because you assume that they will know more about the issue).
- *familism*: Using family as the unit of analysis when the individual might be more appropriate or vice versa (for example, working at the family scale in evaluating the social costs and rewards of in-home care for the elderly rather than exploring the different implications it might have for males and females within the family).
- *sexual dichotomism*: Postulating absolute differences between women and men (for example, women are sometimes considered "naturally" more timid than men).

Objectivity, Subjectivity, and Intersubjectivity in Qualitative Data Collection

Objectivity has traditionally been emphasized in geographic discussions of quantitative research methods. Objectivity has two components. The first relates to the personal involvement between the researcher and other participants in the study. The introduction to this chapter suggested that it is impossible to achieve this sort of objectivity because of the social nature of all research. Objectivity's second component refers to the researcher's independence from the object of research. This implies that there can be no interactive relationship between the researcher and the process of data collection and interpretation. Clearly, however, dispassionate interpretation is difficult, if not impossible, because we all bring personal histories and perspectives to research. **Subjectivity** involves the insertion of personal opinions and characteristics into research practice. Qualitative research gives emphasis to subjectivity because the methods involve social interactions. As will become evident in later chapters, you need to draw on your personal resources to establish rapport and communicate with informants. This is equally the case with online environments where researcher identities affect participation and the data gathered (see Kinsley 2013). Discourse analysis also involves your subjectivity in that your everyday understanding of the world helps you to decipher texts. If subjectivity is important, then so too is intersubjectivity. This refers to the meanings and interpretations of the world created, confirmed, or disconfirmed as a result of interactions (language and action) with other people within specific contexts. Collecting and interpreting qualitative information relies upon a dialogue between you and your informants. In these dialogues, your personal characteristics and social position—elements of your subjectivity—cannot be fully controlled or changed, because such dialogues do not occur in a social vacuum. The ways you are perceived by your informants, the ways you perceive them, and the ways you interact are at least partially determined by societal norms.

Critical reflexivity is the most appropriate strategy for dealing with issues of subjectivity and intersubjectivity. Although you cannot be entirely independent from the object of research, trying to become aware of the nature of your involvement and the influence of social relations is a useful beginning that can help you to identify the implications of subjectivity and intersubjectivity in your research.

Geographers' work on gender provides some good examples of the role of intersubjectivity in research. Gender is important because we often ascribe characteristics to people on the basis of gender. Furthermore, personal interactions vary with the gender of participants; we tend to react differently to men and women. Therefore, gender is a factor that can influence data collection. For instance, Andy

Herod (1993) found that male union officials restricted the sort of information he was given during interviews because he was a man. Specifically, his informants downplayed the role of women in the union's struggles. Herod attributes this not only to the perspectives of the union officials but also to the social (masculine) context of the interviews. The union officials assumed, Herod thinks, either that he was not interested in gender or that it was inappropriate to raise the issue in the context of a male-to-male conversation. By contrast, Hilary Winchester (1996) suggests that being a woman interviewing men aided her research on lone fathers. She found herself adopting a typically feminine role of facilitating conversation with men, which helped considerably in gathering the men's stories.

Your ability to interpret certain situations also depends on your own characteristics. Important here is a debate about the relative merits of being an **insider** or an **outsider**. An insider is someone who is similar to their informants in many respects, while an outsider differs substantially from their informants. Coming from an "out" lesbian interviewing other lesbians, Valentine's research on sexuality could be considered insider research, whereas she was an "outsider" in her interviews with parents and children about childhood (see Valentine 2002). One position in the debate is that for you as an insider, both the information you collect and your interpretations of it are more valid than those of an outsider. People are more likely to talk to you freely, and you are more likely to understand what they are saying, because you share their outlook on the world. If you are not a member of the same social group as your informants, then establishing rapport may be more difficult (see Chapters 3, 4, and 19). And since you do not share their perspective on the world and their experiences, then your interpretations may be less reliable. But being an outsider can also bring benefits to the research. It may mean that people make more of an effort to clearly articulate events, circumstances, and feelings to the researcher. I recently conducted research on the housing needs of PhD students in Australia, where elements of the cultural distance between researchers and participants instigated insightful exchanges of information.

A final perspective to consider is that you are never simply either an insider or an outsider. We have overlapping racial, socio-economic, gender, ethnic, and other characteristics. If we have multiple social qualities and roles, as do our informants, then there are many points of similarity and dissimilarity between ourselves and research participants. Indeed, becoming aware of some of these commonalities and discrepancies can be one of the pleasures and surprises of qualitative research. As a British researcher, Mullings (1999) was far removed from the social worlds of the Jamaican factory workers and employers she interviewed. To her surprise, however, her position as a person of African descent led to shared experiences with local interviewees and subsequently rich interviews.

Intersubjectivity also means that neither you, your participants, nor the nature of your interactions will remain unchanged during the research project. Your general outlook and opinions may indeed change as result of your research. If you are a participant observer, for example, you will be immersing yourself in a situation that will invariably affect you. Robin Kearns (1997) discusses how both he and his research project changed in response to his interactions as an observer. Kearns began his project on Māori health in rural New Zealand in the waiting room of a medical clinic. He did not intend to interact with patients in the waiting room. His attempts to be unobtrusive failed because of the inquisitiveness of patients who talked to him incessantly. These interactions had two immediate effects. First, they forced him to adopt a different stance in his research: he had to be more open about his presence and his research. Second, he modified his method by actively involving the local community in the research design and the findings (Kearns discusses this more fully in Chapter 15).

Critical reflexivity can give rise to new and exciting directions in one's research and is dependent on context. Karen Fisher (2014) uses "autoethnography" to illustrate the very different racialized ways she is considered in different research contexts. As a researcher in both the Philippines and Aotearoa/New Zealand, she was positioned in almost opposite ways: as "insufficient Other" and "sufficient Self."

Summary and Prompts for Critical Reflexivity

Using the qualitative methods described in subsequent chapters will involve you in various social relations and responsibilities. I have advocated critical reflexivity—self-conscious scrutiny of yourself and the social nature of the research. Critical reflexivity means acknowledging rather than denying your own social position and asking how your research interactions and the information you collect are socially conditioned. How, in other words, are your social role and the nature of your research interactions inhibiting or enhancing the information you are gathering? This is not an easy task, since it is not always possible to anticipate or assess accurately the ways in which our personal characteristics affect the information we accumulate.

I shall not end this chapter with answers. Instead, I offer a preliminary set of prompts that might help you reflect critically on issues of ethics, power, and intersubjectivity in different types and phases of research (see Box 2.3).

How to Be Critically Reflexive in Research

On beginning:

- What are some of the power dynamics of the general social situation I am exploring, and what sort of power dynamics do I expect between myself and my informants?
- In what ways am I an insider and/or outsider in respect to this research topic? What problems might my position cause? Will any of them be insurmountable?
- What ethical issues might impinge upon my research (for example, privacy, informed consent, harm, coercion, deception)?

After data collection:

- Did my perspective and opinions change during the research?
- How, if at all, were my interactions with participants informed/constrained by gender or any other social relations?
- How was I perceived by my informants?

Remember to take notes throughout data collection and keep them in a research diary.

During writing and interpretation:

- Am I reproducing racist and/or sexist stereotypes? Why and how?
- What social and conceptual assumptions underlie my interpretations?

Key Terms

asymmetrical power relation
critical reflexivity
fieldwork diary
informed consent
insider
intersubjectivity
objectivity
outsider
potentially exploitative power relation
reciprocal power relation
research diary
research ethics
studying up
subjectivity

Review Questions

1. What is critical reflexivity? Read a study in human geography in your field of interest that uses qualitative methods. Is the researcher reflexive about the research process? If so, how? If not, what sort of questions would you like to ask the researcher about the research process?
2. What forms of power relations may be part of a qualitative research project?
3. Read a piece of qualitative research in human geography. Can you identify any or all of Eichler's (1988) forms of sexism in the way the research has been carried out or in the interpretation?
4. How are qualitative methods intersubjective? How might social relations like gender influence the collection of data?
5. Outline and explain the issues of concern to university ethics committees.

Review Exercises

1. Each university has its own research ethics review processes. Search your university website and find the form required to apply for ethics approval for work in your field. Read the form and identify all the questions that pertain to informed consent, confidentiality, and harm. If you were conducting focus groups involving students and discussing their perceptions of the university campus, which of the ethics approval questions will be particularly important?
2. A research project conducted at Harvard University using Facebook as a data source is a prominent example of the ethical dilemmas faced in conducting online research. The case was reported in *The Chronicle of Higher Education* on 19 July 2011 in a story by Marc Parry entitled "Harvard Researchers Accused of Breaching Students' Privacy" (http://chronicle.com/article/harvards-privacy-meltdown/128166/). Read the report and prepare answers to the following questions:
   a. Which of the ethical concerns discussed in this chapter were raised in the case?
   b. How might the concerns have been remedied at the outset?

Useful Resources

Elizabeth Buchanan (2011) provides a comprehensive account of the ethical issues associated with online research. ("Internet research ethics: past, present, and future" in Mia Consalvo and Chartes Ess, eds. *The Handbook of Internet Studies*, Oxford: Wiley Blackwell, pp. 83–107).

Mark Israel and Iain Hay's highly readable 2006 book *Research Ethics for Social Scientists: Between Ethical Conduct and Regulatory Compliance* (London: Sage) more thoroughly canvasses many of the issues addressed in this chapter.

Samuel Kinsley (2013) nicely interprets online research through a geographical lens: "Beyond the screen: Methods for investigating geographies of life 'online'" *Geography Compass*, 7 [8], 540–555.

Brady Robards's reflections on the process of conducting research on social networking sites are helpful illustrations of critical reflexivity in action (Brady Robards, 2013, "Friending participants: managing the researcher-participant relationship on social network sites" *Young*, 21 [3]: 217–235).

Kate Swanson (2008) offers a novel account of the practice of critical reflexivity that revolves around the place of her pet dog in her research in her paper "Witches, children and Kiva-the-research-dog: Striking problems encountered in the field" (*Area* 40 [1]: 55–64).

Note

1. See Chapter 3 for further discussion of power-related issues.

Cross-Cultural Research: Ethics, Methods, and Relationships

Richie Howitt and Stan Stevens

Chapter Overview: An Explanation of Our Writing Method

Working across the differences that constitute "cultures" is a common challenge for geographical researchers. Ideas about research are profoundly shaped by cultural contexts. As geographers, we are aware of and engaged by the complex and dynamic relationships between culture and geography—our thinking is shaped by where we conduct research and the people we work with. In discussing how we might write about cross-cultural research as part of a wider discussion of qualitative methods, we initially thought that each of us might write sections of the chapter, with a jointly developed introduction and concluding discussion. This approach, we thought, would enable us each to have our own voice (and responsibility for positions that not all might fully share!). Using email allowed us to explore some alternatives. We decided that it might be better to think of our writing in terms of a conversation across cultures and geographies, with each of us framing key questions arising from our readings of each other's work and the wider literature on cross-cultural research and our frustrations with existing practices and their impacts on people we work with. In the end, we have written the chapter as a conversational text. In many ways, the text parallels what we seek to emphasize about cross-cultural field-based research—it involves respectful listening, difficult and challenging engagements, careful attention to nuances in the lives of "others," and a critical, long-term consideration of the implications of methods in the construction of meaning.

Modes of Cross-Cultural Engagement: Colonial, Postcolonial, Decolonizing, and Inclusionary Research

Richie: Can we start our conversation by talking about the intercultural spaces in which we might think about "cross-cultural" research? Conventionally, fieldwork is seen as "the heart of geography" (Stevens 2001, 66), but there has been a lot of discussion about just what we mean by "the field" in "fieldwork" (see, for example, special issues of *Professional Geographer* [1994] and the *Geographical Review* [2001], including Blake 2001; DeLyser and Starrs 2001; DeLyser et al. 2010; Gade 2001; Grimwood et al. 2012; Katz 1994; Kobayashi 1994; Nast 1994; Watson 2012; Zelinsky 2001) and about the multiple versions of "out-there-ness" and "in-hereness" that are constructed around ideas of what constitutes the "cross-cultural" in "cross-cultural research."

As I see it, most human geographic research is cross-cultural, because we are drawn into thinking about other people's constructions of place, other people's ways of reading their cultural landscapes—even when they are the landscapes that we live in ourselves in our everyday lives! I remember as an undergraduate being struck by the idea of a "geographical expedition" into Detroit (for example, Horvath 1971; Pawson and Teather 2002). That was juxtaposed, for me at least, with the whole "boy's own adventure" representation of fieldwork and the ethical issues of interpreting someone else's culture for one's own reasons.[1] The critique of that androcentric-style in geography was simultaneously about its sexism *and* its ethnocentricism.

Stan: I agree with you, Richie. Much geographical research is cross-cultural. Geographers often face similar ethical and methodological issues whether working close to home or on the other side of the planet. I've spent a good deal of time over more than 30 years living and working in the Himalaya with the Khumbu, Pharak, and Katuthanga Sharwa peoples of the Chomolungma region (better known to the rest of the world in British colonial nomenclature as the Sherpas of the Mount Everest region) (Stevens 1993; 1997; 2001; 2003; 2013; 2014; Stevens and Sherpa 1993). But one need not go to the other side of the planet to engage with "others," given the often complex dimensions of diversity created by societal and group constructions of regional, ethnic, linguistic, class, racial, gender, sexual, religious, ideological, and other differences. I suspect that the challenges and moral considerations of cross-cultural communication, learning, and activism might not be much different if I were to undertake research with my neighbours in the rural Massachusetts community I now live in or in the neighbourhoods of nearby cities.[2] Accordingly, I think that there may well be broader applicability and utility to some of the lessons I've learned from Sharwas, my Berkeley advisor Bernard Nietschmann (1973; 1979; 1997; 2001; Maya People of Southern Belize, Toledo Maya Cultural Council, and Toledo Alcades Association 1997) and his work with the Miskitu and the Maya, and postcolonial studies that challenge "conventional"—in the sense of common, long-established, and unexamined—views of fieldwork.[3] These lessons revolve around the importance of rejecting the attitudes, assumptions, purposes, and methodologies of what postcolonial theorists refer to as colonial research in favour of those of "decolonizing," "post-colonial" research.

Richie: Stan, these are issues that fieldworkers have grappled with for a long time. Even in "colonial" modes, people such as Clifford Geertz, Charles Rowley, and Nugget Coombs made valuable and very critical contributions.[4] Among geographers, we might think of the work of people such as Fay Gale, Janice Monk, Elspeth Young, Keith Buchanan, Oskar Spate, and Jim Blaut. You also identify "inclusionary" research as relevant to your work. How do you differentiate these different approaches to research? Do you see them as adopting different positionalities vis-à-vis difference, or do they adopt different methods, conceptual emphases, or perhaps purposes for the research that leads to your particular wording?

Stan: You've written insightfully about these distinctions yourself, Richie (Howitt 2002a; 2002b), which are critical to rethinking research ethics, methods, and relationships. I see colonial and postcolonial research as differing fundamentally from all of the standpoints you've just identified. **Colonial research** reflects and reinforces domination and exploitation through the attitudes and differential power embodied in its research relationships with "others," its dismissal of their rights and knowledge, its intrusive and non-participatory methodologies, and often also its goals and its use of research findings. **Postcolonial** research, to me, is a reaction to and rejection of colonial research and is intended to contribute to the self-determination and welfare of "others" through methodologies and the use of research findings that value their rights, knowledge, perspectives, concerns, and desires and are based on open and more egalitarian relationships. **Decolonizing research** goes further still in attempting to use the research process and research findings to break down the cross-cultural discourses, asymmetrical power relationships, representations, and political, economic, and social structures through which colonialism and neo-colonialism are constructed and maintained. I owe the term **inclusionary research** to your use of it for a particularly revolutionary kind of decolonizing research aimed at helping to empower subordinated, marginalized, and oppressed others to carry out self-defined agendas and projects and providing training and tools they can use to "overturn their world" (Howitt 2002a).

Colonial research has unfortunately long been the dominant mode of cross-cultural research in geography and anthropology and continues to be widespread today despite feminist and postcolonialist critiques. Much research continues to be imposed by outsiders for their own purposes and benefits on Indigenous peoples, on many of the non-Indigenous peoples of Asia, Africa, and Latin America, and on women and ethnic and cultural minorities in many societies throughout the world.[5] So much research has been conducted from such ethnocentric perspectives for morally suspect purposes and through ethically dubious ways that as

Linda Tuhiwai Smith (2012, 1) notes from Māori and other Indigenous peoples' perspectives in her important book, *Decolonising Methodologies: Research and Indigenous Peoples*, "the word itself, 'research' is probably one of the dirtiest words in the Indigenous world's vocabulary." Vine Deloria Jr made the same observation and indictment, declaring that "Indians have been cursed above all other people in history. Indians have anthropologists" (1988 [1969], 78).

Richie: The Australian (Wadi Wadi) poet Barbara Nicholson makes a similar point in her 2000 poem "Something there is . . ." (see Box 3.1).

Something There Is . . .

by Barbara Nicholson

. . . that doesn't like an anthropologist.
You go to a university
and get a bit of paper
that says you are qualified.
Does it also say that you
have unlimited rights
to invade my space?
It seems that you believe your bit of paper
is both passport and visa to my place,
that henceforth you have the right
to scrutinise the bits and pieces
of me.
You have measured my head,
indeed, you preserved it in brine
so that future clones of your kind
can also measure and calculate my cognition.
You've counted my teeth and compared them with
the beasts of the forest,
you've delved into my uterus,
had a morbid fascination with my sacred practices of incision and concision,
with the secret expressions of my rites of passage.
On your bit of paper you record how I dress,

earn my money,
what I eat and drink,
with whom I mix and with whom I don't,
where I go and don't go,
what I spend my money on,
the physical, mental and moral state of my being,
my marriage habits,
my birthing rituals,
my funerary rites,
the position I hold
in my society.
You analyse to a fine point
my art and music,
dance and composition,
horticulture and agriculture,
pharmacology and technology.
Nothing escapes your keen eye
and your pen records it
so that other aspirants to your elevated state
may draw on your findings and further explore
the intricacies
of me . . .
and perpetuate the invasion.

Oh yes . . . something there is.
If I were to go to university
and get a bit of paper that says,
"Wadi Wadi woman, you are an anthropologist,"
will that give me the right to invade your space,
to visit you
in your three-bedroom brick veneer,
note how many rice bubbles
go into your breakfast bowl,
what colour is on the roll
in your bathroom,
and see if the bathroom is clean?
Will I have the right to sit
on the end of your bed
and count every thrust

as you make love?
You will not complain
when I calculate your expenditure
on alcohol and yarndi,[6]
or count the cost when you visit McDonalds?
Remember, because I am an anthropologist,
my bit of paper gives me the right!
From now on I have carte blanche
to all the above
in your society,
and I can invade your space,
and I can record my findings
so that for generations to come
my kin can pursue
a relentless investigation
into the fabric of your existence,
into the bits and pieces
of you
so that you will be better able to fit into.

And resulting from my research
into the common cold and its effects
on you, a representative sample
of a cross-section of the population
of Double Bay, Sydney, 1994,
an avalanche of vultures from the media,
the government,
and the tourism industry
will descend on you
in ever increasing hordes
to see for themselves
if what I said
could really be true.
They will take over your lounge-room
and lay down laws for you to live by
—all for your own good of course;
they will point out to you
the necessity of changing your way of life
the prescriptive patterns of social behaviour

devised by them on your behalf.
I will be to you,
in the guise of humane academic inquiry,
as you have been to me, invader!
something there is . . .

Source: Dr. Hons. Caus. LLD, Wadi Wadi Elder, Barbara Nicholson. Reprinted with permission.

Stan: Exactly! I hadn't come across Barbara Nicholson's poem, but it should be required reading for prospective cross-cultural researchers. These kinds of **subaltern** critiques of colonial research condemn not only the ways in which research has objectified "others," violated their privacy and their humanity, and promoted colonizing agendas but also the ways in which Western science and scholarship have (mis)represented non-Western, Indigenous, and subaltern peoples and groups. Linda Tuhiwai Smith accordingly identifies research as "a significant site of struggle between the interests and ways of knowing of the West and the interests and ways of resisting of the Other" (Smith 2012, 2).[8]

Although colonial research has typically claimed positivistic objectivity, validity, and **reliability**, from the perspectives of the colonized and researched it has instead been seen as subjective, superficial, and misinformed. Smith (2012) argues that this reflects subjectivity, positionality, situated and partial knowledges, and ethnocentrism. She draws on Said's (1978) seminal postcolonial critique of the West's construction of the Orient, Foucault's (1972) concept of Western knowledge as a cultural archive in which many different cultures' knowledges, histories, and artifacts have been collected, categorized, and represented, and feminist critiques of positivism that highlight the partial and situated character of knowledge (Haraway 1991; Reinharz 1992; Stanley and Wise 1993). Colonial research also reflects and embodies unequal power relationships (and associated discourses and methodologies) and the ramifications they have for cross-cultural relationships and interaction. Imposed, exploitative research denies respect for alternative ways of knowing. It undermines trust and sabotages communication and collaborative exploration. By thus shaping the interactions and relationships on which cross-cultural fieldwork relies, this colonial model of unreflective fieldwork produces distorted and ethnocentric findings and analysis, fostering **Orientalism** in its depictions of "others." When it ventures into the realm of activism in the name of development or conservation, such research tends to support paternalism and imposed, external intervention, often with equally regrettable results.

Postcolonial research rejects the assumptions and methods of colonial research. Research is instead envisioned as a means of contesting colonialism and neo-colonialism through fostering self-determination and cultural affirmation (Smith 2012). In my view, postcolonial research aims at being emancipatory not simply through being more culturally sensitive or seeking local research approval but through respect for the **legitimacy** of others' knowledge and their ways of knowing and being and through activism in support of their pursuit and exercise of self-determination. This requires acknowledging and repudiating the dynamics of power that shape colonial research interactions with subordinated and marginalized peoples and groups, attempting to overcome whatever ethnocentrism and paternalism we bring to the research and whatever suspicion we are greeted with, persuading people that we are worthy of being taught and capable of learning, and being willing to put aside preconceptions (and academic and activist preoccupations), to listen, and to be of service to local concerns and projects.

Having drawn such strong contrasts between colonial and postcolonial research, I would like to be clear that I am not suggesting that there has been a simple historical progression from colonial to postcolonial research, nor even that there is always a clear dichotomy between the two. Nor would I advocate that postcolonial researchers casually dismiss and ignore research findings from less critically informed research or assume that all work prior to the development of postcolonial research methodologies was uniformly exploitative, misinformed, and paternalistic. We should recognize that much work today, even work cast within postcolonial frameworks, may remain colonial to some degree. And certainly there is much to learn from the research findings and experience of colonial research (as well as to react to and to revise), despite the ethical issues that it raises and the difficulties of evaluating the degree to which its research findings have been distorted by the goals, attitudes, relationships, and methodologies that shape such fieldwork. There is also much to respect and learn from in the work of earlier researchers whose fieldwork in some ways broke with the prevalent colonial research climate of their time.[9]

Richie: Deborah Rose refers to the need to develop critical perspectives on our own work and to guard against **deep colonizing** (1999, 177). She suggests that even the educated and sensitive "Western" self risks enclosing itself within a "hall of mirrors" in which it mistakes the endless reflections of its own common sense for a universal truth—and truthfulness. And we often see a romantic idealizing of the "Indigenous Other" as a source of some alternative universal truth. How do you see the task of doing research using methods and approaches that foster decolonization of peoples' lives and domains—and even of ourselves and our discipline?

Stan: That raises important issues of attitudes, intentions, and relationships. I'd like to first suggest a few practical strategies for such a decolonizing approach and then discuss the postcolonial perspectives and attitudes towards difference that they reflect and embody. I think we need to work to avoid "deep colonizing" and foster decolonizing in multiple, interactive ways. One vital part of this is "critical self-reflexivity" (England 1994; Rose 1997; see also Chapters 1, 2, 4, 14, and 19 of this volume). Another is fostering relationships with the people we work with that make it possible for them to voice their concerns and other feedback about us and the research project in an open and honest way. This is important on an everyday level, but it can also include seeking formal or informal local authorization for research, as you've written about, Richie (Howitt 2002a) or placing the project under community supervision (Smith 2012). **Local research authorization** or **community supervision** both involve reaching an understanding on appropriate research goals, methodologies (including the **cultural protocols** within which they will be conducted), and how knowledge will be shared and used. If there is no formal process for this, as has been the case in the Sharwa regions I've worked in, one can try to achieve something similar through working with a set of local residents or—still better—with local mentors and **co-researchers**. As I see it, the greatest potential for fieldwork to be decolonizing for all involved is to give up "control" of the research and either develop truly collaborative research or contribute to local or Indigenous research in which we offer our skills as colleagues, consultants, or allies (see Chapters 4 and 17 [for further discussion see also Herlihy and Knapp 2003; Chapin and Threlkeld 2001; Maya People of Southern Belize, Toledo Maya Cultural Council, and Toledo Alcades Association 1997; Park 1993; Nietschmann 1995]).

Decolonizing research thus requires a fundamental shift in research approach and conduct that begins with a perception and response to difference and "others" that is the antithesis of the perspective and practice of colonial research. Colonial research is grounded in the binary, polarizing perception of other peoples as the "Other," a position of distance in opposition to the "Self," "Us," (and often also to the "West") that is fraught with ethnocentrism and hierarchy and undergirds cross-cultural and interpersonal dynamics of dominance and exploitation.[10] To me, postcolonial research is instead grounded in the perception of other peoples as "others" who are different but not intrinsically different or alien, who differ culturally but not in essential humanity and value. It is attracted to difference rather than wary of it; it seeks to interact rather than to remain distant; it coexists with, respects, and honours difference rather than dominating or exploiting it. This difference of attitude and intention makes for very different conceptions of the purposes of research. Colonial research promotes, deliberately or inadvertently, a colonizing (and now globalizing) agenda that includes political, economic, and cultural imperialism and neo-colonialism; national, transnational, and global exploitation and commodification of "resources" (natural and human); Western or other colonial discourses about "civilization," "development," or "conservation"; Western conceptions of "science"; and promotion of a Western over-preoccupation with "self" and self-gratification (not excluding such selfish motivations as the pursuit of academic status or the satisfying of intellectual curiosity). With postcolonial approaches, the purposes of research instead become cross-cultural understanding, the celebration of diversity, and especially empowerment or emancipation—a decolonizing project based on rejecting the ethnocentrism and exploitation of colonialism through cross-cultural respect and through support for self-determination.

These differences in attitudes and intentions in turn tend to be manifest in research topics and conceptual frameworks. I would like to think that postcolonial researchers approach research with a high level of idealism, that they may tend to be less concerned with research topics for their self-serving value in enhancing their careers or bringing financial gain or with scholarly or scientific intellectual pursuits as ends in themselves.[11] Postcolonial researchers are often strongly drawn to research topics that enable them to engage in advocacy and activism and to theory and concepts that enable them to study and affect exploitative or oppressive political, social, and economic relationships, processes, and contexts. This has attracted postcolonial researchers particularly to **postmodernism** (including post-structuralism as well as **postcolonialism** itself), feminist theory, neo-Marxism, and the perspectives and language of independence struggles, "subaltern" social justice movements, and Indigenous rights campaigns.

From these perspectives, values, and conceptual frameworks, postcolonial research has developed methodologies that differ in significant ways from those of colonial research—a topic that will be the focus of much of our remaining discussion in this chapter.

Richie: That makes me think of David Harvey's call for an **applied peoples' geography** (1984, 9; see also Frantz and Howitt 2012) in which the skills of geographical analysis might be harnessed to the service of marginalized "others." That idea always sat comfortably in my mind with the educational philosophies of Paulo Freire (for example, 1972b; 1976) and more recent discussion of using Goulet's ideas of **border pedagogy** in geographical education and research settings (for example, Cook 2000; Howitt 2001) and Dina Abbot's challenge to reconsider whiteness in student field trips to Africa (2006). But universities are themselves a major element in the construction of power and privilege. Academic institutions harness well-intentioned efforts to their own purposes—not all of which are consistent with the liberal values of education and research for the common good. The globalized neo-liberal university often enforces exploitative relationships internally as well as externally, and makes it difficult for researchers to undertake the slow and difficult processes of trust-building, reciprocity, and engagement that these sorts of methodologies require. In some cases, much narrower institutional concerns and specific ideological concerns become dominant. Yet universities also continue to provide opportunities for discursive, and even practical, dissent. Indeed, allowing a small space for difference while reinforcing the status quo of privilege is a defining characteristic of the liberal academy. The task of harnessing these opportunities in pursuing the core values of social justice, economic equity, ecological sustainability, and the acceptance of cultural diversity has been a central element in my projects for inclusionary geographies—they straddle intercultural spaces both discursively and practically (Howitt 2001).

Stan: I would certainly agree with that. I'd add one more point here relative to the discussion of "deep colonizing," though—that prospective researchers should be aware that despite good intentions, even what might be intended as postcolonial applied peoples' geography can still be exploitative, paternalistic, and ethnocentric in practice (Coombes et al. 2014). This is one reason that some Indigenous advocates of postcolonial research have strong reservations about research by outsiders, particularly outsiders who are not under Indigenous guidance (Smith 2012). It is not enough, for example, to *intend* to carry out advocacy and activism-directed research if the means we use reflect our own ethnocentric baggage or if our preconceived notions of what needs to be studied and what needs to be done are considered patronizing, irrelevant, or threatening to the people we wish to work with and for. I am reminded here, Richie, of your call (Howitt 2002a) for "inclusionary research," your accounts of your early experience in Australia, and my own early experiences with the Sharwas of the Chomolungma region. I arrived in Khumbu in 1982 with the idea that I could best "help out" by documenting and exposing the adverse impacts of international tourism. That project was welcomed by local Sharwa leaders, and from the outset of the fieldwork in 1982–3 Sharwas actively participated in and assisted the research. Yet when I returned in 1984 and was taken under the guidance of two Sharwa mentors, I was gently taught that Sharwas had their own perspectives about what kinds of research they needed, their own critique of previous research (much of which had been "colonial" in character), and their own ideas about the kinds of contributions that I could make. I came as a result to work closely with them and with other Sharwa co-researchers on a series of projects that Sharwas thought important, including documenting oral traditions and oral histories at the request of elders who were concerned about their possible loss, reporting the perspectives of Sharwas who wanted to correct outsiders' misimpressions and "orientalizing" of their culture and homeland, and working with Sharwas to document "counter-knowledge" and "counter-history" that they can use in their efforts to redress the Nepal government's seizure of their territory and authority over their land use and management (Stevens 1993; 1997; 2013; 2014).

Getting Started: Research Legitimacy and Local Authorization

Richie: Yes, in late 1978 I was an enthusiastic young student researcher intent on "studying land rights around the Weipa bauxite mine." I visited a regional Aboriginal organization in Cairns on my way to Weipa, expecting to be welcomed with open arms. But I was taken aside and told in no uncertain terms that the last thing local Aboriginal people wanted or needed was a study that told the mining companies about their land rights. "If you want to help, why don't you do something useful?" I was asked. "Why don't you do a study of the companies that we can use?"

This question reoriented not just my first foray into geographical research but my relationship with the discipline. I read Laura Nader on "studying up" (1974) and enthused about "applied peoples' geography" (Harvey 1984; Howitt 2011). I shifted my audience from a scholarly focus to a community focus and began to ask questions that other social scientists hadn't really asked before. When I started working on research projects with Aboriginal people, no formal process of ethics review was required—although Mary Hall, my honours supervisor, required detailed research protocols and critical engagement with participating or affected community groups about ethical concerns. In many ways, she anticipated the shift away from an assumption that academic researchers have a "right to research" and the increased accountability for formal ethical oversight of research. The legal and institutional requirements for ethical oversight and the support for ethical engagement from formal institutional quarters have been quite an important shift in the past 10 years. In that time, most institutions have increasingly put procedures in place that require researchers to establish strong ethical foundations for cross-cultural work, particularly with vulnerable groups, rather than allowing them to assume a right to research non-Western "others." For example, recent guidelines for researchers working with Aboriginal and Torres Strait Islander groups in Australia advocate "building of robust relationships" as the basis for ethical engagement, with no implicit checklist to ensure conformity (NHMRC 2003, 3).[12]

Stan: I've found university human subjects protocols a useful starting point for pursuing decolonizing research. For cross-cultural work generally, and not only with Indigenous peoples, however, I think formal or informal community-based research agreements are critical to creating a space in which non-local researchers are held accountable to local cultural protocols. In breadth and specificity, these protocols may go far beyond the usual university research protocols with "informed consent," confidentiality, and avoiding harm to those with whom we work, identifying appropriate research goals and questions, appropriate ways to seek knowledge (culturally specific, appropriate methodologies), and appropriate ways for research findings and knowledge to be shared. There is usually no place for this in the other kinds of "research permission" that may be required for foreign fieldwork. The process of obtaining research permission from the government of Nepal and its Department of National Parks and Wildlife Conservation, for example, in part involves crafting research proposals and logistical arrangements to meet the self-interest of various government agencies and institutions and requires no consultation with the communities where research would be carried out, no consideration of culturally appropriate research methods, and no concern with the research being co-ordinated with local residents or with knowledge or other research benefits being shared with them. However, regardless of whether or not university, governmental, or local research agreements specify cultural protocols, I would suggest that as researchers we have an ethical responsibility to learn about and respect them. How well we work within them, moreover, will often greatly shape the reception we receive, the kinds and quality of information and insight we obtain, and indeed whether we can carry out research at all.

Richie: Yes. That ethical responsibility can be focused in the process of complying with formal institutional ethics requirements very constructively, I've found. For example, in supervising graduate students, it's helpful to take the student through my university's application form for ethics clearance and ensure that they have considered the ethical and cultural dimensions of various methodological choices as well as the implications of philosophical pluralism and multiple viewpoints in the real-world context of their research interests (for example, Howitt 2001; Howitt and Suchet-Pearson 2003; see also Whyte 2012).

Doing the Work

The Scale Politics of Cross-Cultural Research Projects

Richie: One of the things that I've been drawn to over the years is the ways in which a politics of scale is implicated in the construction of cross-cultural research (for example, Howitt 1992; 1997; 2001; 2002c; 2003; Howitt et al. 2013). In the Indigenous Studies domain, there is an assumption—often rooted in the ways that power relations are constructed between Indigenous and non-Indigenous domains in Australia—that the "correct" entry point for cross-cultural research of any sort is through a "local" or "regional" Aboriginal organization or through a "community." Indeed, many ethical review procedures are predicated on the need for researchers working with Indigenous Australians to demonstrate their accountability to "local" or "community" interests. This implies, however, a relatively naive conceptualization of scale. In contrast, people working in overseas locations often depend on national agency approval as an entry point to their research topic and are forced to conceptualize their study to conform to the window of opportunity offered by such links.

Working on the social impacts of transnational mining company strategies or major development projects in mining, tourism, conservation, and transport over recent years, however, made it impossible for me to restrict my vision to the naively "local." Multi-locational fieldwork in locations linked by various aspects of the mining production cycle (e.g., common ownership, downstream process integration, competition, government policies) meant that the local quickly emerged as a set of particular kinds of relationships that linked to a much wider set of scale relationships rather than as a singularity focused on a bounded location. And any local was always and inescapably contextualized for me by a range of critically important power relations that were constructed at several scales: in a nation that denied its Indigenous populations a right to self-government (or self-determination), and in industries characterized by corporate strategies of integration, cartelization, and financial innovation.

This raises a number of methodological questions, both in relation to how we might conceptualize the places in which we situate ourselves as cross-cultural researchers and in relation to how we conceptualize the range of topics suitable for research in cross-cultural settings. Indeed, in my work, it has even raised serious questions about just what we might mean by "research" and how we might be held ethically responsible for our work as "researchers." How does one engage research participants in "informed consent" for their involvement in, for example, a PhD study, when not a single person from the language group has completed high school and no one has been to a university? What sort of sense might they make of the question of informed consent? And even if one concedes that some sort of consent can be constructed in such circumstances, how is one really held accountable for one's immediate or subsequent actions in relation to the people involved, their representations of their lives and cultures, and one's interpretation of them for other audiences?

Stan, you've raised some important questions that also seem to have implications in terms of scale and the scale politics of research in "foreign" settings, whether they are an unfamiliar part of one's home town or a location on the other side of the world. Your identification of "research affiliations" as an issue is, to my mind, immediately drawing in some critically important scale issues, and I've just had a discussion with some young colleagues about the ways in which Development Studies and Indigenous Studies enter the field via quite different (scaled) windows of "government" (or NGO—non-government organization) and "community" respectively. You also identify many of the key issues of collaboration and accountability.

Stan: The scale issues you've raised, Richie, are quite familiar. They are likely to face researchers carrying out political ecological work with Indigenous peoples anywhere. This often involves research with—and about—multiple actors and processes in order to analyze patterns of regional land use and environmental change. Typically, this requires not only fieldwork with Indigenous communities but also with government agencies and often with regional, national, and transnational NGOs and national entrepreneurs or transnational corporations (Blaikie and Brookfield 1987; Bryant and Bailey 1997; Zimmerer and Bassett 2003). When my Sharwa mentors, co-researchers, and I began to examine forest issues in Khumbu in 1984 and in the Pharak and Katuthanga regions in 1994, for example, they made it clear from the outset that it would be necessary to look at forest use and management by a range of people (Sharwa villagers, non-local Sharwa timber merchants, Nepali foresters and national park staff, and international tourists) and to assess this within the complex politics of contestation over control of territory and the management of forest commons that was being waged within and between villages and between villages and the central government's district forest office and Department of National Parks and Wildlife Conservation. In such situations, the research "site" becomes not only Indigenous settlements but also offices and archives in regional centres, national capitals, and abroad. This also often required conducting more than one kind of cross-cultural research, because in Nepal, as in many countries where Indigenous peoples live in states controlled by other peoples, trying to understand local economic and environmental change involved work with non-Indigenous government officials, NGO staff, and entrepreneurs (Stevens 1993; 1997; 2013; 2014; Stevens and Sherpa 1993).

Cross-cultural research often not only requires work at multiple scales but also requires negotiating relationships at those scales. This takes place at the national level (and with central government officials stationed at the local level), moreover, within a highly charged context, because in many domestic and foreign fieldwork situations research authorization processes are often controlled by state agencies that seek to influence the direction of research, benefit from it, and prescribe procedures to hold researchers accountable to its dictates. Often, research funding agencies and universities expect researchers to adhere to these research authorization requirements. In practice, however, there is often a great range of variation in the degree to which researchers are controlled, how much cooperation and co-ordination is expected, and how well potentially awkward expectations can be defused. Researchers who aim to carry out decolonizing work may be inclined to focus on the ethics of the relationships they develop with Indigenous communities and may well find government requirements inimical. Negotiating these ethical and practical dilemmas may be critical to our work and can derail it before we can begin. Such situations can be particularly intimidating when one lacks familiarity with the internal politics, institutional culture, and bureaucrats involved. It can help to seek advice from other researchers with recent experience.

Richie: In these settings, it is important to contextualize one's work—to recognize that the state and its agencies have ambiguous relationships with diverse and particular Indigenous peoples and local minority groups and their territories and polities, and certainly not with some abstract, singular Indigenous "community" constituted at the scale of the post-colonial nation state, that there is a scale politics in the development discourse itself, and that research is easily drawn into that scale politics in ways that are not obvious to new researchers in a region. There is a substantial literature on the ways in which nation-states, even mature democratic nation-states, respond to developmentalist imperatives to develop remote corners of "their" territories and the people whose homelands they have been since time immemorial (see, for example, Berger 1991; Clarke 2001; Howitt 2001; Wolf 1982; also, more generally, Tully 1995). More recently, Godlewska et al. (2013) have offered a powerful critique of the mobilization of the democratic state to undermine Indigenous rights to self-determination in Canada. Yet there are also elements of accountability, ethics, and responsibility to grapple with in working with state agencies, transnational corporations, and environmental NGOs. In many ways, there is a very different sort of responsibility and accountability in being an informed critic in such complex situations rather than an ignorant critic!

Social/Cultural Transmission and Creation of Knowledge

Richie: We face a tension in how we conceptualize and bring the creation of knowledge to life and negotiate its various purposes. In some situations, research orientation is towards demonstration of a theory or salvaging a "lost" or fragile way of life (for example, Singer and Woodhead 1988). But in many situations, the sorts of research local people themselves prioritize is concerned with exploration of the possibilities of the intercultural domain or explanation of "the other side" of the cross-cultural relationships they experience. For many of my research students, it is their capacity to explain how the institutions, values, and practices of non-Aboriginal society work that is their greatest value for Aboriginal people—not their expertise in cross-cultural matters. Indeed, it is often our lack of facility in listening to and learning from local research participants and collaborators that limits our capacity to frame knowledge and understanding of our own culture and its operations in ways that are accessible and meaningful to local people. It is in listening to and engaging with interpreters about the concepts underpinning the knowledge we create that we make some of the most important realizations about our work. For example, in work on Aboriginal contributions to the central Australian economy in the late 1980s (Howitt, Crough, and Pritchard 1990), we found that Warlpiri interpreters were translating the concept "wealth" with the Warlpiri term for "money." Our analysis was seeking to explore the contributions of non-monetary wealth to Aboriginal futures—including culture, children, health, and so on—and we needed to tackle the terms in which these ideas were being conveyed in workshop discussions with Warlpiri speakers. More recently, in South Australian work (Agius, Howitt, and Jarvis 2003; Agius et al. 2004), Antikarinja interpreters and speakers spent half a day discussing the conceptual differences between ideas of "agreement" and "negotiation" after they realized we were using different words that were being translated into the same Antikarinja term. This discussion of the different implications of a term, one signifying a process and another addressing an outcome, was a careful exploration of how meaning is constructed and the implications for meaning-making in cross-cultural settings. It is through such conversations with the most direct users of our research that I often find my most inspirational collaborators and peers.

Although producing materials for these groups is important, there are other times when producing more conventionally academic publications is important. Ensuring the credibility of our work through peer review procedures, for example, can ensure that work that local or Indigenous collaborators are relying on in their own efforts to change their circumstances has credibility with governments and others. And we have a responsibility to contribute ideas into those academic discourses that reflect the learning made possible by the work we do in our intercultural engagements. Those more academic papers can provide transformational opportunities well beyond the confines we might have previously imagined for our modest endeavours.

Collaborative and Participatory Research[13]

Stan: Yes, we've been talking all along about methodology in the sense of approaches to knowing rather than as only a set of research techniques. As a reaction against colonial research, there has been increased interest in research that is more culturally sensitive and emancipatory in its orientation and often focused on **collaborative research**, locally guided research, and Indigenous research. In some Indigenous societies, as Linda Tuhiwai Smith (2012) describes for the Māori, histories of colonial research have led some people to demand that outsiders cease research on Indigenous peoples and issues, and in such cases some have called instead (as she so powerfully does) for Indigenous research and research methodologies. I feel that research by "insiders" and cross-cultural research by "outsiders" are both important, and that in some situations Indigenous peoples may value decolonizing relationships and friendships with outside researchers, consider outsiders' cross-cultural perspectives and insights into their societies and situations useful, seek to mentor and work with outside researchers out of interest in the research and belief in its significance, and view outside researchers as useful advocates and allies. While Linda Tuhiwai Smith (2012) is critical of research by non-Indigenous researchers, she and Graham Smith identify some approaches to more culturally appropriate, empowering, and emancipatory research that may be valued by Indigenous peoples. These begin with critical awareness of how research is shaped by relationships, power, and ethics. Researchers also can make an effort to work in more culturally sensitive ways, prepare for research by learning the local language, interact with "others" on their terms in their own social/political community venues, become informed about local concerns, seek local support and consent for research, and honour local cultural research protocols and negotiated research agreements. And they can change the nature of their research by making local participation integral to it. This can be done in several ways, as Graham Smith (in Linda Tuhiwai Smith 2012) has suggested for research with Indigenous peoples. Researchers can work under the guidance of local mentors. They can become adopted as members of the community, with all the lifelong obligations and responsibilities this entails. They can establish a "power-sharing" approach in which they seek community support for research. And they can adopt an "empowering outcomes" approach in which research becomes a vehicle through which "others" can obtain information they seek and which they can use to their benefit. To these Linda Tuhiwai Smith (2012,180) adds a more deliberately "bicultural or partnership" approach in which local and outside researchers design and carry out a project together, a process that requires negotiation and agreement on many important aspects of research methodology, design, and use.

In exploring the idea of "bicultural or partnership research," I feel that it is worth drawing a distinction between the attitudes and relationships embodied by participatory research in general and what I would call truly collaborative research. In participatory research, an effort is made to create a space for more involvement by "others" as an integral part of the research approach. Participation itself does not, of course, necessarily represent a break from colonial research, since it can amount to nothing more than enlisting local cooperation in a research project that continues to be driven by outside researchers' definitions of its purposes, methods, and use. Participatory research can also be carried out with an agenda of activism and empowerment by addressing locally relevant issues; supporting self-determination, human rights, and Indigenous rights; providing new knowledge that is of use to subordinated peoples, groups, and individuals; and transferring research skills that they can employ in the future on their own behalf.[14] Even in this case, however, there may still be attitudes of intellectual arrogance, paternalism, and evangelism (Bishop 1994; Smith 2012). Who, it might be asked, is being given the opportunity to participate in whose project?[15]

Richie: Stan, this is important in so many settings, where young, well-intentioned researchers feel that their university credentials give them an expertise in situations they know relatively little about. The risk is in allowing this assumption to frame the collaboration. In native title research in Australia, for example, we often find Aboriginal groups having to tell an expert who they are, how their society works, and what their historical and geographical setting is, because only an accredited "expert" can give evidence on these topics to a court. The claimants themselves are not trusted to represent their societies adequately (or perhaps truthfully), and the intervention of (well-paid) experts is mandatory—an extraordinarily arrogant misreading of the nature and source of such knowledge. Although it is the evidentiary system that drives this situation, many of the researchers accept the misreading without challenging it.

Stan: Collaborative research—truly collaborative research—reflects a sharper break from imposed, colonial research based on different attitudes towards "others" and different relationships among researchers. Ideally collaborators work as equals on a mutual project. This decolonization of the relationships within the research team can generate an interactive, cross-cultural synergy of knowledge and skills through which research conducted with community authorization and within local cultural protocols can be used to address community concerns. Local and non-local researchers conceive and design the research together, including making the key decisions on defining research goals and questions; where and how to seek funding, affiliation, and authorization; who should be on the research team; what methodology should be used; how cultural research protocols should be honoured; how the day-to-day conduct of fieldwork should be handled; what kinds of analyses should be attempted; and how research findings should be shared and used. This requires non-Indigenous researchers to give up "control" over a project and for all involved to contribute their time and efforts in order to work together towards shared goals. This is not easy to do and can only really take place if all researchers can move beyond often quite strong assumptions and behaviours conditioned by their statuses in their own societies and by asymmetrical cross-cultural power relationships. On the basis of my own experience of working within collaborative research relationships with several different Sharwa co-researchers, I would advise not to underestimate the time, care, emotional commitment, self-reflection, learning, and stress it can entail on everyone's part.[16]

Collaborative research approaches and locally guided ones are based on conducting research that works within what are considered culturally appropriate ways of seeking and transmitting knowledge. There are several important dimensions to this. One of the most basic, and often also the most difficult for researchers to accept, is that in many societies, knowledge and information are not necessarily shared openly within communities, much less with outsiders. There may thus be particular types and levels of knowledge that are considered appropriate to share with or withhold from outside researchers (Stevens 2001).[17] Knowledge may also be considered ethnic-, gender-, age-, class-, religion-, or subculture-specific. Researchers should also be aware that there may be types of knowledge that may be shared with outside researchers with the understanding that they are not to be otherwise shared with outsiders. Outside researchers should accept that honouring these concerns may greatly affect the design and conduct of research and the publishing of research findings.

Cultural protocols about acquiring knowledge also may include an understanding that knowledge must be earned by working within culturally sanctioned methodologies. This may apply only to certain kinds of knowledge and may not preclude the use of other culturally sensitive and authorized research methods to learn about local conditions and practices, methods that may produce new knowledge, insight, and community empowerment. But the importance of working within existing local systems of transmission of knowledge should not be ignored. This may mean, for example, that one must be found worthy of being mentored and then undertake a possibly long period of instruction. Researchers whose projects do not have the documentation or translation of a particular type of specialist's knowledge as a goal may be inclined to decline involvement in such a relationship (and knowledge specialists may have their own reasons for declining to take one on as a student). But it may often be the case that a great deal of information that is important to the project can only be obtained through such culturally sanctioned means. Researchers must then weigh whether to make the commitment to the relationship, the process, and the project. Those who persevere may find great personal and research rewards from a kind of cross-cultural communication and learning that ordinary interviewing (much less group meeting-based rapid research) cannot approach.

Cultural protocols also may have considerable ramifications for the use of particular research techniques. The use of questionnaires or formal, structured interviews, for example, may be considered intrusive, rigid, and exploitative by some peoples. Informal, semi-structured interviews, on the other hand, may not conflict with local etiquette about social interaction and communication because they can be interactive discussions or conversations in which there can be reciprocal exchanges of information. Interviewees may indeed value such interviews as social occasions that provide an opportunity to get to know the researchers, inquire about research findings, and learn about the outside world. Researchers may find that elders and knowledge specialists do not wish to be interviewed, or may think that requests for assistance are appropriate only after a process of acquaintance and relationship, or may initially speak only in generalities. Yet in other cases, if properly approached, they may be willing to act as mentors. People may be uncomfortable with group discussions or consider them essential. Mapping may be highly suspect if one is not well known and respected, while in other contexts it may be invited and welcomed. Cultural protocols and negotiated local research agreements may also influence research design in terms of whom it is appropriate to interview, what topics are considered suitable, and when interviews are and are not appropriate (such as not being appropriate during festivals or times when people are busy with subsistence activities). Local etiquette may also influence conversation in very specific ways, such as when there are taboos against using the names of the dead.

Another aspect of honouring local cultural protocols that can significantly affect research arises in cases in which societal discourses and practices promote relationships and interactions of inequality and domination between women and men or among people of different ethnicity, race, class, religion, age, or other socially defined categories and groups. Such cultural protocols can pose a formidable dilemma for outside researchers who wish to respect local concerns but object to the character of the interactions and relationships created by societal discourses and asymmetries of power and do not wish to engage in or to tacitly accept them. Local societal attitudes towards difference may also complicate the interactions and relationships of team members and affect the degree to which their research is fully collaborative. And cultural protocols that discourage or constrain interaction between women and men or between people of different ethnicities, classes, castes, or other social groups can make research across socially defined boundaries of difference difficult or impossible for both local and outsider researchers.[18]

Constructing Legitimacy

Stan: The reception that we and our research projects receive depends in part on local perception of the projects' legitimacy in terms of whether or not they address local concerns, needs, and interests, partly on how the projects are conducted and whether they meet cultural protocols, and partly on local perception of our character and those of our co-researchers. The importance of the perception of our character should not be underestimated. Often, it will be based primarily on our personal qualities, the evaluation of which can vary considerably among cultures but may well include whether we are considered to have a good heart or spirit, whether we can be trusted, how we treat and interact with people, how well we listen, and what skills and resources we bring to the community (see Smith 2012). While our academic achievements and status may matter little, we should be aware that how we are perceived can be very much affected by the company we keep or are perceived to keep. Research teams will often also be judged by the character, reputation, and actions of all of their members, including local ones (see, for example, Berreman 1972). And we may need to work hard to counter suspicions that we are agents of the central government, a transnational corporation, or a locally unpopular NGO and to reassure people that research findings will not find their way into such hands. However, affiliation with local, regional, national, or transnational NGOs or with particular government programs can also enhance one's legitimacy when they are well regarded locally, and this can sometimes both smooth the process of negotiating for research authorization from government agencies and serve as an easy entree into communities. The problem is that it can be difficult to know how agencies, organizations, and programs are locally perceived until one has spent time in an area and people trust one well enough to be candid. This is one of many reasons why it is often ideal to carry out an extended reconnaissance of a potential research site before establishing affiliations and research authorizations. In any event, we need to be clear with community members about our relationships (if any) with government agencies and NGOs and to be prepared to rethink those relationships based on community concerns.

Richie: Of course, this means that many students have to rely on the credibility and experience of a supervisor. That can open doors, but in my experience it is ultimately the students' own credibility and qualities that carry them through into worthwhile working relationships with people. Because those relationships are so dependent on personal integrity rather than on status, it becomes important for students and new researchers to think about what this integrity might mean in the setting they hope to work in. But it also means that the time frames involved in negotiating these matters can be inconsistent with the bureaucratic expectations around a student's candidature.

Stan: Legitimacy is also created through social relationships. Working in places and within communities fosters the development of relationships, and these relationships bring with them expectations of reciprocity and diverse responsibilities. Meeting these obligations over time can greatly enhance our acceptance in a community and attitudes towards our work. And over time, we also establish networks of friends and allies who will vouch for our character and intentions.

Making Sense, Reaching Conclusions

Richie: One of the really important issues that we've hardly touched so far is the question of audiences in cross-cultural research. Working in an academic environment, we are inevitably influenced by the publication culture of our institutions. Our own legitimacy as intellectuals is constructed in our "public" work. But in many cross-cultural settings, the primary audience for our work is neither academic nor reached by academic publications. "Publishing" to community-based audiences is often the most important element of cross-cultural research. It is the point at which we become accountable to our participants and collaborators for the ideas and knowledge our work produces. And it often involves exceptionally rigorous scrutiny, with multilingual discussion, careful (re)contextualization of ideas and information, checking of facts and interpretations, challenges, and debate. While this is some of the most demanding peer reviewing one experiences as an academic, it is not acknowledged as such by the institutional academic community. Many of the publication formats it involves (for example, community newsletters, comic books, radio broadcasts, long debates in community settings, joint submissions to inquiry processes, manifestos, and community statements) will not count for credit when it comes to thesis examinations, academic tenure, and promotion. Yet if we are in the business of producing ideas that change the world, this can be our most effective and influential work, and it is powerfully tested in communities' efforts to actually change their own circumstances!

Of course, we cannot escape the requirement to publish in more academic and professional settings, but that also raises important questions about how one represents the intercultural domain for other audiences. It is easy to be cast in a role as an "expert" on another culture—however limited our grasp of or engagement with that culture in all its complexity. In some cases, a range of outputs or different types and formats for publications can reframe quite sophisticated academic outputs for multiple audiences (Coombs et al. [1989] and Rose [1996b] offer impressive examples).

Stan: Yes. Researchers have a responsibility to ensure that co-produced knowledge and research findings are widely shared within communities. This is more than the courtesy of ensuring that copies of our academic publications are readily available. We need to adopt culturally appropriate ways of sharing knowledge. Academic publication is often not a very effective way of doing this. In societies where knowledge is shared face to face and orally, it may be effective to discuss research findings in group meetings, community meetings, or workshops. It can also be very effective and rewarding to discuss findings directly with individuals, although this takes considerable time to do widely. These methods also have the major benefit that they are interactive and provide opportunities for feedback that may correct misimpressions and overgeneralizations, provide alternative information and analyses, clarify concerns with the communication of some information or particular portrayals to the outside world, and provide opportunities to discuss how findings from the research can be used by individuals and communities in their lives and actions. Co-researchers also play a major role in disseminating research findings, since they become in-place, living repositories of that knowledge and the skills and experiences through which it was created. They are able to use these directly in their own lives and to inform community discussions and action. This can be much more powerful than anything an outside researcher would be able to contribute.

Finally, for me, another important aspect of making sense of the research is coming to terms with how that experience reshapes our lives and the lives of those with whom we work. Cross-cultural research can lead to long-term relationships between researchers and "others," and with them come obligations and responsibilities that can far transcend the sphere of the research itself and the relatively brief time that most researchers devote to living in places and carrying out fieldwork. The closeness that we develop with co-researchers, mentors, and friends can lead to continuing, lifelong affection and interaction, and in these relationships there can be strong cultural expectations of generosity and mutual aid that extend to each other's relatives, children, and associates. For some researchers (as Barney Nietschmann and I both found), field experiences and relationships can create a strong sense of commitment to a people, particular communities, and a region that leads us to return repeatedly to the place, the people, and further rounds of research and activism.

Richie: Yes, what might be conceptualized academically as "cross-cultural" is simultaneously interpersonal, and it has profound implications for just what sort of human beings we imagine ourselves to be or to be capable of becoming! To some extent, one is drawn into more activist and advocate roles than many of one's colleagues and even into framing our academic roles as "teacher" and "researcher" somewhat differently. I have long considered the links between my own intellectual nourishment from research and teaching and the construction of critical engagements with students and the wider societies of which I am part to be a central element of my responsibility as a public intellectual. That sort of engagement is not limited to the classroom. It happens in a wide range of places where one tries to make new sorts of sense that might disrupt the certainties of colonial (and deep colonizing) practices. For many researchers, finding the balance between scholarship and activism is far from easy.

Postscript

Stan: Richie, resuming our dialogue for the revised edition of the book, there are a few points I would like to underscore in support of postcolonial, decolonizing, and inclusionary research. First, Indigenous peoples and other marginalized and oppressed peoples, communities, and groups continue to struggle against colonial research in many parts of the world. Colonial research continues despite increased attention in some academic fields and in some countries to research ethics, including codes of conduct and research protocol requirements implemented by universities and funders (Israel and Hay 2006). Some peoples and communities are responding by developing their own research authorization and conduct protocols and requirements. The Inuit, for example, have developed such procedures for research in Nunavut territory, Canada (the Nunavut Research Institute's website, http://www.nri.nu.ca, includes research authorization forms and a number of other useful materials) and have produced a booklet, *Negotiating Research Relationships with Inuit Communities: A Guide for Researchers* (Nunavut Research

Institute and Inuit Tapiriit Kanatami n.d.), that is useful for researchers and communities more generally and includes a sample negotiated research relationship agreement. Yet many peoples and communities lack the recognized self-governance and self-determination to be able to legally require that researchers obtain their prior, free, informed consent for research on their lives, lands, and waters, much less to ensure that research is collaborative, respects local cultural protocols, and is decolonizing. The Sharwas of the Mount Everest region are one of the many peoples who find themselves in this circumstance. Under an onslaught of recent, largely unwelcome research conducted without their consent, collaboration, or respect, Sharwa leaders have attempted to confront and reform (with modest success) egregious research programs and recently have been developing an informal, voluntary research authorization and conduct protocol. Sharwa leaders hope that most researchers will welcome working together with them once they are aware that procedures are in place.

Richie: That's a familiar story—but it's often the case that the demands and expectations of researchers exceed the capacities of and opportunities for local groups to intervene informally. Strategies such as the Nunavut website and disciplinary mentoring can also go a long way in addressing the colonizing consequences of otherwise well-intended research. But opportunistic and exploitative research is much harder to address. The persistence of colonial imperatives of possession, erasure, and denial is reinforced by many of the elements of exploitation that are glossed as "globalization" and seems to fly in the face of the powerful idea of the "stickiness" of place, which makes those relationships of governance, accountability, and ethics at the local scale that we emphasized previously so important. It reminds me of just how important Nader's idea of "studying up" is in cross-cultural research (Nader 1974). We often need to train the focus of our research on the structures of privilege, power, and marginalization to really get at the questions of what local groups can or can't do. In many cases, it is the lack of capacity among the government agencies and corporate interests to actually work on the ground with people in their community groups and organizations that frustrates people so profoundly (Howitt et al. 2013)—and that can be investigated by social researchers. In many of the bureaucratic systems that dominate local peoples' lives, there is a genuinely wicked complexity that develops from the fragmented, conflicting, and competitive systems of agency "silos," which communicate with each other only by imposing contradictory requirements on Indigenous or local groups unlucky enough to receive funding support from multiple agencies.

Stan: In seconding your call for "studying up," Richie, I would add that research on the varied intercultural relationships and dynamics between marginalized "others" and national and international NGOs can also be a contribution. Much of what you've observed about the relationships between communities and corporate interests or government agencies can also apply to interactions with outside NGOs. Indigenous peoples and other communities and groups may find "scaling up" through alliances with national and international NGOs a useful strategy in their struggles for recognition, respect, and self-determination. But in other cases, NGO interventions can create significant problems and indeed perpetuate or accentuate subjugation, dispossession, and assimilation. Affected communities may value research that illuminates these dynamics and contributes to transforming existing relationships and interactions. Researchers should not assume that affiliation with NGOs is a straightforward way of facilitating local authorization, legitimacy, and collaboration.

Before we conclude our dialogue, I also want to highlight that although Indigenous peoples are not well represented in academe in many countries, some Indigenous geographers are actively engaging with issues of research methods, including both **Indigenous methodologies** and the methodological and ethical issues raised in cross-cultural, collaborative research. I would like to call attention to Jay Johnson, Renee Pualani Louis, and Albertus Pramono's (2005) discussions and Renee Pualani Louis's paper (2007), as well as your reviews in *Progress in Human Geography* with Coombes and Johnson (Coombes, Johnson, and Howitt 2012, 2013, 2014). These issues are also being much discussed within professional associations, including the Indigenous Peoples Specialty Group of the Association of American Geographers, the Canadian Association of Geographers Native Canadians Study Group, the Indigenous Peoples' Knowledges and Rights Study Group of the Institute of Australian Geographers, and the Indigenous Peoples' Knowledges and Rights Commission of the International Geographical Union. I hope that these voices will be heard and listened to more widely in geography.

Richie: The paper by Renee Pualani Louis (2007) is a terrific piece. I think that the development of Indigenous methodologies is moving rapidly at the moment. We are seeing the development of collaborative networks, research management frameworks, and ethical oversight and accountability arrangements that have been taken up in the new edition of this book (Chapter 4). It is also an important reminder of the need for always being open to challenge from new ideas, approaches, and concerns, rather than assuming that one's good intentions will be enough to produce justice and good research.

Key Terms

applied peoples' geography
border pedagogy
collaborative research
colonial research
community supervision
co-researcher

cultural protocols
decolonizing research
deep colonizing
inclusionary research
Indigenous methodologies
legitimacy
local research authorization
Orientalism
postcolonial research
postcolonialism
postmodernism
reliability
subaltern

Review Questions

1. What are some key differences between colonial and postcolonial research in assumptions, attitudes, relationships, and methodologies?
2. How can researchers go about attempting to carry out decolonizing research and inclusionary research?
3. How can research be made more truly collaborative in all aspects of a project, and how might this affect the time, energy, and outputs involved and affect a student's progress towards her or his degree?
4. Why might university research protocols often need to be supplemented by local cultural protocols?
5. How might issues about research with Indigenous peoples raised in this chapter be relevant to cross-cultural research in your own situation?

Review Exercises

1. Over the years, I (Richie Howitt) have used a number of role play and simulation exercises, particularly negotiation simulations, to encourage students to put themselves into the role of "other" (both as Indigenous community roles and mining company management roles, for example) and used debriefing to explore how difficult certain important ideas are to get across in such situations. These exercises have often developed out of case studies students are studying in their classes and push students to think critically and differently about all the positionalities involved in a range of settings.
2. For shorter exercises, using a brief period for pairs of students to interview each other and then reverse roles before reporting "something interesting" about each other to the class can provide opportunities to demonstrate how difficult it is to start conversations in various situations. I have found the following "Five-Minute Scenarios" particularly useful both to "break the ice" in new class groups and to encourage reflective engagement with issues such as those we have raised in this chapter.

Negotiations and its problems

Students divide into groups of three (two negotiators and an observer).

1. In each five-minute session, the students work through the set situation, with one student acting as observer/rapporteur for the group.
2. After each session, the groups will change to allow students to get to know a range of people in the class, and a new task will be set.

Scenarios for negotiation

1. Teenage child asking parent to use vehicle to attend a party
2. Parent confronting teenage child with suspicions of drug use
3. Adult child seeking to invite new girlfriend/boyfriend to family celebration
4. Same as (3) but new partner is same sex or a different religious or racial background
5. Local community member seeking to secure rejection of an unwanted local planning proposal in over-the-counter discussion with a council planning officer
6. Person taking a protected sea creature for family consumption under local customary law seeking to secure understanding from a government fisheries officer enforcing the species protection order

Debrief issues

- Negotiation is an everyday event.
- Certain issues (e.g., cultural values and expectations and intercultural differences) make negotiations harder (or easier). What are they?
- Certain factors influencing negotiations are invisible to participants. Can you think of examples?
- Negotiations are easily diverted into confrontation, particularly by cultural miscues.
- It is easy for participants to get locked into a "position" rather than dealing with the problems at issue.
- What are the implications for researchers involved in negotiating access to research opportunities in unfamiliar settings?

Useful Resources

Johnson, J.T., and S.C. Larsen. 2013. *A deeper sense of place: Stories and journeys of Indigenous-academic collaboration.* Corvallis: Oregon State University Press.

Lavallée, L. 2009. "Practical application of an Indigenous research framework and two qualitative Indigenous research methods: Sharing circles and Anishnaabe symbol-based reflection." *International Journal of Qualitative Methods* 8: 20–40.

Louis, R.P. 2007. "Can you hear us now? Voices from the margin: Using Indigenous methodologies in geographic research." *Geographical Research* 45 (2): 130–9.

Mullings, B. 1999. "Insider or outsider, both or neither: Some dilemmas of interviewing in a cross-cultural setting." *Geoforum* 30: 337–50.

Nunavut Research Institute and Inuit Tapiriit Kanatami. n.d. "Negotiating research relationships with Inuit communities: A guide for researchers." http://www.nri.nu.ca/pdf/06-068%20ITK%20NRR%20booklet.pdf.

Smith, L.T. 2012. *Decolonising Methodologies: Research and Indigenous Peoples.* Second Edition, London: Zed Books.

Notes

1. A useful review of Bunge's expeditions is available at: http://civic.mit.edu/blog/kanarinka/the-detroit-geographic-expedition-and-institute-a-case-study-in-civic-mapping. For me (Richie Howitt), it was my awareness of the US government sponsorship of "research" on social movements in Latin America and hill tribes in Indo-China in the 1960s and the disputes it produced within anthropology (e.g., Horowitz 1967; also Gough 1968; Jorgensen 1971) that helped push me into engaging with intercultural research ethics early in my strange career! It also drew me into a critical consideration of our own discipline's implication in colonizing efforts (e.g., Howitt and Jackson 1998; Howitt 2011).
2. One example of a similar approach close to home is Herman and Mattingly's (1999) effort to negotiate "reciprocal research relations," cultivate "the authority of self-representation," and implement "ethical responsibility" through community action as well as research methodology in their work in the most culturally diverse neighbourhood in San Diego, California. Also see Chapter 17, Katz (1994), Kobayashi (1994), and Nast (1994).
3. "Postcolonial" can signify both "after" colonialism and "rejecting," "against," or "anti" colonialism. I use the term in the sense of "rejecting" colonialism. Even in cross-cultural settings that are not appropriately characterized in these terms, rethinking research relationships in terms of participatory action frameworks (see Chapter 17) can redefine research outcomes towards mutual respect and benefit.
4. In Australian public policy, the work of Charles Rowley (e.g., 1970; 1971a; 1971b) and Nugget Coombs (e.g., 1978; but see also Rowse 2000) reflected a deeply reflective and non-paternalistic approach within a predominantly "colonial" period. Reynolds (1998) refers to such dissent as a "whispering in our hearts" and notes that "In each generation people have expressed their concern about the ethics of colonisation, the incidence of racial violence, the taking of the land and the suffering, deprivation and poverty of Aboriginal society in the wake of settlement" (1998, xiv). Geertz (e.g., 1973; 1980; 1984) similarly throws a different light on "colonial" efforts to understand cultural difference.
5. Many of the criticisms of colonial research can also be applied to much cross-cultural research with subordinated and marginalized peoples and groups in non-colonial contexts (although some postcolonial scholars would prefer to define the term more narrowly), as in Nietschmann's (2001, 183) remark that "Who studies and who gets studied reflects power, economics, status, class, color, and identity."
6. Yarndi: marijuana.

7. For another powerful perspective on similar issues of academic engagement with Indigenous issues, see Emma Larocque's 1994 poem "Long Way From Home," reproduced accessibly in Larocque, E. (2013).
8. **Stan:** I prefer to use "others" in preference to "the Other" to acknowledge cultural diversity, although I remain uncomfortable with the distance and attitudes implied in any form of the word and prefer less polarized language.
   Richie: Yes, I find that too, but there is something important in acknowledging that the language of power reflects something significant about the relationships it represents. As the philosopher Levinas suggests, that awkward singularity of a generalized "Other" annihilates something very significant in a relationship "whose terms do not form a totality" (Levinas 1969, 39). In the case of the "self–other relation," there is no larger concept higher in some implied hierarchy that encompasses these two terms. So for Levinas, aggregation of the self and the other does not produce a new, larger singularity because to do so would be to deny the ethical (or unethical) power relations that create differences. While the more generalized singularity "human" might encompass such differences, the risk is that in seeking to challenge the realities of the relationships of power, we use language that obscures it. It's a good reminder of the need to problematize language and to reflect on and challenge many aspects of the hidden constructs of injustice in the work we do.
9. In the case of Sharwa studies, this has included research by pioneering anthropologists Christoph von Furer-Haimendorf (1964; 1975; 1984), Sherry Ortner (1978; 1989; 1999), and James Fisher (1990).
10. Said (1978) notes that such views of "others" go back to the ancient Greek delineation of themselves and "barbarian" others, although the ancient Chinese developed a similar perception early on and it might be argued that such ethnocentrism is very old, very widespread, and common outside of imperial and colonizing situations as well as in them.
11. In practice, many of us find we must negotiate multiple personal motives in our work, and I would suggest that while our ideals and commitments may often call for a measure of self-sacrifice in our work, it is also legitimate to be concerned with such matters as completing one's degree or attaining tenure.
12. These guidelines seek to establish grounds for developing ethical relationships between Indigenous communities and research groups rather than offering a checklist of how to make a project application look as though it conformed to institutional ethics requirements. This approach shifts the orientation of ethics oversight away from formal legal concerns about risks to the institution and constructs a framework for higher levels of accountability to those participating in or affected by the research and will require some degree of rethinking in institutional compliance structures—as does the recognition of native title.
13. Readers are encouraged to consult Chapter 17 for a complementary discussion of participatory approaches to research.
14. The term "participatory research" is often used to refer to exactly this kind of collaborative, empowering research (see Park 1993).
15. This also brings up the issue of local "research assistants," the term that it has long been conventional in geography and anthropology to use to refer to local members of a research team. In recent years, I have come to feel that this practice needs to be examined critically because of the way it can narrow local project members' participation in research by fostering hierarchical relationships among team members and restricting—and in some cases not properly acknowledging—local project members' contributions and roles.
16. On collaborative research, see Park (1993), Herlihy and Knapp (2003), Chapin and Threlkeld (2001), and the approach that Barney Nietschmann and Berkeley geography students working for GeoMap developed with the Maya of Belize (Maya People of Southern Belize, Toledo Maya Cultural Council, and Toledo Alcades Association 1997). This "community-based cartography" is both more collaborative and more empowering than most participatory mapping (see Wainwright 2008, however, for a critique of some aspects of the Maya Atlas project).
17. This is somewhat different from the issue of local perceptions of what levels of knowledge

outside researchers are capable of understanding. Local residents may often over-generalize and simplify in response to what they perceive as an outside researcher's rather basic level of understanding of their homeland and ways of life. This is a major problem for short-term research and for researchers who do not realize that there are multiple levels of explanation and understanding.

18. For reflections on the issues raised by these situations, see the 2001 special issue of the *Geographical Review* on "Doing fieldwork," the 1994 special issue of the *Professional Geographer* on "Women in the field," and several other collections of essays by feminist geographers working within and outside of their own societies and communities (Jones III, Nast, and Roberts 1997; Moss 2002; Wolf 1996).

Empowering Methodologies: Feminist and Indigenous Approaches

Jay T. Johnson and Clare Madge

Chapter Overview

This chapter explores empowering methodologies through an examination of feminist and Indigenous research approaches. First, we consider what might be involved in employing empowering methodologies and suggest that such methodologies have deep roots in feminist research practice, which is discussed in section two. Through time, features of feminist research practice have been incorporated into many other research approaches. In section three we examine one of these approaches—that of Indigenous geographies—and outline some of the specific empowerment strategies currently being employed by Indigenous researchers. Having briefly outlined some of the potentials and challenges of feminist and Indigenous approaches for qualitative researchers, in section four we outline some key issues to consider when using empowering methodologies. The chapter finishes with a brief conclusion in section five.

Introduction

In this chapter we are concerned with the empowering and transformative potential and mechanisms of qualitative research. We consider first what is meant by *empowering research*. Such research is not simply about studying "something"; through its objectives and day-to-day research practices, empowering research holds significant transformative potential for those involved. Thus, according to Raju (2005, 194), **empowerment** is "a process of undoing internalized oppression" and, in the case of women's empowerment, "it is also about changing social and cultural forms of patriarchy that remain the sites of women's domination and oppression." Women are recognized as proactive agents who can exercise power to alter the process of empowerment and participate in social change. However, for Louis (2007, 131), empowering research with Indigenous communities must "be conducted respectfully, from an Indigenous point of view" and should have "meaning that contributes to the community. If research does not benefit the community by extending the quality of life for those in the community then it should not be done."

At its simplest, empowerment therefore refers to the process of increasing the social, political, spiritual, economic, and/or psychological potential of individuals and communities. It often concerns groups that have been marginalized from hegemonic decision-making processes through discrimination based on historically constructed unequal relations of power (on the basis of age, gender, race, ethnicity, sexuality, class, religion, nation, dis/ability). The process of empowerment aims to undo or overcome oppression and increase opportunities, knowledge, skills, collective action, and choices for those groups routinely pushed to the margins of society. It can also disrupt further attempts to deny improvement to their opportunities. Empowering research aims to support these groups to (re)shape organizations, policies, institutions, and everyday encounters affecting their lives. It can also challenge normative assumptions and negative stereotypes about these groups to promote greater justice or equity. Such research can be instigated and developed by marginalized groups, or through collaboration with researchers as "allies" who have access to useful resources, knowledge, and skills (see Chapter 17). In sum, according to Scheyvens (2009, 464) "empowerment means activation of the confidence and capabilities of previously disadvantaged or disenfranchised individuals or groups so that they can exert greater control over their lives, mobilize resources to meet their needs, and work to achieve social justice."

However, doing empowering research is not easy. Research that aims to transform the people it is working with, or to challenge hegemonic power relations and promote social justice, cannot be simply and quickly achieved. It can involve hard work, frustrations, contradictions, uncomfortable reflexivity, reinforced power relations (as well as successes, progressive change, and satisfaction). It should not be assumed, for example, that such communities wish to "be" transformed, nor that there are shared meanings about what empowerment might entail or how it might be attained. Rather, empowerment unfolds, is resisted, and is transformed through the process of research, and there are no straightforward guarantees of liberatory research (de Leeuw et al. 2012). Indeed, in a recent critical appraisal of empowerment, Ansell (2014) argues for the need to adopt a relational approach that recognizes the need to transform power relations at multiple levels, while Wijnendaele (2014) suggests that emotions and embodied knowledges are crucial elements in bringing about such social transformation. An understanding of these types of complexities involved in the process of doing empowering research has been deeply influenced by feminist research practice.

Feminist Research Practice

There is no single route to conducting empowering feminist research because there are many different approaches to feminist geography. As Moss and Al Hindi (2010, 1) note, there are "myriad ways of being feminist, engaging in feminist praxis and producing feminist geographies." Moreover, feminist research practice has changed over time and has been conducted in different ways in different places. There is therefore no single "story" or totalizing account about empowerment and feminist research practice: it is diverse, sometimes contradictory, and overlaps with, draws on, and influences other bodies of geographical work, as we will discuss.

Early debates about feminist research practice concerned many questions. What makes geography research feminist? Are there any distinctly feminist methods? To what use should such methods be put? Answers circulated in a range of special journal issues and edited collections around the political goals that might be afforded through attention to the design, analysis, and dissemination of a research project and to the diverse methods that might be used to achieve these feminist aims (see *Antipode* 1995; *Canadian Geographer* 1993; Jones et al. 1997; *Professional Geographer* 1994, 1995; WGSG 1997). There was (contested) understanding that no particular research methods were distinctly feminist. Rather, it was more important to consider the "work" to which the methods were put and to choose methods appropriate for answering the research questions and addressing the aims of the research. In other words, it was the epistemological stance taken towards the methods that was important in achieving feminist goals. However, what many studies of feminist geography did have in common was their political and intellectual goal of changing the world they sought to research, in other words, engaging in social and political change. Such transformative feminist research often initially foregrounded women and/or gender as the primary social relation (see Box 4.1). So feminist geography research involved politicizing a methodology through feminism to conduct research that was often pro-women, anti-oppression, or based on social justice (Moss 2002a, 3, 12), which challenged male dominance, made women's lives visible, and exposed gender inequalities (England 2006, 286). It was recognized that this involved the whole project, from the initial decision to undertake research on a specific topic to presenting the final outcomes.

As feminist geography asserted itself (*ACME* 2003; Bondi et al. 2002; *Gender, Place and Culture* 2002; Moss 2002b; Moss and Al Hindi 2008), there was increasing recognition that there was no one feminist geography political project, precisely because understandings of feminism were grounded in specific histories, cultures, places, and biographies. As feminism became reconstituted into diverse feminisms that challenged the idea of a "universal female identity" (often based on the unstated assumptions of white, heterosexual, able-bodied "Western" norms), so too came increasing awareness of the multiple oppressions affecting women's lives, which demanded that feminist geography go beyond gender as the central construct, to recognize its intersection with lifecourse, class, race, ethnicity, sexuality, dis/ability, place, nation, and religion (see, for example, Blanch 2013). Drawing on influences from queer, postcolonial, and ecofeminist theory and anti-racist and transgender politics, there was increased recognition of the multiplicity of social

Example of Research Empowering Women

Based on the Dangme West district of Ghana, this paper explores how poverty reduction programs (PRPs) with credit components can reduce women's vulnerability to poverty and significantly improve their socio-economic status through access to financial and non-financial resources. This has, in some cases, improved gender relations at the household level, with women being recognized as earners of income and contributors to household budget. However, other women still regard their spouses as "heads" and require their consent in decisions even related to their own personal lives while for other women improved economic status has created confrontation between spouses. The paper recommends that assisting organizations must address "power relations" at the household level, otherwise socio-cultural norms and practices, underpinned by patriarchal structures, will remain "cages" for rural women.

Source: Charlotte Wrigley-Asante. 2012. "Out of the dark but not out of the cage: women's empowerment and gender relations in the Dangme West district of Ghana." *Gender, Place & Culture: A Journal of Feminist Geography* 19 (3): 344–63.

relations of difference and the myriad hierarchies of power that were involved in research with diverse social groups. This recognition of the differences between women (and men) is sometimes referred to as "third wave feminism." It led to some demanding questions for feminist geographers through the deconstruction of the category of gender that had initially formed such an important foundation of feminist geography enquiry. Thus, Jenkins et al. (2003) asked, if gender inequalities were no longer privileged, what made a feminist project distinct from other critical human geography projects? In response, Raghuram and Madge (2007, 221) suggested that diverse feminist geographers still shared a (polyvocal) interest "in challenging the varied forms and effects of gendered power differentials as they intersect with a host of other factors, such as race, class and nation, and in a commitment to dialogic, pedagogic, research and political practices." Sharp (2005, 305) concurred that it was "not just the processes through which data is collected . . . that makes it feminist, but also the way in which projects are conceptualized and how we as researchers act as people (ethically, politically, emotionally) while engaged in the process."

As Sharp (2005) was intimating, three key themes were emerging out of constantly evolving feminist geography research practice. First, feminist geography research critiqued *ways of knowing* (epistemology) by contesting objectivity and validating subjective experience, acknowledging the situatedness and non-universality of knowledge creation and demanding awareness of the importance of context in producing what could only ever be a partial understanding of any research situation. This created new understandings of what counted as "knowledge." But it also presented challenges to *ways of researching* (methodology, the second theme): accounts were replete with discussions of the complexity of power relations and ethical issues involved throughout the research process, the multiple and shifting identities of all those involved, and how these influenced both the knowledge created and contested the boundaries of "the field" of research. Feminist geographers were at the forefront of discussions surrounding reflexivity, positionality, politics, and accountability. Emerging out of these debates were reflections about the ambivalent yet embodied nature of the field (Parr 2001; Sundberg 2003); the emotional entanglements involved in feminist research, in which research subjects were viewed as knowledge agents (Bondi 2003; Chacko 2004); and the complexities of establishing collaborative ways of presenting findings through alternative writing strategies (Sharp et al. 2004). And thirdly, feminist geography research was also significant in stimulating debate surrounding *the politics of research*, promoting ideas about conducting research that allowed "silenced" voices to be heard; recognizing the multiplicity of viewpoints, voices, and locations that geographical investigations might entail; and retaining a sharp focus on the social and political empowering potentials of research.

These features of feminist geography research practice have been highly influential beyond feminist geography: they have formed important elements of debates and developments in qualitative research in cultural, queer, sexual, emotional, children's, participatory, postcolonial, and Indigenous geographies, for example. Perhaps this should come as no surprise, for feminist geography has always been in iterative dialogue and contestation, both changing and being changed, by other sub-disciplinary ideas, languages, and political visions. Through this process Sharp (2005) argues that although an awareness of, and sensitivity to, gender has been mainstreamed in geography, the feminist political project still operates on the discipline's margins (Sharp 2009, 77). Despite this, many feminist research practices have become commonly accepted aspects of qualitative research (for example, consideration of reflexivity, positionality, research power relations, situated knowledges, emotions), although feminist geographers are also still at the forefront of troubling over and unsettling those very practices (see Billo and Hiemstra 2013; Chattopadhyay 2013). Thus feminist geographers have highlighted the tricky and messy nature of empowering research in practice (see Attanapola et al. 2013; Nagar 2013).

So perhaps feminist geography research practice can best be envisaged as "rhizomic" (Moss and Al Hindi 2010), constantly changing and being changed. The political visions of feminist geography are not sedimented into simple stages of historical progression, nor do feminist research practices seeking empowerment remain unaltered. Rather, there is a situation of both continuity and change, with some tried and tested research methods and political intentions coexisting alongside newer ideas and practices. Thus today feminist geography research projects employing empowering methodologies are greatly varied. Some projects focus on empowering women (Buang and Momsen 2013); others on exposing "naturalized" gender power relations or highlighting patriarchal assumptions, practices, and male bias (Bee 2013; Zanotti 2013); others on considering boys/men, manhood, and on challenging hegemonic masculinity (Faria 2013; Lahiri-Dutt 2013). Other projects interrogate heteronormativity through a focus on sexualities routed through queer theory or transgender politics (Dominey-Howes et al. 2013; Selen 2012) while others are more concerned with sexual geographies of blackness, questioning the universalism of Western gender theories and feminist readings of gender and sexuality (Bailey and Shabazz 2013; He 2013). More recent works explore how feminist politics might be "revitalized" through the dynamic networks of new media (McLean and Maalsen 2013) or through reinvigorated discussions surrounding patriarchy and everyday sexism (Valentine et al. 2014). However, as Moss (2005, 42) summarizes, what still distinguishes research as feminist is that "it deals with power in some way—whether conceived as something to be held, exerted, deployed, mobilized, sought after, or refused, or as something structural and inevitable, despotic and concentrated, or dispersed and everywhere." Next we examine one research approach that has drawn on, and influenced, feminist research practices: that of Indigenous geographies.

Indigenous Research

Geography has a long history of supporting the colonial expansion of Europeans into Africa, Asia, the Americas, and the Pacific. Wherever explorers ventured, the grid of the cartographer was soon to follow. Geography's colonial history has been documented by a number of geographers since the mid-1970s. Godlewska and Smith's (1994) *Geography and Empire* detailed the history of geography's complicity in European colonialism and laid a foundation for geographers interested in shifting the focus, or decolonizing, contemporary research within the discipline. One part of decolonizing the discipline has been focused on examining how research is undertaken.

At the same time that geographers, and other social scientists, began to reflect on their role in constructing and perpetuating colonialism, Indigenous peoples around the globe were coming together to redefine their tribal or regional conflicts with colonial and settler-state powers within a new, global anti-colonial narrative. This articulation of a global Indigenism, or **Indigeneity** (see Niezen 2003) has resulted in relocating Indigenous–state conflicts from the national to international scale. The establishment of the United Nations Permanent Forum on Indigenous Issues (2000) and the Declaration on the Rights of Indigenous Peoples (2007) are two recent milestones within this global narrative around Indigeneity.

As Indigenous peoples have been articulating their struggles within international fora, they have also been rejecting their long history as the "natural" subjects of Western research. This push back by Indigenous peoples has led to a significant shift in Indigenous-focused research by geographers since 2000. While Indigenous peoples have been the focus of research by geographers since the rise of geography as a modern discipline in the early nineteenth century (see von Humboldt 1811), much of this work has been exploitative in nature. It was research "on" and "about" Indigenous peoples and geared toward controlling and dominating populations who had only recently come under the jurisdiction of European crowns.

This method of doing research "on" Indigenous peoples has remained the dominant paradigm for the past two centuries and only began to shift once Indigenous communities began to articulate their own research methodologies and agendas. The seminal moment of paradigm shift within the academy is marked by many as the publication of Linda Tuhiwai Smith's (1999) influential book *Decolonizing Methodologies*. Smith's book provided an opening for dialogue not only on how research methodologies have harmed Indigenous communities around the globe by aiding the Euro-American colonial enterprise, but also by asserting that research could and should be conducted "for" and "with," instead of "on" and "about" Indigenous communities. While it may seem like a straightforward and common-sense notion that Indigenous communities should be active collaborators and participants, deciding what research is done in and on their communities and homelands, this is a relatively new idea within the academy (see Box 4.2 for an example).

The other crucial shift Smith's (1999) work has brought to the fore is even more significant. For Smith, the logical progression of decolonized research methodologies is an inevitable assertion of **Indigenous methods**—methods conceived and articulated from non-Western world views. Smith describes a set of research projects within a framework based on her experience as a Māori woman. However, Smith did not simply imply that all Indigenous research need originate within a Māori perspective but instead she opened the door for different Indigenous communities to articulate their own specific Indigenous research methodologies. Hence, Indigenous communities need no longer be constrained by the Western research paradigm that has laboured to colonize them for centuries. Today, Indigenous communities frequently control not only the research agenda in their own communities, but that research can also take place using methodologies conceived within their own ontology.

Smith's groundbreaking work has opened space, albeit at the margins of the academy, for a deeper dialogue on how Indigenous methodologies might be

Example of Research Empowering Indigenous Peoples

Based on a research partnership involving the James Bay Cree community of Wemindji, northern Quebec, and academic researchers at Canadian universities, this research for a conservation project that used protected areas documents the process of applying community-based participatory research principles as a political strategy to redefine relations with governments in terms of a shared responsibility to care for land and sea. The authors describe how empowering methods, including collaborative, equitable partnerships in all phases of the research; promotion of co-learning and capacity building among all partners; emphasis on local relevance; and commitment to long-term engagement, can provide the basis for a revamped community-based conservation that supports environmental protection while strengthening local institutions and contributing to cultural survival.

Source: Monica E. Mulrennan, Rodney Mark, and Colin H. Scott. 2012. "Revamping community-based conservation through participatory research." *The Canadian Geographer* 56 (2): 243–59.

conceived and articulated. Recent books by Shawn Wilson (2008) and Margaret Kovach (2009) have begun to push these boundaries forward for a broader Indigenous studies audience. Recent journal special editions have brought this discussion within the sphere of the geographic community (*American Indian Culture and Research Journal* 2008; *Canadian Geographer* 2012; *Geografiska Annaler B* 2006; *Geographical Research* 2007). Although the specific articulations concerning how Indigenous research methodologies should operate are as varied as the individuals and communities voicing their opinions, a few key concepts seem to have reached consensus status.

These key concepts, first identified by Harris and Wasilewski (2004) through their work with Americans for Indian Opportunity, articulate an Indigeneity that cuts across the obvious differences between Indigenous groups throughout the Americas and beyond. These key concepts, commonly referred to as the 4 Rs—relationship, responsibility, reciprocity, and redistribution—were determined to be core shared values articulated through two decades worth of meetings and discussion among diverse Indigenous groups in the 1980s and 1990s. They have also been adopted by Indigenous academics and are now commonly referenced as the core ethical values that should govern an Indigenous research methodology. The drive to articulate Indigenous research methodologies is, in part, as Louis (2007) has identified, because there is a significant difference between research done with Indigenous communities using Western methodologies and that using Indigenous methodologies.

The first stage in any research project is the establishment of a relationship between the researcher and those with whom she or he intends to work. This is the same for those intent on working with Indigenous communities, although the establishment of a relationship in this context implies a deeper sense of responsibility than might be expected in many research relationships. Russell Bishop (2005, 118) describes this establishment of an ongoing relationship as "the process of establishing an (extended family) relationship, literally by means of identifying, through culturally appropriate means, your bodily linkage, your engagement, your connectedness, and, therefore, an unspoken but implicit commitment to other people." Building such a research relationship by creating an extended family commitment around your shared interests requires showing your face to the community. This form of relationship cannot be negotiated through emails and phone calls. Many a non-Indigenous researcher has been confounded by a lack of response from Indigenous communities to their research queries. For many Indigenous communities, no response is a way of saying "no" without creating disharmony. This raises the fundamental question of whether a no response should be conceived as meaning "no" to the research being proposed or whether it merely means "no, not now." Sometimes developing a research relationship with an Indigenous community or individual may require many hours sitting in an office or home, perhaps drinking tea or coffee, and talking. This dialogue is not only about establishing the research and agreeing to its parameters, but it is also about developing trust across the complex power inequalities inherent in any relationship between academics and non-academics, particularly where research has played or continues to play a role in the colonial relationship.

The myriad responsibilities one takes on in establishing such a relationship are founded within acts of reciprocity. This reciprocity, while predicated on both acts of giving and receiving, is motivated by giving: not giving as charity, but giving as honouring. As Harris and Wasilewski (2004, 493) describe, "at any given moment the exchanges going on in a relationship may be uneven. The Indigenous idea of reciprocity is based on very long relational dynamics in which we are all seen as 'kin' to each other." Building a research relationship, then, with an Indigenous community cannot be based on a "helicopter" approach where you drop into their lives for a short stay and then disappear with the information you need, never to return. These extended familial, research relationships require lasting and durable commitments; they require not only being hosted in the community but they also require a reciprocal hosting in your own home or institution. It is a cyclical reciprocity predicated on a continual renewal and sharing (see also Chapter 17).

Redistribution inherent in sharing serves to balance or rebalance relationships. As researchers, we are disciplined to view research as our possession: knowledge we have created and own. The truth, though, is that research never takes place in a vacuum free from the influence of those we work with. Those with whom we "do" research are aiding us in creating new knowledge. Frequently, it is just a translation of knowledge already commonly held by Indigenous communities to a non-Indigenous audience. Sharing gifts, whether they be material wealth, information, time, talent, or knowledge, is all a part of this obligation of redistribution. Within many Indigenous communities, this is referred to as a "give away" or potlatch. Central to this obligation is the ethos "to whom much is given, much is expected." It is through this redistribution that everyone in the community is valued.

While increasingly influenced by Indigenous world views in their formation, Indigenous research methodologies remain in dialogue with critical, participatory, and feminist methodologies. It is through this ongoing exchange of ideas, through geography and the social sciences more broadly, that Indigenous methodologies are starting to influence the ways in which research is conceived and conducted. This transformation of geographical methodologies from colonial to postcolonial sensibilities requires vigilance and determination.

Having discussed some of the potentials and challenges feminist and Indigenous research practices raise for qualitative researchers, in the next section we outline some key issues to consider in making the move towards using more empowering methodologies. This discussion is only a starting point—there are numerous other issues we have not had space to mention—and it is structured around approaching the research, doing the research, and the politics of the research, although of course these three processes are constantly interacting.

Using Empowering Methodologies

1. Approaching the Research

Early Beginnings: Creating a Long-Term Dialogic Relationship

According to Raghuram and Madge (2006, 275), the initial framing of research might be in terms of why the research is being conducted in the first place, an approach that forefronts the ethical issues of who gains from the research and why. The researcher must start to think about and work through in dialogue the power relations, inequalities, and injustices that enable and allow the research to occur and must be committed to working towards challenging these at different scales—the personal, the institutional, and the global-political. This might be attempted through a process of engaged pluralism (DeLyser and Sui 2013, 10) where various "views are engaged, divergences openly tolerated, and differences dialogically embraced. Differences may not be resolved, but genuine engagement can lead to enhanced . . . creativity on all sides, stimulating new thought." This might, for example, be undertaken through the use of "talking circles" (see Evans et al. 2009, 903), in which the opportunity to speak is distributed sequentially around the circle and confrontational style argument is discouraged.

However, from the outset, it is also important to recognize that the research project may not always be instigated by the researcher. It is imperative to "make space" to listen for and respond to the self-determination of women's or Indigenous groups who might articulate a need for a specific research project or who might initiate the research in the first instance using their conceptual notions, research designs, political intentions, and ethical review practices, which may differ from those of the researcher. Here the researcher might become an "academic ally," acting as a conduit between the research community and academic institutions and public funding organizations through all stages of the research process. This process of "walking with" (Sundberg 2014, 39) the research community might take the form of supporting and fostering the group's "capabilities" to undertake research themselves and advocating for the institutional and structural changes necessary to make this possible. Thus the researcher might actively work with (or in response to) the community or group from the inception of the project, including articulating initial research questions, writing grant proposals, agreeing on shared responsibilities in the implementation of the research, discussing redistribution of the resources for carrying out the research, and considering how results might be analyzed, written, and reported to produce different types of research products that may be differentially beneficial to the various groups involved in the research. This process is likely to involve lasting and sustained relationships, commitments, and obligations. It is also involves recognition of the active political subjectivity of women's and Indigenous groups, who may have their own structures of power that shape research agendas, designs, and relations, as well as potential harmony and/or dissent within their group.

Moving the Centre: Making Space for Multiple Ontologies and Polycentric Epistemologies

A second key issue in approaching empowering research is the value of developing a research sensibility that is open and hesitant, that refuses "to allow the taken-for-granted to be granted" (Ahmed 2004, 182, quoted by Sharp 2009, 78). An example of this refusal is given by Mishuana Goeman (2013). She argues that it is vital to refocus the efforts of Native nations beyond replicating settler models of territory, jurisdiction, and race to remapping settler geographies and centring Native knowledges. This appreciation that different societies (or groups and individuals in society) might have distinct views of the world, or have diverse ways of being in the world, involves being receptive to the idea of **multiple ontologies** (Hunt 2014). Furthermore, different groups may have diverse ways of knowing, asking different types of questions about the world and transmitting them in varied ways, signifying the need to validate **polycentric epistemologies** (Harding 2011, 154). In other words, if we can start to understand and value that there are multiple world views and many different ways of conceptualizing knowledge (although these might initially be unfamiliar and difficult to comprehend), we can start to appreciate that the world is made up of manifold, heterogeneous, dynamic ways of being and knowing. This is a vision of a **pluriversal world**, in which many worlds belong (see Sundberg 2014, 34).

Thinking about the world as pluriversal involves advocating and making space for multiple knowledge systems and life worlds that are legitimated *on their own terms*. In this process, "Western" knowledge loses its central and universal position and becomes one of a range of competing and contested knowledge systems. This suggests that Western knowledge might start to be regarded as a local or provincial knowledge (Chakrabarty 2000)—knowledge that is locally produced but has gained its apparent universality through being projected outwards throughout the world through colonial and neocolonial power relations. Thus, according to Escobar (1995), the domination of Western knowledge is explained not through a privileged proximity to the truth, but as a set of historical and geographical conditions tied up with the geopolitics of power. This move forces recognition that so-called "powerful" Western discourses are also partial and fragmentary, often involving knowledges and practices emanating from "Indigenous informants." In turn, this challenges the idea of a precise dichotomy between Indigenous–Western knowledge formations, moving us towards a position of multiepistemic literacy (see Sundberg 2014, 34). In making this conceptual relocation that unsettles the hegemony of Western knowledge and challenges the strict Indigenous–Western binary, space is cleared for Indigenous knowledge to be relocated as one of many legitimate and valid (albeit sometimes competing) knowledge formations. This enables moving beyond "anthropological particularism"—in which Indigenous knowledge is seen as unchanging, pristine, traditional, or local, in opposition to modern, universal, global, Western knowledge—towards a position in which *all* knowledge might include mysticism, spiritual ontologies, and ritualistic methodologies and in which *all* knowledge is considered partial and emerging, but at the same time also place-specific or situated.

Research with No Guarantees: Troubling over the Research Process

However, this creation of a long-term dialogic relationship, which identifies and validates multiple ontologies and polycentric epistemologies, is not easy to achieve. It will include a commitment to respond to issues raised by research communities, a willingness to engage in continuing dialogue that takes into account the conceptual landscape of all those involved in the research process, and an awareness that, despite a shared desire to participate in empowerment politics, there may be contested meanings about what empowerment might entail or how it might be achieved. In other words, from the outset of the research it is important that all parties involved acknowledge that there are *no guarantees* of successful emancipatory outcomes (de Leeuw et al. 2012; Noxolo et al. 2012). Rather, these outcomes must be carefully worked towards through everyday research practices and intimate research relations.

This process of conducting empowering research is not likely to be straightforward; indeed, it can create a range of complicated practical issues in institutions (e.g. universities and grant agencies) where a more limited model of research is espoused and reinforced. Moreover, it may be wrought with contradictory and potentially refuted relations and complex emotional investments because the creation of knowledge is never "innocent"—it is always entwined with differentiated relations of power. This constitutes what Smith (2005) has termed "tricky ground." This tricky ground concerns the troubling methodological, ethical, and political issues and inter-subjective relations that require continual communal reflexivity in the process of developing workable research relationships. For example, as researchers grappling with the production of academic work, we should be acutely aware of the limits to our understanding and acknowledge our limitations in "speaking for others." This involves being mindful of the risks of appropriation of knowledge creation while always being open to new ways of thinking about and understanding the world. (For examples of the complexities of cross-cultural dialogue see Desbiens and Rivard 2014; Eshun and Madge 2012; and Hunt 2014). This open and hesitant approach is also important during the process of doing the research, as explored next.

2. Doing the Research

Employing a Multi-Layered Reflexivity

As we have outlined earlier, the development of a relationship is the first and primary component of any methodology that aims at empowerment within collaborative anti-colonial research. The relationship-building component of the research process can take many forms. Bishop (2005) has described the process as one of developing an extended family whose common interests are the agreed-upon research goals and objectives. De Leeuw et al. (2012) have described a process centred upon friendships that extend beyond the research framework, allowing for a more profound critique of the research process and greater reflexivity. Reflexivity has, first through feminist and now also through Indigenous research approaches, become key to any collaborative, empowering research methodology.

The reflexivity we outline here, adapted from Ruth Nicholls's (2010) work, encourages a multi-layered approach. This first layer, or self-reflexivity, asks the researcher to explore the hidden assumptions about the research that originate within disciplinary structures or funding streams that enable the work to proceed.

It also involves being self-reflexive about the epistemological and ontological assumptions that the researcher brings to the research project, a process that may well involve "unlearning" what one has already learned (see Sundberg 2014, 39). This might be in terms of rethinking the questions asked, or delving deep into analytical and interpretative understanding in the field through ongoing dialogue, or making room for redefining terms of representations or conceptual framings. The researcher should also attempt to become cognizant of the complex and changing power relations inherent throughout the research process, particularly the (almost inevitably privileged) position researchers bring to the relationship. This is especially important for non-Indigenous researchers in establishing a critical and dynamic relationship with Indigenous collaborators but is also true for women from the global north working with women from other parts of the globe or across other axes of difference.

The development of a research relationship requires the researcher to carry this self-awareness into dialogue with others. This second layer of reflexivity, termed *interpersonal reflexivity* by Nicholls (2010), implies a relationality that necessitates evaluation of interpersonal encounters within particular institutional, geopolitical, and material situations. Recognizing one's role within the (changing) research relationship necessitates that researchers reflect on their ability to collaborate as opposed to lead, control, or delegate. The researcher is commonly placed in between the expectations of academic institutions and the community, navigating the intersection of ethical demands. As de Leeuw et al. (2012, 188) observe, "researchers who carry out participatory projects quickly confront the mismatch between demands of the institutions within which they operate and their own commitment to build meaningful relationships with the people and places about which they care."

The third layer of reflexivity, termed *collective reflexivity* by Nicholls (2010), requires all participants to engage in a dialogue about the process of doing research together. What are the terms of participation? Who initiated the research project, and why? Who involved themselves, and why? Whose voices have been heard and what form has this taken? How was the research conceived and carried out, and how did this affect social change and practical knowing? Has the research process been transformative, affirming, cathartic, empowering, and if not, why not, and for whom? This third layer of reflexivity entails a shift in the researcher's positionality, "a ceding of research control beyond the initial phase of negotiation, and extending participation into data collection, analysis and distribution" (Nicholls 2010, 25). This approach, founded within a radical pedagogy, pushes beyond mere information transfer towards a **critical consciousness** that Freire (2000) argued provides the foundation of empowerment. Conducting empowering research may also be promoted through the employment of dialogic research tools.

Dialogic Research Tools: Reworking the Field as a Methodological Site of Agency
Raghuram and Madge (2006, 276) argue for the need to explore methods that will make research questions more dialogic. But in advocating dialogue they do not presume that difference can be simply "dissolved" to attain complete understanding, "for there will always be degrees of incomprehensibility and continuing spaces 'in between.'" However, they do suggest that working through these in-between spaces can "bring moments of enlightenment; glimpses of the world through someone else's reality and a sense of the losses associated with privilege. From such moments of deep personal and political change more relevant research questions can arise, questions that can potentially challenge the 'master narrative' of northern-centred research."

One research method that might enable such dialogue is storytelling—a research tool "wherein personal, experiential geographies are conveyed in narrative form" (Cameron 2012, 575). This approach to narrating research experiences presents expressive and affective methods that can uncover new understandings and perspectives, or be a means to express different world views or expressions of being in the world. There is also a political potential to storytelling to construct counter-narratives that test dominant discourses and produce social change (Gibson-Graham 2006). A storytelling approach is particularly well suited to collaborative research with Indigenous communities as there are many synergies with Indigenous forms of knowledge creation and sharing (Christensen 2012). However, such an approach does raise questions concerning authorship, first- or third-person narration, and issues arising from translating oral stories into text. Storytelling also requires considerable attention to concerns of power and representation (Eshun and Madge 2012; Garvin and Wilson 1999). Despite the complexities of employing more dialogic methods, they can enable inclusion of community research agendas, thus having potential to rework the field as a methodological site of agency.

The importance of thinking critically about the field—the place and the specific context in which research occurs—has been stressed for some time by feminist geographers (see Moss 2005, for example). As research is place-specific and all knowledge production is situated, the critical consciousness advocated by Freire (2000) develops through everyday lived experiences (Johnson 2012). To understand the place-based struggles of different communities necessitates engagement with the experiences, conflicts, languages, and histories they rely upon to construct their collective identity. Here the place of research (or the field) has potential to become an active location of empowerment. How such emancipatory change might be achieved in a particular place involves not only thinking about and doing the research, but also includes consideration of the political outcomes of the research process, as we explore next.

3. The Politics of the Research

Changing Ourselves: Breaking Out of the "Hall of Mirrors"

As is probably clear from the preceding discussion, the political outcomes of the research process can take many forms. Initially this might be considered in terms of changing ourselves by interrogating, destabilizing, and reconfiguring our underlying epistemological and ontological assumptions. This is an important process in recognizing but also challenging the links between geography and imperialism and moving towards more decolonized versions and visions. Rose (1999, 177) observes that Western science "sets itself within a hall of mirrors . . . mistakes its reflection for the world, sees its own reflections endlessly, talks endlessly to itself, and, not surprisingly, finds continual verification of itself and its world view." So how might researchers break free of this "hall of mirrors" in their research to promote the process of empowerment and participate in social change?

For research outcomes to be empowering, we need to embrace the uncanny realization that multiple ontologies are not only possible but are also the lived reality of most of our fellow humans. Putting ourselves into a space within which, as researchers, we can begin to glimpse these alternative but equally valuable ontologies requires us to challenge commonly accepted frameworks and hidden assumptions in a shift towards breaking out of the "hall of mirrors." This will involve the relentless need for a rigorous interrogation of the politics of speaking and writing, and being open to opportunities to criticize and challenge dominant world views and propose alternative agendas rather than adding to existing ways of thinking about the world and conducting research. This process is likely to be demanding. Avoiding co-option of groups on the margins of society and appropriation of their knowledges and world views will involve constant vigilance and an active political agency on behalf of all those involved in the research process. It will take courage and determination to avoid propagating an underlying neocolonial "business as usual with the odd tweak," to move beyond existing knowledge formations, experts, and institutional structures of power.

Changing Institutional Structures and Processes: From Research "For" and "With" to Research "By"

It is clear that there will be no easy and definitive answers, but a careful working towards dismantling, or at minimum acknowledging the complexity of, the historically produced power geometries (of imperialistic, white supremacist, capitalist, heteronormative patriarchy) upon which geographical research is based is crucial. It is only by "stepping outside" hegemonic systems of knowledge production that a shift in the paradigms of research can begin. As Kuhn's (1962) work has demonstrated, these shifts are frequently concurrent with social and political upheavals that not only upset the rationalized frameworks of science and research, but also question fundamental social structures that perpetuate colonialism, imperialism, homophobia, racism, and sexism. Following the manner in which feminists have occupied and changed the academy and its knowledges, everyday practices, and politics, Indigeneity as a social movement operating both inside and outside of the academy is now also placing pressure on these hegemonic societal structures of control.

By placing pressure on the academy, Indigenous geographers are beginning to uncover "the spaces between intellectual and lived expressions" of Indigeneity, prising open gaps in regimes of knowledge production and providing "sites where ontological shifts are possible" (Hunt 2014, 30). The trick to identifying these "gaps" to facilitate ontological shifts requires that researchers respect the autonomy and independence of Indigenous organizations (Sundberg 2014). By serving as "allies" in support of the self-determination of Indigenous nations, and as collaborative partners focused on the research agendas of those communities, we also serve to broaden the ontological foundations of our discipline and the academy. Serving as allies in research and struggle, though, is only one step in the process of aiding Indigenous research agendas. As Coombes (2012, 290) identifies, a Freirian approach to research identifies collaboration as a "mere intermediary step towards the democratization *and* dissemination of knowledge production itself." The final step is the fostering of "communities' capacities to complete research for themselves" (Coombes 2012, 291). This can be seen as the final prepositional shift, from research "for" and "with" to research "by" Indigenous peoples.

Summary: A Critical Reflection

Qualitative research methodologies in geography have been significantly influenced by feminist and Indigenous research practices in the past few decades. Both feminist and Indigenous approaches have moved geographical research towards more empowering methodologies. However, as this chapter has illustrated, employing empowering methodologies is not easy and involves careful reflection regarding approaching the research (for example, the development of long-term dialogic relationships and the validation of multiple world views), doing the research (for example, employing a troubling multi-layered reflexivity and using dialogic research tools), and the politics of the research (for example, "stepping outside" hegemonic systems of knowledge production and challenging academic institutional structures). Nevertheless, if employed thoughtfully and compassionately, empowering methodologies can move qualitative research towards more inspiring, meaningful, and potentially transformatory and equitable outcomes.

Key Terms

critical consciousness
empowerment
Indigeneity
Indigenous methods
multiple ontologies
pluriversal world
polycentric epistemologies

Review Questions

1. How would you define "empowerment"? In what circumstances might you use it as a strategy for qualitative geographical research, and why?
2. Critically discuss the potentials and limitations of empowering methodologies. Consider the ways in which these limitations might be ameliorated.
3. Find an example of geographic research that has used an empowering methodology. How was the project initiated? What were the outcomes of the research for the different individuals/groups involved? How would you evaluate the "success" of the project?
4. Do you need to be a woman to do feminist research? Can only an Indigenous researcher do effective research with Indigenous communities? What skills and attributes do you need to have or develop to be able to undertake research with such (diverse) groups using empowering methodologies?
5. Outline some of the ethical issues involved in using empowering methodologies in geography. How might you negotiate these issues through the research process?
6. Does place matter when using empowering methodologies? Justify your viewpoint.

Review Exercise

Read the following paper:

Eshun G. and Madge C. 2012. "Now let me share this with you: Exploring poetry for postcolonial geography research." *Antipode* 44 (4): 1395–1428.

In small groups, debate the following issues:

1. How was poetry used as an empowering methodology?
2. Consider some of the potentials and problems of using poetry as a research method. Do you consider this approach was successful? Justify your viewpoint.
3. How and why did the use of poetry allow the researcher to "be (de)centred" in the research process? How and why did it (re)inscribe marginality?

4. In what ways were issues of relationship, responsibility, reciprocity, and redistribution raised in this project?

Useful Resources

Johnson J., and S. Larsen, eds. 2013. *A Deeper Sense of Place: Stories and Journeys of Collaboration in Indigenous Research*. Oregon State University Press. Corvallis, OR. The stories, essays, and reflections collected in this book offer insight into the challenges and rewards encountered by geographers in their academic and personal approaches to research when working collaboratively with Indigenous communities across the globe.

Louis R. 2007. "Can you hear us now? Voices from the margin: Using Indigenous methodologies in geographic research." *Geographical Research* 45 (2): 130–9. This article argues for participatory and Indigenous-led research agendas and methodologies so that research can be carried out in a respectful and ethically sound manner.

Moss, P., ed. 2002. *Feminist Geography in Practice: Research and Methods*. Oxford: Blackwell. This book gives a thorough introduction to taking on, thinking about, and doing feminist geography research.

Moss, P. 2005. "A bodily notion of research: Power, difference and specificity in feminist methodology." In L. Nelson and J. Seager, eds., *A Companion to Feminist Geography*. Blackwell. Oxford. pp. 41–59. A thought-provoking chapter that urges researchers to take seriously the principles of feminist methodology they use and to be critical about the context in which research takes place.

Parpart J.L., Rai S.M., and K.A. Staudt, eds. 2013. *Rethinking Empowerment: Gender and Development in a Global/Local World*. London: Routledge. This book explores the uneven and highly contested terrain of gender and empowerment, illustrated through case studies from a variety of places. Includes chapters on education, cyberspace, the nation state, micro-credit, and NGOs.

Acknowledgement

Clare would like to thank Katy Bennett for her useful suggestions about the chapter.

5 Writing a Compelling Research Proposal

Janice Monk and Richard Bedford

Chapter Overview

This chapter is intended to help you with the challenging task of initiating research by writing a proposal that will focus your thoughts, plan your approach, and convince others that your project is important. We address how ideas for research originate, how to specify research questions, how to demonstrate your ideas' significance, and how to define your research methods. We discuss the process of writing research proposals and comment briefly on how reviewers evaluate them. Our concluding comment sums up why we think writing compelling research proposals is so challenging while at the same time being one of the most enjoyable parts of a geographer's post-secondary training. Learning to write such a proposal is also one of the most valuable contributions you can make to the development of your skills as a researcher, especially if you choose a career that involves defining research priorities or bidding for funding to support research activity.

The Research Challenge

Writing a compelling research proposal is a real challenge. In our view, it is probably the most demanding task that any undergraduate or graduate student experiences. It is also one of the most exciting, because it is a chance for you to define your research questions, frame them in the context of existing knowledge, identify appropriate methods of inquiry to address the questions, negotiate ethical issues, and establish data collection and analysis procedures.

Writing a research proposal requires you to bring into focus all of the essential elements of your university education. It demonstrates your ability to synthesize and question existing knowledge on a topic and your understanding of how ideas are used to formulate theoretical frameworks. It draws on your knowledge of and experience with different ways of approaching problems and asking questions, as well as your ability to construct a coherent design for new research. It indicates the extent to which you are stimulated by the "cutting edges" of knowledge and inquiry in the geographical questions that really interest you.

As we show in this chapter, a proposal offers a roadmap for the research journey, but as on many journeys, you are likely to encounter crossroads, roadblocks, and many twists and turns that you need to negotiate. You need to be flexible enough to accommodate these changes. Your proposal is not a static, definitive statement; it is a vibrant, living component of your research. It is subject to revision, extension, and amendment as the research unfolds.

Writing research proposals is stimulating. Carrying out the research to address the questions in the proposal is an adventure. Research is to be enjoyed, not endured, even if at times it is difficult to see how you will ever address all the research questions you started with. No worthwhile journey is easy; research is not easy, but research makes university-based study really rewarding.

Where Do Research Ideas Come From?

The first thing to establish when developing a research proposal is a topic. In our experience, ideas for research come from at least three sources: personal experience, reading, or conversations with other geographers and scholars. In most cases, some combination of these sources is involved. It is important to appreciate that you need to have a strong personal interest in the research you are planning to do. The work has to be something that you care about and have the skills to carry out (or know where you can get these skills).

If you are writing a thesis, you will be spending a lot of time on the research, so you want to be sure you have a topic that really interests you. Do not be surprised if it takes time to refine the topic; the initial idea may come from some experiences you had a long time before you actually became involved in research. In the course of writing a proposal, the ideas will be clarified, and you may find that by the time the proposal is written, the original idea you started with is substantially altered and refined.

To illustrate how research ideas can evolve, we draw on our own experiences as graduate and post-graduate students embarking on our first major research exercises (see Boxes 5.1 and 5.2). While these examples relate mainly to writing theses for Master's or doctoral degrees, the situations and experiences they describe have relevance for undergraduate research projects and honours dissertations as well. There is a lot about writing a compelling research proposal that is generic, in addition to the specific requirements of projects or theses for particular degrees.

Something we both found highly motivating about our research was that the ideas that contributed to the development of our research proposals were of interest to participants, local government officials, and other academics studying contemporary social, cultural, and economic transformation (for an elaboration, see Chapters 3 and 4). Personal as well as broader interest in what you are researching

The Origins of a Dissertation in Social Geography

Jan Monk grew up in a fairly low-income family in Australia at a time when non-English-speaking immigrants were arriving in substantial numbers. She was conscious of social inequalities, ethnic differences, and stereotyping of those we would now label "Other." Shortly after graduating with her BA, she participated in a work camp that built a house for an Aboriginal family so that they could move from a reserve into a small town in New South Wales. The project reflected the state's policy of assimilating Aboriginal people into the white community. The government paid for the building materials, and a church group brought together young professional men and women who donated their labour.

The project raised geographic questions about who lives where and who has the power to shape those residential patterns. The hands-on experience, and some of the project's ethical and political implications, made a strong impression on Jan. At that time, Australian geographers had not been writing about Aboriginal affairs,[1] and anthropologists were mainly interested in "traditional" cultural patterns or in psychological questions about Aboriginal life on reserves. There was really no geographic precedent for Jan to link her personal interests and potential research to at this time. The ideas for possible research were stimulated by personal experience; further reading in the social sciences while she undertook post-graduate studies was needed before the research proposal could be developed.

Jan moved on to doctoral study in geography in the United States. Attention to relations between groups was rare in geography, but she took courses in sociology and anthropology that spoke to her interests. An anthropology professor's ecological framework took into account demography, economics, and politics. With this new perspective, Jan saw a way to address what had been on her mind in Australia. It so happened that Australian policies towards Aboriginal people were also coming under scrutiny at the time she embarked on doctoral research, so her ideas happened to coincide with an emerging area of policy concern. This was a bonus—government officials were also interested in the research, and this made the whole project seem more useful and relevant.

She now had context that identified her question as one of policy importance, she was more familiar with relevant interdisciplinary literature, and she had a set of research tools to apply to questions grounded in personal experience. Imbued with the geographical perspective that place matters, Jan designed a comparative analysis of the social and economic lives of Aboriginal communities in six small towns in the state of New South Wales and went on to explore how the lives of these Aboriginal people were influenced by white social, political, and economic history.

BOX 5.2

Beginning a Career in Research on Migration

Serendipity can play a major role in the selection of research topics. Richard Bedford's undergraduate training in geography, history, geology, and Pacific studies at the University of Auckland led him to consider two quite different directions for a graduate research degree: fluvial geomorphology and population geography. In the end, it was an invitation to visit a central Pacific atoll country (Kiribati) for a holiday that swung the balance towards research about people rather than rivers. There are no rivers on atolls, but there are some very interesting people–environment relationship issues, especially apparent to someone from a much larger, mountainous country who is visiting coral atolls for the first time.

A combination of reading for a graduate paper on the geography of the Pacific Islands, discussions with senior government officials in Kiribati, and several months of fieldwork in the islands provided the ideas that were to become the basis for a Master's thesis on migration in an atoll environment. As was the case with Jan Monk's research on Aboriginal social geography, Richard Bedford's research on migration as a response to population pressure on coral atolls was of interest to the colonial government of the day. The concerns the government had about population change thus fed into informing the ideas that underpinned the research proposal that was developed to guide the research over the subsequent 12 months. This interest in the migration of Pacific peoples then took Richard to the Australian National University's Research School of Pacific Studies (now the Research School of Asia and the Pacific) where he completed his doctorate.

Doctoral research in Vanuatu and post-doctoral research in Papua New Guinea and Fiji during the 1970s and 1980s, as these countries moved from colonial rule to independence, was exciting and very challenging for a "white" male New Zealander. A mix of research methods and sources of information was always required: archival research, analysis of census data, questionnaire surveys, in-depth interviews, group discussions. It was not a question of "either quantitative or qualitative research methods and approaches"; it was always necessary to use a mix of both, especially if the analysis of population movement, as described in the reports of government officials and the statistics collected in surveys and censuses, was ever to be informed by the personal experiences of those who moved.

can be critically important, both for sustaining your own engagement with the project as well as for stimulating new research ideas and opportunities.

One of the great benefits of much research done by geographers is that it seems "relevant" to others: that is often very important for researchers when they are trying to define a topic on which they will work for a substantial period. However, it should be noted that research does not have to be "relevant" to particular interest groups to be worthwhile; much theoretical inquiry is driven by curiosity, not by a concern with addressing a question that is of interest to a particular group. As we have already said, the key thing about doing research is to be genuinely interested in the topic you choose to work on.

Getting Started

When and how you start work on a research proposal will depend on the stage of your post-secondary education, what your department and supervisor expect, the nature of your project, and whether or not you need funding for the research. If you are an undergraduate, your department may offer a course that includes some training in writing a research proposal. At this level, it is likely that you will be expected to develop a full research proposal, but you may not be required to actually carry out the proposed research or to seek outside funding to support your research. If you are an honours, Master's, or doctoral student, however, you will usually be expected to write a proposal and present it in a colloquium prior to having it approved. Once it is approved, you will then undertake the proposed research. You may also need to prepare a proposal to a granting body to obtain financial support for the research.

Even if you are not required to write a formal proposal along the lines of what is suggested in Box 5.3, we suggest that you do—it will clarify your ideas, taking them from a broad theme to specific questions. The proposal will provide you with a roadmap for the research journey. It will help you to say why your project is important, how you will carry out the work, what resources (financial and other) you will need, and what timetable you will follow for collecting data, analyzing the materials, and writing up your results.

In Box 5.3, we summarize some of the key components of a research proposal. Several of the items that are bulleted in this summary are discussed in greater detail in subsequent sections.

In the next three sections, we use examples from a research proposal prepared by Ranjana Chakrabarti, a doctoral student in the United States in the early 2000s who has given us approval to quote from her work, to illustrate the major parts of the proposal. These parts are specifying the research questions, framing the research in terms of its theoretical and empirical contexts, and

Components of a Research Proposal

A good proposal will enable the reader to appreciate immediately:

- the *question/problem* that is being addressed in the research;
- the current state of *knowledge* on the topic that is the subject of inquiry;
- the *methods* that will be used to collect the information required;
- the *ethical issues* that have to be addressed before undertaking any fieldwork;
- the *resources* that will be required to collect the required information and complete the research report;
- the *intended outcomes* of the research.

The proposal will have several sections in which the points raised above will be covered. The key sections are:

- introduction to and background of the proposed research, including a statement as to why the topic is worth studying;
- the key research questions/problems (which might be specified in the form of hypotheses, but this is not always appropriate);
- the wider theoretical and empirical context for the research;
- research methods and associated ethical issues;
- research plan (timetable outlining the various stages of the research);
- budget and sources of funding for the research;
- references cited.

developing the methodology for the project. We then provide some suggestions about the way successful research proposals are written and return briefly to the issue of funding.

Specifying Your Research Question

One of the most challenging aspects of writing a proposal is articulating your major research question and the sub-questions that flow from it. Everything else follows. You have to keep the "big picture" in mind, but think through how to identify its component parts so that you have a manageable project. Suppose you are interested in policy aspects of housing and homelessness in Australia. You know, stereotypically, that homelessness is associated with urban areas and "vagrant" men; remedies are couched in terms of providing shelters, mostly in inner-city neighbourhoods. Then you see a report that young people make up almost one-third of the homeless.[2] You grew up in a rural area and know of students who dropped out of your high school and ended up homeless within the local community. This knowledge prompts your first decision: you will conduct research on aspects of youth homelessness in country towns.

Now you have more decisions to make. Will you focus on the causes of the problem, the experiences of homeless youth themselves, the types of services provided or needed, or on some combination of these aspects? Will you undertake a case study of a single community or of several? What criteria will you use to select the study area(s)? Will you focus on homeless young men, young women, or both? How will you connect your research back to policy concerns? There is no "right" answer to these questions. Which questions you address will depend partly on the time and resources you have but also on what you learn about existing related research so that you can show that yours will make a new contribution.

Your choice of research question will also be heavily influenced by your knowledge of and skills with different research methods. Some geography students have a strong interest in statistical analysis, especially if they work with geographic information systems (GIS) or have subjects like psychology and economics in their degrees. Other students have a strong preference for qualitative modes of inquiry, particularly if they have studied cultural geography, critical perspectives, gender and ethnicity, and so on. There is no "right" set of skills that everyone doing research must have, aside from an ability to think and write clearly. The particular research and analytical skills needed will depend very much on the nature of the research questions being addressed.

Research offers the best opportunity there is to use the skills you have already developed in the course of getting your degree, while at the same time enabling you to gain new skills. You should not feel constrained to work just with methods you are familiar with; research for a thesis will often encourage you to extend your understanding of research methods. But be realistic when developing your research questions; make sure when you are specifying questions that you have or can develop the skills to address them. Over-ambitious research proposals can make for very frustrating research experiences.

The sorts of questions raised in the example of homelessness among rural youth lend themselves to analysis using qualitative research methods, and students with a good understanding of such methods may be both more interested in and better placed than others to undertake research on this topic. We discuss the development of the methodology section of a research proposal later in this chapter, but it is important to appreciate that the definition of research questions is very much influenced by your preferences for and understanding of different research methods and modes of analysis.

It is useful to cite a further example to illustrate how research questions are formulated and incorporated into a research proposal. Ranjana Chakrabarti went to the University of Illinois at Urbana-Champaign in the United States for study after completing a Master's degree in India with a focus on medical/health geography. She was interested in how place shaped women's health-care practices. Given her background, she chose to study South Asian immigrant women in the United States, specifically in New York where she had personal connections. Her research addressed their low rates of prenatal care utilization, since this had consequences for their health and that of their babies. In the rest of this chapter, we use sections of her dissertation proposal, written in 2004, to illustrate some of the key points we seek to make. Box 5.4 summarizes how Ranjana defined the research question and the sub-questions for her dissertation research on the roles of place, context, and culture in shaping South Asian immigrant women's prenatal health-care practices in New York.

Once the research questions have been defined in draft (they will be refined as the proposal is developed), the **literature review** and the identification of an appropriate theoretical framework and methodology for conducting the research can be completed. We now turn to the critically important component of all research proposals that addresses the theoretical and empirical contexts within which the research questions are situated.

Framing Your Research

As you develop your research questions, it is important to think about how you will convey their significance to others and what theoretical and **conceptual frameworks** you will employ. If your audience and study are primarily academic, you will mostly rely on the research literature to justify your project's approach and significance. If your work has a strong applied component, you should also study the perspectives and needs of the group(s) with which you will work. Among the various questions that need to be addressed while framing your research, the following four are especially important:

Research Questions for a Health Geography Study

Much of public health research on prenatal care has focused on the role of individual-level maternal risk factors in explaining low levels of utilization of prenatal care. A large number of socio-demographic and structural as well as attitudinal and psychological factors have been identified in previous studies. However, the ways these barriers are experienced and expressed differ among women from different ethnic backgrounds. This difference can be understood through an appreciation of the complex interactions between place, culture, and health (Chakrabarti 2004).

Ranjana's primary research question was "What causes the low rates of utilization of prenatal care facilities by South Asian immigrant women in New York?"

The sub-questions she identified to assist her frame the study and develop the research proposal were:

- How and why does prenatal care use by South Asian immigrant women vary within New York City?
- How do South Asian women gain access to prenatal services?
- How does the experience of place mediate the prenatal care practices of these women?
- How do these women relate to and gain knowledge from the formal and informal prenatal care resources in their local environment? What geographical, economic, and cultural barriers do they face?
- How does culture affect the prenatal care knowledge and practices of these women? Are such practices place-based and at what scale?
- How do South Asian women create and utilize place-based social networks at the local, national, or transnational scales to gain access to formal and informal prenatal care services and advice? (Chakrabarti 2004)

- Does the project deal with a significant and meaningful problem that lends itself to a substantial research effort?
- Why is the problem of interest to other scholars or practitioners in the field?
- Has a persuasive case been made as to why the problem is worth solving?
- Is it clear who or what will be aided by the research findings? (Dissertation Proposal Writing Tutorial n.d.)

We know that identifying the "right" project is a difficult task and that you may agonize over it. Peggy Hawley (2003, 36) reminds us that there are dozens of "right" topics and that "what makes the choice succeed or fail depends mostly on what you do about it after the choice has been made." As you think about the questions above, it is useful to also follow Hawley's suggestion that you avoid a topic that is broad and unwieldy and ask yourself whether it will be manageable, be in the range of your competence, draw on sound sources of data, and be of interest to you. She also recommends avoiding research on extremely controversial issues early in your career. This is also the time when the advice of your supervisor(s) can be very helpful. While it is your job to define and justify the topic, it is the responsibility of research mentors (and that is what supervisors are) to assist you in developing confidence in the topic you have chosen. They are likely to have views about the "significance" of the problem or question you are thinking of addressing, why this problem/issue might be of interest to other scholars in the field, and how to make a case for research on the topic. Do seek advice from your supervisor(s) as you begin to frame your research, but always keep your own interest in the topic to the fore. After all, this is *your* research project.

Writing a review of the literature is essential for preparing the proposal. To situate the research in an appropriate theoretical framework will require careful study of some of the key journals in the discipline (see Box 5.5). This literature review should not be confused with an annotated bibliography or a summary of all you have read. It should be both constructive and critical in tone, identifying the strengths of the pieces you select in order to show how they support your own work and the weaknesses or gaps that demonstrate why your work is fresh and significant. The final report or thesis that you write will always include reference to earlier research on the topic and, if it is a thesis, a substantive assessment of previous findings and how your proposed research furthers understanding of the topic.

Using a literature review to help establish the context for your research is a demanding and time-consuming task. It is not something that is done overnight! It is also not something that you will complete until you are writing up the results of your research. Literature reviews evolve during the course of inquiry, so the summary you provide in the research proposal is really just the initial review of key findings by others. You will add to this as the research progresses, especially if you are writing a thesis for a Master's or doctoral degree. Box 5.5 contains some suggestions on how you can approach this very important component of a research proposal.

A useful source of examples of research contexts and the literature reviews associated with them is the introductory sections of published articles. An article by Marianna Pavlovskaya (2004) on her doctoral research on household economies in post-communist Moscow is one such example. She begins by noting that the existing literature on the transition from communism tends to emphasize macroeconomic themes. Next, she draws attention to the theoretical writing by

The Research Context

The key questions that are addressed when describing the context for your research are:

1. What does your work offer to the available field of knowledge that helps us to better understand the issue or topic you have chosen to address in your research?
2. How do existing studies inform your research topic?

In answering these questions, keep in mind that there are two important dimensions to the context of any research proposal:

1. the theoretical ideas that inform research on the topic (these ideas will tend to come mainly from the existing literature);
2. the empirical or "real-world" situation that relates to your research topic (there will usually be examples of research on your topic or a closely related one that you can find in the literature).

Make sure that you address both of these dimensions when framing your research. Heath (1997) suggests that when describing the context for your research, you should:

- Use specific language to name and describe the conceptual foundation for your research—that is, the research perspective that informs your approach to the topic. Show clearly why this perspective is appropriate and relevant for your study. For example, you may be using insights from a post-structuralist perspective to provide a theoretical context for your research. Make sure that you explain clearly and simply what the approach means for the way you will do your research and how you will interpret your results. Useful definitions of most of the main theoretical approaches used in human geography can be found in the various editions of *The Dictionary of Human Geography* (Gregory et al. 2009).
- Cite authors who have already used this approach to address questions and problems similar to the ones you are researching. An excellent starting point for recent reviews of the theoretical and the empirical literature in most of the main areas of research currently being addressed by human geographers is the journal *Progress in Human Geography*. This journal and its companion, *Progress in Physical Geography*, contain regular updates on new ideas, methods, and findings in contemporary geography.

geographers on the importance of scale and connections over space in shaping the social and spatial relations of people and the environments within which they conduct their lives. This geographical literature is used to justify her approach to the topic through an analysis of households and the ways in which they link formal and informal economies in the city. This theoretical perspective leads her to a research design that will enable her to integrate in-depth interviews with households in selected neighbourhoods of inner Moscow (taking into account their gender and class dimensions) with a GIS-based reporting of spatial patterns of neighbourhood characteristics and interactions.

The research context thus establishes the rationale for the particular research methods that will be used to undertake the study. We now turn to this very important component of all research proposals.

Developing Your Methodology

Identifying how you plan to carry out your research is one of the most demanding aspects of writing a proposal. Reviewers of proposals repeatedly find that the discussion of the methodology is the least satisfactory part of the document. This seems to be a particular problem with research that will involve qualitative methods. The researcher may not go much beyond indicating that data will be collected through focus groups or in-depth interviews, perhaps naming software that will be used for analysis but failing to *show why and how these methods will allow the specific research questions to be answered fully and ethically.*[3]

In Box 5.6, we summarize some of the most important considerations that need to be borne in mind when developing a research methodology. We do not discuss specific methods in this chapter in any detail; these methods are the subjects of other chapters in this volume.

In Boxes 5.7 and 5.8, we draw again on Ranjana Chakrabarti's doctoral research proposal to provide a specific example of a research timetable and a statement of research methodology. These ideas about research methods should be read in the context of her key research question and sub-questions outlined in Box 5.4.

Writing the Proposal

Exactly how you organize the final proposal will depend on who will read it. If you are writing it for your supervisor or a class in research methods, you may be given guidelines to follow. If you are applying to a funding agency, it will usually have guidelines that should be followed very carefully. A cover sheet or letter and an abstract, for example, may be required. You will probably write these items last to make sure that they are accurate and compelling statements about your project.

Thinking about a Research Methodology

- Be realistic. You will have limited time and resources to conduct the project. If your work involves individual interviews, how many, for example, will you be able to complete while still getting sufficient information to explore the range of views people may hold? Consider that you will have to recruit people, schedule the interviews, and deal with some refusals or cancelled meetings. How do you know people will talk to you? What strategies will you use to recruit them? Be careful not to over-commit yourself.
- Will your work be served best by combining methods? Researchers frequently use a combination of quantitative and qualitative approaches (see, for example, Tashakkori and Teddlie [1998] and Chapter 1 of this book). How can you do this and be realistic in relation to your resources and context?
- Assess the strengths of the approach you propose over alternatives. Would focus groups serve your needs better than individual interviews? Why or why not? What are the strengths and limits of the methods you are proposing?
- Show that you know what data you may need to draw from (e.g., public records) and that you will be able to gain access to them and gain that access in time to complete the project.
- Consider what you will do if you run into data collection problems. Strong proposals identify potential obstacles and show that you have thought of alternatives. The best-laid plans are not always feasible at the time and place you want to implement them.
- Prepare and include a timeline showing when you will undertake specific tasks. One (but not the only) format for a timeline is a matrix, also known as a Gantt chart, that lists tasks on one axis and time periods (for example, months) on the other. This allows you to show that some components of your work will overlap in time (for example, you may be arranging for interviews at the same time that you are researching background information from other sources on the context). Box 5.7 provides an example of a timeline based on Ranjana Chakrabarti's PhD dissertation.
- If possible, undertake pilot research to demonstrate the feasibility of your study, and refer to it in your proposal.
- Finally, do not forget to allow time to obtain approval from the departmental or university committee responsible for overseeing ethical issues in research.

BOX 5.7

Example of a PhD Timetable

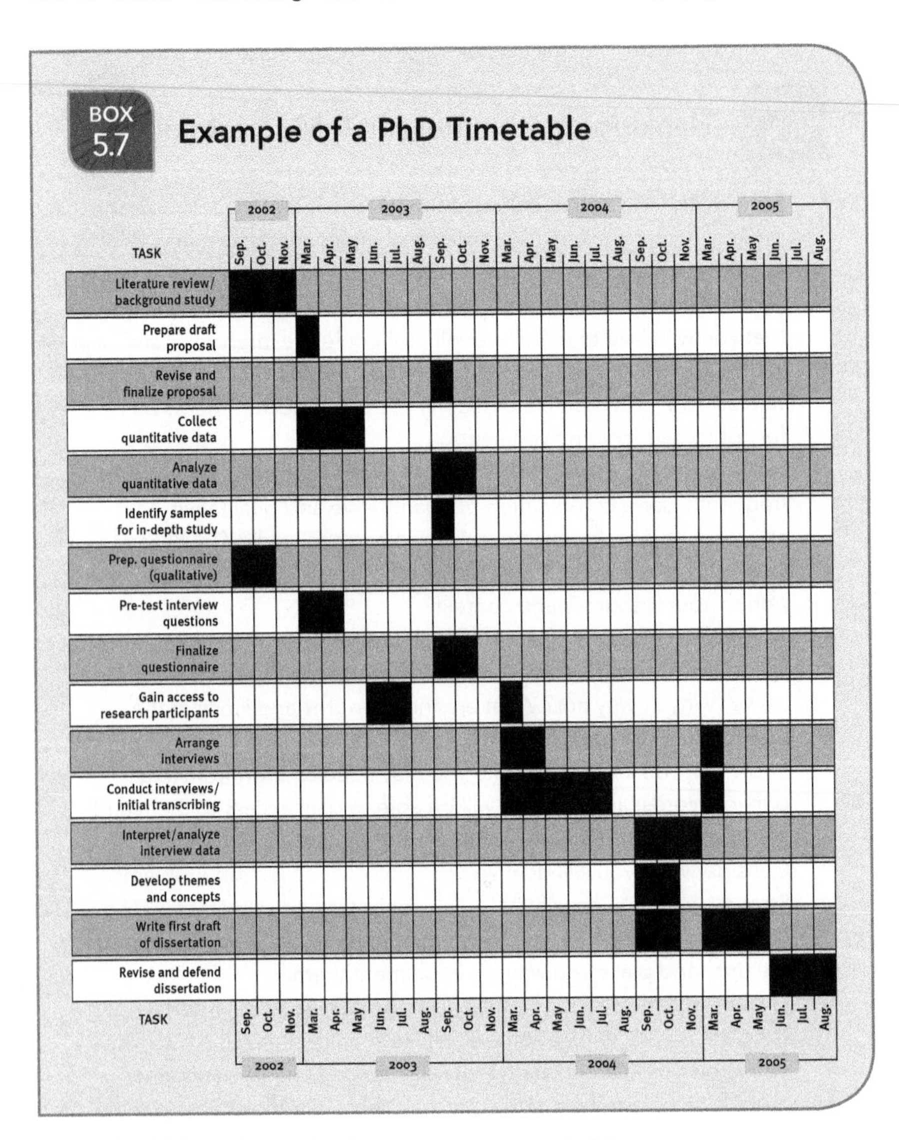

BOX 5.8

Research Methodology for a Health Geography Study

"In this study, a mix of qualitative and quantitative methods will be used, and these relate to two distinctive, but related, parts of the proposed research into the use of prenatal care services by South Asian mothers. . . . The first phase involves a quantitative analysis of spatial variation in the use of formal prenatal care services by South Asian mothers in New York City (NYC). This work uses data from the births master file of the NYC Department of Health. A geographical information system (GIS) will be used for mapping and analyzing the geographic distribution of South Asian mothers and for exploring geographic variation in prenatal care use for the major South Asian groups. Logistic regression techniques will be used to shed light on the influence of factors such as education and insurance coverage on the use of prenatal care.

The second phase of the research involves an extensive field-based qualitative study to understand how the everyday lives of immigrant South Asian women shape their prenatal care practices. Interviews will be carried out with approximately 60 immigrant South Asian women in New York City. The main goal is to understand the type of barriers women face in using formal prenatal care and how women utilize social networks, both national and transnational, to gain knowledge, social support, and prenatal services during pregnancy. The types of questions that will be used to obtain this information are discussed in the proposal.

The sample women for the in-depth interviews will be chosen to reflect the geographic, economic and cultural diversity amongst South Asian immigrant population in New York City. . . . Based on the quantitative GIS analysis, place-based stratification is proposed to identify neighborhoods for intensive study. The initial research participants in each neighborhood are to be contacted through South Asian organizations, nongovernmental organizations working for immigrant South Asian women, and through informal networks and social gatherings. . . . and through 'snowball-sampling' procedures." (Chakrabarti 2004).

The qualitative phase of the project began in February, 2004 and lasted for around 8 months. Interviews were tape-recorded and transcribed, and the interview data were analyzed using interpretive methods. The qualitative analysis software, **NUD*IST**, was used to assist in identifying major themes from the interviews and in aiding the data interpretation. The proposal makes reference to the fact that Ranjana conducted pilot interviews with a small sample of South Asian women in Champaign, Illinois, to pre-test the format and questions for the in-depth interviews. Based on that experience, a detailed set of interview questions were formulated.

The Institutional Review Board [i.e., ethics committee] at the University of Illinois granted final approval for the project in December 2003.

continued

In whatever form the proposal is written, it should address the key questions of what, why, how, and when, as this chapter has discussed and, if you are applying for funding, it will need to identify who you are, your qualifications, and how much (a justified budget) the research will cost. Ancillary information may well include items such as a list of the references cited in your text, the ethics/institutional review clearance, letters of recommendation, and your credentials in the form of a biographical statement or curriculum vitae.

As we noted earlier when outlining the key components of a research proposal, we think that the opening one or two paragraphs of the research proposal should capture the reader's attention by succinctly answering the what, why, how questions. Anthony Heath (1997) provides a useful outline of the content of the introduction to a proposal advocating that you address what it is you want to know or understand, how you became interested in the topic, why the research is needed, referring briefly, for example, to the lack of existing knowledge or limits of other work and to whom the work will be of value. We also suggest that you indicate briefly what conceptual or methodological approach the research will adopt. In Box 5.9, we offer the opening paragraphs from Ranjana Chakrabarti's research proposal for illustrative purposes.

This kind of approach is more effective than beginning with a lengthy "background" statement and failing to define your project until several pages into the proposal. The latter failing is one of the most common among inexperienced writers. It is especially important to have a strong opening if you are applying for funds in a national competition. Reviewers have many proposals to evaluate in a short time. The easier you make their task, the better chance you have of being successful. In this regard, it is a good idea to specify in the opening paragraphs what is innovative about your research.

We observed at the outset of this chapter that you should allow plenty of time to write your proposal. Just how much time depends on your career stage and the context in which you are writing. Six months preparation is not unusual for a doctoral proposal. Six weeks might be typical for an honours proposal. You should expect to write multiple drafts, to seek critiques from peers, your supervisor(s), and other relevant academic staff members, and then to think carefully about how to address their comments. Solís (2009) provides some useful examples of how geographers have revised their proposals to respond to such critiques. You should seek out models of other successful proposals. Websites that provide helpful examples exist, and we list some of them at the end of this chapter. If program officers

Introducing the Proposal

Ranjana Chakrabarti's (2004) proposal began with the following two paragraphs:

> The main objective of this PhD dissertation is to understand the roles of place, context and culture in shaping prenatal health-care practices of South Asian immigrant women in New York City. These women have very low levels of prenatal care utilization compared to US-born women and women from other immigrant backgrounds. . . . This puts at risk the maternal and infant health of one of the most vibrant immigrant populations in the country, because low and inadequate prenatal care use can result in low birth weight babies and other adverse pregnancy outcomes. . . .
>
> Little is known about the complexity of circumstances that leads to inadequate utilization of formal prenatal care by South Asian women in New York City. This research is intended to fill this gap by exploring how the geographical contexts of everyday life influence their prenatal care practices. The primary focus is to understand how the experiences of place, culture and context mediate prenatal health-care practices. Using quantitative methods and qualitative in-depth interviews I seek to understand how women's situatedness in local and transnational social and geographical networks constrains access to prenatal care and how women draw on such networks in creating new spaces of prenatal care access.

in a relevant funding agency are willing to discuss your proposal with you, or if they offer workshops at professional meetings, by all means take advantage of such opportunities.

While your ideas are clearly extremely important, your writing style and the format of the argument are critical features of a successful proposal. Use clear language rather than excessive technical jargon. Using discipline-specific technical jargon is especially problematic if you are applying for funds from agencies that use multi-disciplinary panels to assess the merits of the proposed research. Statements that readers with a good education in any of the sciences can read are much more likely to succeed in competitive funding rounds than ones written in complex prose that is really only understood by researchers working from particular theoretical or ideological perspectives.

Make sure the typeface is easily legible. Make judicious use of underlining, bold, or italic type where appropriate. Proofread carefully, and double-check all details. If you have written a proposal for external funding, be especially careful about the funding agency's requirements. Does the proposal have to be mailed by or received by a specific date? How many copies must you send? Is the proposal to be submitted electronically or as paper copy? Do you require approval by your institution before you submit your proposal to an external agency? What does that process involve? How much time does it take? Failure to follow the required guidelines for research proposals can lead to significant delays in getting the project approved or funded or, more likely, to rejection of the proposal.

Seeking Funding for Research

Obtaining funding for research operates very differently in different settings, and for this reason we will deal with the topic only briefly. We suggest that you talk with your supervisor and other students who may have had grants, check the Internet and libraries for information sources, and consult offices in your university that assist in identifying funding sources. You may also wish to check whether funding is available for a **pilot study** that will enable you to demonstrate that your main project is feasible.

If you need funding, start the application process well before you expect to begin the research. You may wait as long as six months to learn whether you have been successful in a national competition. It is sometimes argued that research involving qualitative methods—for instance, collecting detailed information from small numbers of interviewees—has a lower chance of getting financial support from external sources than, say, large quantitative surveys. This was much more of an issue a decade ago than it is today. Well-argued research proposals, even if they rely heavily on qualitative rather than quantitative data, are likely to be funded. A critical issue with regard to funding research is the nature of the research question that is being addressed, the relevance it has for the priorities of the funding agency, and the extent to which the panel assessing your application is excited by and understands your research proposal.

Before applying for funding, check the agency's guidelines *very carefully*: are you eligible for an award, what is required in the proposal (for example, format, page length, type style, budget, bibliography, letters of reference, your credentials), and what is the deadline for receipt of proposals? Writing your research proposal and associated funding applications is time-consuming. Experienced supervisors often suggest beginning your draft up to six months before submission. So start early!

A Concluding Comment

We pointed out in the introduction to this chapter that writing a research proposal is probably the most demanding task any undergraduate or graduate student will experience. We have found that students invariably tell us that the requirement to design and carry out their own research project was the highlight of their geography training at university. One of the hardest things to do is to decide on what you think is a good research topic. As we noted when drawing on our own experiences of graduate and post-graduate research, there is no single source of inspiration for topics. With the wisdom of hindsight, we offer one useful piece of advice: choose a topic that can be specified and explained simply and clearly. A real trap when choosing topics for research is trying to be very clever in order to convince the reader that yours is really an original idea. In reality, very few research ideas are completely novel or "original"; most research builds on existing ideas and extends them in interesting ways.

The key to success in writing a compelling research proposal is your own excitement about doing the research. If you are interested in the topic, you will think more clearly about all of the issues we have raised in this chapter. Strong commitment to the topic on your part will tend to ensure that the research questions will be well specified, the research will be framed effectively, the methodology will be appropriate, and the proposal will be well written. Undertaking research does not appeal to everyone, and a lack of interest in this aspect of university training usually shows up very quickly in the content of a research proposal. We hope that you will want to carry out research on a topic that is of interest to you; in both our cases, it was a transformative experience that led us on to careers in which research is at the heart of the job. We can also say that our research questions, posed over 40 years ago, still remain very relevant in the twenty-first century and remain at the heart of debates about development of the Aboriginal people of Australia and the inhabitants of the coral atolls and reef islands of the central Pacific. We can think of no better or more rewarding challenge in a university program than the challenge of writing a compelling research proposal and then carrying out and writing up the research.

Key Terms

conceptual framework
literature review
NUD*IST
pilot study
snowball sampling

Review Questions

1. Identify two or three possible topics you will consider for your research project, thesis, or dissertation. Describe how these ideas originated. Why

do you think they will sustain your interest? What preparation will you need in order to address them?

2. Select a published research article in your area of interest. What does the author identify as the main research questions? Write an opening paragraph that you think would be a strong introduction for a research proposal that will address these research questions.
3. List the information that you will include in a proposal for research you will undertake to demonstrate that you have considered how your methodology will be implemented (for example, how you will select people to interview).
4. Outline the "big picture" question and up to four related sub-questions for a research topic of your choice. Identify the main elements of the methodology you would need to employ to address these questions.

Review Exercises

1. Framing a Research Project

This exercise is meant to help you clarify and express what your research project will address. It is important to be able to explain to others the goals of your project and to draw on their feedback as you refine your research question.

One way to accomplish this task is to work in small groups (e.g., with two or three of your fellow students) in or out of class.

a. In a single sentence, identify your main research question. Be prepared to explain to members of the group (a) why you chose this topic and (b) how you think it will contribute to knowledge or society (i.e., what is the significance of the work).
b. Ask participants in the group for comments on your proposed question, for example, do they think it is clear? Do they think it is feasible to conduct research in the setting you propose, in the time you have available?
c. Drawing on the group's feedback, how might you revise your original research question so as to clarify or strengthen the proposed project?

If this exercise is done in class time, it is useful to have two or three groups report back on the main points that emerged in their discussions. What kinds of comments or suggestions did members of the group offer? For example, did they see the question as too narrow or too broad? Did they think it feasible to research? Based on the discussion, how might the proposed question be revised?

2. Situating the Research in the Wider Literature

This exercise is designed to get you more familiar with checking for relevant concepts that address aspects of the research question you are interested in exploring. It is not sufficient to simply Google the topic and work from the list of hundreds (or thousands) of entries that might come up. It is also not sufficient to go to Wikipedia and start there.

One way you might work collectively with a small group of peers on this aspect of your research proposal is to form a reading group that might meet weekly to discuss key concepts and sources of information relating to each of the group member's research topics. A suggested timetable of meetings is as follows:

a. At the initial meeting, work collectively to share ideas on the key concepts and authors that you know of from your geographical education and wider reading to date that might inform each group member's research question. Assign to each member of the group a series of concepts and ask them to copy and circulate relevant extracts from the *Dictionary of Human Geography* for discussion at the next meeting.
b. At a second meeting, discuss the relevant extracts from the latest edition of *The Dictionary of Human Geography*. By sharing information and your own insights in this way you will gain confidence in discussing conceptual ideas in geography (often something students are reluctant to do openly) and extend your understanding of ideas that might be relevant for addressing your research question.
c. At subsequent meetings discuss the last three "progress reports" in the different topic areas selected by group members that are published in the journal *Progress in Human Geography*. Using insights obtained from these "progress reports" and their bibliographies that will contain recent studies in your chosen research field, you will be able to start developing the literature review that accompanies the conceptual framework for your research.
d. Further meetings, devoted to discussing particularly exciting and challenging articles in the contemporary (and earlier) geographical literature may be useful, depending on how much the group enjoys working together to explore conceptual ideas.

As we noted in our chapter, preparing research proposals takes time, and one of the most difficult parts of the preparation is sorting out the conceptual framework. It is often easier to make progress with this by working collectively and becoming comfortable with verbally articulating ideas that can seem quite complex when you read them. A similar approach can be adopted to reviewing possible research methods you might use in your research—discussions among peers who have different skills and backgrounds can be very useful in this regard.

Useful Resources

Bouma, G.D., and R. Ling. 2004. *The Research Process*. 5th edn. New York: Oxford University Press.

Gregory, D., Johnston, R. Pratt, G. Watts, M., and Watmore, S., eds. 2009. *The Dictionary of Human Geography*. Revised 5th edn. London: Wiley Blackwell.

Hay, I. 2012. *Communicating in Geography and the Environmental Sciences*. 4th edn. Melbourne: Oxford University Press.

Heath, A.W. 1997. "The proposal in qualitative research." *The Qualitative Report* 3 (1). http://www.nova.edu/ssss/QR/QR3-1/heath.html.

Institute of International Studies, University of California, Berkeley. 2001. "Dissertation proposal workshop." http://globetrotter.berkeley.edu/DissPropWorkshop. In addition to sections on conceptualizing, writing, and revising proposals, this site includes sample proposals and comments by the authors reflecting on how they regarded the proposal after they had completed the research.

Punch, K.F. 2006. *Developing Effective Research Proposals*. 2nd edn. London: Sage.

Sarantakos, S. 2012. *Social Research*. 4th edn. Melbourne: Palgrave.

Sides, C.H. 1999. *How to Write and Present Technical Information*. 3rd edn. Phoenix: Oryx Press.

Notes

1. Although Fay Gale was shortly to complete her doctoral research in South Australia on regional patterns of Aboriginal communities in South Australia (Gale 1964).
2. Andrew Beer drew our attention to this research idea.
3. Keep in mind that any research involving collection of information from human subjects will necessitate a process of ethical approval for the project. Writing the ethics approval application can be a time-consuming task that should be factored into your timetable. See Chapter 2 for a fuller discussion of ethical issues.

Qualitative Research Design and Rigour

Elaine Stratford and Matt Bradshaw

Chapter Overview

Careful design and rigour are crucial to the dependability of any research. Research that is well conceived results in research that is well executed and in findings that stand up to scrutiny. Thoughtful planning and the use of procedures to ensure that studies are rigorous should therefore be central concerns for qualitative researchers. The questions we ask, the cases and participants we involve in our studies, and the ways in which we ensure the rigour of our work need to be considered; these are hallmarks of any dependable research.

Introduction

In this chapter, we focus on some matters of design and **rigour** that qualitative researchers need to consider over the life of a study to ensure that the work satisfies its purpose and critical audiences. We outline various principles of qualitative **research design** and specific means by which rigour can be achieved in our work.

The chapter is organized into three sections. First, we make a link between the **interpretive communities** in which we work and the sorts of issues that are raised when we begin a study. Second, we elaborate on several steps that map how to select suitable cases and participants for study. In qualitative research, the number of people we interview, communities we observe or with whom we collaborate, or texts we read are important considerations. Nevertheless, these matters are secondary to the *quality* of who or what we involve in our research and subordinate also to *how* we conduct that research. Third, we outline some of the ways in which it is possible to ensure rigour in qualitative research to produce work that is dependable.

Careful research design is an important part of ensuring rigour in qualitative research. While texts and topics on research methods and design often imply that studies should be conducted in a particular way, no single correct approach to research design can be prescribed (Gould 1988; Mason 2004). For certain kinds of work, the order and arrangement of stages may be different, stages can overlap, other stages might well be included, and the combination of qualitative and quantitative research is possible and not uncommon. Nevertheless, by the end of the chapter we will have described in a particular order a number of stages of qualitative research design and summarized them in three diagrams. We believe that you will find this movement through stages helpful in approaching your own qualitative research work, though we must stress it is not intended to be prescriptive.

Asking Research Questions

Each of us needs to acknowledge that our fellow geographers and other colleagues are *already* involved in our studies (Box 6.1). None of us ever formulates research questions or undertakes research in a vacuum. We are all members of interpretive communities that involve established *disciplines* with relatively defined and stable areas of interest, theory, and research methods and techniques (Butler 1997; Fish 1980). Increasingly, too, we are members of *interdisciplinary* research communities that cross over discipline boundaries in ways that enable collective consideration of "**wicked problems**" that are difficult to solve because of their complexity. Such collaborative engagements are often deeply rewarding, and that compensates for the particular challenges they pose, not least among

BOX 6.1 **Considerations in Research Design, Stage 1: Asking Research Questions***

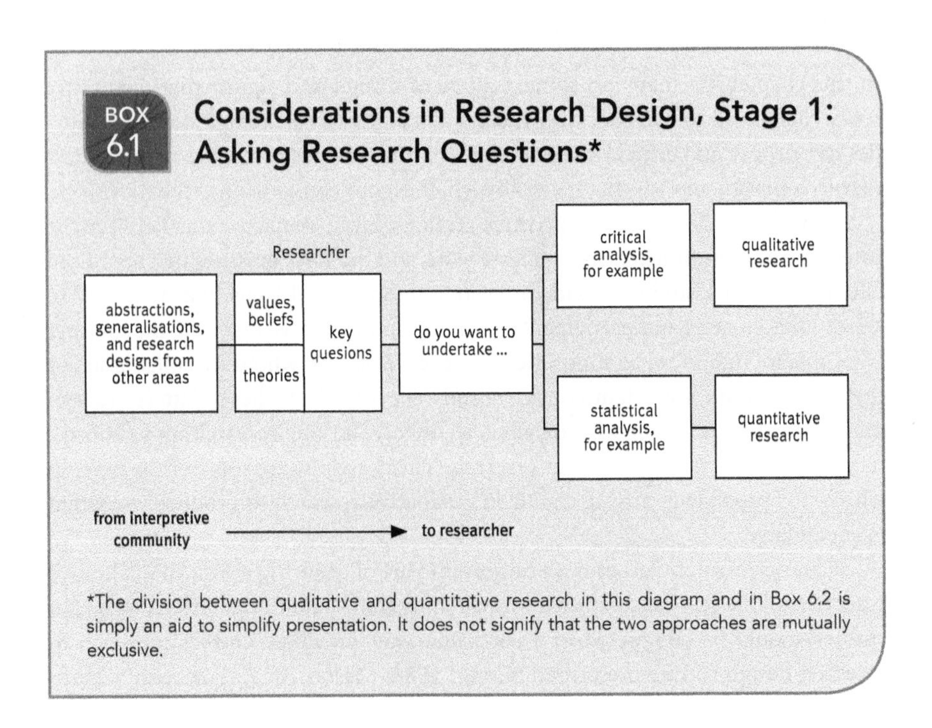

*The division between qualitative and quantitative research in this diagram and in Box 6.2 is simply an aid to simplify presentation. It does not signify that the two approaches are mutually exclusive.

them having to learn each other's distinct vocabularies, methods, values, and conceptual frameworks. These disciplinary and interdisciplinary interpretive communities influence our choice of topic and our approach to and conduct of study; this is because of what Livingstone (2005, 395) describes as "the inescapably collective character of interpretation . . . [which, to] the extent that interpretive communities occupy material or metaphorical spaces, they fall within the arc of the cultural geography of reading" and, we would add, other geographical actions and analyses. We also fold our own values and beliefs into research, and they can influence both what we study and how we interpret our studies and findings (see Flowerdew and Martin 2005; Jacobs 1999; and Chapter 19 of this volume for more detail).

From Asking Research Questions to Conducting Research

Research aims affect research design. For example, asking "how many skateboarders frequent a particular public place compared with other types of users?" will involve a research design different from one that aims to answer the question "why do skateboarders use this place, and how do they interact with other types of users?" The first question focuses on quantification and statistical analysis, and the second is more concerned with the qualitative investigation of skateboarders' behaviours and practices, and with questions about how such behaviours and practices inform various socio-spatial relations—with other skaters, peers, members of the public, private security guards, police and municipal officers, and so on.

In considering the conduct of research, we also need to ask what to do with the information collected. Answering this question will influence the kind of research we do and the ways in which we write our findings. Before making such a decision, we need to be aware of some of the differences between quantitative (extensive) and qualitative (intensive) research. As Sayer (2010) has suggested, each approach helps us to answer different research questions, employs different research methods, has different limitations, and ensures rigour differently.

Extensive research is characterized by identifying regularities, patterns, and distinguishing features of a population, often by reference to a sample that has been selected using a random procedure to maximize the possibility of generalizing to a larger population from which it is drawn. Extensive methods are designed to establish statistical relations of similarity and difference among members of a population. We may be able to determine that "N" number of respondents in a sample think "P" in relation to an issue. So, it might be possible to establish that 86 per cent of a randomly sampled group of daily users of an inner-city public park are in favour of the infrastructure in that park being redesigned to allow skateboarding to be done safely so that multiple uses of the park could include young people's pursuits. Wood and Williamson (1996) distributed a standardized questionnaire to a random sample of the users of Franklin Square, an inner-city square in the city of Hobart, Tasmania, that had been partly claimed through day-to-day use by skateboarders. The data from their study were aggregated, and statements were made about the degree to which these data were likely to reflect the opinions of all the square's users about the presence of skateboarders. This extensive approach produced useful information suggesting the existence of common characteristics and patterns; for instance, skateboarders used certain parts of the square, while other users avoided them. Yet such findings did not account for the shifting quality of various people's different experiences of Franklin Square and of each other, or the reasons behind their opinions.

So how are we to determine *why* respondents held the opinions they did? Such determination is often best advanced by means of **intensive research**, which requires that we ask how processes work or opinions are held or actions are taken in a particular case (Platt 1988; see also Chapter 7). This kind of open-ended questioning will often illuminate the reasons for something—the *how and why* components. In short, intensive research is a powerful tool when we need to establish what actors do in a case, when we need to understand why they act as they do, and when we need to establish what produces change in actors and the contexts in which they are located. Also, in selecting Franklin Square as a case by which to examine multiple-use conflict in public places, Stratford (1998; 2002) and Stratford and Harwood (2001) used intensive methods such as in-depth interviews and observation to understand various responses to skateboarding in the square and around Tasmania more generally. Indeed, qualitative research methods and intensive modes of investigation characterize most studies of multiple-use "conflict" in inner cities because researchers and policy makers see pressing need to understand the motivations, values, and positions of all those involved in the occupation and use of spaces and places. By implication, in various disciplinary and interdisciplinary collaborations wherever such need exists, intensive methods of work are likely to be highly relevant.

Clearly, both extensive and intensive approaches have merit, and are often used together in multi-method combinations, with Sui and DeLyser (2012, 119–20) noting that all methods "simultaneously enable and disable, and mixing methods is not the only way to approach methodological challenge . . . [and that] balanced specialist-synthesis approaches adapted to each new situation may be the most successful methodological framework." However, if we are mainly interested in working through the elements of structure and process that arise from analyzing responses rather than in generating data that make statistical analysis possible, then we are pointed in the direction of intensive research. In opting for a qualitative research design, we are influenced by the theories we are concerned with, by studies undertaken by other researchers in our interpretive communities that we have found interesting, and by the research questions we wish to ask—all of which are interrelated.

Selecting Cases and Participants

Definitions of the terms **case** and **participant** will differ among interpretive communities. However, it is our view that cases are examples of more general processes or structures that can be theorized. Researchers should be able to ask "that categorical question of any study: 'What is this case a case of?'" (Flyvbjerg 1998, 8). In our view, they should resist any anxiety about questions related to the validity of case-based research. As Flyvbjerg (2006, 219) has noted, "a scientific discipline without a large number of thoroughly executed case studies is a discipline without systematic production of exemplars, and a discipline without exemplars is an ineffective one."

In the example just cited, Franklin Square is a *case* of multiple-use conflict in a public place, but it resonates outward to embrace a number of general social and spatial processes involving, for example, the privilege of consumerism, the ways in which citizenship has come to "attach" to acts of consumption, and the relations of government to capital (Stratford 2002). *Participants* make up some of the elements of the case in question and can be widely divergent—for example, parkour practitioners, the elderly, the business community, city councils, law enforcement agencies, the international media, or the health professions (Carr 2010; Dumas and Laforest 2009; Kidder 2012; Stratford 2015). In some theoretical dispositions, participants can also include non-human elements, or actants, whose effects on the case are profound (Callon 1986; Murdoch 2006; Brunner 2011).

Selecting Cases

Sometimes we find a case, and sometimes a case finds us. In both instances, selection combines purpose and serendipity (Box 6.2). On the one hand, we may read about multiple-use conflict in public places in other cities and want to see whether explanations advanced there have merit in—or inform our understanding of—situations with which we are familiar. In this instance, the general or theoretical interest "drives" the research, and we must narrow the field, selecting cases and participants for research. On the other hand, a local government parks manager might draw attention to conflict among various groups in one site in the city and want an investigation of the options available to manage this conflict. In this situation, the case has "found" the researcher—and theories about multiple-use conflict in public spaces are subsequently woven into it. It is worth noting that if, for example, a community development officer rather than a parks manager had contacted researchers about the same issue, the researchers might well have been presented with a different brief that would in some ways make for a "different" case.

Regardless of how a case is selected, it is usually advisable to work at sites or on cases that are both practical and appropriate. In our example of skateboarders' use of public places, ambiguous sites—such as shopping malls, which are generally

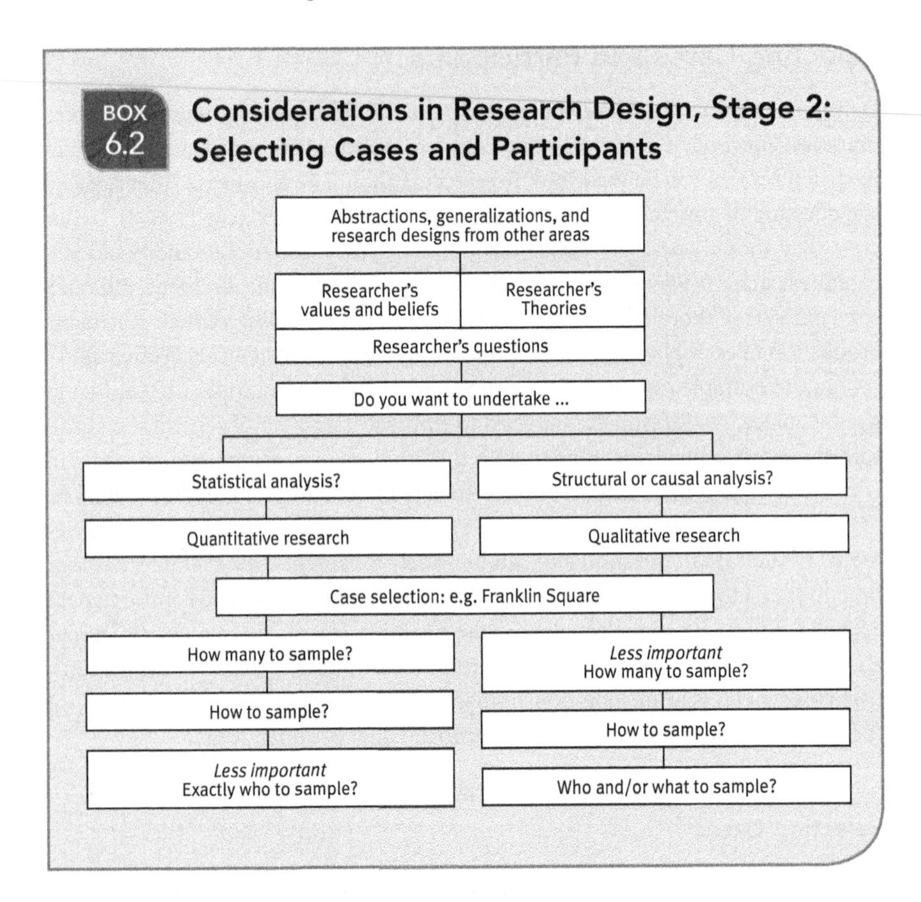

private places behind public facades—add a layer of complexity to investigation and often require researchers to secure a range of permissions to gain access to property, customers, and employees. One final issue needs to be considered in case selection. On the one hand, we might choose to work with so-called typical cases on the grounds that they will provide useful insights into processes evident in other contexts. Alternatively, we might deliberately seek out **disconfirming cases**. Such cases might include individuals or observations that challenge a researcher's interpretations or do not confirm ways in which others portray an issue. For example, we may have studied media reports suggesting that there is unmitigated conflict between youths and the elderly in a public square. However, interviews with elderly pensioners might lead us to understand that some aged people regularly frequent the square when youths are present because they enjoy the company of young people. Indeed, *The Guardian* reports that, in south London, young members of Parkour Dance are working with people over sixty years of age teaching them the principles of parkour (Jenkins 2013), which is a highly disciplined and extremely physically demanding mode of movement (Chow 2010; Mould 2009), and one that reveals much about qualitative and playful engagements with human geography (see, for example, Woodyer 2012).

Such disconfirming cases can be important in the research process, making us think through the way that different institutions and the practices they use (such as the media and their tendency to sensationalize events) create stereotypes. Such cases also require us to ask how various actors are represented (and for what reasons) and how they represent themselves.

Selecting Participants

Exploratory and/or background work (for example, reading, observation, viewing documentaries, conducting preliminary interviews) will often give researchers the capacity to begin to comprehend the perspectives of participants with whom we think we want to interact. Understanding their perspectives in complex cultural situations usually requires some form of in-depth interviewing (see Chapter 8 for details) or observational method (see Chapter 15 for details) that, though time-consuming, often result in a deep and detailed appreciation of the complicated issues involved (Geertz 1973; Herod 1993).

Generally speaking, the more focused our research interest becomes and the more comprehensive our background information and understanding, the more confident we are about who we wish to involve in our research and why. Nevertheless, this confidence needs to be underpinned by a rigorous process of justification. As Mason (2004, 129) points out,

> Your answers to questions about which people to sample should therefore be driven by an interpretive logic which questions and evaluates different ways of classifying people in the light of the particular concerns of your study. Underlying all of this must be a concern to identify who it is that has, does or is the experiences, perspectives, behaviours, practices, identities, personalities, and so on, that your research questions will require you to investigate.

In this respect, it is conceivable that conducting in-depth interviews with a small number of the "right" people will provide significant insights into a research issue. The choice of technique may be dictated by logistics and financial and human resources, or by the needs of participants, among other factors. That choice has expanded significantly as a result of web-based technologies. Stratford, one of the authors of this chapter, has, for example, undertaken extended and multiple-return semi-structured conversations using email and Skype as well as telephone and face-to-face modes of engagement. In one study, she and a colleague sent a list of questions to professionals engaged in urban planning and design (Stratford and Henderson in review). Respondents were able to choose the time at which they responded, and several commented that they valued having that choice. A number also suggested that the email "interview" allowed them to consider their answers, including the use of emoticons, bolded or underlined words, and forms of punctuation for emphasis. This method also provided instant transcription but the researchers were not privy to important information such as non-verbal cues and intonations that attend "normal" interviews.

Participant Selection

Michael Patton's (2002) work on **purposive sampling** is among the more useful summaries of the topic available to researchers. Patton refers to various forms of purposive sampling, including the following seven commonly employed strategies. *Extreme* or **deviant case sampling** is designed to help researchers learn from highly unusual cases of the issue of interest, such as outstanding successes or notable failures, top-of-the-class students or dropouts, exotic events, or crises. **Typical case sampling** illustrates or highlights what is considered "typical," "normal," or "average." **Maximum variation sampling** documents unique or diverse variations that have emerged through adaptation to different conditions and identifies important common patterns that cut across variations (Williams and Round 2007). **Snowball** (or *chain*) **sampling** identifies cases of interest reported by people who know other people involved in similar cases (Kirby and Hay 1997; Stratford 2008). **Criterion sampling** involves selecting all cases that meet some criterion, such as involvement in natural resource management governance in Australia (Lockwood et al. 2010). **Opportunistic sampling** requires that the researcher be flexible and follow new leads during fieldwork, taking advantage of the unexpected (Clough et al. 2004). **Convenience sampling** involves selecting cases or participants on the basis of access (for example, interviewing passers-by on the street). While this final strategy saves time, money, and effort, it often produces the lowest level of dependability and can yield information-poor cases. Much purposive sampling combines a number of these strategies.

How Many Participants?

In both qualitative and quantitative research, it is usual that only a subgroup of people or phenomena associated with a case is studied. The size of the group is more relevant in quantitative research because representativeness is important. In qualitative research, the "emphasis is usually upon an analysis of meanings in specific contexts" (Robinson 1998, 409) and the sample is not intended to be representative.

The following analogy between a case and an island may help to explain the distinction. Suppose you are looking at a special kind of aerial photograph of an island, so detailed that you can see all of its inhabitants.

> Clearly, if the population of the island were ten thousand instead of ten, enumeration would count for a great deal. . . . But this is because of the investigator's limitations: [she or he] cannot really get to know ten thousand people and the various ways in which each interacts with others. The use of formalist techniques is a second-best approach to this problem because the ideal technique is no longer feasible. Even on this big island, the old technique will count for a great deal, but that is not the main point. The point is that counting and model building and statistical estimation are not the primary methods of scientific research in dealing with human interaction: they are rather crude second-best substitutes for the primary technique, storytelling. [Ward 1972, 185]

Numbers *do* tell us things about the island, and if what interests us happens to be the frequency and geographic distribution of the island's population, then we need no more than the photograph. If we are interested in a particular "story," such as might revolve around an aspect of the cultural geography of the island—for example, multiple-use conflict in public places—then we will need more than the photograph to go on, and might start by thinking about how circumstances on the island compare with others elsewhere (see, for example, Stratford 2012).

One way to then conduct our specific investigation will be to talk with the island's inhabitants. We could also engage in participant observation and consult relevant texts such as submissions to government, letters to the editor of the island's newspaper, or television news stories that might give us an insight into multiple-use conflict in the island's public places. As researchers, however, we are usually resource limited, both in terms of funding and time, and we must make decisions about what and whom to include and what and whom to exclude from our study. It is clear, however, that we still face the issue of how many people to talk with, how many texts to read, and so forth. While it may seem disconcertingly imprecise, Patton's (2002) brutally simple advice remains accurate: there are few if any rules in qualitative inquiry related to sample size, and it depends on what is needed in the way of knowledge, on the purpose of the research, on its significance and for whom, and on logistics and resources. The richness of information, and its validity and meaning, is more dependent on the abilities of the researcher than on size of sample. In the final analysis, then, it is you as the researcher who must be able to justify matters of case and participant selection to yourself, your supervisor, your interpretive community, and the readers and users of your work.

Ensuring Rigour

It is no frivolous matter to share, interpret, and represent others' experiences. We need to take seriously "the privilege and responsibility of interpretation" (Stake

1995, 12). This responsibility to informants and colleagues means that it must be possible for our research to be evaluated. It is important that others using our research have good reason to believe that it has been conducted dependably. (An extensive literature on these matters, undertaken over many years, includes Anfara, Brown, and Mangione 2002; Baxter and Eyles 1997; Bogdan and Biklen 1992; Ceglowski 1997; Denzin 1978; Dey 1993; Flick 2009; Geertz 1973; Jick 1979; Keen and Packwood 1995; Kirk and Miller 1986; Lincoln and Guba 2000; Manning 1997; Mays and Pope 1997; Seidman 2013.)

Ensuring rigour in qualitative research (Box 6.3) means establishing the **trustworthiness** of our work (Bailey, White, and Pain 1999a; 1999b; Baxter and Eyles 1999a; 1999b; Golafshani 2003). Research can be construed as a kind of **hermeneutic circle**, starting from our interpretive community and involving our research **participant community** and ourselves, before returning to our interpretive community for assessment (Burawoy et al. 1991; Jacobs 1999; Reason and Rowan 1981). This circle is a key part of ensuring rigour in qualitative research; our participant and interpretive communities check our work for credibility and good practice. In other words, trust in our work is not assumed and has to be earned.

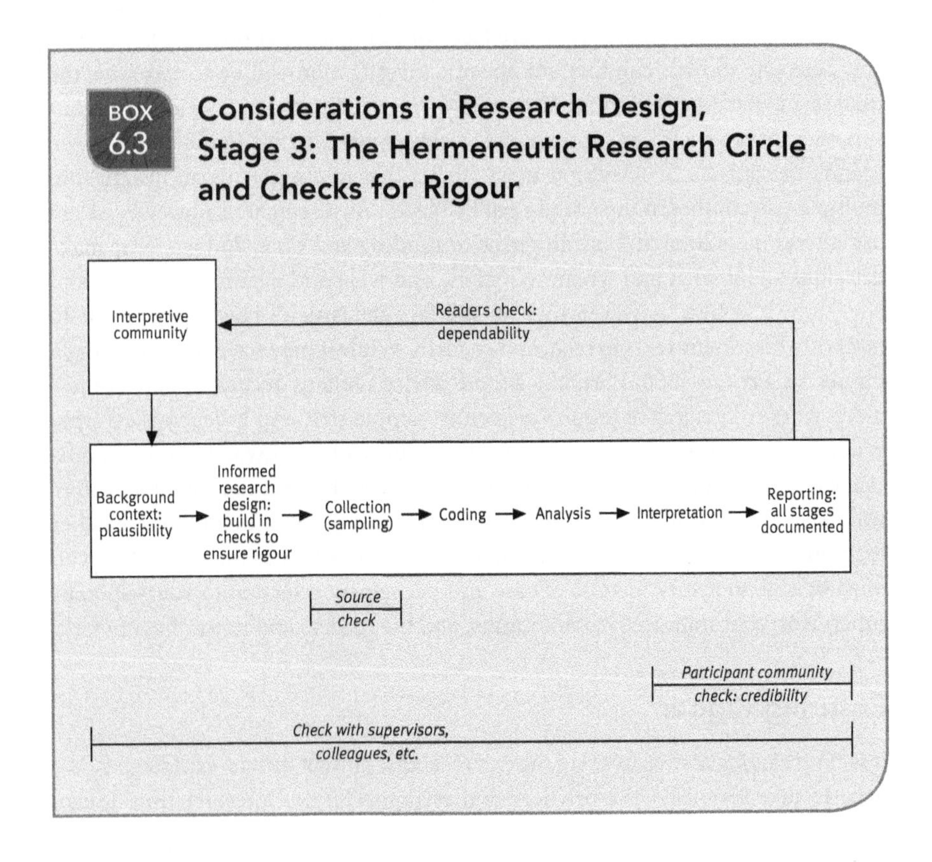

BOX 6.3 **Considerations in Research Design, Stage 3: The Hermeneutic Research Circle and Checks for Rigour**

Two steps in particular need to be followed to ensure and defend the rigour of our research for our interpretive communities. First, strategies for ensuring trustworthiness need to be formulated in the early stages of research design and applied at each stage in the research process (Baxter and Eyles 1997, 1999a; Lincoln and Guba 1985). These strategies should include *appropriate* checking procedures in which our work is opened to the scrutiny of interpretive and participant communities (Mason 2004, but see also Bradshaw 2001 on some of the possible perils involved in these procedures). Second, we need to document each stage of our research carefully so that we can report our work to our interpretive community for checking; "we should focus on producing analyses that are as open to scrutiny as possible" (Fielding 1999, 526).

Rigour must be considered from the outset of our research, underpinning the early stages of research design. It is important to incorporate appropriate *checking* procedures into our research process. These procedures were outlined in foundational qualitative research works by Denzin (1978) and Baxter and Eyles (1997) as the four major types of **triangulation**: multiple sources, methods, investigators, and theories. For example, as we move through various research stages, we might check: (a) our sources against others (re-search); (b) our process and interpretations with our supervisors and colleagues; and (c) our text with our research participant community to enhance the credibility of our research. We note that this last check can be problematic if that community has considerable power, such as might be the case with a multinational corporation if its managers refuse permission for us to publish work related to findings derived from the corporation.

As indicated in the research stages in Box 6.3—which often overlap as they become a whole research composition—we also need to document our work fully: how we came to be interested in the research, why we chose to do it, and for what purpose. We may declare our own philosophical, theoretical, and political dispositions, and we will almost certainly review literature dealing with both the general area of our research and the research methods we intend to use. This elaboration of context permits us to establish the plausibility of our research by demonstrating that we embarked on our work adequately informed by relevant literature and for intellectually and ethically justifiable reasons. We will most likely have checked the plausibility of our research with supervisors and/or colleagues before embarking on detailed research design. At the final stage of reporting research, we should also attempt to acknowledge limits to the **transferability** of our research due to particularities of the research topic, the research methods used, and the researcher. In this way, we confirm that the methods we use and the interpretations we invoke influence our research outcome. Thus, it is vital that we document all stages of our research process. Such documentation allows members of our interpretive and participant communities to check all of these stages and confirm that our work can be considered dependable.

Final Comments

We began this chapter by suggesting that consideration of research design and rigour is essential to the conduct of dependable qualitative inquiry. We have addressed issues of case selection and participant selection and outlined some reasons to be concerned with rigour, as well as some means by which rigour might be achieved.

Most research is undertaken to be shared with others. We therefore need to ensure that our research can stand up to the critical scrutiny of our interpretive and participant communities. The work presented in this chapter provides some of the conceptual and practical tools by which this outcome of sharing plausible, credible, and dependable work can be achieved.

Key Terms

case
convenience sampling
criterion sampling
deviant case sampling
disconfirming case
extensive research
hermeneutic circle
intensive research
interpretive community
maximum variation sampling
opportunistic sampling
participant
participant community
purposive sampling
research design
rigour
snowball sampling
transferability
triangulation
trustworthiness
typical case sampling
"wicked problem"

Review Questions

1. Why is rigour important in qualitative research? Is it an ethical consideration?
2. What is an "interpretive community"?
3. What is meant by the phrase "participant community"?
4. What are some ways we might check our research to establish its dependability to members of our interpretive community?

Review Exercises

1. Imagine that you are working as a geographer/planner in a consulting firm. You have been approached by a municipal government's general manager and asked to investigate ways in which to provide for the civic needs and aspirations of diverse groups and individuals in an inner-city square. The square has heritage values—including a central fountain dedicated to one of the city founders; one side of it is part of the central bus interchange; it is adjacent to a shopping mall and government and office buildings; and there are large areas of lawn and established trees and benches, as well as intersecting paths throughout the park. Among your first tasks is to determine whether you will use a quantitative, qualitative, or mixed methods approach. Working with your colleagues, draw up a comprehensive table to consider the advantages and disadvantages of each approach. Does a clear decision emerge? What do you learn by this exercise?
2. Imagine that you and your colleagues in the consulting firm mentioned in exercise 1 decided that you would adopt a mixed-method approach to your investigation for the municipal government. Before you begin any primary data collection, you have decided to develop a checklist of other types of information you will need in order to ensure rigour. Working together, tabulate the other forms of information you would seek to draw upon and justify why you think those forms of information will assist triangulation.

Useful Resources

Bradshaw, M. 2001. "Contracts and member checks in qualitative research in human geography: Reason for caution?" *Area* 33 (2): 202–11.

Hennick, M., I. Hutter, and A. Bailey. 2010. *Qualitative Research Methods*. London: Sage.

Mason, J. 2004. *Qualitative Researching*. 2nd edn. London: Sage.

Patton, M.Q. 2002. *Qualitative Evaluation and Research Methods*. 3rd edn. Beverly Hills, CA: Sage.

Platt, J. 1988. "What can case studies do?" *Studies in Qualitative Methodology* 1: 1–23.

Robson, C. 2011. *Real World Research: A Resource for Users of Social Research Methods in Applied Settings*. Chichester: Wiley.

Sayer, A. 2010. *Method in Social Science: A Realist Approach*. Revised 2nd edn. London: Routledge.

Case Studies in Qualitative Research

Jamie Baxter

Chapter Overview

This chapter defines the case study as a broad methodology or approach to research design rather than as a method. Much of the chapter clarifies precisely what a case study is, describes the different types of case studies, and addresses some misplaced depictions (*N*=1) and criticisms (lack of generalizability) of case study research. Although most case study research is cross-sectional, conducted typically on one case at one point in time, this chapter reviews two major variants of multiple case studies that might appeal to geographers in particular: the within-case *temporal* comparison and the *spatial (place-oriented)* cross-case comparison. Much of the "how-to" of case study methods and fieldwork is covered in companion chapters within Part 2 of this volume. This chapter instead focuses more on broader research design issues specific to case studies.

What Is a Case Study?

Gerring (2004, 342) provides a very concise and useful base definition of the **case study** as "an intensive study of a single unit for the purpose of understanding a larger class of (similar) units." However, we must be careful not to conflate sample size with the quality of case study research—a point that is addressed near the end of this chapter. Further, Gerring (2007) later amended his definition to consider the notion of multi-case studies associated with temporal and spatial comparisons. Thus, case study research involves the study of a single instance or small number of instances of a phenomenon in order to explore in-depth nuances of the phenomenon and the contextual influences on and explanations of that phenomenon. Some examples of phenomena researched as case studies might include an event (e.g., a protest rally, a disaster), a process (e.g., immigration, discrimination, risk amplification, deforestation), or a particular place (e.g., a neighbourhood with a high crime rate, a community hosting a hazardous waste facility). Case studies are often used to better understand and sometimes directly resolve concrete problems (e.g., why is uptake of immunization so low in town X?). From an academic point of view, though, case studies are well suited to corroborating existing explanatory concepts ("theory"), falsifying existing explanatory concepts, or developing new explanatory concepts. Perhaps most important, the case study provides detailed analysis of *why* theoretical concepts or explanations are not inherent in the context of the case.

Thus, a case study is perhaps most appropriately categorized as an approach to research design or **methodology** (a theory of what can be researched, how it can be researched, and to what advantage) rather than as a **method** (a mechanism to collect "data"). It is more an approach or methodology than a method because there are important philosophical assumptions about the nature of research that support the value of case research. The primary guiding philosophical assumption is that in-depth understanding about one manifestation of a phenomenon (a case) is valuable on its own without specific regard to how the phenomenon is manifest in cases that are not studied. This depth of understanding may concern solving practical/concrete problems associated with the case or broadening academic understanding (theory) about the phenomenon in general, or a case study may do both of these things. Other philosophical assumptions are described below as clarifications particularly concerning N=1 and **transferability/generalizability**.

It is worth examining where case studies fit relative to other approaches to social research. Case studies are often considered equivalent to field research, participant observation, **ethnographic research**, or even qualitative research. Although case study research certainly intersects with all of these, it deserves more attention than most social science textbooks provide. That is, case studies are typically mentioned within chapters that mainly concern these other aspects of qualitative research (this text and chapter notwithstanding). Further, the terms "qualitative" and "case study" are not entirely interchangeable, largely because quantitative researchers also conduct case studies and many use a combination of methods and methodologies within case study research. In fact, the relationship between qualitative, quantitative, and case study research looks more like the Venn diagram in Figure 7.1 whereby much of case study work is indeed qualitative but some is quantitative or a mixture of the two. Thus, researchers need to specify up front precisely what type of case study is being conducted (quantitative, qualitative, or mixed) so that the quality of the work is fairly judged according to the methodology used (see Chapter 6). However, in keeping with the theme of this text, the focus in this chapter is on the qualitative case study and on qualitative aspects of mixed-method case studies.

The Historical Development of the Case Study

The popularity of case study research has perhaps never been higher. Although the case study has a long history in the social sciences, its popularity faded during

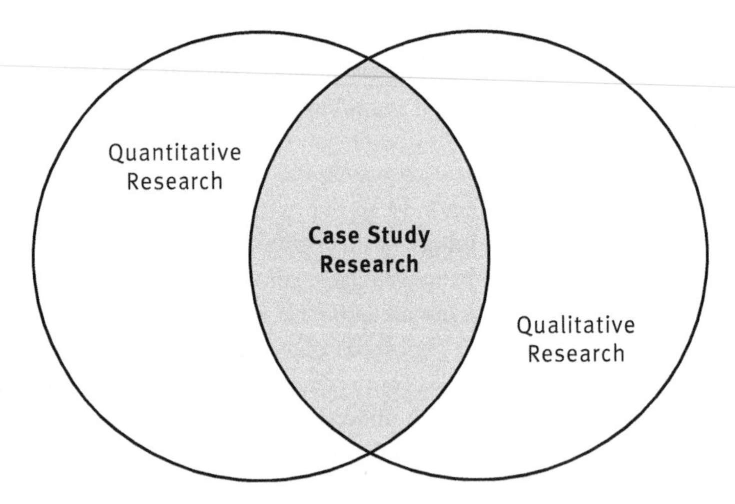

Figure 7.1 Intersecting Domains of Inquiry

Although there is much overlap, case study research is *not* synonymous exclusively with qualitative research. Rather, case study is a methodology or approach to research that can be either predominantly qualitative or predominantly quantitative or a mixture of both of these approaches.

the quantitative revolution of the post–World War II era. However, its status has dramatically resurged during the past few decades (Platt 1992).

Most writers trace the origin of modern-day case study research to the **Chicago School of Sociology**. The so-called Chicago School emerged in the 1920s and 1930s during a time of rapid industrialization in North America. Industrialization in turn spawned the need to better understand the workings of rapidly growing cities. Major writers from the Chicago School include William Thomas (1863–1947), Robert Park (1864–1944), Ernest Burgess (1886–1966), and Louis Wirth (1897–1952), and their works have been heavily influential on a diverse array of qualitative researchers in areas such as urban and cultural sociology and geography. Much of the work was and continues to be ethnographic, combining quantitative survey work and qualitative participant observation, semi-structured interviews, and unstructured interviews (Platt 1992). For example, Thomas and Znaniecki (1918) produced a massive four-volume account of Polish immigration to America as a case study of the "immigration experience." Such work and much to follow was heavy on detail and conceptual development that included empathetic accounts of human experience. Thus, the Chicago School approach was appealing to humanistic geographers interested in the manifestation of various phenomena in "places" imbued with contextualized meaning rather than conceptualized simply as "locations" (Tuan 1977). However, the density and sheer length of the early Chicago School ethnographies made them largely inaccessible except through second-hand interpretation—in geography textbooks, for example. Indeed, the same issue remains a challenge for qualitative case study researchers today who want to provide rich detail but must produce work brief enough to appeal to a wide audience.

During the post–World War II period, quantitative aspects of case study work began to gain more prominence among scholars than its qualitative aspects. Nevertheless, the qualitative/ethnographic methodologies used in case studies continued to be honed throughout this era through such well-known breakthrough case studies as Whyte's (1943) *Street Corner Society* (Boston) and Liebow's (1967) *Tally's Corner* (Washington, DC). Both studies concerned men living in poor neighbourhoods, and the authors produced meticulous accounts of structural social issues about such phenomena as unemployment and gender relations in predominantly Italian and Black working-class neighbourhoods respectively. The fieldwork was measured in months and years rather than weeks. These works are lauded as much more concise and accessible (though still book-length) accounts than the early Chicago School works while retaining the conceptual/theoretical richness of the Chicago School studies. Further, they helped to develop both participant observation and less structured interview methods more fully.[1] Whyte and Liebow drew out concepts grounded in the case context in concrete terms (e.g. "manliness") rather than academically abstract concepts. Although well-grounded in the case, such concepts did and still do resonate in other contexts that are similar to the ones in the Whyte and Liebow studies. What is important here is that good case studies are so richly described (theorized) that one generally finds it quite easy to draw parallels with contexts outside the case. Indeed, this detail or richness is one of the best strategies for creating credible and trustworthy (rigorous) qualitative case study work (see Chapter 6).

It is important to note that one thread of the case study's history is the ongoing use of multiple methods, including the combination of qualitative and quantitative methods in the same case studies. Platt (1992) provides a detailed account of the history of the rise, fall, and rebirth of the case study, which is beyond the scope of this chapter. Yet in her account, she carefully shows how the Chicago School studies ushered in a long era during which researchers highlighted and, more detrimentally, exaggerated *differences* between quantitative ("objective") research and qualitative ("subjective") research, a tendency that still exists today. In the early years of the Chicago School, however, researchers were more inclined to highlight that quantitative and qualitative approaches were both powerful on their own but also complemented each other (Platt 1992). That perspective on combining methods has now resurged (Hesse-Biber and Leavy 2004), which means that there is a refreshingly diverse set of research design possibilities from which the case study practitioner can choose. Before we turn to some common ways that geographers design case studies, it is worth clarifying the importance of depth and context as opposed to sample size.

N=1 and the Importance of Depth and Context

The term "N=1" is often used to succinctly describe case study research, but as explained in Chapter 6, there is far more to case studies than the number of units studied. In fact, case study researchers tend not to think in terms of sample size. Thus, it is important to clarify what N=1 means in the context of case study research and to highlight that depth of understanding and contextualized understanding are far more important.

The use of the term "N=1" to describe case studies can be confusing, because it grafts quantitative/statistical terminology onto non-statistical research. In statistics, *N* refers to the population (the group about whom conclusions are drawn), while *n* refers to a sample, or subset, of that population. Strictly speaking then, *n*=1 is more appropriate statistical terminology—one case from the population of *N*. Moreover, there is also selection within the case itself that must be accounted for—the sub-units within the case. That is, case study researchers often study large numbers of "things" within their case. Yet qualitative researchers tend to think of their cases as wholes that cannot be fully understood when lumped together with a large number of other cases. That is, the context of the case is important, since it more often than not substantially influences the phenomenon in question (e.g., a change in national employment policy can affect local crime rates). That is why multiple sub-units of "things" (e.g., people, newspapers, policies) are often studied—to get at these contextual influences. Thus, rather than study a few of each sub-unit across a wide array of communities (e.g., 10 people, one newspaper, and two policies each from 10 or 100 communities), the qualitative case study researcher instead prefers to study one carefully selected community intensively and holistically to understand how the various things studied interact with one another in, for example, one place. There is no statistical notation to account adequately for the importance of context, and any use of *N* and *n* does not do justice to the value of case study research.

Further, it is important to underscore that case study research is intensive rather than extensive research (see Chapter 6). Social scientists use the terms **idiographic** research and **nomothetic** research to describe this difference. Idiographic research is depth oriented, since it tends to focus on the particular to understand a phenomenon in more detail. Nomothetic is breadth oriented, since it focuses less on the details and more on investigating a limited number of things across several units (cases) simultaneously. Nomothetic researchers typically use probabilistic sampling to select those units. Although qualitative case study researchers do not typically choose sub-units based on probabilistic sampling, few case study researchers concede that their findings are exclusive *only* to the case. Indeed, qualitative case study researchers working in an idiographic frame expect and look for what might be common *among* cases. A case is viewed as neither entirely unique nor entirely representative of a phenomenon (i.e., a "population" of cases, according to statistical terminology). This point is taken up again in the discussion of generalizability and transferability later in the chapter.

To explain the points about sub-unit selection and how to study context more fully, Figure 7.2 offers an example of how a case study can be based on multiples of different sub-units. Suppose a researcher is interested in the phenomenon of community hazard risk perception—specifically, why groups of residents vehemently oppose potentially hazardous or polluting facilities (e.g., industrial, waste, nuclear power) being located near their community.[2] This is a problem of practical importance, since at the very least, failed facility siting efforts cost millions of dollars, not to mention the potential negative impacts of such facilities on local residents. The researcher could look for a community that faces having a facility located in their midst. However, for the purpose of identifying a single valuable case relevant to the problem at hand, the researcher might instead flip the problem on its head and ask the related question "why are some communities *un*concerned and not opposed to these facilities?" Thus, the researcher needs to identify a town where residents seem to be supportive of a local hazardous waste treatment facility—a case study. What the researcher might actually do is use multiple methods whereby she might interview several different types of residents (councillors, facility workers, opposition group members) to understand their risk perceptions and conduct a discourse analysis on various types of facility-related media coverage to understand the influences on (a lack of) risk perception. These are some of the many design and selection considerations that are documented elsewhere in this book (e.g., Chapter 6). The intensive, holistic aspect of the study comes in part from trying to understand how the various contexts—which are conceptualized in Figure 7.2 as "local," "regional," and "broad"—are involved with facility hazard risk perception. In this example, the analysis of local, regional, and national media coverage helps the researcher to better understand these contextual influences. The analysis could include other influences, such as recent national economic changes and policies at various scales. In the interviews, residents would be asked how they view these external influences, such as the reporting by various types of media. We will revisit this example later in the chapter.

Types of Case Studies

Theory-Testing and Theory-Generating Cases

As outlined above, case studies play two key but not necessarily mutually exclusive roles: to test theory and to generate or expand theory. The former may involve the search for negative or falsifying cases, while the latter concerns cases that are more typical.

For some, what distinguishes case studies from other approaches, such as grounded theory and ethnography, is that in case studies, theoretical propositions

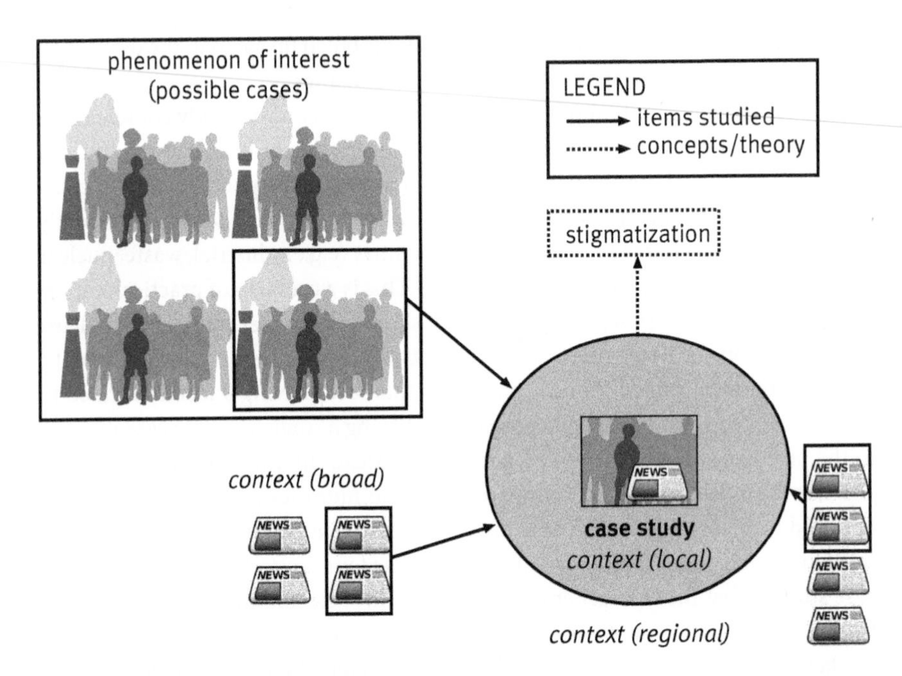

Figure 7.2 Case Study Selection, Sub-Units, and Context

A case study is one manifestation of a broader phenomenon. Researchers carefully select the case to understand the practical/concrete aspects of the case itself but also to better understand the broader phenomenon. In this example, one community living with a hazardous facility (n=1) is selected from all possible communities with hazardous facilities (N). Further, multiple people will be studied within the case community, along with multiple newspapers both inside (local) and outside the community (regional and extra-regional). This helps to understand an array of contextual influences on the phenomenon (e.g., low risk perception) within the case to develop contextualized theory (e.g., stigmatization).

should be stated prior to entering the field (Yin 2003). If those propositions are particularly well developed through, for example, past case study work, the researcher may choose to operate in a **theory-testing mode**, in their new case study. This means emphasising **deductive logic** whereby the researcher looks for data that supports or falsifies the concept under consideration. Yet many tend to view qualitative case studies as primarily **theory-generating** endeavours such that ethnography and grounded theory can be easily incorporated within a case study design. In practical terms, both positions are mainly a matter of degree, since most practitioners of grounded theory and ethnography do not commence field research without adequate knowledge of *some* theory. Nevertheless, if one decides to follow Yin's recommendation that formal propositions need to be stated upfront, two cautions are necessary.

First, qualitative researchers presume that propositions are contingent or context-dependent such that concepts describing relationships are only "true" *under*

Studying One Person: A Case Study of "Praxis"

Sometimes cases are not identified as places or institutions with multiple people as "sub-units" within to be studied. Instead, they may concern multiple manifestations of a phenomenon with one person only. Yet findings from such work can still be transferable beyond the single person. For example, Wakefield (2007) studies the phenomenon praxis in critical geography. Praxis concerns the way that researchers may use research itself to make positive change outside academia. The "case" in this study is Wakefield's own experiences as an activist researcher in the "food movement" operating in such contexts as her own household and Toronto food movement organizations, as well as within broader environmental justice organizations and "globalized, corporate industrial agriculture" (Wakefield 2007, 332). That is, she studies how she uses research in all of these different contexts—how research praxis is manifest in each of these contexts. Wakefield draws on her personal experiences to expand on the underappreciated concept of "activism at home." This concept moves beyond academic teaching and writing, which she argues are the dominant forms of academic praxis. She highlights the value of taking *personal* action in the household and community, partly to set an example for others but also to better understand the challenges of alternative food consumption. Thus, most of her research is reflexive in the sense that she systematically studies her own experiences and interactions with others in relation to alternative types and sources of food. She has made a contribution to the general theory of academic praxis in terms of expanding the theoretical concept "activism at home." Further, one can easily imagine that activism at home resonates beyond the case of food movements and Wakefield alone. That is, it is relatively easy to see how the concept might apply to other cases, such as waste generation or fuel consumption.

certain conditions (Sayer 2000). That said, concepts are still "true when . . . ," and accounting for the context or contingencies within which a "truth" happens certainly falls within the realm of what most qualitative researchers do in practice. Using our example of the low risk-perceiving, hazardous facility-hosting community depicted in Figure 7.2, the researcher might focus on the concept "stigma" prior to field work. She might propose that such communities feel that the outside (non-local) media stigmatize the community, especially *when* accidents happen at the facilities and *when* these accidents are widely reported and *when* that reporting does not simultaneously report the social good the community is doing by hosting a facility that solves a widespread social problem (i.e., disposal of hazardous waste). Further, this stigma may minimize or redirect the concern or anger that residents once directed at the facility onto the media instead. The effects of stigma are not always "true," but they seem to be under these specific conditions (and perhaps others).

The second caution is that there is a potential logical flaw in *ever* stating propositions up front at the beginning of a study. That is, formal propositions typically require well-developed theory as their basis. Yet qualitative case studies are often used to delve into under-explored and thus under-theorized phenomena. Moreover, researchers tend to borrow from related areas of inquiry. Often, it is not necessary to "re-invent the (conceptual/theoretical) wheel." In the case of our Figure 7.2 example, there might already be a well-developed theory on the negative effects of stigma on hazardous facility–hosting communities, but it might not go as far as suggesting that stigma can redirect or minimize concern about the facility itself. The concept of stigma, which may itself have been developed from case study work, is thus borrowed from the literature and elaborated upon.

This raises the question of how theory is actually *generated* using case studies. This question is covered in greater detail in Chapter 18 on coding data, but it is worth pointing out that qualitative research in practice is rarely a purely deductive or purely inductive endeavour. Rather, it tends to be more cyclical in the sense that theory stated initially either formally as hypotheses or loosely as budding ideas is explored (deductively) by studying the real world of the case and then that information is used to generate new concepts (theory) to explain what is observed (inductively) (Figure 7.3). These refined or new concepts are then further scrutinized by ongoing analysis of the real world of the case. All the while, the good researcher will remain aware of existing academic literature that might contribute to the explanations. Grounded theory—described in Chapter 18—is one approach to moving through this deductive/inductive cycle to form in-depth understanding of the entire case.

Case Studies across Time and Space

Case studies need not be one-off, single studies of one case at one particular point in time. However, multiple case studies are generally not approached with the purpose of establishing *statistical* generalizability (see the discussion later in this chapter). Instead, it is better to view multiple cases in one of two ways. In the first instance, multiple case studies provide a broader basis for exploring theoretical concepts and explanations of phenomena. In the second instance, longer-term study of the case may be useful to **corroborate** and further explore theory as it relates to the case at hand. Both enhance the credibility and trustworthiness of

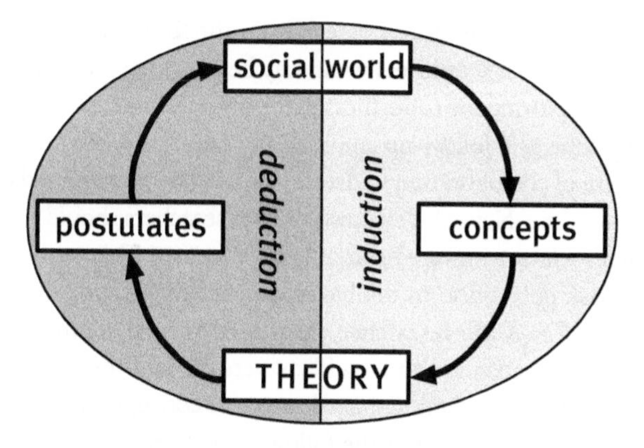

Figure 7.3 Cyclical Modes of Exploration in Case Studies

Often, qualitative case study researchers tend to emphasize the *inductive* mode of inquiry, moving from empirical observation to concepts/theory. In some cases, the main purpose may instead be to test an existing postulate (hypothesis). However, both inevitably involve "multiple loops" of reasoning as the researcher tentatively develops concepts (induction) and then compares them to the details of the social world that comprise the case (deduction).

the concepts and explanations by exposing them to different scenarios between cases in the first instance and within the case as it evolves over time in the second. Neither one is replication in the statistical or experimental sense of the term, but they may instead be seen as ways of both deepening and expanding the theoretical concepts (see the examples of "stigma" and "activism at home" above). Thus, for multiple cases, little attention needs to be paid to the sequence of the case studies—they can be studied in parallel. However, for longitudinal cases, more attention must be paid to the temporal sequencing of the research.

Time: Cross-Sectional and Longitudinal Case Studies

The most common form of social research is **cross-sectional**—that is, it is conducted at one point in time (Bryman 2006). The definition of "one point in time" can be somewhat fuzzy for the qualitative researcher, however, since he or she often spends extended time in the field collecting and analyzing data. Operationally then, a study may be considered cross-sectional if fieldwork is conducted in one block of time regardless of how long it takes. It would only be considered **longitudinal** if there were a revisit, with the researcher returning to the case after an intervening time period during which no appreciable research was done. Similarly, for studies that do not involve fieldwork or face-to-face interaction (e.g., discourse analysis of media), identifying collections of data at specific time periods and analyzing them for changes from one period to the next would make them longitudinal.

A key advantage of longitudinal research on the *same* case is that it makes it possible for the researcher to address what may be considered the enduring versus the ephemeral by exploring the robustness of the original concepts and explanations (theory). When done as a follow-up study on the same case, longitudinal research amounts to a form of corroboration to determine whether the original explanations have endured over time. Nevertheless, case study researchers should not be so naive as to assume that social settings are invariant over time (see Chapter 1). For example, in our study of risk perception in the hazardous facility–hosting community case depicted in Figure 7.2, if the researcher were to revisit the community two years later and find that stigmatization and its effects were the same or stronger, then there would be good grounds for arguing that the original concepts and theory are credible, are trustworthy, and endure. If the follow-up research is done with the same participants as in the previous study, it serves as a tactic for guarding against such threats to rigour as **member checking** (Lincoln and Guba 2002).

However, a more challenging scenario is when the phenomena and the concepts that explain them seem to make only *some* sense still, but less so, or perhaps in a changed form. This condition may signal that the concepts or even the phenomenon itself is ephemeral and may have been relevant for only a brief period. For example, residents might talk less about the "stigmatizing effect of non-local media" during the follow-up. This might threaten the original theory about stigmatization, leaving the researcher to explore the reasons for the apparent change of view. Such an exploration is potentially immensely valuable, because new insights may be gained. Nevertheless, if the changes regarding stigma were due, for example, to the success of the community's legal action against a non-local newspaper, the theory would remain credible. That is, the community might have moved on, knowing that the "stigmatizer" had been appropriately dealt with. If residents were *more* concerned about the facility (e.g., because of a recent accident) and tended to sympathize more with non-local media reporting on the community, it would pose a serious threat to the initial conceptualization. Recall that the original concept of the effect of stigma was that harsh criticism of the facility by outside media stigmatized the community and residents reacted by outwardly expressing little concern about the facility. That is, low concern about facility risks was a reaction to stigma that served to defend the community and enhance community pride. An apparent rise in concern, together with more sympathy towards outside media, go against the original thesis and would require either (1) the reconceptualization of "stigma" or (2) abandoning it as an explanation of low concern altogether.

It is obviously very problematic to assume that social phenomena are static over time. The internal dynamics of a community typically change. For example, new people come into positions of power. Although the research context of the case may change, this does not *necessarily* invalidate (threaten the credibility or trustworthiness of) the original theory in relation to the overall phenomenon.

Researchers need to be careful about identifying what aspects of their theory and concepts seem less or more relevant over time and, most important, *why*. Thus, longitudinal case studies are very good for tracking how the phenomenon changes over time in the case in question. Ultimately, this should lead to the development of well-rounded explanations. In this regard, the timing of follow-up research should be theoretically informed. For example, in our community with the hazardous waste facility, if stigmatization of the community by the media is important, then future research might solicit residents' opinions after the publication of key articles and reports that were highly negative—or highly positive—towards the local waste facility and/or the community.

Unlike the prospective longitudinal case studies just described, retrospective studies go back in time. They constitute the other major class of longitudinal study and may at first appear problematic for the qualitative case study researcher intent on primary data collection. For example, if semi-structured interviews require participants to "think back" to the way things were and how they "felt then," recall bias becomes a serious and unpredictable problem. Nevertheless, in a design that depends on the triangulation of methods and data including such things as diaries, letters, policy documents, and meeting minutes, it may be a far less serious threat.

Space: Comparative Analysis

Not all comparative analyses are spatial, but most conducted by geographers are spatial in one way or another. For example, there is a long tradition in human geography that emphasizes how phenomena may present very differently from one case to the next because of the place itself—i.e., the meanings derived from the interactions among residents. When conducted at the same point in time, case studies of multiple instances of a phenomenon are commonly known as **comparative analysis**, **comparative case study**, or **parallel case study**. For example, a comparative analysis might involve a parallel study of three of the communities shown in Figure 7.2. Comparative studies tend to share many of the same advantages as longitudinal case studies in that there are opportunities to generate and modify concepts and theory so that they explain commonalities across cases *despite* being embedded in different contexts. Although research phenomena (e.g., risk perception) need not be place specific, it is often useful for geographers to focus on place as an initial basis for comparison. It is also important to clarify that we are talking about field comparisons of two or more instances of a phenomenon rather than merely the comparison of a case study with existing, published case studies. Although situating one's research within the literature is essential for good academic work, the focus here is on multiple cases built into a study at the design stage. For example, in a special issue on comparative studies of transnationalism (e.g., remittances of money or visits back to the country of origin), Dunn (2008) suggests that much can be gained by comparing the impact of transnationalism on the same cultural groups in different places or on different cultural groups in the same place, as well as the impact of transnationalism on the places themselves. The emphasis is not on the number of cases per se but on understanding how the phenomena are manifest in different contexts.

Castree (2005) underscores the value of comparative analysis by making a call for more explicitly comparative case study work in geography concerning research on "nature's neoliberalisation" (2005, 541). He complains that a growing number of such case studies focus too much on the particular and not enough on what is common *across* case studies. He wants to explore some general impacts (manifestations) of neo-liberalization on nature, irrespective of context. This is an ongoing tension in qualitative research generally—balancing the particular with the more abstract when developing explanations. We see this tension in what Castree himself writes in the sense that elsewhere in the same paper, he claims that he is also frustrated by *too much* abstraction. Despite his call for some theory that cuts across case studies, he complains that existing concepts quoted in the literature are "clearly so abstract that [they] fail to tell us how and with what effects otherwise different neoliberalisations work," (2005, 543). These over-general concepts include privatization and deregulation. Castree's frustration is directed towards the literature and not towards any particular **cross-case comparison**. Indeed, a formal cross-case comparison of neo-liberalization's influence on nature in a single study is one solution to Castree's concern, as long as the concepts are sufficiently specific (but of course not *too* specific). These ideas are taken up further in the following discussion on generalizability.

Are Case Studies Generalizable?

The short answer is "yes." Quite probably, the most common criticism of case study research is its supposed lack of generalizability (Campbell and Stanley 1966; see also Flyvbjerg 2006). Yet such concern may be exaggerated. That is, generalizability should not be a problem if case study research is designed appropriately and the analysis is attentive to the tension between concrete and abstract concepts.

Generalizability (or **external validity**) is a term used by quantitative social scientists, but many qualitative researchers prefer the term **transferability** (Lincoln and Guba 2002). Generalizability or transferability concerns the degree to which findings apply to other cases of the phenomenon in question. It may be interpreted as "the more cases the theory applies to, the better." However, this is only one way to look at generalizability, and qualitative case study researchers tend not to emphasize the more-cases-are-better approach. Instead, qualitative researchers are more concerned that explanations of the phenomenon as manifest in the case are credible. This distinction is partially explained as the difference between statistical generalization and **analytical (theoretical) generalization**. In the former case, generalization is achieved through large probability samples, while in the latter, transferability is accomplished by (1) carefully selecting cases and (2) creating useful theory that is neither too abstract nor too case specific (Flyvbjerg 2006; Yin 2003).

Flyvbjerg (2006) provides some good examples of how generalizable, or transferable, theoretical concepts and explanations can be generated from a single case study (e.g., a single experiment). Perhaps the most famous is Galileo's debunking of Aristotle's theory of gravity. Aristotle claimed that heavier objects will accelerate faster than lighter ones. However, with a single experiment (case study) involving only a few different types of balls rolled down inclined planes, Galileo showed that balls of different weights accelerated at the same rate. This case study disproved the theory that weight is a determinant of gravitational acceleration, which paved the way for others like Sir Isaac Newton to suggest that the determinant was something else (i.e., the relationship between mass and distance). Thus, the choice of case (in the Galileo example, the choice of a ball and ramp experiment) can be critical. Further, it is important to recognize that falsifying existing theory simultaneously opens up avenues for new theory. Similarly, Popper (1959) argued that it takes only one black swan to falsify the theory that all swans are white. Keep in mind that although these examples serve to show both the value of single cases and the strategy of **falsification**, social scientists rarely expect any concept or theory to apply in all cases—i.e., to constitute a theoretical law. What is important is to describe *why* a theory does or does not apply in a particular case.

There are numerous examples of case studies in geography that falsify, particularly in areas like political economy and political ecology that use case studies to show how social and environmental problems are connected to flaws in the global socio-economic system. For example, Weis (2000, 300) provides an explanation of unsustainable deforestation in the blue mountains of Jamaica that shifts blame for the problem away from the "poor stewardship" of the peasants towards the "grossly inequitable land regime" within a globalized food system. Thus, Weis's empathetic account of these Jamaican farmers is a form of falsification as it attempts to supplant a long-standing and traditional explanation of deforestation with a more compelling alternative one.

Finding cases that falsify existing theory is but one of many examples of the role case studies can play in terms of addressing generalizability or transferability. For most qualitative researchers, the development of a coherent theory itself is of primary concern, not necessarily whether the findings challenge hegemonic wisdom or whether findings adhere in *all* or even *most* cases at the time they are studied (George and Bennett 2005). In practical terms, it is too great a burden for one study not only to understand the context, contingencies, and details of the case but at the same time to know all contingencies and contexts for all other cases of the phenomenon in question—regardless of whether the phenomenon is risk perception, academic praxis, transnationalism, neo-liberalization, or deforestation.

Strictly speaking, the only way to truly know whether a theory applies in other cases is to study those other cases—but perhaps not in as much detail as the cases that were used to develop the theory in the first place. The true value of a qualitative case study, then, may not be known until several years after the research or policy community has had a chance to digest the concepts and theory. The theory may only *eventually* prove useful for explaining similar phenomena in different cases. Indeed, this is one of the main reasons for the enduring fame of Whyte's (1943) and Liebow's (1967) studies. Although they focused on Italian and Black men in Boston and Washington respectively, the studies' insights apply in numerous cities among a wide array of ethnic and minority groups. Their concepts have resonated in several contexts over space and time. They seemed to strike the right balance between describing concepts that explain the concrete details of the case and creating concepts that are sufficiently abstract to apply to similar phenomena in different places and times (i.e., living in a poor ethnic neighbourhood in a large urban centre). The take-home message is to generate theoretical concepts and explanations that potentially resonate in other (as yet unstudied) contexts.

Conclusion

Case studies have a long and rich history in the social sciences and geography. Case study methodology is a powerful means by which to both (1) understand the concrete and practical aspects of a phenomenon or place and (2) develop theory. That is, case studies may be used to understand and solve practical problems relating to the case alone, and they may be used to test, falsify, expand, or generate explanatory theoretical concepts. Case studies are valuable because, when done well, they produce deep, concrete explanations of social phenomena that are attentive to a variety of contextual influences at various scales. Most of what qualitative researchers do involves a case study methodology. This research often uses various combinations of qualitative (and quantitative) *methods* to support data collection (e.g., interviews, focus groups, participant observation) as well as a variety of analytical *strategies* (e.g., grounded theory, discourse analysis). These methods and analytical strategies may be used in longitudinal analyses of the same case, cross-case comparisons of different cases, or one-time analysis of a single case. It is important to recognize that although a case study may only involve a sample of one, a carefully chosen and well-studied case can be used to produce very robust, credible, and trustworthy theoretical explanations. These explanations are generalizable, or transferable, in the analytical sense rather than in the statistical sense. Good theoretical explanations are those that are well rooted in the concrete aspects of the case yet sufficiently abstract that others in similar situations can see how they might apply to their own context.

Key Terms

analytical generalization
case study
Chicago School of Sociology
comparative analysis
comparative case study
corroboration
cross-case comparison
cross-sectional case study
deduction
ethnographic research
external validity
falsification
generalizability
idiographic
induction
longitudinal case study
member checking
method
methodology
nomothetic
parallel case study
praxis
theoretical (analytical) generalization
theory generation
theory testing
transferability

Review Questions

1. How can a study based on N=1 produce useful research results? What role does generalizability play in evaluating the quality of such results?
2. What is the historical role of multiple methods in case study research?
3. Outline some of the key advantages and challenges of temporal or spatial comparative case studies.
4. What are the roles of induction and deduction in case study research? How do they relate to nomothetic and idiographic research?

Review Exercises

1. Break into groups of two to four and design a qualitative case study of a phenomenon important to students at your university. Discuss what you will measure and consider how you will select units to study. Try designing the units you study with and without people.
2. Break into groups and then divide the group in half to have a debate. One half of the group should take the position that qualitative case studies are generalizable/transferable while the other half take the opposite position.

Useful Resources

Colorado State University. 1993–2009. "Writing guides—Case studies." http://writing.colostate.edu/guides/research/casestudy. A helpful guide to the history and conduct of case studies as well as to the presentation of their results.

Flyvbjerg, B. 2006. "Five misunderstandings about case-study research". *Qualitative Inquiry* 12 (2): 219–45.

George, A., and A. Bennett. 2005. *Case Studies and Theory Development in the Social Sciences.* Cambridge, MA: MIT Press.

Gerring, J. 2007. *Case study research: Principles and practices.* Cambridge: Cambridge University Press.

Hamel, J., S. Dufour, and D. Fortin. 1993. *Case Study Methods.* Newbury Park: Sage.

Hesse-Biber, S., and P. Leavy. 2004. *Approaches to Qualitative Research: A Reader on Theory and Practice.* New York: Oxford University Press.

Platt, J. 1992. "'Case study' in American methodological thought." *Current Sociology* 40: 17–48.

Stake, R. 2006. *Multiple Case Study Analysis.* New York: Guilford Press.

Yin, R. 2003. *Case Study Research: Design and Methods.* Los Angeles: Sage.

Notes

1. At the time, structured surveys were by far the most popular means of collecting information in social research.
2. The example is modelled on my own research largely to avoid any misinterpretations about research intentions and interpretations. See Baxter and Lee (2004).

PART

"Doing" Qualitative Research in Human Geography

Interviewing

Kevin Dunn

Chapter Overview

This chapter provides advice on interview design, practice, transcription, data analysis, and presentation. I describe the characteristics of each of the three major forms of interviewing and critically assess what I see as the relative strengths and weaknesses of each. I outline applications of interviewing by referring to examples from economic, social, and environmental geography. Finally, some of the unique issues regarding Internet-based interviews are reviewed.

Interviewing in Geography

Interviewing in geography is so much more than "having a chat." Successful interviewing requires careful planning and detailed preparation. A recorded hour-long interview involves days of preliminary background work and question formulation. It requires diplomacy in contacting **informants** and negotiating "research deals." A 60-minute interview will require at least four hours of transcription if you are a fast typist, and verification of the record of interview could stretch over a couple of weeks. After all that, you have still to analyze the interview material. These are time-consuming activities. Is it all worth it? In this chapter, I outline some of the benefits of interviewing and provide a range of tips for good interviewing practice.

An interview was once defined as "a face-to-face verbal interchange in which one person, the interviewer, attempts to elicit information or expressions of opinion or belief from another person or persons" (Maccoby and Maccoby 1954, 499). An interview is a data-gathering method in which there is a spoken exchange of information. While this exchange has traditionally been face to face, researchers have also used telephone interviews (Groves 1990). However, the new millennium has seen the emergence of **computer-mediated communications** (CMC) **interviewing**, a mode in which there is no direct access to the informant.

Types of Interviewing

There are three major forms of interviewing: structured, unstructured, and semi-structured. These three forms can be placed along a continuum, with the **structured interview** at one end and the **unstructured interview** at the other (see also Chapter 1, Box 1.1). Structured interviews follow a predetermined and standardized list of questions. The questions are asked in almost the same way and in the same order in each interview. These interviews are much more like questionnaires (see Chapter 12). At the other end of the interviewing continuum are unstructured forms of interviewing such as **oral histories** (discussed fully in Chapter 9). The conversation in these interviews is actually directed by the informant rather than by set questions. In the middle of this continuum are **semi-structured interviews**. This form of interviewing has some degree of predetermined order but maintains flexibility in the way issues are addressed by the informant. Different forms of interview have varying strengths and weaknesses that should be clear to you by the end of this chapter.

Strengths of Interviewing

Research interviews are used for four main reasons (see also Krueger 1994; Minichiello et al. 1995, 70–4; Valentine 1997, 110–12):

1. To fill a gap in knowledge that other methods, such as observation or the use of census data, are unable to bridge efficaciously.
2. To investigate complex behaviours and motivations.
3. To collect a diversity of meaning, opinion, and experiences. Interviews provide insights into the differing opinions or debates within a group, but they can also reveal consensus on some issues.
4. When a method is required that shows respect for and empowers the people who provide the data. In an interview, the informant's view of the world should be valued and treated with respect. The interview may also give informants cause to reflect on their experiences and the opportunity to find out more about the research project than if they were simply being observed or if they were completing a questionnaire.

Interviews are an excellent method of gaining access to information about events, opinions, and experiences. Opinions and experiences vary enormously among people of different class, ethnicity, age, and sexuality. Interviews have allowed me to understand how meanings differ among people. Geographers who use interviewing should be careful to resist claims that they have discovered the *truth* about a series of events or that they have distilled *the* public opinion (Goss and Leinbach 1996, 116; Kong 1998, 80). Interviews can also be used to counter the claims of those who presume to have discovered *the* public opinion. This can be done by seeking out the opinions of different groups, often marginalized or subaltern groups, whose opinions are rarely heard.

Most of the questions posed in an interview allow for an open response as opposed to a closed set of response options such as "yes" or "no." In this way, each informant can describe events or offer opinions in her or his own words. One of the major strengths of interviewing is that it allows you to discover what is relevant to the informant.

Because the face-to-face verbal interchange is used in most interviewing, the informant can tell you if a question is misplaced (Box 8.1). Furthermore, your own opinions and tentative conclusions can be checked, verified, and scrutinized. This may disclose significant misunderstandings on your part or issues that you had not previously identified (Schoenberger 1991, 187).

Asking the Wrong Question: A Tale from Cabramatta

On 23 June 1990, I began my first formal research interview. The informant was a senior office-bearer from one of the Indo-Chinese cultural associations in New South Wales, Australia. My research interest was in the social origins of the residential concentration of Indo-Chinese Australians around the outer Sydney suburb Cabramatta and the experiences of these immigrants. The political context of the time was still heavy with the racialized and anti-Asian overtones of the 1988 "immigration debate" in which mainstream politicians and academics such as John Howard (later Australian prime minister from 1996 to 2008) and Geoffrey Blainey (professor of history) had expressed concern about "Asian" immigration and settlement patterns in Australian cities. Specifically, Vietnamese immigrants were accused of congregating in places like Cabramatta (Sydney) and Richmond and Springvale (both in Melbourne) and of purposefully doing so in order to avoid participating with the rest of Australia. I had hypothesized that Vietnamese Australians did not congregate voluntarily but that they were forcibly segregated by the economic and social constraints of discrimination in housing and labour markets. Indeed, the geographic literature supported my assertion at the time. I was a somewhat naive and colonialist investigator who saw his role as "valiant champion" of an ethnic minority.

But back to my first interview. One of my first questions to these Indo-Chinese-Australian leaders was: "Please explain the ways in which discrimination has forced you, and members of the community you represent, to reside in this area?" Their answer: "I wouldn't live anywhere else." This informant, and most subsequent informants, described the great benefits and pleasures of living in Cabramatta. They also explained how residing in Cabramatta had eased their expanding participation in Australian life (Dunn 1993). I had asked the wrong questions and had been told so by my informants. I decided to focus the project on the advantages and pleasures that residence in "Cab" brought to Indo-Chinese Australians. The face-to-face nature of the exchange, and the informed subject, make interviews a remarkable method. The participants can tell the researcher, "You're on the wrong track!"

Interview Design

It is not possible to formulate a strict guide to good practice for every interview context. Every interview and every research issue demands its own preparation and practice. However, researchers should heed certain procedures. Much of the rest of this chapter focuses on strategies for enhancing the credibility of data collected using rigorous interview practice. In the next section, we look at the organization of interview schedules and the formulation of questions.

The Interview Schedule or Guide

Even the most competent researcher needs to be reminded during the interview of the issues or events they had intended to discuss. You cannot be expected to recall all of the specific questions or issues you wish to address, and you will benefit from some written reminder of the intended scope of the interview. These reminders can take the form of **interview schedules** or **interview guides.**

An interview guide or **aide-mémoire** (Burgess 1982c) is a list of general issues you want to cover in an interview. Guides are usually associated with semi-structured forms of interviewing. The guide may be a simple list of key words or concepts intended to remind you of discussion topics. The topics initially listed in a guide are often drawn from existing literature on an issue. The identification of key concepts and the isolation of themes is a preliminary part of any research project (see Babbie 1992, 88–164).

One of the advantages of the interview guide is its flexibility. As the interviewer, you may allow the conversation to follow as "natural" a direction as possible, but you will have to redirect the discussion to cover issues that may still be outstanding. Questions can be crafted in situ, drawing on themes already broached and from the tone of the discussion. The major disadvantage of using an interview guide is that you must formulate coherent question wordings "on the spot." This requires good communication skills and a great deal of confidence. Any loss of confidence or concentration may lead to an inarticulate and ambiguous wording of questions. Accordingly, a guide is inadvisable for first-time interviewers. Guides are more appropriate for very skilled interviewers and for particular forms of interview, such as oral history.

Interview schedules are used in structured and sometimes semi-structured forms of interviewing. They are also called question schedules or "question routes" (Krueger 1994). An interview schedule is a list of carefully worded questions (see Box 8.2).

I have found that a half-hour interview will usually cover between six and eight primary questions. Under each of these central questions I nest at least two detailed questions or **prompts**. In some research, it may be necessary to ask each question in the same way and in the same order to each informant. In others, you might ask questions at whatever stage of the interview seems appropriate. The benefits of using interview schedules mirror the disadvantages of interview guides. They provide greater confidence to researchers in the enunciation of their questions and allow better comparisons between informant answers. However, questions that are prepared before the interview and then read out formally may sound insincere, stilted, and out of place.

A mix of carefully worded questions and topic areas capitalizes on the strengths of both guides and schedules. Indeed, a fully worded question can be placed in a guide and yet be used as a topic area. The predetermined wording can

Formulating Good Interview Questions

- Use easily understood language that is appropriate to your informant.
- Use non-offensive language.
- Use words with commonly and uniformly accepted meanings.
- Avoid ambiguity.
- Phrase each question carefully.
- Avoid leading questions as much as possible (i.e., questions that encourage a particular response).

be kept as a "fall-back" in case you find yourself unable to articulate a question "on the spot." I find it useful to begin an interview with a prepared question. It is damaging to one's confidence if an informant asks, "What do you mean by that?" or "I don't know what you mean" in response to your first question.

Interview design should be dynamic throughout the research (Tremblay 1982, 99, 104). As a research project progresses, you can make changes to the order and wording of questions or topics as new information and experiences are fed back into the research design. Some issues may be revealed as unimportant, offensive, or silly after the initial interview. They can be dropped from subsequent interviews. The interview schedule or guide should also seek information in a way that is appropriate to each informant.

While the primary purpose of interview schedules and guides is to jog your memory and to ensure that all issues are covered as appropriately as possible, it is also useful to provide informants with a copy of the questions or issues before the interview to prompt thought on the matters to be discussed. Interview guides and schedules are also useful note-taking sheets (Webb 1982, 195).

Types of Questions

Interviews utilize **primary** (or original) **questions** and **secondary questions**. Primary questions are opening questions used to initiate discussion on a new theme or topic. Secondary questions are prompts that encourage the informant to follow up or expand on an issue already discussed. An interview schedule, and even an interview guide, can have a mix of types of original questions, including descriptive questions, storytelling prompts, structural questions, contrast and opinion questions, and devil's advocate propositions (see Box 8.3). Since different types of primary questions produce very different sorts of responses, a good interview schedule will generally comprise a mix of question types.

On an interview guide or schedule, you might have a list of secondary questions or prompts (Box 8.5). There are a number of different types of prompts, ranging from formal secondary questions to nudging-type comments that encourage the informant to continue speaking (Whyte 1982, 112). Sometimes prompts are listed in the interview guide or schedule, but often they are deployed, when appropriate, without prior planning.

Ordering Questions and Topics

It is important to consider carefully the order of questions or topics in an interview guide or schedule. Minichiello et al. (1995, 84) advise that the most important consideration in the ordering of questions is preserving **rapport** between you and your informant. An interview might begin with a discussion of the general

BOX 8.3

Primary Question Types

Type of Question	Example	Type of Data and Benefits
Descriptive (knowledge)	What is the full name of your organization? What is your role within the organization? How many brothers or sisters do you have?	Details on events, places, people, and experiences. Easy-to-answer opening questions.
Storytelling	Can you tell me about the formation and history of this organization and your involvement in it?	Identifies a series of players, an ordering of events, or causative links. Encourages sustained input from the informant.
Opinion	Is Canadian society sexist? What do you consider to be the appropriate size for a functional family?	Impressions, feelings, assertions, and guesses.
Structural	How do you think you came to hold that opnion? What do you think the average family size is for people like yourself?	Taps into people's ideology and assumptions. Encourages reflection on experiences and social expectations may have influenced opinions and perspectives.
Contrast (hypothetical)	Would your career opportunities have been different if you were a man? Or if you grew up in a poorer suburb?	Comparison of experience by place, time, gender, and so forth. Encourages reflection on (dis)advantage.
Devil's advocate	Many practising town planners are voicing concern about the lack of transparency in the new development assessment process.	Controversial/sensitive issues broached without associating the researcher with people who are not prepared to make their opinion public.

problem of homophobia or racism: how widespread it is, how it varies from place to place, and how legal and institutional responses to those forms of oppression have emerged. Following a **funnel structure**, after broaching these general or macro-level aspects of oppression, the interview might then turn to the particular experiences of the informant.

Asking the Tough Questions without Sounding Tough

My work with Indo-Chinese communities in Cabramatta occurred within a political context in which Vietnamese Australians were being publicly harangued by academics and politicians (Dunn 1993). I felt it was important for informants to respond to their critics. My interview schedule had the following two devil's advocate propositions. I also used a preamble to dissociate myself from the statements.

> In my research so far, I have come across two general explanations for Vietnamese residential concentration. I would like you to comment on two separate statements that to me represent these two explanations:
>
> First: that Vietnamese people have concentrated here because they don't want to participate in the wider society.
>
> Second: that the Vietnamese are segregated into particular residential areas through social, economic, and political forces imposed upon them by the wider society.

The aim was to gather people's responses to both statements. In most cases, informants were critical of both views. Some informants took my question as a request for them to select the explanation that they thought was the most appropriate. Those people selected the second statement. Others were in no doubt that I disagreed with both views. Either way, devil's advocate propositions are often leading. My political views were noticeable in the question preamble and wording as well as in the preliminary discussions held to arrange the interviews. It is fairer that the researcher's motives and political orientation are obvious to the informant rather than hidden until after the research is published (see Chapters 2, 4, 6, and 19 of this volume).

In an interview with a **pyramid structure**, the more abstract and general questions are asked at the end. The interview starts with easy-to-answer questions about an informant's duties or responsibilities or their involvement in an issue. This allows the informant to become accustomed to the interview, interviewer, and topics before they are asked questions that require deeper reflection. For example, to gather views on changes to urban governance, you might find

BOX 8.5

Types of Prompt

Prompt Type	Example	Type of Data and Benefits
Formal secondary question	Primary Q: What social benefits do you derive from residing in an area of ethnic concentration?	Extends the scope or depth of treatment on an issue.
	Secondary Q: What about informal child care?	Can also help to explain or rephrase a misunderstood primary question.
Clarification	What do you mean by that?	Used when an answer is vague or incomplete.
Nudging	And how did that make you feel? Repeat an informant's last statement.	Used to continue a line of conversation.
Summary (categorizing)	So let me get this straight: your view, as just outlined to me, is that people should not watch shows like "Big Brother"?	Outlines in-progress for verification. Elicits succinct statements (for example, "quotable quotes").
Receptive cues	Audible: Yes, I see. Uh-huh. Non-audible: nodding and smiling.	Provides receptive cues, encourages an informant to continue speaking.

it necessary to first ask an informant from an urban planning agency to outline his or her roles and duties. Following that, you might ask your informant to outline the actions of the informant's own agency and how those actions may have changed in recent times. Once the "doings and goings-on" have been outlined, it may then make sense to ask the informant why agency actions and roles have changed, whether that change has been resisted, and how the informant views the transformation of urban governance.

A final question-ordering option is to use a **hybrid** of the funnel and pyramid structures. The interview might start with simple-to-answer, non-threatening questions, then move to more abstract and reflective aspects, before gradually progressing towards sensitive issues. This sort of structure may offer the benefits of both funnel and pyramid ordering.

When thinking about question and topic ordering, it can be helpful to have key informants comment on the interview guide or schedule (Kearns 1991a, 2). Key informants are often initial or primary contacts in a project. They are usually the first informants, and they often possess the expertise to liaise between the researcher and the communities being researched. Key informant review can be a useful litmus test of interview design, since these representatives are "culturally qualified." They have empathy with the study population and can be comprehensively briefed on the goals and background of the research (Tremblay 1982, 98–100).

Structured Interviewing

A structured interview uses an interview schedule that typically comprises a list of carefully worded and ordered questions (see Boxes 8.3 and 8.5 and the earlier discussion on ordering questions and topics). Each respondent or informant is asked exactly the same questions in exactly the same order. The interview process is question focused.

It is a good idea to **pre-test** a structured interview schedule on a subset (say two to four) of the group of people you plan to interview for your study to ensure that your questions are not ambiguous, offensive, or difficult to understand. Though helpful, pre-testing is of less importance in semi-structured and unstructured interviews in which ambiguities (but not offensive questions!) can be clarified by the interviewer.

Structured interviews have been used with great effect throughout the sub-disciplines of geography, including economic geography (Box 8.6).

Semi-Structured Interviewing

Semi-structured interviews employ an interview guide. The questions asked in the interview focus on content and deal with the issues or areas judged by the researcher to be relevant to the research question. Alternatively, an interview schedule might be prepared with fully worded questions for a semi-structured interview, but the interviewer would not be restricted to deploying those questions. The semi-structured interview is organized around ordered but flexible questioning. In semi-structured forms of interview, the role of the researcher (interviewer or facilitator) is recognized as being more interventionist than in unstructured interviews. This requires that the researcher redirect the conversation if it has moved too far from the research topics.

Unstructured Interviewing

Various forms of unstructured interviewing exist. They include oral history, **life history**, and some types of group interviewing and in-depth interviewing. Unstructured interviewing focuses on personal perceptions and personal

Interviewing—An Economic Geography Application

In 1991, Erica Schoenberger argued that most industrial geography research had been on the outside looking in, deducing strategic behaviour from its locational outcomes rather than investigating it directly (1991, 182). One of the assumptions challenged by the use of structured interviews was that the location of firms was strongly associated with proximity to industry-specific inputs and markets. Using structured interviews with managers, Schoenberger was able to show that the location of foreign chemical firms in North America was as much, if not more, related to historical and strategic contingencies than to contemporary location preferences.

For example, one of Schoenberger's case studies was a German-owned chemical firm. Her interviews revealed that the firm's board of directors had decided on a major expansion in the US market. But the board had been split between establishing a "greenfields" site, which would be purpose-built to company needs, and acquiring an already established chemical plant, which would hasten their expanded presence in the market. Plans to establish a greenfields facility were foiled by organized community opposition. The directors who argued for an acquisition then gained the upper hand, and at about the same time a US chemical firm came up for sale. For many decades, German chemical firms had agreed among themselves to specialize in certain parts of the chemical sector. These agreements were about to end, and Schoenberger's case-study firm was keen to expand horizontally into another speciality. The US firm that came up for sale happened to specialize in that area. A host of historical and strategic events had combined to produce a particular locational result.

The historical and strategic contingencies that accounted for the location of the German chemical company were revealed through structured interviews. Their location was in fact quite at odds with the apparent preferences of the firm and revealed nothing about the firm's location preferences (Schoenberger 1991, 185). The US chemical sector has a high level of foreign ownership, and most of it was established through acquisition. Replicated interviews with managers and directors across the chemical sector revealed the prevalence of location choices being determined by historical and strategic contingency. Interviewing was therefore an essential method for unravelling the location determinants of chemical plants in the North America.

histories. Rather than being question-focused like a structured interview or content-focused as in a semi-structured format, the unstructured interview is informant-focused. Life history and oral history interviews seek personal accounts of significant events and perceptions, as determined by the informants and in their own words (see Chapter 9 and also McKay 2002). Each unstructured interview is unique. The questions you ask are almost entirely determined by the informant's responses. These interviews approximate normal conversational interaction and give the informant some scope to direct the interview. Nonetheless, an unstructured interview requires as much if not more preparation than its structured counterpart. You must spend time sitting in musty archive rooms or scrolling through digital archives gaining a solid understanding of past events, people, and places related to the interview. But through these interviews we can "find out about" events and places that had been kept out of the news or that had been deemed of no consequence to the rich and powerful (Box 8.7).

Interviewing Practice

Rapport with another person is basically a matter of understanding their model of the world and communicating your understanding symmetrically. This can be done effectively by matching the perceptual language; the images of the world; the speech patterns, pitch, tone, speed; the overall posture; and the breathing patterns of the informant (Minichiello et al. 1995, 80).

Achieving and maintaining rapport, or a productive interpersonal climate, can be critical to the success of an interview. Rapport is particularly important if you need to have repeat sessions with an informant. Even the first steps of arranging an interview are significant, including the initial contact by telephone and other preliminaries that might occur before the first interview. Interviews in which both the interviewer and informant feel at ease usually generate more insightful and more valid data than might otherwise be the case. In the following paragraphs, I outline a set of tips that can help you to enhance rapport before, during, and while closing an interview.

Contact

Informants are usually chosen purposefully on the basis of the issues and themes that have emerged from a review of previous literature or from other background work. This involves choosing people who can communicate aspects of their experiences and ideas relevant to the phenomena under investigation (Minichiello et al. 1995, 168). Once you have identified a potential informant, you must then negotiate permission for the interview. This means getting the consent of the informants themselves, and in some circumstances, it will also involve gaining the sanction of "gatekeepers" like employers, parents, or teachers. This might occur,

Oral Environmental Histories

Oral history interviews can collect data about environmental history. This type of interviewing helps produce a more comprehensive picture of the cause and process of environmental change than is available through physical methods of inquiry. Data collected might include people's memories of changes in local land use, biodiversity, hydrology, and climate.

Lane (1997) used oral history interviews to reveal changes in watercourses, weeds, and climate in the Tumut Region high country of the Australian Alps. Interviews were conducted with five main informants, first in their homes and then while driving and walking through the countryside where they had resided. The informants told of the waterholes and deep parts of creeks where they would fish and swim and where they and their children had learned to swim. One informant commented that one of the creeks used to be almost a river—and now you could step over it (Lane 1997, 197). The same informant noted the change in colour and quality of the water. Lane's informants described how the water level and quality had steadily degraded since pine plantations had been planted in the 1960s. This description was consistent with "scientific" understandings of the impact of pine plantations in which there is an ever-decreasing level of run-off as the pines grow.

Such specific observations from local residents may often be the only detailed evidence on environmental change that is available. Oral history can fill gaps in the "scientific record," or it can be used to complement data gathered using physical or quantitative methods. More important, with the use of oral history, environmental change can be set in a human context and related to the history of people who lived in the region (Lane 1997, 204).

for example, if you wanted to interview schoolchildren, prisoners, or employees in some workplaces.

Your first contact with an informant will often be by telephone or by some form of correspondence such as email. In this preliminary phase, you should do at least four things (Robertson 1994, 9):

1. Introduce yourself and establish your bona fides. For example: "My name is Juan Folger, and I am an honours student from Java State University."
2. Make it clear how you obtained the informant's name and telephone

number or address. If you do not explain this, people may be suspicious and are likely to ask how you got their name or email address. If you are asked this question, rapport between you and your informant has already been compromised.

3. Outline why you would like to conduct the interview with this particular informant. Indicate the significance of the research, and explain why the informant's views and experiences are valued. For instance, you may believe that they have important things to say, that they have been key players in an issue, or that they have experienced something specific that others have not. On the whole, I have found that most people are flattered to be asked for an interview, although they are often nervous or hesitant about the procedure itself.
4. Indicate how long the interview and any follow-up is likely to take.

Making an informant feel relaxed involves dealing with all of the issues mentioned above and in addition spelling out the mechanics of the interview and negotiating elements of the interview process. All of these matters can be outlined in a "letter of introduction" that may be sent to an informant once they have agreed to an interview or while agreement is still being negotiated. This formal communication should be under the letterhead of your organization (for example, your university) and should spell out your bona fides, the topic of the research, the manner in which the interview will be conducted, and any rules or boundaries regarding confidentiality. You must, of course, seek permission from your supervisor to use the letterhead of an organization such as a university, although ethics procedures within your institution (see Chapter 2) are likely to have made this mandatory. In the absence of a letter of introduction, informants should be made aware of their rights during the interview. This procedure is sometimes referred to as brokering a "research deal" or a "research bargain." The research deal may be agreed to over the telephone or by email or orally just before an interview begins. The deal can be set out in written form. (See Box 8.8 for some of the rights of informants that can be established. Chapter 2 of this volume includes material relevant to the ethics of interviewing. See also Hay [1998] and Israel and Hay [2006].) These preliminary discussions are important to the success of an interview. Indeed, they set the tone of the relationship between interviewer and informant.

The Interview Relation

The relationship established between interviewer and informant is often critical to the collection of opinions and insights. If you and your informant are at ease with each other, then the informant is likely to be communicative. However, there are competing views on the nature of the interviewer–informant relationship. On the one hand is an insistence on **professional interviewing** and on the other "creative" or empathetic interviewing. Goode and Hatt (in Oakley 1981, 309–10) warned that interviewers should remain detached and aloof from their informants: "the interviewer cannot merely lose himself [*sic*] in being friendly. He must introduce himself as though beginning a conversation, but from the beginning the additional element of respect, of professional competence, should be maintained. . . . He is a professional in this situation, and he must demand and obtain respect for the task he is trying to perform."

Codifying the Rights of Informants

In their research on the Carrington community in Newcastle, New South Wales, Winchester, Dunn, and McGuirk (1997) decided to codify informants' rights in the oral histories and semi-structured interviews that were to be conducted. They included the following list of informants' rights on university letterhead, and they gave a copy to each of the informants:

- Permission to record the interview must be given in advance.
- All transcribed material will be anonymous.
- Tapes and **transcripts** will be made available to informants who request them.
- Informants have the right to change an answer.
- Informants can contact us at any time in the future to alter or delete any statements made.
- Informants can discontinue the interview at any stage.
- Informants can request that the audio recorder be paused at any stage during the interview.

To this list one might add further statements (for example, that informants can expect information about the ways in which their contributions to the research may be used). A codification of rights was deemed necessary for two reasons. First, it was done to empower the informants and assure them that they could pause or terminate the interview process whenever they deemed it necessary to do so. Second, the researchers had employed an articulate local resident to conduct the interviews, and so it was important that the interviewer was also constantly reminded of the informants' rights.

A very different, indeed opposite, sort of relationship was proposed by Oakley (1981) and Douglas (1985). In their view, a researcher who remains aloof would undermine the development of an intimate and non-threatening relationship (Oakley 1981, 310). Rather than demanding respect from the informant, Douglas's model of **creative interviewing** insists that each informant must be treated as a "Goddess" of information and insight. Douglas recommends that researchers humble themselves before the Goddess. The creative or empathetic model of interviewing thus advocates a very different sort of relationship between the informant and interviewer than that recommended for professional interview relations. Overall, there is a range of interview practice that lies between the poles of professional and creative interview relationships.

Decisions about the interview relationship will vary according to the characteristics of both the informant and the interviewer. The cultural nuances of a study group will at times necessitate variations in the intended interaction. However, it is wise to remember that despite any empathy or relationships that are established, the interview is still a formal process of data gathering for research. Furthermore, there is usually a complex and uneven power relationship involved in which information, and the power to deploy that information, flows mostly one way: from the informant to the interviewer (see Chapters 2, 4, and also McKay 2002 and Stacey 1988).

Rapport may increase the level of understanding you have about the informant and what they are saying. There are a number of strategies for enhancing rapport. The first is through the use of respectful preliminary work. The second involves the use of a **warm-up** period just before an interview commences. Douglas (1985, 79) advises that rather than getting "right down to business," it is better to engage in some "small talk and chit-chat [which] are vital first steps." This warm-up discussion with an informant could be a chat about the weather, matters of shared personal interest, or "catching-up" talk. In their surveys and interviews of Vietnamese Australians in Melbourne, Gardner, Neville, and Snell (1983, 131) found that "[t]he success of an interview (when measured by the degree of relaxation of all those present and the ease of conversation) generally depended on the amount of 'warm-up' (chit-chat, introductions, etc.)."

My own warm-up techniques for face-to-face interviews have included giving the informant an overview of the questions I plan to ask and presenting relevant diagrams or maps, as well as discussing historical documents (see also Tremblay 1982, 99, 103). Maps, diagrams, tables of statistics, and other documents can also be used as references or stimuli throughout an interview. If an informant offers you food or drink before an interview, it would be courteous to accept. As we shall see in more detail later in this chapter, many of these forms of warm-up are not available in CMC interviews.

You should also have acquainted yourself with the cultural context of the informants before the interview. As Robin Kearns pointed out, "If we are to engage someone in conversation and sustain the interaction, we need to use the right words. Without the right words our speech is empty. Language matters" (Kearns 1991a, 2). For instance, you must be able to recognize the jargon or slang and frequently used acronyms of institutions or corporations as well as the language of particular professions or cultural groups.

Listening strategies can improve rapport and the productivity of the interview. Your role as interviewer is not passive but requires constant focus on the information being divulged by informants and the use of cues and responses to encourage them. Your role as an active participant in the interview extends well beyond simply asking predetermined questions or broaching predetermined topics. You must maintain an active focus on the conversation. This will help to prevent lapses of concentration. You must also avoid "mental wandering"—otherwise, you may miss unexpected leads. Moreover, it is irritating to the informant, and a threat to rapport, if you ask a question they have already answered (Robertson 1994, 44).

Adelman (1981) advises researchers to maintain a **critical inner dialogue** during an interview. This requires that you constantly analyze what is being said and simultaneously formulate the next question or prompt. You should be asking yourself whether you understand what the informant is saying. Do not let something pass by that you do not understand with the expectation that you will be able to make sense of it afterwards. Minichiello et al. (1995, 103) provide a demonstration of how critical inner dialogue might occur: "What is the informant saying that I can use? Have I fully understood what this person is saying? Maybe, maybe not. I had better use a probe. Oh yes, I did understand. Now I can go on with a follow-up question."

Strategies to enhance rapport continue throughout the interview, through verbal and non-verbal techniques that indicate that the responses are valued. When undertaking repeat interviews it is essential that the researcher does not appear to have forgotten the story of, as told so far by, the informant. In some professional fields, such as police, forensic, or cognitive interviewing, it is recommended not to re-interview because of the enhanced opportunity for "interviewer introduced" perspectives and suggestibility (Burgwyn-Bailes et al. 2001). However, repeat interviews can be used to successfully build rapport and enhance the communicativeness of the informant, expanding the breadth and depth of the data collected (Hershkowitz and Terner 2007). Vincent (2013) found from interviews with pregnant schoolgirls and schoolgirls' mothers in Britain that these advantages were particularly salient with vulnerable groups and when dealing with sensitive topics. Researchers should be well aware of the information provided in the previous interview, and they can be ready to test apparent inconsistencies in the accounts of events and feelings between interviews, although the aim in most social science interviews should not be to "test" or "trip-up" an informant.

Informants may sometimes recount experiences that upset them or stir other emotions. When an informant is becoming distressed, try pausing the interview or changing the topic and possibly returning to the sensitive issue at a later point. If the informant is clearly becoming very distressed, you should probably terminate the interview.

There may be a stage in an interview when your informant does not answer a question. If there is a silence or if they shake their head, the informant may be indicating that they have not understood your question or simply do not know the answer. They might be confused as to the format of the answer expected: is it a "yes–no" or something else (Minichiello et al. 1995, 93)? In these cases, try restating the question, perhaps using alternative wording or providing an example. You should always be prepared to elaborate on a question. It is important to remember, however, that choosing not to respond is the informant's right. If the informant refuses to answer and says so, you should not usually press them. They may have chosen not to answer because the question was asked clumsily or insensitively (or for some other reason—if the question dealt with sensitive commercial matters, for instance). If you prepare your questions carefully, you should avoid this sort of problem and the consequent loss of empathy and data.

As an interviewer, you should also learn to distinguish between reflective silence and non-answering. Robertson cautions, "Do not be afraid of silences. Interviewers who consciously delay interrupting a pause often find that a few seconds of reflection leads interviewees to provide the most rewarding parts of an interview. . . . There is no surer way of inhibiting interviewees than to interrupt, talk too much, argue, or show off your knowledge" (Robertson 1994, 44).

It is important to allow time for informants to think, meditate, and reflect before they answer a question (see Box 8.9). It is also important to be patient with slow speakers or people who are not entirely host language–fluent. Resist the temptation to finish people's sentences for them. Supplying the word that an informant is struggling to find may seem helpful at the time, but it interrupts them and inserts a term they might not have ordinarily used.

Closing the Interview

Do not allow rapport to dissipate at the close of an interview. It is critical to maintain rapport—especially if you intend to re-interview the informant. You must prepare for the closure of an interview—otherwise, the ending can be clumsy. Because an interview establishes a relationship within which certain expectations are created, it is better to indicate a sense of continuation and of feedback and clarification than to end the interview with an air of finality. Indeed, near the end of an interview, or after the recording device is switched off, an informant may continue to divulge very interesting information. The ethical and political protocols of the interview should be maintained in these closing moments.

Finishing Sentences, Interrupting, and "Rushing-On"

During February 2001, Minelle Mahtani (then of the University of British Columbia) joined me in Sydney to undertake joint, and comparative, research on the media representations of ethnic minorities in Australia and Canada. This involved interviews with managers and employees within newsrooms. They were powerful and confident informants. Dr Mahtani had a wealth of expertise in such environments, having been a producer with the flagship Canadian Broadcasting Corporation's *The National*, a news television program. Our first field interview was with a network news editor for one of the commercial networks in Australia. Our questions included themes such as media representations of ethnic minorities, attempts by the organization to improve the portrayal of ethnic minorities, the presence of "minority" journalists, and circumstances in which they or their staff had challenged stereotypical storylines. The questions had been developed and agreed to in advance, but what very different styles we had! The informants would sometimes provide very short and dismissive responses to some questions. When it was clear that they had answered, I would probe or move to another question. Dr Mahtani would wait, however. The silence would hang heavy over the interview. I felt uncomfortable, but these powerful informants got the idea that we wanted a fuller response. They would then justify the view they had briefly given, or they would admit that there were alternative viewpoints to the one they had expressed. My rushing-on was a strategy vastly inferior to the "sounds of silence" for uncovering richer insight into ethnic minority representations and the dynamics of the newsroom (Dunn and Mahtani 2001).

Try not to rush the end of an interview. At the same time, do not let an interview "drag on." There is an array of verbal and non-verbal techniques for closing interviews (Box 8.10). Of course, non-verbal versions should be accompanied by appropriate verbal cues; otherwise, you could appear quite rude. The most critical issue in closing an interview is to express not only thanks but also satisfaction with the material that was collected. For example, you might say, "Thanks for your time. I've got some really useful, insightful information from this interview." Not only is gratitude expressed this way, but informants are made aware that the process has been useful and that their opinions and experiences have been valued.

Techniques for Closing Interviews

Four types of verbal cue:

- direct announcement: "Well, I have no more questions just now."
- clearinghouse questions: "Is there anything else you would like to add?"
- summarizing the interview: "So, would you agree that the main issues according to you are . . . ?"
- making personal inquiries and comments: "How are the kids?" or "If you want any advice on how to oppose . . . just ring me."

Six types of non-verbal cue:

- looking at your watch
- putting the cap on your pen
- stopping or unplugging the audio recorder
- straightening your chair
- closing your notebook
- standing up and offering to shake hands

Source: Adapted from Minichiello et al. (1995, 94–8).

Recording and Transcribing Interviews

Interview recording, transcription, and field note assembly are referred to as the mechanical phases of the interview method. These are the steps through which the data are collected, transformed, and organized for the final stages of analysis.

Recording

Audio recording and note-taking are the two main techniques for recording face-to-face and telephone interviews. Other less commonly used techniques in geography include video recording, compiling records of the interview after the session has ended, and using cognitive maps. Both audio recording and note-taking have associated advantages and disadvantages, as will become clear in the discussion to follow. Therefore, a useful strategy of record-keeping is to combine note-taking and audio recording.

The records of an interview should be as close to complete as possible. An audio recorder will help to compile the fullest recording (Whyte 1982, 117–18). Interviewers who use note-taking need excellent shorthand writing skills to produce verbatim records. However, the primary aim in note-taking is to capture the gist of what was said.

Audio or video recording can allow for a natural conversational interview style because the interviewer is not preoccupied with taking notes. Instead, you can be a more attentive and critical listener. Audio recording is also preferable to note-taking because it allows you more time to organize the next prompt or question and to maintain the conversational nature of the interview. Note-taking researchers can be so engrossed in taking notes that they may find themselves unprepared to ask the next question. Note-takers can also miss important movements, expressions, and gestures of the informant while they are hunched over, scribbling at a furious pace (Whyte 1982, 118). This undermines rapport and detracts from attentive listening.

On the other hand, an audio recorder may sometimes inhibit an informant's responses. The presence of the mobile telephone or other recording device serves as a reminder of the formal situation of the interview (Douglas 1985, 83). Informants may feel particularly vulnerable if they think that someone might recognize their voice if the recording were aired publicly. Opinions given by the informant on the "spur of the moment" become fixed indelibly on the hard drive (or memory stick) and have the potential to become a permanent public record of the informant's views. This may make the informant less forthcoming than they would have been if note-taking had been used. Some informants become comfortable with an audio recorder as the interview progresses, but others do not. If you find the latter to be the case, consider stopping the recorder and reverting to note-taking.

If you use an audio recorder, place it somewhere that is not too obvious without compromising the recording quality. Modern digital audio recorders have long recording capacities and can be turned on and then left alone. But take care when using an audio recorder not to be lulled into a loss of concentration by the feeling that everything is being recorded safely. There may be a technical failure. You can maintain concentration and avoid the problem associated with recorder failure by taking some written notes. If you are taking notes, there is little likelihood of mental wandering. Everything is being listened to and interpreted, and parts of it are being written down, demanding that you maintain concentration. I find this particularly important when I am conducting the second or third interview in a long day's fieldwork.

Because an audio recorder does not keep a record of non-verbal data, non-audible occurrences such as gestures and body language will be lost unless you are also using a video recorder or taking notes. If an informant points to a wall map and says, "I used to live there," or if they say, "The river was the colour of that cushion," then the audio recording will be largely meaningless without some written record. These written notes can be woven into the verbal record during the transcription phase (described later in this chapter). Written notes also serve as a back-up record in case of technical failures. Overall then, a strategic combination of audio recording and note-taking can provide the most complete record of an interview with the least threat to the interview relationship.

Transcribing the Data

The record of an interview is usually written up to facilitate analysis. Interviews produce vast data sets that are next to impossible to analyze if they have not been converted to text. A transcript is a written "reproduction of the formal interview which took place between researcher and informant" (Minichiello et al. 1995, 220). The transcript should be the best possible record of the interview, including descriptions of gestures and tone as well as the words spoken (although see Box 8.13). The name or initials of each speaker should precede all text in order to identify the interviewer(s) and informant(s). Counter numbers at the top and bottom of each page of the transcript enable quick cross-referencing between the transcript file and digital records or tapes of the interview. Converting interview to text is done through either a reconstruction from handwritten notes, a transcription of an audio or video recording, the use of **voice recognition software** (see Box 8.11), or editing CMC interview correspondence.

Interview notes should be converted into a typed format preferably on the same day as the interview. If there were two or more interviewers, it is a good idea to compile a combined reconstruction of what was said using each researcher's notebook. This will improve the breadth and depth of coverage. The final typed record will normally comprise some material recalled verbatim as well as summaries or approximations of what was said.

Recorded interviews should also be transcribed as soon as possible after the interview. Transcription is a very time-consuming and therefore resource-intensive task (Whyte 1982, 118). On average, most interviews take four hours of typing per hour of interview. Transcription rates vary according to a host of variables, such as typist skill, the type of interview, the informant, and the subject matter. You can facilitate transcription by using a purpose-built transcribing recorder. You should transcribe your own interviews for two main reasons. First, since you were present at the interview, you are best placed to reconstruct the interchange. You are aware of non-audible occurrences and therefore know where such events should be inserted into the speech record. You are also better able to understand the meaning of what was said and less likely to misinterpret the spoken words. Second, transcription, although time-consuming, does enable you to engage with the data again. Immersion in the data provides a preliminary form of analysis.

BOX 8.11 Voice Recognition Software and Interviews

Computing packages have been developed that convert the spoken word into computer text. Packages such as NaturallySpeaking by Dragon Systems/Nuance can help you convert text at rates above 160 words per minute, and faster than most researchers can type (see Matheson 2007). "Free" packages associated with search engines and other software giants are also available. However, these systems will only convert the speech of a single speaker. Each system has to be "trained" to understand a single "master's voice." The success of these packages for converting interview data has been mixed. The researcher has to simultaneously listen to a recording and orally repeat the informant's contributions to enable the system to convert the data. This technique has been referred to as "re-speaking" or "parroting." Gestures and indications of intonation have to be typed into the word processing document manually, and the systems will not automatically provide punctuation. Nonetheless, typists can train the software and then re-speak the interview, typing corrections (using the "correct that" command) and inserting notations as they go along (using hot-keys for such things as speaker initials). Fatigue, illness, and alcohol have all been reported as reducing recognition accuracy. The limited comparative assessments to date—of Voice Recognition Software (VRS) versus traditional audio-to-text transcription—do not suggest that there is a substantive efficiency (see the assessments by the American urban geographer Brian Johnson [2011]). However, there is some agreement that VRS can ease the physical and mental stress of transcription, especially those for whom typing is difficult.

Listserver reports and blogs lodged by researchers have claimed accuracy rates as high as 95 per cent once the program has been "trained." I have read a claim that the new age of software requires only 15 minutes of training, although my experience was that even after four hours of coaching, the software was still getting every third word "wrong" and some of the conversions were hilarious. It is critical that you save your speech files (the training) after each use, and it is advisable that you use a very good quality microphone (see Stockdale's 2002 useful technical tips). It should also be noted that serious investigation of the methodological issues that may surround voice recognition software has barely begun.

There is no accepted standard for symbols used in transcripts, but some of the symbols commonly used are set out in Box 8.12. CMC interviews do not require transcription, since each informant's answers are already in a text format. The text may well include some **emoticons** that are popular in SMS (short message service) and Internet-based communication.

BOX 8.12 Symbols Commonly Used in Interview Transcripts

Symbol	Meaning
//	Speaker interrupted by another speaker or event: //phone rings//
:	Also used to indicate an interruption
KMD	The initials of the speaker, usually in CAPS and bold
—	When used at the left margin, refers to an unidentified speaker
Ss	Several informants who said the same thing
E	All informants made the same comment simultaneously
. . .	A self-initiated pause by a speaker
. . . . or	Longer self-initiated pauses by a speaker
-	Speech that ended abruptly but without interruption
( )	Sections of speech, or a word, that cannot be deciphered
(jaunty)	A best guess at what was said
(jaunty/journey)	Two alternative best guesses at what was said
*	Precedes a reconstruction of speech that was not recorded
(. . .)	Material that has been edited out
But I didn't want to	Underlined text indicates stressed discourse
I got nothing	Italicized text indicates louder discourse
[sustained laughter]	Non-verbal actions, gestures, facial expressions
[hesitantly]	Background information on the intonation of discourse
Emoticon	(see http://en.wikipedia.org/wiki/List_of_emoticons)
:) or ☺ or :-) or :-D	Smiling, joke marker, happy, laughing hard
:(or :-(	Frowning, sad
;) or *)	Wink
:'(or :: or :,(	Tears, shedding a tear, crying
lol	Laughing out loud

Once completed, the transcript should be given a title page stating the informant's name (or a code if there are confidentiality concerns), the number of the interview (for example, first or third session), the name(s) of the researcher(s) (i.e., who carried out the interview), the date of the session, the location, duration of the interview, and any important background information on the informant or special circumstances of the interview. Quotations that demonstrate a particular point and that could be presented as evidence in a final report on the research might be circled or underlined.

The transcript can be given to the informant for vetting or authorizing. This will normally improve the quality of your record (see Box 8.13). This process of **participant checking** continues the involvement of the informants in the research process and provides them with their own record of the interview.

Assembling Field Note Files

Assembling interview records marks the beginning of the analysis proper. It begins with a critical assessment of the interview content and practice and is followed by formal preparation of interview logs. To my mind, the best advice on assembling field note files comes from Minichiello et al. (1995, 214–46). In the wide margins of the transcript file, you can make written annotations. Comments that relate to the practice of the interview, such as the wording of questions and missed opportunities to prompt, should be placed in the left margin. These annotations and other issues concerned with contact, access, ethics, and overall method should be elaborated upon in a **personal log** (Box 8.14). The right margin of the transcript file can be used for annotations on the substantive issues of the research project. These comments, which generally use the language and jargon of social science, are then elaborated upon in the **analytical log.** The analytical log is an exploration and speculation about what the interview has found in relation to the research question (Box 8.14). It should refer to links between the data gathered in each interview and the established literature or theory.

Analyzing Interview Data

Researchers analyze interview data to seek meaning from the data. We construct themes, relations between variables, and patterns in the data through content analysis (see Chapter 18). Content analysis can be based on a search of either manifest or latent content (Babbie 1992, 318–19). **Manifest content analysis** assesses the visible, surface content of documents such as interview transcripts. An example would be a tally of the number of times the words "cute" and "cuddly" are used to describe koalas in interviews with members of the public. This might be important to understanding public opinion and the politics of culling in areas of koala

BOX 8.13 Debates about Changing the Words: Vetting and Correcting

In general, it has been thought that a transcription should be a verbatim record of the interview. This would include poor grammar, false starts, "ers," and "umms." There are a number of good reasons advanced for this position. A verbatim record will include the nuances of accent and **vernacular**, it will maintain any sense of hesitancy, and it could demonstrate an embarrassment that was present. For example, Sarah Nelson (2003, 16) reflected on how the "humming and hawing" of Ulster politicians when asked about sectarian killings was reflective of their hesitancy and hypocritical stances on sectarianism. Transcripts that are not exact textual replications of an interview will lose the ethnographic moment of the interview itself. However, it may be difficult to search for key terms if they are "misspelled" in a transcript (misspelled as a means of indicating accent or mispronunciation).

A range of researchers working in different disciplines and countries have expressed some concern about the political effects of exact transcription. Many have reflected on the embarrassment that many informants articulate when they receive the transcript of their interview. They express anxiety about the grammar, the false starts to their sentences, repetition, and the "ers" and "umms" and "you knows." This anxiety is even more strongly felt by informants who live in societies where the dominant language is not their first language. Informants might be so concerned as to withdraw their interview and avoid any future ones (McCoyd and Kerson 2006, 397). Moreover, research reports on the less powerful in society (the poor, single mothers, youth groups) that use the real language of informants and are largely sympathetic to those people can often portray them in a way that reproduces negative images and stereotypes. Nelson (2003) reflected on the way such quoted material reconstructs images of illiteracy, powerlessness, and inferiority. As stated earlier, transcription is a transformation of verbal encounter into text; it is a constructed document that is of the researcher's making (Green, Frauquiz, and Dixon 1997). As bell hooks (1990, 152) famously stated, "I want to know your story. And then I will tell it back to you in a new way . . . . Rewriting you, I write myself anew. I am still author, authority." Informants are more interested in the impact of their words than in the nuances of expression. Many researchers recommend sending informants summaries or interpretations of the interview rather than transcripts. It is certainly a good idea to send informants the eventual publications and reports.

BOX 8.14 **Field Note Files**

Transcript File	Personal Log	Analytical Log
Includes the record of speech and the interviewer's observations of non-audible data and intonation. Also includes written annotations in the margins on the practice and content of the interview.	Reflection on the practice of the interview. Includes comments on the questions asked and their wording, the appropriateness of the informant, recruitment and access, ethical concerns, and the method generally.	Exploration of the content of the interview. A critical outline of the substantive matters that have arisen. Identification of themes. Reference to the literature and theory. In-progress commentary on the research aims and findings.

overpopulation (for example, Muller 1999). Searching interview data for manifest content often involves tallying the appearance of a word or phrase. Computer programs such as **NVivo** or QSR N6 are particularly effective at undertaking these sorts of manifest searches.

Latent content analysis involves searching the document for themes. For example, you might keep a tally of each instance in which a female has been portrayed in a passive or active role. Latent content analysis of interview texts requires a determination of the underlying meanings of what was said. This determination of meanings within the text is a form of coding.

A coding system is used to sort and then retrieve data. For example, the text in transcripts of interviews with urban development authorities could be coded based on the following categories: structures of governance (for example, legislation, party political shifts), cultures of governance (with sub-codes like "managerialist perspective" and "entrepreneurial perspective"), coalitions and networks (of various types and agendas), the mechanisms through which coalitions operate, and the various scales at which power and influence emanate and are deployed (see McGuirk 2002). Once the sections of all the interviews have been coded, it is then possible to retrieve all similarly coded sections. These sections of text can be amalgamated and reread as a single file (Box 8.15). This might allow a researcher to grasp the varying opinions on a certain issue and to begin to unravel the general feeling about an issue.

Not every section of text needs to be coded. An interview will include material that is not relevant to the research question, particularly warm-up and closing sections and other speech focused on improving rapport rather than on gathering

BOX 8.15 **Coding Interview Data: Five Suggested Steps**

Coding Step	Specific Operations: Computing/Manual Versions
Develop preliminary coding system	Prepare a list of emergent themes in the research. Draw on the literature, your past findings, as well as your memos and log comments. Amend throughout.
Prepare the transcript for analysis	Meet the formatting requirements for the computing package being used. / Print out a fresh copy of the transcript for manual coding.
Ascribe codes to text	Allocate coding annotations using the "code text" function of computing packages. / Place handwritten annotations on transcript.
Retrieve similarly coded text	Use the "retrieve text" function of computing packages to produce reports on themes. / Extract and amalgamate sections of text that are similarly coded.
Review the data by themes	Assess the diversity of opinion under each theme. Cross-referencing themes allows you to review instances where two themes are discussed together. Begin to speculate on relations between themes.

data. Some sections of text will be multiple-coded. For example, in one sentence an informant may list a number of causes of fish kills, including open-cut mine run-off, super phosphates, acid sulphate soils, and town sewage. This may require that the sentence is given four different coding values. Coding is discussed more fully in Chapter 18.

Presenting Interview Data

Material collected from interviews is rarely presented in its entirety. The emerging exceptions include the increasing trend for transcripts to be available for review in digital repositories, and for some journals requesting that data files (e.g. interview recordings or transcripts) be "published" online as an appendix to published papers. These affirm the importance of the records being good transformations of the interview, and the importance of those vignettes on ordinary life being respectful. However, most interview data must be edited and (re)presented selectively in research publications. While it is difficult to locate a "genuinely representative" statement (see Connell 1991, 144–5; Minichiello et al. 1995, 114–15), it is usually possible to indicate the general sense and range of opinion and experience expressed in interviews. One way to indicate this is to present summary statistics of what was said. Computing packages such as NVivo can help you to calculate the frequency with which a particular term or phrase appeared in a document or section of text (see Chapter 18). However, the more common method is through a literal description of the themes that emerged in the interviews (see, for example, Boxes 8.6 and 18.1 and the discussion in Chapters 19 and 20 on presenting results).

When describing interview data, you must cite transcript files appropriately. For example, in her interview-based honours research on the changing identity of the Australian industrial city of Wollongong, Pearson (1996, 62) noted that "Several respondents asserted that elements excluded by the new identity were of little significance to the overall vernacular identity of Wollongong (Int. #1, Int. #6, and Int. #7)." The transcript citations provided here indicate which of the informants expressed a particular type of opinion. In research publications, the transcript citations can indicate the informant's name, number, code, or recorder count. Whenever a direct quotation from an informant is presented, then a transcript page reference or recorder count can be provided.

Transcript material should be treated as data. A quotation, for example, ought to be treated in much the same way as a table of statistics. That is, it should be introduced and then interpreted by the author. The introduction to a quotation should offer, if it has not already been provided, some background on the informant. It is important that readers have some idea of where an informant is "coming from"; information about their role, occupation, or status is important in this regard. Also important, as Baxter and Eyles (1997, 508) point out, is "some discussion of why particular voices are heard and others are silenced through the selection of quotes." Quotations should be discussed in relation to, and contrasted with, the experiences or opinions of other informants. Statements of opinion by an informant should also be assessed for internal contradiction. Finally, a quotation cannot replace a researcher's own words and interpretation. As the author, you must explain clearly what theme or issue a quotation demonstrates.

Knowledge is a form of power. The accumulation and ownership of knowledge is an accumulation of power: power to support arguments and power to effect change. In most interviews, information and knowledge flow from the informant to the researcher. The researcher accumulates this knowledge and ultimately controls it. There is a host of strategies and guidelines to which researchers can adhere to reduce the potential political and ethical inequities of this relationship (see Chapters 2, 3, 4, 17, and 19).

In terms of data presentation, it will sometimes be important that an informant's identity be concealed. Pseudonyms or interviewee numbers have been used by geographers to disguise the identity of their informants when it has been thought that **disclosure** could be harmful. Informants can be given the opportunity to select their own pseudonym. Robina Mohammad (1999, 238–9) used this technique with some success in her interviews with young Pakistani Muslim women in England. The interviews included discussion of patriarchal authority, "English cultures," and the cultures and dynamics of the Pakistani Muslim community in Britain. Some informants selected Pakistani Muslim pseudonyms for themselves, others chose very English names. These selections were themselves very interesting and provided further insight into the cultural perspectives and resistances of these women. Other researchers allocate pseudonyms that reflect the ethnicity of the informants (McCoyd and Kerson 2006, 396, 404). Gill Valentine (1993) felt it necessary to disguise the name of the town in which her interviews with British lesbians had taken place. Similarly, Mariastella Pulvirenti (1997, 37) disguised the street names that were mentioned by female Italian Australians when discussing their housing and settlement experiences in Melbourne.

Naming an informant (or locating them in any detailed way) and directly associating them to a quotation could be personally, professionally, or politically harmful. Researchers must be very careful when they deploy data they have collected. Interviewers are privileged with insights into people's lives. Some researchers recommend instituting an alias or pseudonym for informants very early in the mechanical phase so that no digital records will bear the informant's real identity. However, it can prove difficult to remember who the real people behind the aliases are, and some researchers only impose the pseudonym in the presentation phase of the research.

The presentation of interview-based research must contain an accessible and transparent account of how the data were collected and analyzed (Baxter and Eyles 1997, 518). This account should outline the subjectivity of the researcher, including their biases or "positioned subjectivity" (see Chapters 2, 19, and 20). As we have already seen from the discussion in Chapters 3 and 6, it is only through transparent accounts of how interview-based research was undertaken that the trustworthiness and wider applicability of the findings can be assessed by other researchers.

Interviews Using Computer-Mediated Communication

Computer-mediated communication (CMC) interviewing can include interviews with individuals and groups. They can be either **asynchronous** or **synchronous** (see Meho 2006, 1284; Mann and Stewart 2002, 604). The most common form of CMC interviewing has been asynchronous, using the to-and-fro of email exchanges. Synchronous CMC interviewing has mirrored the environment of online chat. Meho's (2006, 1284–5) review of literature on email interviewing concluded that since the early 2000s, this mode had moved from being experimental and is now considered an established format for research-oriented interviewing. Most early uses of CMC for social research involved attaching questionnaires (and sometimes interview schedules) to emails. However, email interviewing now commonly takes the form of an ongoing set of exchanges between a researcher and an informant. In the next few pages, I review advantages and disadvantages of CMC interviewing and then provide a series of tips for better practices specific to this mode.

Advantages of CMC Interviewing

CMC-delivered interviews offer five general sets of advantages: (1) an expanded sample; (2) reduced **interviewer effects**; (3) enhanced convenience; (4) more reflective informant responses; and (5) cost savings (see Chapter 12 for additional discussion of some of these issues).

First, Internet delivery of questions allows a researcher to overcome spatial, temporal, and social barriers that would restrict access to informants for face-to-face interviews. Interviews can be more easily facilitated with people living overseas or in inaccessible locations (remote places or war zones) or who have mobility limitations (e.g., differently abled), and emailed exchanges can be much more convenient for people who are shift workers or those who are based at home with small children (Bampton and Cowton 2002; Mann and Stewart 2002, 604–6; Meho 2006, 1288). Email interviews can also allow researchers to transgress social hurdles to gain access to informants: they will suit people who are shy, who are cautious about their identity being revealed, or whose cultural context is too disparate from that of the researcher; they are also appropriate when the topic is very sensitive. Good examples in the literature include interviews with political and religious dissidents, criminals and criminalized people, oppressed minorities, and deviant subcultures. McCoyd and Kerson (2006) found from their research with women in North America who had had medical terminations (after learning of a fetal anomaly) that email interviews were very effective for discussions on a topic that generated strong emotions among the informants and broached issues that are not normally discussed in public. Meho's (2006, 1286–7) review of the uses of email interviewing indicates that response rates among these "difficult to reach" samples were on average about 40 per cent.

The absence of "face-to-face" in email interviewing means that researchers can interview people who dress and look very differently from themselves—such as subcultural groups like Goths or bikers (Mann and Stewart 2002, 606). More broadly, a second advantage of email interviews is a reduction in interviewer effects as a result of the researcher's visual anonymity. Many of the cues we use to make judgments about people are based on visual appearance (e.g., dress, body shape, skin colour, jewellery, hair styles), and these cues are much more limited in CMC interviews in which informants can adopt pseudonyms and even de-gender themselves. This anonymity comes at the cost of the non-audio data that are usually gathered in a face-to-face interview. Contextual effects also dissipate. At the instrumental level, this means that the interview is not interrupted by telephone calls, colleagues, small children, or spouses. A more complicated matter is the absence of the researcher from the informant's own ethnographic setting and the loss of those observations. However, email interviewing offers great convenience to informants, allowing them to choose the time of their responses, to consider their answers at their leisure, and to do so in the comfort of their own home (Bampton and Cowton 2002; Mann and Stewart 2002, 607; Meho 2006, 1290). An informant in McCoyd and Kerson's (2006, 397) research with women who had medical terminations reflected that "I'm looking forward to doing the interview . . . it is a much more relaxed and productive way to do it [through email]. This way, I can do it when things are quiet and I'm in the right frame of mind." The informant has much more control over the pace and flow of the interview, more so than if it were a telephone or face-to-face interview during which they might feel rushed to offer an answer.

A fourth advantage of email interviewing is that the answers that informants offer can be more detailed, reflective, and well considered than those in other formats. James and Busher's (2006) email interviews with people in British university settings (and beyond) found that there was a "richness of reflection among the participants." Informants who took their time in responding to a question "tended to generate more thoughtful answers" (2006, 414). James and Busher gave the example of an informant who began one of their answers with "I didn't email you straight back, because I was thinking about my answer. So my responses were more carefully thought through and probably longer than if I'd tackled the whole thing in a face-to-face interview" (university-based informant, quoted in James and Busher 2006, 415). Informants can rethink, proofread, and re-craft their responses so that they most accurately represent their views and experiences (Bampton and Cowton 2002). Many of these advantages apply to one-on-one email interviews rather than to group discussions.

A final set of advantages of CMC interviewing is the reduced cost relative to face-to-face interviewing. The obvious savings are associated with travel costs and time—and carbon footprints. Another saving comes in the mechanical phase, since there is no need for a conversion of spoken word to text. The answers are already in text form, and this also removes issues of transcription error and interpretation (Chen and Hinton 1999; Mann and Stewart 2002, 608–9). This obviates the need to engage one of the debates about transcription referred to earlier (Box 8.13). Informants have "cleaned" their own responses to a level they are satisfied with before they post them (McCoyd and Kerson 2006, 397). Of course, researchers do have to edit out the unnecessary email symbols, signature sections, and line returns.

Challenges of CMC Interviewing

The weaknesses or limitations of CMC interviews, relative to face-to-face interviewing, stem mostly from the spatial and temporal displacements between the informants and the researcher. These issues include concerns about the authenticity of the informant, the loss of visual cues that assist rapport-building, and the "clunkiness" of the interview relationship. There are also issues of uneven Internet access and comfort with the medium, as well as ethical issues having to do with privacy and anonymity.

The advantages that stem from the visual anonymity and the use of pseudonyms have the negative effect of reducing our ability to know who we are interviewing. The identity dynamism facilitated in Internet correspondence means that people can make misleading claims about who they are (James and Busher 2006). However, Mann and Stewart (2002, 210) noted that maintaining a false identity in a substantive correspondence is not so easy to do, and researchers ought be wary of inconsistencies and contradictions that reveal such false personae. More subtly, informants are more able to embellish or be bombastic in emailed answers, because the researcher does not have the visual cues to help detect and address such tendencies (Meho 2006, 1289). Similarly, prompts are delayed until the next email interchange, at which point they may be considered less important than new primary questions, crowding out space for such clarifications (Chen and Hinton 1999; Meho 2006, 1289–90). Researchers using email interviews must instead rely heavily on "reading between the lines" of answers.

The absence of **paralinguistic clues** in CMC interviews raises broader issues. It is much more difficult for researchers to tell whether an informant is becoming distressed or uncomfortable when reflecting on or writing an answer to a question (Bampton and Cowton 2002). Their answers are in a "narrower bandwidth," and this has dramatic consequences on the potential for rapport. It is difficult to communicate empathy and sympathy (e.g., in regard to grief) to informants via email without it sounding banal or insincere (Mann and Stewart 2002, 617–19). The generally truncated process of email interviewing means that interviewing a single informant can extend over weeks and months. There will be substantial gaps at times between responses. These issues are related to the processes of reflection and consideration discussed earlier, but they are also linked to the personal availability of the informants and to the ability of the researcher to properly assimilate previous answers and to consider necessary probes and relevant primary questions. These gaps can be interpreted, incorrectly, as disinterest on the part of the informant or as a sign that the researcher is underwhelmed or disgusted by previous responses. It is also difficult for researchers to know why an informant might be delaying a response (Bampton and Cowton 2002). But the resulting form of rapport that develops is very different from what has been described for face-to-face interviews.

CMC interviewing is really only appropriate for study groups with widespread Internet access and literacy. This means that it would be inappropriate for homeless people or those who lack basic information technology skills and familiarity (Bampton and Cowton 2002; Mann and Stewart 2002, 605). Participating in an interview through email, blogs, or online in a chat format is quite physically demanding, requiring a lot of time spent hunched over a computer screen and typing (Chen and Hinton 1999). It is more onerous for the informant than answering a question orally over the telephone or face-to-face. Interviews that require informants to download materials and open attachments shift a good deal of technical work to the informant, raising the concern that it may divert attention from the questions and research topic to the technology. One general lesson is that CMC interviews are most appropriate for technically savvy study groups, including, for example, "online communities," which are natural groups that already exist within the Internet environment (Mann and Stewart 2002, 615).

Decisions about good ethical practices surrounding CMC interviews are still evolving. One concern surrounds the privacy of informant comments. On the one hand, people often feel an inflated sense of anonymity within the Internet, using pseudonyms to avoid attribution. But emailed messages and posted comments can be traced (Cho and LaRose 1999). A bigger issue concerns the usual promises that researchers offer to informants about the confidentiality of their comments. Internet communications are intercepted, and server hosts do retain copies of emails, as do institutions like universities. Informants' answers are kept by, and are accessible to, the researcher. As well, many universities and most research corporations have "My Documents" directories on all researchers' computers through which files are stored centrally on a common server. Emails should not be considered ephemera that disappear as soon as you delete them from your own inbox: they have potential longevity and circulation that is almost limitless, and the ethical risks entailed in that require careful consideration (Crystal 2006, 132; Spinks, Wells, and Meche 1999, 148–9). Although many universities currently require informants to sign consent forms before they participate in interviews, the Internet environment has different protocols for signatures and approvals, and some researchers consider that an email from the informant detailing their informed consent should be sufficient. Some argue that such correspondence must occur before and separate from any discussion of substantive matters; others are more sanguine about the separation, suggesting that consent statements can immediately precede the first set of answers (Cho and LaRose 1999, 429; McCoyd and Kerson 2006, 394).

A final set of concerns regarding CMC interviews has to do with the way they add to Internet clutter. Requests that a person be involved in email interviews or post responses in an online setting can be confused with corporate and commercial correspondence with which Internet users are increasingly bombarded. Forecasts by the Radicati Group (2011, 3) determined that in 2015, those who use email for work will receive 125 messages per day on average (as opposed to 99 per day in 2005). The rate of increase has slowed lately with the rise of other modes of messaging (e.g. small message service [SMS] and social media platforms). Nonetheless, this Internet clutter could pose substantial problems for engaging CMC interview informants in the future.

Strategies for Good CMC Interviews

As with face to-face interviews, rapport between researchers and informants can be enhanced or jeopardized in various phases of CMC interviews. In the contact phase, you should make quite clear how you obtained the potential informant's email address. Internet users are increasingly suspicious of how people obtained their address and fed-up with the email spam they receive. The sweeping up of email addresses has been referred to as **trolling**, and any research that carries a sense that informant contacts were collected in that way will be treated with suspicion. Emails also need to be sent in an individualized way to informants, not as a group email (James and Busher 2006, 408). Contacting online groups or networks and asking them participate in online interviews should be preceded by some interaction with the group and some sanction from gatekeepers. A researcher who launches into a group unannounced will likely receive antipathetic reactions. Similarly, a long period of non-participation in a group followed by a request for interviews that demonstrates a good deal of knowledge about the group will create the impression that the researcher has been **lurking**—another form of Internet anti-social behaviour (Cho and LaRose 1999, 422–4, 430). Further, informants should receive individually tailored responses and follow-ups, not formally worded emails that anaemically state, "Here is the next question." This suggested recursiveness means that most CMC interviews take on a semi-structured or unstructured form.

One of the threats to rapport once an email interview is underway is the silences and gaps in communication. The longer the gaps between exchanges, the higher the rates of drop-out and non-completion (Meho 2006, 1288). Researchers will need to remind informants about questions for which an answer has not yet been forthcoming. Bampton and Cowton (2002) advise that this asynchronous aspect (clunkiness) of interviews (and other differences from face-to-face format) should be made clear to the informants at an early stage. In general, informants need to be given a clear sense of the time frame of the whole process, the expected amount of response, the number of parcels of questions they will receive, and that there will be courteous reminders at times (see also Mann and Stewart 2002, 616). Nonetheless, patience is likely to be a virtue when one is waiting respectfully for answers, and pestering could spoil the field for future researchers (Cho and

LaRose 1999, 425). When an answer from an informant is received, it is wise to send a prompt acknowledgment, and more deeply considered follow-ups can be delayed until you have been through a process of inner critical dialogue. James and Busher (2006, 412) recommend "strategies of visibility" to maintain contact while informants are considering their answers: "Haven't heard from you for a while," "How is it going?" These types of individualized prompts will remind your informants that you are still keen to hear from them, and it can provide an opportunity for informants to raise any issues they have about the question or make clear their desire to skip that question. These are ways to be vigilant to potential drop-outs and distress (Mann and Stewart 2002, 616–19).

Some researchers recommend that initial emails from the researcher should contain some element of personal disclosure relevant to the topic of the interview. This has the effect of building up a sense of trust and sharing (Mann and Stewart 2002, 615–17). The Internet is an environment with a spirit of information-sharing, so CMC interviews should commit to releasing the general findings and recommendations to the World Wide Web (Cho and LaRose 1999, 432).

Online cultures may include the use of acronyms, abbreviations, and emoticons. Some researchers suggest that these textual cultures should be encouraged (Meho 2006, 1293), but others think that they should be discouraged in formal correspondence (Dumbrava and Koronka 2006, 62), and still others believe that they should be tolerated. Emoticons, for example, could be used by informants to indicate the kind of tone and emotion that would be apparent in face-to-face interviews (Bampton and Cowton 2002). McCoyd and Kerson (2006, 396) reported an informant placing the emoticon for tears "::" at the end of a sentence to indicate her emotional context as she was telling the researchers about her feelings regarding the medical termination of her fetus. Common emoticons could enhance rapport between researchers and informants once an interview is well underway, and the use of extra letters and dots can be a good indication of pauses ("hmmmmmm, I'm not sure about that," "weeellll maybe") (see Box 8.12). There are debates regarding the extent to which researchers should ensure proper grammar and spelling in CMC interviewing. In general, some latitude is granted for contractions that imitate speech (e.g., I'm, don't) as well as some abbreviations (txt, fwd) (Crystal 2006, 113–17). However, rich and well-expressed formal text provides a clearer and more usable form of data than text that is full of emoticons and SMS-like abbreviations (Mann and Stewart 2002, 15; or see Spinks, Wells, and Meche 1999, 149, on business communications).

Many of the rules and cultures of Internet communication carry over into CMC interviews. Researchers should be cautious about attempting to use humour and sarcasm. Emails should have appropriate subject lines, and university email addresses and URLs should be used to provide informants with a stronger sense of the project's credibility (Chen and Hinton 1999; Cho and LaRose 1999, 431).

Capitalized and bold text should be avoided, since it is considered online shouting. Each email should have only enough text to fit into a single screen, or the first screen should have a summary of all that follows (see Bampton and Cowton 2002; Crystal 2006, ch. 4; Hassini 2006, 33; Spinks, Wells, and Meche 1999, 147–9). These rules of "netiquette" flow from long-standing principles established in the mid-1990s (see Scheuermann and Taylor's [1997, 270] reference to the Ten Commandments of Etiquette or Rinaldi's [1994] often-cited guidelines).

The ethical issues surrounding CMC interviews mean that the threats to privacy should be prominent in consent agreements. Researchers can make the compiled records of interviews anonymous and store all such files off-line on external hard drives, and they can attempt to erase all emails and destroy all print-outs (Chen and Hinton 1999; McCoyd and Kerson 2006, 394; Mann and Stewart 2002). These actions can provide the informant with some assurance of confidentiality.

Because of the limitations and challenges of CMC interviewing, some have concluded that it is best seen as a complementary mode and that face-to-face interviewing remains the gold standard method (see Meho 2006). According to this view, CMC interviewing is appropriately used for informants who are difficult to physically reach or for those who already have a presence within some form of online community or network. However, the mode has special advantages of its own, such as the opportunity for more reflective responses, and it represents an appropriate method for the Internet age.

Conclusion

The rigour of interview-based research is enhanced through adequate preparation, diverse input, and verification of interpretation. Being well informed and prepared will give you a deeper understanding of the "culture" and discourse of the group(s) you study. You can then formulate good questions and enhance levels of rapport between you and your informants. You should also purposely seek out diversity of opinion. By interviewing more than one informant from each study group, you can begin to draw out and invite controversy or tensions. An opinion from one informant should never be accepted as demonstrative of group opinion unless it is shown to be so. Finally, some means of verifying your interpretations of interview data are necessary (for example, participant checking, peer checking, and cross-referencing to documentary material).

Interviews bring people "into" the research process. They provide data on people's behaviour and experiences. They capture informants' views of life. Informants use their own words or vernacular to describe their own experiences and perceptions. Kearns (1991a, 2) made the point that "there is no better introduction to a population than the people themselves." This is what I find the most refreshing aspect of interview material. Transcribed interviews are wholly unlike other forms of data. The informant's non-academic text reminds both researcher and reader of the lived experience that has been divulged. It reminds geographers that there are real people behind the data.

Key Terms

aide-mémoire
analytical log
asynchronous interviewing
computer-mediated communications (CMC) interviewing
creative interviewing
critical inner dialogue
disclosure
emoticon
funnel structure
informant
interviewer effects
interview guide
interview schedule
latent content analysis
life history
lurking
manifest content analysis
NVivo
oral history
paralinguistic clues
participant checking
personal log
pre-testing
primary question
professional interviewing
prompt
pyramid structure
rapport
secondary question
semi-structured interview
structured interview
synchronous interviewing
transcript
trolling
unstructured interview
vernacular
voice recognition software
warm-up
word processing

Review Questions

1. Match the three categories of interviewing (unstructured interviewing; structured interviewing; semi-structured interviewing) with the following three descriptions (question focused; informant focused; content focused). Explain what is meant by each of these three descriptions.
2. List some rapport strategies you could use if you were to conduct a face-to-face interview with an older relative not well known to you. Provide a rapport strategy for the preliminary, contact, warm-up, and closing phases of such an interview.
3. List five benefits and five disadvantages or challenges of computer-mediated communication (CMC) interviewing.

Review Exercises

1. Select one of the four questions below, and spend about 15 minutes constructing an interview schedule for a hypothetical five-minute interview with one of your colleagues. Use a mix of primary question types and prompts. Think about the overall structure of your schedule, and provide a sense of order in the way the issues are covered. Try to imagine how you will cope if the interviewee is aggressive, very talkative, or non-communicative. Will your schedule still work?
   a. Most of us would agree that a greater use of public transport is an environmentally and economically sound goal. However, most of us would personally prefer to use a private car and only pay lip service to such noble goals. Why?
   b. Beach activity is decidedly spatial. Performances are expressive, and behaviour is at times territorial.
   c. The re-integration of the differently abled into "normal society" is a noble ideal. However, this integration will always be confounded by the organization of public space and the reactions of the able-bodied when the differently abled are in public space.
   d. The local environment plan (LEP) of every local council should allocate a specific area for sex industry uses.
2. Conduct two semi-structured in-depth interviews with someone of an older generation than yours. It could be an older relative (however, do not interview a sibling or parent). Limit both interviews to approximately 30 minutes. Construct an interview guide that operationalizes key concepts in the following research question: "Ours is a patriarchal society. We are often told, however, that the society of our parents and grandparents was structured by an even more restrictive and oppressive system of sexism and compulsory heterosexuality. Investigate how the opportunities, resources, and experiences differed according to gender for earlier generations. Pay particular attention to gender variations in the use of and access to space."
3. The absence of direct contact in CMC interviewing has implications for rapport. Map out a list of strategies specifically tailored to maintain rapport in an interview comprised of email exchanges. You can imagine that the topic of the interview is one of those outlined in Questions 1 and 2. In making your list, reflect on the means by which you could maintain "listening strategies" over the web, as well as "strategies of visibility."

Useful Resources

Baxter, J., and J. Eyles. 1997. "Evaluating qualitative research in social geography: Establishing 'rigour' in interview analysis." *Transactions of the Institute of British Geographers* 22 (4): 505–25.

Bennett, K. 2002. "Interviews and focus groups." In P. Shurmer-Smith, ed., *Doing Cultural Geography*. London: Sage.

Blunt, A. 2003. "Home and identity." In A. Blunt et al., eds., *Cultural Geography in Practice*. Euston: Arnold.

Cloke, P.J., I. Cook, P. Crang, M.A. Goodwin, J. Painter, and C. Philo. "Talking to people." In *Practising Human Geography*. London: Sage.

Douglas, J.D. 1985. *Creative Interviewing*. Beverley Hills, CA: Sage.

Edwards, J.A., and M.D. Lampert, eds. 1993. *Talking Data: Transcription and Coding in Discourse Research*. Hillsdale, NJ: Lawrence Erlbaum Associates.

Findlay, A.M., and F.L.N. Li. 1997. "An auto-biographical approach to understanding migration: The case of Hong Kong emigrants." *Area* 29 (1): 34–44.

Kearns, R. 1991. "Talking and listening: Avenues to geographical understanding." *New Zealand Journal of Geography* 92: 2–3.

Mann, C. 2000. *Internet Communication and Qualitative Research: A Handbook for Researching Online*. London: Sage.

Minichiello, V., et al. 1995. *In-Depth Interviewing: Principles, Techniques, Analysis*. 2nd edn. Melbourne: Longman Cheshire.

Oakley, A. 1981. *From Here to Maternity: Becoming a Mother*. Harmondsworth: Penguin.

Robertson, B.M. 2000. *Oral History Handbook*. 4th edn. Adelaide: Oral History Association of Australia SA Branch Inc.

Schoenberger, E. 1991. "The corporate interview as a research method in economic geography." *Professional Geographer* 43 (2): 180–9.

Stockdale, A. 2002. "Tools for Digital Audio Recording in Qualitative Research," Social Research Update, 38, see also http://sru.soc.surrey.ac.uk/SRU38.html (accessed 28th November, 2013).

Tremblay, M.A. 1982. "The key informant technique: A non-ethnographic application." In R.G. Burgess, ed., *Field Research: A Sourcebook and Field Manual*. London: Allen and Unwin.

Valentine, G. 1997. "Tell me about . . . : Using interviews as a research methodology". In R. Flowerdew and D. Martin, eds, *Methods in Human Geography: A Guide for Students Doing a Research Project*, 110–26. Harlow: Longman, Harlow.

Whyte, W.F. 1982. "Interviewing in field research." In R.G. Burgess, ed., *Field Research: A Sourcebook and Field Manual*. London: Allen and Unwin.

Oral History and Human Geography

Karen George and Elaine Stratford

Chapter Overview

This chapter describes how oral history can be used in geographical research. After defining the meaning and scope of oral history as a term and a method of approach to qualitative research, we outline the aspects that distinguish it from other forms of interviewing. These features include establishing rapport, dealing with sensitive issues, understanding the ethics of interviewing, knowing how to ask questions, and appreciating the importance of recorded sound quality.

Introduction

> It was a blinking dust storm. Every time you come up to Loxton there was you got off the track oh well that's where you was until you got yourself out again and it was always blowing dust. I thought "Gawd" I always used to say, "Fancy living up in this hole." [Ruth Scadden, in George 1999a, 161]

These are the words of Ruth Scadden, the wife of a soldier settler who was eye-witness to a wave of major environmental change in the horticultural town of Loxton in South Australia's Riverland after World War II. Over subsequent decades, this "dust bowl" farming area was gradually transformed into an irrigated settlement producing citrus fruits, grapes, and stone fruits for Australian and overseas markets. When interviewed in 1999, Ruth had begun to witness a further wave of change as long-term irrigation revealed its impact on the River Murray. More recently this situation has become critical, leading to further remaking of the Riverland environment. Ruth's perceptions, understanding of rural life at a particular instance in time, and her representations of that life to others are threads of everyday and ordinary existence whose cumulative weavings constitute a rich tapestry of local geographical knowledge.

> Ultimately what has always struck me as being remarkably interesting about how one is influenced is that there is a local geography involved.

> For example, when I was working in Bougainville, there was one other academic working on the island, an anthropologist, and simply because we met fairly frequently we managed to produce two or three joint articles together, and that situation seems to me to have always continued. So there are these local factors which no one can actually build into an intellectual trajectory or even practical planning, have been incredibly important at how one actually shapes what it is one does. [John Connell, geographer]

These are the recollections of a scholar in the field, recorded during an interview in 2000 for the Institute of Australian Geographers' Millennium Project on Australian Geography and Geographers. His words trace just some of the complex lines that ultimately form the web of an individual's life experiences and locales.

This chapter outlines the basic scope of oral history as a technique to gather information, insights, and knowledge from participants in social research—people such as Ruth Scadden and John Connell. It describes how oral history can be a powerful source of **situated learning** that can facilitate enhanced understandings of space, place, region, landscape, and environment—five central filaments of human geography (see Stratford 1997). Importantly, the chapter also summarizes a range of ethical, technical, and communicative guidelines for the effective conduct of oral history.

What Is Oral History and Why Use It in Geographical Research?

The practice of oral history involves a prepared interviewer recording a particular kind of interview. The interview is usually conducted in an informal question-and-answer format with a person who has first-hand knowledge of a subject of interest. Background preparation allows the interviewer to follow up responses and prompt further information. Oral history interviews may concern a very specific subject, cover an entire lifespan, or trace a complex issue that unfolds over time.

Historian Allan Nevins first used the term "oral history" in the 1940s to describe a project at Columbia University in which the memories of a group of *significant* Americans were recorded (Robertson 2006, 3). While "oral *tradition*" as a method of passing stories down through generations has existed for centuries, "oral *history*" was defined differently because its aim was to record the first-hand knowledge and experience of participants. During the 1960s and 1970s, the value of oral history in documenting and preserving the experiences of *ordinary* people was recognized. Since then, oral history has become an important tool for studying hidden histories and geographies, the place-based lives and memories of disadvantaged people, minority groups, and others whose views have been ignored or whose lives pass quietly, producing few if any written records (for example, Kwan 2008). In short, there are insights to be gained from oral histories to better understand space, place, landscape, region, and environment in ways that are sensitive to context (see, for example, Andrews et al. 2006; Stratford 1997).

American geographer Isaiah Bowman suggested that "Geography tells what is where, why and what of it" (in Rivera 1997). Acknowledging that this definition no longer fully represents the complexity of the discipline but borrowing Bowman's phraseology in any case, oral history tells what happened, how, and why, and does so from a personal perspective. For this reason, it has become a useful tool in human geography, illuminating how recollections and representations are *placed* over extended periods, and enabling researchers and participants to track and understand changes across spatial scales as well as temporal ones. In this respect, oral history has been described as the "voice of the past" (Thompson 2000) and as "a picture of the past in people's own words" (Robertson 2006, 2). As a research method, it provides a means to step back to the mix of past times and places *as they are mediated* through the words and memories of another person in the present.

Another way of thinking about oral history in relation to human geography is to acknowledge that people witness and engage in all manner of change, including environmental change—and here we mean "environment" broadly as "that which surrounds." Documents, maps, illustrations, and photographs tell part of a story—hence, too, the importance to geography of writing (Hones 2011; Hulme 2012; Saunders 2010) and visual culture (Hawkins 2011; Rogoff 2000; Stratford and Langridge 2012). However, eyewitness accounts can deepen the image and provide unique or specific detail from many different perspectives (Sackett 2005). Take, for example, the ways in which oral history helps to uncover people's experiences of the built environment and to trace the narratives of their geographical engagements within it. As oral historian for the Adelaide City Council, Karen George learned from statistical records in annual reports that during the 1950s and 1960s the population of the city declined markedly. Many buildings were declared unfit for human habitation, and residents were forced out of the city as it was being transformed into a business district. Only when she interviewed such former residents did Karen understand the significance of this phenomenon. The city was the home of individual families *and* also a community and a support network; destruction of homes compromised both. People described meeting places that no longer existed and reminisced about people they used to see every day. From such stories Karen reconstructed an image of the city before the exodus. Interviews with a health inspector who had declared many of the houses unfit then allowed her to see the event from another perspective. As often happens in the built environment, which has a life cycle of its own, events have now gone full circle. With vigorous encouragement from the council, people have moved back to the city to live, resulting in a new generation of city dwellers with different community networks and a reborn sense of place (George, 1999b; City of Adelaide Oral History Collection, Adelaide City Archives).

A second example shows that oral history can uncover the way that geographers themselves understand their professional contributions to how space, place, region, landscape, and environment are constituted (Matless, Oldfield, and Swain 2007; Powell 2008). As coordinator of the Institute of Australian Geographers Millennium Project between 1996 and 2001, Elaine Stratford encouraged members of the profession to undertake oral histories with eminent geographers, and the idea was taken up in New Zealand thereafter (Stratford 2001; see also Hay 2003; Roche 2011; Rugendyke 2005; Sheridan 2001). Participants were asked to give some thought to the contributions that geography has made or may make to Australian society. Two responses begin to hint at the wealth of disciplinary knowledge that can be gained by the extended and in-depth interview style of oral history.

> I think that geography could continue its contribution to the development of Australian society by expanding the public imagination and . . . values about [the public's] . . . relationship with the environment and so on. Because it is a very special place here, a very special environment with special needs and so on. [Joseph Powell]
>
> I think geography has made an enormous contribution. It would be difficult to look at that in its totality because different geographers are obviously looking at that component from the perspective of the work that they themselves have done. I think geography has contributed or could contribute again enormously to the understanding of the habitation of this country and what this means for the future. [Elspeth Young]

How Is Oral History Different from Interviewing?

In Chapter 8, Kevin Dunn describes in detail how to conduct research interviews. Most of his guidelines hold true for oral history. However, oral history practice does differ from interview practice in a number of ways, which we set out in the sections that follow.

Perhaps one of the clearer differences between interviewing for research and conducting oral history interviews is that many elements in the oral history process position the participant differently from the interview respondent. Through preparations and techniques that make a participant comfortable with recording an interview, oral historians aim to record as natural, rounded, and complete a story as possible.

Starting Ethically

Elaine Stratford and Matt Bradshaw note in Chapter 6 that it is a significant matter to engage with research participants and share, interpret, and represent their experiences. Therefore, these acts require an ethical approach. Oral history work involving researchers from post-secondary institutions must be assessed and approved by those institutions' ethics committees (see Robyn Dowling's comments in this regard in Chapter 2). Similarly, private practitioners must be ethical in their approach, and membership of oral history associations demands this. Thus, in describing below the various stages and techniques of oral history, we assume that ethical considerations and/or clearances are in place before research commences—a matter that has parallels with interviewing more generally and on which Kevin Dunn elaborates in Chapter 8.

Oral History Australia (http://oralhistoryaustraliasant.org.au/) was established in 1978 to promote the practice and methods of oral history, educate in the ethical use of oral history methods, encourage discussion on all aspects of oral history, and foster the preservation of oral history records. There are branches of the association in each state. As well as providing advice and training in oral history, the association has drawn up "Guidelines for ethical practice" (http://oralhistoryaustraliasant.org.au/about-us/). It strongly advises that these guidelines, which protect the rights of both participant and interviewer, are followed by anyone involved in oral or life history. A similar organization exists in Canada (http://www.canoha.ca), and others thrive internationally: in the UK (http://www.ohs.org.uk), the US (http://www.oralhistory.org/), and New Zealand (http://www.oralhistory.org.nz). Many oral historians belong to the International Oral History Association (http://www.iohanet.org/).

Getting to Know Your Participant—The Preliminary Meeting

Establishing rapport with a participant is integral to success. Oral historians use a particular approach to help establish rapport, and the **preliminary meeting** is a key part of it. After making contact by letter, telephone, Internet, social media, email, or other appropriate means, the interviewer arranges a meeting, usually at the home of the participant or another place of his or her choice. This orientation session is not recorded. Rather, the time is used to establish a relationship with the participant, gather background information, and assess the interview environment. Some interviewers use an information sheet to make brief biographical notes about participants, such as their birthplace, school, employment, and personal background. They might also discuss other issues or information that might be pertinent to the interview. Although much of this material will be covered again in more detail in the recorded interview, preliminary notes establish context and ensure accuracy—for instance, in spelling and pronunciation. These details are important because oral history recordings should be preserved as key "primary documents" in and of their own right and not simply as data repositories for further analysis.

During the preliminary meeting the interviewer can ask whether the participant has other materials that may enhance research. News clippings, letters, diaries, or photographs may suggest new questions not previously considered. If novel topics emerge in a preliminary session, the interviewer then has time to conduct further research about them before the interview. The preliminary meeting also offers an opportunity for the researcher/interviewer to explain the purpose of the interview and to explain how it may be used in research. Any questions the participant might have about the interview process can be addressed, and the conditions of use form, discussed in detail later in this chapter, can be provided.

Finally, memory is individual and some people remember and can talk about experiences more readily than others. You may find at this first meeting that someone you thought would provide great material remembers very little. If this occurs, you may be able to defer or cancel the interview, telling a participant that he or she has already provided what was needed.

Sensitive Issues

If a project deals with sensitive issues (including personal, political, or professional details that may require the researcher to guarantee the confidentiality of parts of the oral history transcript), a preliminary meeting will allow you to discover how your participant feels about answering particular questions or exploring aspects or phases of their life or the subject under investigation. Even when interviewing in a subject area that seems straightforward and non-contentious, it is sound practice to assure participants of their right to refuse to answer a question or to withdraw from the research altogether without prejudice. Discussing these matters in advance avoids discomfort during an interview and allows participants to think in advance about what they might wish to say about difficult subjects.

Sensitive subjects can arise unexpectedly during interviews. If this happens, exercise care and consideration. Ask participants if they wish to continue or would rather stop. If they wish to continue, of course let them speak. If the subject matter becomes very personal and you think that it should be excluded, let participants know this. University ethics committees (sometimes known as research ethics boards or institutional review boards) may make the useful suggestion that if participants show distress, you might ask whether they would like you to call a friend or family member on their behalf or refer them to a helpful counsellor.

After the interview is over, allow time for winding down. Winding down can be as simple as accepting another cup of tea or listening to other stories not related to the research subject. As an interviewer you should also be aware of the cumulative impact of listening to sensitive or traumatic memories and consider **debriefing** or seeking counselling after the interview if you feel affected.

Multiple Interviews

Unless there are unavoidable resource constraints, oral history recordings can be completed over several sessions. A first session of one to two hours could be followed up with subsequent sessions over a number of weeks.

Multiple interviews can be very valuable. There is time between appointments for you to listen to the initial recording and note responses to enlarge upon later. Participants may also reflect on their answers: remembering triggers further memories to be shared at ensuing meetings. By a third or even fourth meeting, trust between interviewer and participant has usually developed, which can result in very high quality interviews.

The Question of Questions

Open questions are integral to effective oral history. These questions begin with words such as "who," "what," "where," "when," "why," and "how." They reveal who was involved in an event, what happened, where and when it happened, how it felt, and why that was so. They yield the details that make oral history such an effective source of nuanced recollections.

In Chapter 8, Kevin Dunn refers to secondary questions or "prompts," which oral historians often label **follow-up questions**. They often comprise the body of an interview. Most are prompted by a participant's response to an initial question. If a participant says that his or her first day on the job was "frightening," the logical follow-up question is "why?" or "in what way?" If he or she responds by saying that "fellow workers were aggressive," a logical follow-up question is "can you give me an example?" Through follow-up questions, great depth may be added to the detail of information being sought.

Interview Structure

Oral histories can appear to be "unstructured," but such is not really the case. Interview guides or *aides-mémoire* are often used. Interviews in the tradition of oral histories can often be divided into a three-part format, comprising orientation, common, and specific questions. As Robertson (2006, 22) points out, this

> three part structure provides an excellent framework for interviews. It helps you to avoid aimless or superficial interviews and it can lead to

recordings that are easier to use for research, publication or broadcast because of their well-defined structure and focus.

Orientation questions establish the participant's background. **Common questions** are those asked of each participant in a project. They build up varying views and information about certain themes. **Specific questions** relate to individual experiences and are developed through follow-up work. The flow of an interview is determined by participant responses, so follow-up questions are always different across interviews.

Questioning the Source

As the previous sections suggest, oral history is active and shared; you can question the source. For example, as part of Karen George's South Australian research on War Service Land Settlement after 1945, she consulted written records of applicant interviews with the Land Board. Although these documents included board members' notes about applicant responses, she could not ask them, "What do you mean by that?" When conducting oral histories with people who had appeared before the board, she could. Karen then created as complete a picture of those board interviews as memory allowed. She was able to ask what happened that particular day, what the interviewers were like, and how respondents felt about the questions when they were asked and after the experience (George 1999a).

Sound Quality, Interview Sites, and Other Technical Issues

Oral history is *always* recorded via audio, and sometimes audio-visual, technologies—these days that is predominantly using digital recorders. "Solid state" recorders that use compact flash memory cards, or that record onto a hard disc, are highly recommended (Robertson 2006, 43–4), because a significant aim of recording interviews is to create enduring **sound documents**. Oral history preserves the sound of participants' voices and the content of their interviews. High-quality sound allows the emotion, inflections, and tone of each voice to be heard. It is also important because material may be used for broadcast. Background noises, interruptions, substandard recording equipment, and a too-talkative interviewer diminish sound quality.

Because you are recording a unique sound document, it is important to record interviews in a location that is as quiet as possible and to use the best equipment available. Oral history associations in all countries are the best places to approach with inquiries about hiring equipment. High-quality recorders are often available on loan from libraries and, in Australia, from branches of Oral History Australia.

Some of these organizations run workshops on conducting oral history and the correct use of recording equipment.

Robertson's *Oral History Handbook* remains an excellent source of information and advice on all recording equipment, from budget-priced to super high-quality (Robertson 2006, chapter 4). Recording technology is constantly undergoing change, and new digital recording devices are regularly becoming available, so it is advisable to seek professional advice from organizations such as your national oral history association before investing in new equipment. If you use digital recording technology, you must remember to download sound files after each interview and ensure that backup copies of work are saved in secure settings, especially if you have promised to do so as part of the ethical commitments of the project. It is often very valuable to work in conjunction with a major university library or state/provincial library. Some libraries offer access to high-end recording equipment if original recordings are deposited and preserved in their repositories.

No matter what type of recorder you are using, always try to find a quiet place free from interruptions to conduct an interview. Recording high-quality sound means that extracts from your recordings can easily be broadcast or used in museum interpretations, on websites or CDs, or for audio walking tours. This quality also makes any transcription and analysis much easier. Note that workplaces are among the worst locations, whereas private homes usually offer a dining or living room, both of which can be quieter. Avoid kitchens whenever possible. Refrigerators are renowned for droning away in the background or cutting in and out with a thump. Ticking or chiming clocks should be stopped or removed because their regular pulse in the background is distracting. If you cannot avoid background noises completely, such as traffic sounds on a busy road, direct the interviewee's microphone away from the noise. Close doors and windows to minimize background noises. If you talk about these things at your preliminary meeting, participants will usually help out and not think you are rude when you ask them to stop great-grandmother's cuckoo clock or turn off the fridge!

Interpreting non-verbal responses and gestures is common to oral histories and other interviews. The comment "It was about this big" needs to be translated by the interviewer into "about a metre high." It is best to convey these details on the recording, since you might forget them later. Recall, too, that an interviewer's verbal responses can be detrimental to a sound recording, particularly one for broadcast purposes. A litany of "yes, yes," "mmm," "oh really," or "wow" remarks that commonly occur in a conversation interrupts the recorded flow of a story. Respond with a nod or a smile instead. Let participants know you will remain quiet *and* involved. Listen to your own recordings to gauge how silent you actually are. A pause, a moment of quiet may be the instant before the best story. Oral history also can be a demanding process for participants, and they may need time to stop and reflect. Be sure to allow them this time. Silence on tape can also be very emotive. Long seconds of silence recorded in the midst of a painful story reveal a struggle with strong emotions better than words ever could.

Why and How to Make Oral History Accessible

Interviews conducted for research often have very limited circulation. However, in oral history interviewers are encouraged to deposit their recordings in libraries or archives. In conjunction with published work on oral histories, scholarly journals such as *Geographical Research* are now also making such recordings available through their websites. These steps ensure the preservation of master or original recordings, and if the participant has agreed, allows recordings to be made available to other researchers. It is always worthwhile to search oral history collections *before* you begin an interview to make sure that your informant has not already been interviewed and to check whether there are other recordings that might provide data for your project. Even interviews concerning completely different topics may contain useful information. For example, since the majority of the men and women Karen George (1999a) interviewed about soldier settlement in South Australia grew up during the Depression of the 1930s, a researcher interested in that period of history could glean a lot of information from their answers to questions about their backgrounds and childhoods.

Another advantage of depositing recordings in a library is that some larger repositories offer limited assistance with the transcription of interviews. Whether you produce full transcripts, **timed tape logs** (which note subjects discussed at different time points in the recording), or broad interview summaries is normally dependent on the project's aim and on funding. Professional transcription is expensive but worthwhile if material from interviews is to be reproduced in a publication. It is worth noting here that professional transcribers are trained to reflect pauses and the unique cadences of the spoken word through punctuation and layout. Transcripts are rarely completely verbatim because it is common for participants to feel concerned about their poor grammar, repetitions, and crutch phrases (such as "you know") when they see them in print.

Making interviews, both recordings and transcripts, accessible to others should only be done with the signed agreement of your participants. It is essential that you draft a **conditions of use form** outlining what will happen to the material they share with you—what their rights are, who will own copyright, where the recorded interview will be stored and for how long, and what it will be used for. Although participants should be encouraged to share their stories with a wide audience, they should also be allowed to add conditions to this agreement, such as to restrict portions of the recording—or indeed the entire interview—until after their death if they wish.

Uses of Oral History—Spreading the Word

As well as depositing oral history recordings into libraries or archives and using information from interviews in published and unpublished writing, there are other ways to share the results of your work. If you have made high-quality recordings, the possibilities are extensive. While quotations from interviews can be presented in displays and on the Internet in written form, it can be even more effective to use sound excerpts. "Sound bytes" or even complete interviews accessed through a webpage allow users to hear as well as read about your interviews. For example, the South Australian branch of Oral History Australia has a web gateway to a series of interviews recorded by its members (http://oralhistoryaustraliasant.org.au/category/interviews/). **Listening posts**—that is, posts or other structures from which audio can be heard—may be used as part of exhibitions, or portions of interviews can be used in audio commentary.

Well-recorded interviews offer much scope for presentation to groups in radio, film, and video. For example, the voices of long-term employees of Balfours, one of the last city-based factories in Adelaide, the capital city of South Australia, were used effectively in conjunction with video recording of the working of the bakery (Starkey and George 2003). Using images and oral history excerpts, the project recorded and preserved images and descriptions of the original factory and of the processes that have not been used since the factory relocated to modern suburban premises. Certainly, narration may be made much more engaging when excerpts from oral histories are played in conjunction with visual materials. Similarly, when Karen George conducted an interview with a long-time printer, Rob Wilson, in 2008, she also recorded the sounds of early linotype and printing machines. As Rob died in 2012, it is unlikely that these machines will operate again. The sounds of the machines have therefore been included with oral history, photographs, and video footage in an interactive touchscreen at the Pinnaroo Letterpress Printing Museum in South Australia's Murray Mallee. Excerpts from oral history can also be used effectively as part of an audio tours or walks (see, for example, Butler 2007).

Last Words

Geography's central concern is to understand people in place, spatial relations, landscapes, regions, and environments. It also aims to contribute to research-based outcomes that advance well-being. For many geographers, these composite tasks involve philosophical and political investments in learning about—rather than appropriating—marginal, informal, and otherwise undocumented perspectives as well as in comprehending those that are central, formal, and documented. Like those of the interview or focus group, oral history techniques allow both researchers and participants to explore the nuances of social and spatial interactions, events, and processes in ways that can make these goals possible. However, in the pursuit of these goals, never forget that your "source" is another human being, a person sharing with you a distinctive and valuable gift—their memory.

Key Terms

common questions
conditions of use form
debrief
follow-up question
listening posts
multiple interviews
open questions
orientation questions
preliminary meeting
situated learning
sound document
specific questions
timed tape logs

Review Questions

1. What are some of the ethical issues associated with the use of oral history?
2. What are some of the relationships between oral history and human geography?
3. What are some of the ways in which oral history might help you explore an area of human geography in which you are interested?
4. Develop an idea for an oral history project, prepare a list of potential participants, and search existing oral history collections for previously recorded interviews on related topics.
5. Record an interview, paying particular attention to sound quality. Develop a multimedia spoken or web-based presentation using excerpts from the interview combined with other media—such as photographs and documents. You could present this as a talk or on the web.
6. Develop a conditions of use form to be used in conjunction with one of the two projects above.

Review Exercises

1. You have been commissioned to record interviews for the fiftieth anniversary of a geography department. Two of your participants are prominent graduates of the department and were advisors to senior political figures involved in some major developments and/or environmental campaigns.
   a. What ethical constraints and/or other issues might you need to consider in planning these interviews?

   b. Write a list of questions for one of the interviews. Ask yourself about, and test with fellow students, the ethical quality, sense, and merit of those questions. How might you be assured that those questions will assist you to derive rich responses that help you understand significant parts of a person's experiences and life?

2. You have completed a series of interviews with the senior policy advisors/university geography department graduates mentioned in exercise 1, and you have been asked to make the results of your project public.
   a. In what different ways might you use oral history in a public forum?
   b. Describe how you could you use oral history as part of (1) a display to be located in the main foyer of the university department's building for 12 months, and (2) at the site where the fiftieth anniversary celebrations are to be launched.
   c. How would you evaluate the effect and effectiveness of the project?

Useful Resources

Baylor University, Institute for Oral History. 2008. "Introduction to oral history." http://www.baylor.edu/oralhistory/index.php?id=23566.

Butler, T. 2007. "Memoryscape: How audio walks can deepen our sense of place by integrating art, oral history and cultural geography." *Geography Compass* 1: 350–72.

George, K. 1999. *A Place of Their Own: The Men and Women of War Service Land Settlement at Loxton after the Second World War*. Adelaide: Wakefield Press.

Jenkins, A., and A. Ward. 2001. "Moving with the times: An oral history of a geography department." *Journal of Geography in Higher Education* 25 (2): 191–208.

Kwan, M.-P. 2008. "From oral histories to visual narratives: re-presenting the post-September 11 experiences of the Muslim women in the USA." *Social & Cultural Geography* 9 (6): 653–669.

Perks, R., and A. Thomson, eds. 2006. *The Oral History Reader*. 2nd edn. London: Routledge.

Powell, R.C. 2008. "Becoming a geographical scientist: Oral histories of Arctic fieldwork." *Transactions of the Institute of British Geographers* 33 (4): 548–65.

Robertson, B.M. 2006. *Oral History Handbook*. 5th edn. Adelaide: Oral History Association of Australia SA Branch Inc.

Roche, M. 2011. "New Zealand geography, biography and autobiography." *New Zealand Geographer* 67 (2): 73–78.

Sheridan, G. 2001. "Dennis Norman Jeans: Historical geographer and landscape interpreter extraordinaire." *Australian Geographical Studies* 39 (1): 96–106.

Sommer, B.W., and M.K. Quinlan. 2009. "Capturing the living past: An oral history primer." Nebraska State Historical Society. http://www.nebraskahistory.org/lib-arch/research/audiovis/oral_history/.

Stratford, E. 1997. "Memory work, geography and environmental studies: Some suggestions for teaching and research." *Australian Geographical Studies* 35 (2): 206–19

Stratford, E. 2001. "The Millennium Project on Australian Geography and Geographers: An introduction." *Australian Geographical Studies* 39 (1): 91–5.

Thompson, P. 2000. *The Voice of the Past: Oral History*. 3rd edn. New York: Oxford University Press.

University of California, Berkeley. The Oral History Centre. http://bancroft.berkeley.edu/researchprograms/roho.html.

Focusing on the Focus Group

Jenny Cameron

Chapter Overview

An investigation of options for the post-mining uses of industrial peatlands in Ireland (Collier and Scott 2010), an exploration of emotions experienced by volunteers participating in conservation programs in South Africa (Cousins, Evans, and Sadler 2009), and a study of the barriers to Latinos using parks in Los Angeles (Byrne 2012)—all are examples of research projects that employ focus groups to disentangle the complex web of relations and processes, meaning, and representation that comprise the social world. With the shift to more nuanced explorations of people–place relationships in geography, the focus group method has been increasingly recognized as a valuable research tool.

Focus groups can be exhilarating and exciting, with people responding to the ideas and viewpoints expressed by others and introducing you, the researcher, and other group members to new ways of thinking about an issue or topic. This chapter discusses the diverse research potential of focus groups in geography, outlines key issues to consider when planning and conducting successful focus groups, and offers strategies for analyzing and presenting results.

What Are Focus Groups?

The focus group method involves a small group of people discussing a topic or issue defined by a researcher. Briefly, a group of between 6 and 10 people sit facing each other around a table; the researcher introduces the topic for discussion and then invites and moderates discussion from group members. A session usually lasts between one and two hours (you might see parallels here with university tutorial group meetings).

The focus group is one of the group techniques used in research. As shown in Figure 10.1, these techniques range from group interviews in which each participant is asked the same question in turn and there is little or no interaction between

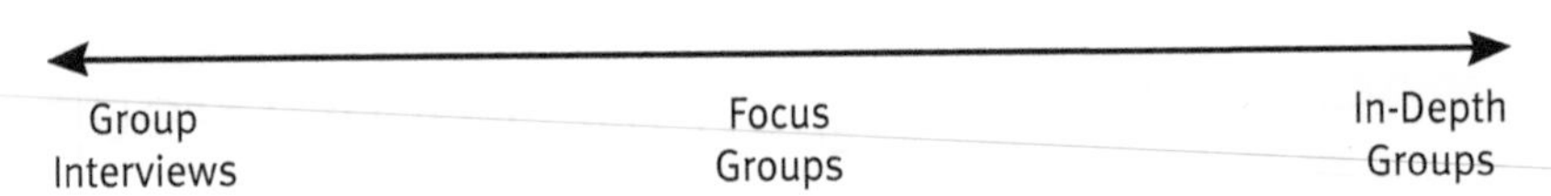

Figure 10.1 Relationship between Focus Groups, Group Interviews, and In-Depth Groups

participants (Barbour 2007) through to in-depth groups in which the emphasis is on the interaction between participants, with participants sometimes even deciding on discussion topics (Kneale 2001). In-depth groups also meet regularly for extended periods of time (sometimes months).

As with in-depth groups, interaction between participants is a key characteristic of focus groups. The group setting is generally characterized by dynamism and energy as people respond to the contributions of others (see Box 10.1). One comment can trigger a chain of responses. This type of interaction has been described as the **synergistic effect** of focus groups, and some propose that it results in far more information being generated than in other research methods (Berg 1989; Stewart, Shamdasani, and Rook 2007). In the focus group excerpt in Box 10.1, for example, three social enterprise practitioners discuss their motivations. It starts with Practitioner A discussing how they are motivated by the people with whom their social enterprise works. Practitioner B then discusses the challenge of balancing the social goals with the business operations. Practitioner C brings the discussion back to the personal motivations, and makes the point that they are "in it for me." There's then discussion about what this means with Practitioners A and C refining what being "in it for me" means in the context of a social enterprise.

The interactive aspect of focus groups also provides an opportunity for people to explore different points of view and to formulate and reconsider their own ideas and understandings. Kitzinger (1994, 113) describes this form of interaction in the following terms: "[p]articipants do not just agree with each other. They also misunderstand one another, question one another, try to persuade each other of the justice of their own point of view and sometimes they vehemently disagree." For researchers who are interested in the socially constructed nature of knowledge, this aspect of focus groups makes them an ideal research method; the multiple meanings that people attribute to places, relationships, processes, and events are expressed and negotiated, thereby providing important insights into the practice of knowledge production.

As in group interviews, the researcher plays a pivotal role. He or she promotes group interaction and focuses the discussion on the topic or issue. The researcher draws out the range of views and understandings held within the group and manages—sometimes even encourages—disagreement among participants (Myers 1998).

Initially, focus groups can be extremely challenging for researchers who are new to the process. They are, however, well worth it. In focus groups, the diversity

The Synergistic Effect of Focus Groups

Practitioner A: When I first started [the social enterprise] we didn't have anyone who could look after the business side and because the other guy was even less good at it than me, I ended up falling into that role. I learnt. To me it's not about money, it's about "Why are you doing it?" I'm not doing it for me. I'm doing it for them. To me that's the fundamental distinction between what I do and what a businessman does. The businessman is doing it ultimately for themselves. I'm doing it for the people that I'm working with. That's my motivation.

Facilitator: What you're saying is that you're clear about what your motivation is. You are adapting yourself to the context in order to make sure you can act on that motivation.

Practitioner A: The money doesn't scare me.

Practitioner B: Social enterprise should have a social mission as its ultimate purpose for existing. Sometimes the only way you can make that mission happen is by having the business feed money into it. But the more money you make the less social mission. I can remember someone saying we could make more money but we would probably have to replace some of the students with more full-time chefs. We could make more, but then we can put less students through. Then our social mission drops and our profits go up.

Practitioner A: Why are you doing it? Why are you doing it?

Practitioner B: So it's that very fine balance.

Practitioner C: Yeah, but just going back to what you said. I'm actually in it for me too. I have no bones about that.

Facilitator: Because of the challenge?

Practitioner C: Absolutely. I don't want to sound like I think I'm good. You've really have got to have your heart in it and you really have to love it. You have to be really selfish like that. I'm in it for me.

Practitioner A: Yeah, yeah, in it for you—in terms of your own motivations. You're not in it for you to see how many bucks you can make for yourself so you can retire at whatever age or buy the big house or whatever.

Practitioner C: That's right.

Source: Audio excerpt from focus group conducted by Jenny Cameron and Sherelle Hendriks, Newcastle, Australia, 23 July 2012 (see Cameron and Hendriks 2013).

of processes and practices that make up the social world and the richness of the relationships between people and places can be addressed and explored explicitly. Furthermore, group members almost invariably enjoy interacting with each other, offering their points of view, and learning from each other. Researchers also find the process refreshing (for example, see the discussion by two skeptical anthropologists in Agar and MacDonald 1995).

Using Focus Groups in Geography

Focus groups were originally used by sociologists in the US during World War II to examine the impact of wartime propaganda and the effectiveness of military training materials (Merton 1987; Morgan 1997). Although this work resulted in several sociological publications on the technique, focus groups were neglected by social scientists in the post–World War II period in favour of one-to-one interviews and participant observation (Johnson 1996). It was in the field of market research that the focus group method found a home. Since the 1980s, there has been renewed interest in the technique among social scientists, and this has led to considerable diversity in focus group research (Lunt and Livingstone 1996; Morgan 1997). Focus groups can be a highly efficient data-gathering tool, but they are also appropriate in "more critical, politicized, and more theoretically driven research contexts" (Lunt and Livingstone 1996, 80), exploring, for instance, the discourses that shape practices of everyday life, the ways in which meanings are reworked and subverted, and the creation of new knowledges out of seemingly familiar understandings. The range of uses and purposes of focus groups is evident in geographic research employing the technique.

Geographers have used focus groups to collect information. Zeigler, Brunn, and Johnson (1996) used them to find out about people's responses to emergency procedures during a major hurricane. They claim that the focus group technique provided insights that might not have been revealed through methods like questionnaires or individual interviews. As a consequence, they were able to recommend important refinements to disaster plans. Barr et al. (2010) used focus groups to explore what people who are environmentally conscious at home thought about air travel for holiday purposes. They found while people were aware of the contribution of air travel to greenhouse gas emissions they were not prepared to reduce their holiday air travel, though some would consider paying for carbon offsets. The research highlighted how difficult it is to change air travel behaviour particularly as low-cost airfares have become so readily available. As a data-gathering technique, focus groups are ideal for investigating not just *what* people think and do but *why* people think and behave as they do. Barbour (2007, 24) notes that focus groups are particularly useful for understanding people's beliefs and practices. In focus groups, "participants are given scope to justify and expand on their views in a non-judgemental environment," giving researchers a chance to learn how people rationalize what might seem illogical to researchers. For example, in the research by Barr et al. (2010), people justified their use of air travel for holidays for a range of reasons that included "trading-off" sustainable practices when at home for less sustainable practices when on holidays, claiming that people living in other parts of the world were more of a problem than they were and even expressing skepticism about the contribution of air travel to climate change.

One concern of some researchers involved in collecting information is that because of the relatively small number of participants, findings are not applicable to a wider population (for discussion of this issue, see Chapters 6 and 7). Combining focus groups with quantitative techniques is an extremely useful way of dealing with this issue. A survey questionnaire, for instance, might be administered to a random sample of the population from which the focus group was drawn to test the generalizability of the insights gained from the group discussions.

Quantitative methods can be combined with focus groups in other ways. Preliminary surveys are sometimes helpful in identifying focus group members or the topics for detailed focus group discussion (e.g. Barr et al. 2010). Focus groups have been used to generate questions and theories to be tested in surveys (Goss and Leinbach 1996), to refine the design of survey questionnaires (Jackson and Holbrook 1995), and to follow up the interpretation of survey findings (Goss and Leinbach 1996), particularly when there seem to be contradictory results (Morgan 1996).

Focus groups can also be combined with other qualitative methods to collect information. For example, Cousins et al. (2009) sought to investigate the experiences of conservation volunteers from the UK in South Africa. Participant observation was supplemented with focus groups held in the third week of volunteering, giving participants time to be exposed to the volunteering experience and time to feel comfortable with each other so they might share their views. It is, however, entirely appropriate to use focus groups as the sole research method rather than in combination with other research techniques.

For geographers interested in the process of producing knowledge, the focus group is an excellent research tool. Robyn Longhurst (1996) employed focus groups as a forum in which pregnant women could converse and interact. The narratives, accounts, **anecdotes**, and explanations offered by these women provided Longhurst (1996) with insights into a new discursive landscape of pregnancy. Similarly, Gibson, Cameron, and Veno (1999) have been concerned not just with reproducing knowledge of the problems and difficulties confronting rural and non-metropolitan communities in Australia but with reshaping understandings so that new responses might be engendered. The seemingly isolated instances of innovation that several focus group members could readily recall provoked other participants to think of additional examples. The beginnings of a body of knowledge on regional initiative began to emerge through these discussions.

In a report of their Indonesian research on individual and household strategies related to the allocation of land, labour, and capital, Goss and Leinbach (1996) also highlight the collective rather than individual nature of knowledge production. By interacting with other focus group members, Javanese villagers developed new understandings of their social conditions. Indeed, Goss and Leinbach argue that "the main advantage of focus group discussions is that both the researcher and the research subjects may simultaneously obtain insights and understanding of particular social situations *during* the process of research" (Goss and Leinbach 1996, 116–17, emphasis in original). For geographers who are committed to the idea that research can be used to effect social change and empower "the researched," the potential for focus groups to create and transform knowledges and understandings of researchers and participants is compelling (see also Johnson 1996).

The focus group method has an important contribution to make to geographic research. It is a highly effective vehicle for exploring the nuances and complexities associated with people–place relationships. The material generated in focus groups can add important insights to work that seeks to describe, document, and explain the social world. But focus groups serve not just to "mine," "uncover," and "extract" existing knowledges (Gibson-Graham 1994); they can also provide insights into how people construct their world views through interactions with others (Pratt 2002), and they can provide opportunities for researchers and participants to jointly develop new knowledges and understandings.

Planning and Conducting Focus Groups

Given that the focus group method can be used for a range of research purposes in geography, there will be some variation in how groups are organized and conducted. There are, however, basic principles and methodological issues that need to be considered. To be sure, the success of a focus group depends largely on the care taken in the initial planning stage.

Selecting Participants

Selecting participants is critically important. Generally, participants are chosen on the basis of their experience related to the research topic. J. Burgess's (1996) study is a good example of this **purposive sampling** technique (see Chapter 6 for a discussion of participant selection). In work intended to ascertain the perceptions of crime and risk in woodlands among different social and cultural groups, she selected women and men of varying age, stage in the life cycle, and ethnicity to participate in focus groups. In their study of home-based and travel-based environmental behaviour, Barr et al. (2010) recruited participants who displayed what could be called strong to medium to weak environmental behaviours at home in order to investigate each group's attitude to air travel.

Composition of Focus Groups

Should people with similar characteristics participate in the same group, or should groups comprise members with different characteristics? This decision will be largely determined by the purpose of your research project.

Holbrook and Jackson (1996), for example, sought to address issues of identity, community, and locality by grouping together people with characteristics like age and ethnicity in common. In their research on environmental responsibility, Bedford and Burgess (2002) had people with similar experiences in each focus group but a range of different focus groups—suppliers, retailers, regulators, consumers, and advocates. They describe this as "ensur[ing] homogeneity within the group and heterogeneity between them" (2002, 124). Other researchers have noted that discussion of sensitive or controversial topics can be enhanced when groups comprise participants who share key characteristics (Hoppe et al. 1995; O'Brien 1993). In some projects, it may be more appropriate to have groups made up of different types of people. Goss and Leinbach (1996) were interested in the social relations involved in family decision-making and deliberately chose to conduct mixed-gender groups. The different knowledges, experiences, and perspectives expressed by women and men in that study became an important point of discussion, whereas in the context of forest fringe communities in Ghana, Teye (2012) conducted separate focus groups with women and men, as the patriarchal nature of these traditional communities meant that women would be unlikely to discuss and debate issues with men.

Another consideration is whether people already known to each other should participate in the same group. Generally, it is best not to have people who are acquainted in the same group, but in some research, particularly place-based research, it may be unavoidable. Researchers need to be aware of the limitations that this can produce. One is peer pressure, with participants not wanting to appear "out of step" with their acquaintances. Similarly, some participants may **under-disclose** or selectively disclose details of their lives, as Pratt (2002) found in her research.

A different problem arises when participants **over-disclose** information about themselves. One strategy for dealing with this is to outline fictional examples and ask group members to speculate on them. In groups they ran in Indonesia, Goss and Leinbach (1996) provided details of three fictional families and asked group members to discuss which of the families would be most likely to accumulate capital. Participants did not have to disclose information about their own situations but could still discuss family strategies. Participants can also be asked to treat discussions as confidential. Since this confidentiality cannot be guaranteed, it is appropriate to remind people to disclose only the things they would feel comfortable about being repeated outside the group.

Of course, you should always weigh whether a topic is too controversial or sensitive for discussion in a focus group and would be better handled through another technique such as individual in-depth interviews. (Most universities now have ethics committees, institutional review boards, or research ethics boards to help ensure that researchers carefully manage material from focus groups and other qualitative research methods. For more on this, see Chapter 2.)

Size and Number of Groups

The size of each group and the number of groups are other factors to be considered. Too few participants per group—fewer than four—limits the discussion, while too many—more than 10—restricts the time available for individual participants to contribute.

In terms of the number of groups, one rule of thumb is to hold three to five groups, but this guideline will be mediated by factors such as the purpose and scale of the research and the heterogeneity of the participants. A diverse range of participants is likely to require a larger number of groups. For instance, J. Burgess (1996) was interested in perceptions of crime and risk in woodlands among different social and cultural groups so she conducted thirteen focus groups with people of varying age, stage in the life cycle, and ethnicity, whereas in Byrne's study of why Latinos do not use parks in Los Angeles only two focus groups were conducted, each comprising Latinos who were non–park users.

The structure of the focus group is also a factor to consider. When fewer **standardized questions** are used and when there is a relatively low level of researcher intervention and moderation, more groups are needed, since both these factors tend to produce greater variability among groups (Morgan 1997). Time, cost, and availability of participants may also limit the number of groups that can be held. The overall research plan—especially whether focus groups are the sole research tool or one of a number of tools—will also affect decisions about the number of groups convened. Finally, another guide to the number of focus groups is to use the concept of **saturation** (Krueger 1998, 72). This means that you continue to conduct focus groups until you can gather no new information or insights.

Recruiting Participants

The strategy used to recruit participants will depend on the type of participants you require for your study. Gibson, Cameron, and Veno (1999) recruited business and community leaders in two regions by initially contacting local people who were featured in local newspapers and targeting managers of key government and non-government agencies. These initial contacts were asked to suggest other people who would make interesting contributions to the study, including those who would have a different point of view from them (this snowball **recruitment** technique is also discussed in Chapter 6). A preliminary phone conversation quickly established whether nominees were interested and able to attend. This was followed by a letter with more information about the project. A few days before the focus groups were held, participants were telephoned again to reconfirm their participation. Twelve people were invited to attend each group to allow for cancellations due to illness, last-minute change of plans, and so on (several people from each group did drop out).

After an unsuccessful attempt to recruit participants through strategies such as web-based bulletin board postings, Byrne (2012) worked directly with two community-based organizations that recruited participants on his behalf. When Eden, Bear, and Walker (2008) wanted a more diverse range of food consumers, they made contact with a whole-food retailer and a vegan group, among others. Like these researchers, it is important to think strategically about how best to locate potential participants.

Questions and Topics

Before conducting focus groups, give thought to the questions or topics for discussion. This involves not only the general content of questions or selection of issues for discussion but also the wording of questions and issues, identification of key phrases that might be useful, the sequencing and grouping of questions (see Chapter 8 for additional material on question order), strategies for introducing issues, and the links that it might be important to make between different questions or issues. Also give thought to using others sorts of stimuli to generate discussion. Barbour (2007, 84–8), for example, discusses the use of cartoons, snippets from television shows, photographs, newspaper clippings, advertising materials, and vignettes. However, it is important to select such material carefully and, where possible, to check that it will not offend participants.

One way to proceed is to devise a list of questions. Hares, Dickinson, and Wilkes (2010) had a list of six open-ended questions that acted as **probes** and that moved from the general to the specific. Holbrook and Jackson (1996) identified six themes related to the experience of shopping and then used these themes to develop questions that were raised spontaneously and that fitted with the flow of the discussion. J. Burgess (1996) first took the participants in each focus group on a walk through a woodland and then introduced for discussion five primary themes related to elements of the walk. As part of the recruitment process, Cameron and

Hendriks (2014) asked each participant to prepare a brief two- to three-minute statement about her or his background and current role working in a social enterprise. The themes that emerged from these statements provided the initial basis for discussion.

Take care when letting people know in advance what the questions or topics will be. If attendance or discussion is likely to be enhanced by providing this information, then it may be appropriate. Sometimes, however, it might be necessary for you to paint a very broad picture. For example, even though Hares, Dickinson, and Wilkes's (2010) research was on people's awareness of the impacts of tourism on climate change, they told potential participants that the research was about climate change and everyday lives. They deliberatively avoided mentioning tourism in the recruitment process as they did not want to create a connection in participant's minds between tourism and climate change. (See Chapter 2 for a consideration of the ethical dimensions of this sort of approach.) When recruiting and explaining the research, it is also important to use language that participants will understand.

Generally, questions or topics should be suited for a discussion of between one and two hours. With very talkative groups, it may be necessary to intervene and move the discussion on to new topics. Alternatively, if you have planned a hierarchy of questions or themes, then it may be appropriate to allow the group to focus on the more important areas of discussion. With less talkative groups, you may need to introduce additional or rephrased questions and prompts to help draw information out and open up the discussion. You should think about such questions and prompts during the preparation stage.

Another issue to consider is whether questions and topics should be standardized across all focus groups involved in your study or whether new insights from one group should be introduced into the discussions of the next. In many qualitative research situations, it may be appropriate to incorporate material from earlier groups, but this issue should be determined with reference to the project aim. Information that might identify people who attended earlier groups should not be revealed to subsequent groups.

As well as conducting meetings with several groups, you may find it useful and appropriate to have each group meet more than once. Multiple focus groups may be a particularly useful strategy when participants are being asked to explore new and unfamiliar topics or to think about an apparently familiar topic in a new way. Multiple groups may also be appropriate as a way of developing trust between the researcher and research participants. For instance, when researching the experiences of single mothers, I met several times with one group of teenagers who were very wary of talking with people associated with educational, medical, and media institutions (Cameron 1992).

Conducting Focus Groups

Generally, focus groups are best held in an informal setting that is easily accessible to all participants. The rooms of local community centres, libraries, churches, schools, and so on are usually ideal. The setting should also be relatively neutral: for example, it would not be advisable to convene a focus group about the quality of service provided by an agency in that agency's offices. Food and drink can be offered to participants when they arrive to help them relax, but alcohol should never be provided. It is also helpful to give out nametags as participants arrive.

Much has been written about the ideal focus group **facilitator** or **moderator** (for example, Morgan 1997; Stewart, Shamdasani, and Rook 2007). In academic research, the researcher, who is familiar with the research, is often best positioned to fill this role. To gain some confidence and familiarity with the process, a less experienced researcher might initially take the role of note-taker while a more experienced researcher facilitates the first groups. Focus groups can also be run with more than one facilitator, and a less experienced researcher might invite a more experienced researcher to take the lead. But beginning researchers should not be afraid to try facilitation; as Bedford and Burgess (2002, 129) note, "the desired qualities are those possessed by the average undergraduate—the ability to listen, the ability to think on your feet, and a knowledge of and interest in the subject the group is discussing."

When a note-taker is present, that person should sit discreetly to one side. The notes, particularly a list of who speaks in what order and a brief description of what each talks about, can be helpful when transcribing audio recordings of the discussion. A seating plan is also essential. Because the facilitator has to attend to what participants are saying and monitor the mood of the group, he or she should not take extensive notes, although the facilitator may want to jot down a point or two to return to in discussion.

It is highly advisable to audio-record focus groups. The group will usually cover so much material that it will be impossible to recall everything that was discussed. In addition, because presentation of focus group results generally includes direct quotes to illustrate key points, a transcribable audio recording can be very helpful. The quality of the recorder and microphone is crucial. (See Chapters 8 and 9 for a fuller discussion.) Most recorders come with a built-in microphone, but several flat "desk" microphones placed around the table will ensure much better sound quality and that quieter voices are recorded. The audiovisual departments of universities can be excellent places for you to get advice. Ensure that you test the equipment long before the focus group as well as later in the room before group members arrive. Make sure that any devices you need are fully charged (or that you have spare batteries). Take care that the setting for the group meeting is quiet enough for discussion to be recorded clearly.

The facilitator usually initiates discussion by giving an overview of the research and the role of the focus group in the project. The themes or questions for discussion can then be introduced. Since group members may be unfamiliar with the focus group technique, a brief summary of how focus groups operate should also be given. Box 10.2 provides an example of a focus group introduction.

BOX 10.2 A Sample Introduction to a Focus Group Session

In this focus group, two researchers acted as facilitators. As people arrived, the researchers greeted them, introduced them to other participants, and offered them tea or coffee. Participants had already been emailed a consent form, so the researchers checked with each participant whether they had any questions and collected the signed form. When all participants had arrived and were seated around the table, the researcher acting as the primary facilitator introduced the project:

> Well, thank you everyone for being here and making the time to come along today. So as you know from your invitation the research today is looking at the motivations of people who work in the social enterprise sector. And you've been asked to prepare a little two or three minute statement that gives us a bit of a snapshot of your background and how you came to be working in the role that you now have in a social enterprise. So maybe we could just go round the group and hear everyone's story. And can I just ask that if you've not already turned off your mobile phone that you do that now, just so we can stay focused on what each other is saying. Thanks. And then once we've gone around the group we'll talk more about your motivations for working in the sector. So we might start here . . .

Once all the group members had made their presentations, the primary facilitator opened the discussion up:

> Great, thanks for that everyone. And just listening to everybody a couple of things immediately spring to mind and that's the, even in this relatively small group, the incredible diversity of what people

The facilitator moderates discussion by encouraging exploration of a topic, introducing new topics, keeping discussion on track, encouraging agreement and disagreement, curbing talkative group members, and encouraging quiet participants. Examples of the sorts of phrases used by facilitators are outlined in Box 10.3.

are doing from the manufacturing types of activities to creatively focused enterprises to youth work and to things like urban agriculture. And the other thing is that people are talking about the relationship between the social focus and the more business or commercial focus and it seems some people are pretty comfortable with the more business side of things but for others there's a more tricky balance there. So from here we've got a series of questions but these are really just some prompts as we are happy for the discussion to go in whatever direction it's going to go. And we are really happy for you to jump in and ask each other questions especially now we've all heard a little bit about each other, we've broken the ice. And if things start to go too much off-topic we'll just jump in and we'll bring us back on-topic, but it's meant to be a conversation where people can ask each other questions, clarify things and even disagree with each other cause it's really important that we hear different points of view and if things get a bit noisy we'll jump in and we can move the discussion on if we need to. So perhaps we could just start by talking about the role of social enterprises in Australia today and to get people's thoughts on that and then come back and talk about how that matches up with the motivations you've got. So let's start with your thoughts on the role of social enterprise are at the moment?

From this point on, different group members responded to questions from the researchers, asked each other questions, agreed, and disagreed with each other. The topics for discussion flowed as people each contributed, adding a slightly different perspective and introducing new ideas. The researchers also asked questions, made points of clarification, and introduced new areas of discussion.

Source: Audio excerpt from focus group conducted by Jenny Cameron and Sherelle Hendriks, Newcastle, Australia, 23 July 2012 (see Cameron and Hendriks 2014). An example of an introduction is also provided by Myers (1998, 90).

BOX 10.3 Examples of Phrases Used in Focus Group Facilitation

- Encouraging exploration of an idea:
  "Does anyone have anything they'd like to add to that?"
  "How do you think that relates to what was said earlier about . . . ?"
  "Can we talk about this idea a bit further?"
- Moving onto a different topic:
  "This is probably a good point to move on to talk about . . ."
  "Just following on from that, I'd like to bring up something we've not talked about yet."
  "This is an important point because it really picks up on another issue."
- Keeping on track:
  "There was an important point made over here a moment ago, can we just come back to that?'"
- Inviting agreement:
  "Has anyone else had a similar experience?"
  "Does anyone else share that view?"
- Inviting disagreement:
  "Does anyone have a different reaction?"
  "We've been hearing about one point of view, but I think there might be other ways of looking at this. Would anyone like to comment on other sorts of views that they think other people might have?"
  "There seem to be some differences in what's been said, and I think it's really important to get a sense of why we have such different views."
- Clarifying:
  "Can you give me an example of what you mean?"
  "Can you say this again, but use different words?"
  "Earlier you said that you thought . . . now you're saying . . . can you tell us more about what you think/feel about this topic/issue?"
- Curbing a talkative person:
  "There's a few people who've got something to add at this point. We'll just move onto them."
  "We need to move onto the next topic. We'll come back to that idea if we have time."
- Encouraging a very quiet person:
  "Do you have anything you'd like to add at this point?"

Source: Drawn from discussions in Carey (1994), Krueger (1998), and Myers (1998) and from personal experience.

Some aspects of facilitation require special comment. Expressing and exploring different points of view is important in focus groups, yet research shows that groups have a preference for agreement (Myers 1998). The facilitator plays a central role in creating the context for disagreement. This can be done by stating in the introduction that there is no correct answer and that disagreement is normal and expected, by asking directly for different points of view, and by making explicit implied disagreement and introducing it as a topic for discussion (Myers 1998, 97). Watch for non-verbal signs of disagreement such as folded arms, movement away from the table, and a shaking or downcast head. You might ask the whole group or target the disagreeing member to give a different point of view. Of course, as facilitator, never state that someone is wrong or display a preference for one position. In the unlikely event that the discussion becomes heated, then intervene immediately, suggest that there is no right answer, and move the group on to the next question.

Very talkative or very quiet participants can be a challenge. Talkative people need to be gently curbed, while quiet ones need to be encouraged to participate. Along with the sorts of phrases listed in Box 10.3, your non-verbal signals can be useful. Pointing to someone who is waiting to speak indicates to the talkative person that there are others who need to have a turn. Making frequent eye contact with the quieter person and offering encouraging signs, like nodding and smiling when they do speak, is important. Remember, though, that silence gives people time to reflect and gather their thoughts. Do not feel that you have to fill silences; give people time to respond.

At the conclusion of a focus group, you might review key points of the discussion, providing a sense of completion and allowing participants to clarify and correct your summary. Group members should always be thanked for taking the time to attend and for their contributions. You can do this again with a personal letter to each participant.

Online Focus Groups

With developments in online technology, some researchers find **online focus groups** useful, particularly for bringing together people who are extremely busy or geographically dispersed (even in different parts of the world). They also save time, since "saved" discussions eliminate the need for transcribing. Some researchers have used online focus groups to broach sensitive topics because anonymity is possible—either participants devise their own pseudonyms or, in some computer programs, participants have the option of submitting both anonymous and identified postings (Kenny 2005; O'Connor and Madge 2003). However, participants need online access and some familiarity with computers.

There are two forms of online focus groups—**synchronous** (or real-time) **groups** such as chat rooms and **asynchronous groups** such as **bulletin boards** or blogs that may run for days, weeks, or even months. Like face-to-face focus groups, synchronous groups tend to have 6 to 10 participants, whereas asynchronous groups can be much larger. Sweet (2001) recommends 12 to 20 participants, but Kenny (2005) reports successfully running one with 40. In both settings, the role of the facilitator is critical. They post questions for participants to respond to, they seek clarification, they probe for more information, they bring the discussion back "on track," and they can even use computer logs to monitor when someone is not contributing or is dominating. Facilitating can be demanding, particularly in synchronous settings. The discussion tends to be as lively as it is in face-to-face groups; however, because participants cannot be seen (although the use of webcam technology could change this), the facilitator cannot "read" the visual cues: is a participant disagreeing, bored, asleep, out of the room making a cup of tea, distracted by someone else in the room? Nevertheless, those who have used online focus groups find them a useful addition to the qualitative researcher's toolbox.

Analyzing and Presenting Results

Krueger (1998, 46) importantly reminds us that "analysis begins during the first focus group." Listen carefully to responses, and clarify any unclear or contradictory contributions, since this information may be critical later when you present the results. For example, if young people say that they would watch television news and current affairs if the coverage were more relevant to them, it is probably important to get them to explain or give examples of how news items could be made more relevant (see also Box 10.4).

Since there is always a richness of material, analyzing focus group discussions can be as time-consuming as it is interesting. The first step involves transcribing the audio recording. A complete transcript of the entire discussion takes a great deal of time, since one hour of recording usually requires more than four hours to transcribe. When a detailed comparison of groups is to be undertaken, full transcripts may be necessary. Generally, a partial or abridged **transcript** (which involves transcribing only key sections of the discussion) will suffice. This is best done as soon as possible after the focus group, with the facilitator(s) and note-taker working in collaboration to decide which sections should be transcribed. A record of the seating plan and running order of speakers and a brief description of what was said are extremely helpful at this point. If you as the researcher have the time, it is also advantageous for you to transcribe the audio recording, because it will give you an opportunity to become more fully immersed in the content (and if you are new to focus group research, you will be able to reflect on your facilitation style and identify your strengths and weaknesses). (For a full discussion on transcribing interviews, see Chapter 8.)

It is advisable to transcribe and undertake a preliminary analysis of the first focus group before conducting any others. This is a way of checking that your questions are understood by participants and are eliciting the type of information you need for your research. It is also a way of checking that you understand and can interpret the responses of participants. For example, in an initial focus group, you might not think to ask young people to clarify what they mean by relevant news coverage, but by carefully reading the transcript, you are likely to pick up this omission.

Once you have the complete set of focus group transcripts available, read the material over several times to help make yourself very familiar with the discussion. One relatively straightforward strategy for proceeding draws from the questions or themes on which the discussion focused. Write each question or theme on the top of a separate sheet of paper, and then list on each sheet the relevant points made. Finally, make a note of key quotes that might be used in written material (Bertrand, Brown, and Ward 1992). This is an approach that works well when the discussion did not deviate widely from the questions or themes set by the researcher or when comparisons are to be made between focus groups (Bertrand, Brown, and Ward 1992). For example, in a research project comparing the land management strategies for dealing with salinity preferred by farmers, policy-makers, and researchers, the sheets with the responses of the different groups to each question or theme can be easily compared.

When the purpose of the research project is to identify key themes or processes associated with a particular issue or topic, it may be more appropriate to use **margin coding** (Bertrand, Brown, and Ward 1992). To do this, read through the transcripts, identify key themes or categories, and devise a simple colour-, number-, letter-, or symbol-based coding system to represent the themes or categories. You should then reread the transcripts and highlight the words, sentences, and paragraphs related to each category or theme by writing the appropriate code in the margin. Once transcripts have been coded, a cut-and-paste technique—completed either on a computer or manually—can be used to group the discussion related to each theme or process (see Chapters 8 and 18 for more information on this). Always keep an original of the transcripts for future reference. A variation of this thematic analysis is to develop a list of key words and, in a word-processing package, type two or more key words beside each comment. Using the search function, you can then locate related points of discussion. Computer programs specifically designed for qualitative analysis, like NVivo, can also be used and are particularly helpful when you have a large amount of transcribed material to analyze (see Chapter 18 for a discussion of this).

The ability to find material quickly is an important consideration, since analysis and writing rarely proceed in a linear fashion. During the writing process, new insights unfold (see Chapter 19), and frequently you may find it necessary to return to the original transcripts to refine and reformulate ideas. Sometimes it will be necessary to listen to and make additional transcripts of sections of the recordings.

When reporting on focus group research, present your results only in terms of the discussion within the groups. As noted earlier, focus groups do not produce findings that can be generalized to a wider population. Focus group results are also expressed in impressionistic rather than numerical terms. Instead of precise numbers or percentages, the general trends or strength of feeling about an issue are typically given. As Ward, Bertrand, and Brown (1991, 271) have noted, focus group reports are "replete with statements such as 'many participants mentioned . . .,' 'two distinct positions were observed among the participants,' and 'almost no one had ever . . . .'" Reporting on their study into people's responses to emergency procedures, Zeigler, Brunn, and Johnson (1996), for example, noted that the people in their focus groups generally responded with either compliant behaviour or under-reaction. Zeigler, Brunn, and Johnson then used

Example of How Focus Group Results Can Be Written Up

The following is an extract from a journal article reporting on findings from Australian focus group research on the impact of television news and current affairs on young people's political participation and active citizenship. The analysis of the focus group discussion starts by highlighting the way that young people find the reporting of news and current affairs too complex, particularly because of the sophisticated language and the absence of background information. The analysis continues:

> Political current affairs television was viewed by respondents as too complex to incorporate into their everyday viewing habits, but young people also feel it is not worth investing time in television current affairs because any political information received from the programs is usually trivialized and played for entertainment value. For example, *A Current Affair* [a news magazine show] was described by Debra, a nineteen-year-old university student, as "Hey Hey It's Ray" after the celebrity of its host Ray Martin, and was seen by her focus group as a form of populist emotional exploitation. As the following responses suggest, there was a strong feeling amongst the groups that television current affairs portrays politicians as being "full of it."

direct quotes to illustrate the different ways that the responses were expressed (see Box 10.4 for an example of a focus group analysis, and see also the ways that Barr et al. [2010], Byrne [2012], and Cousins, Evans, and Sadler [2009] discuss focus group findings).

In some projects, it might not be the general trends but the ambiguous or contradictory remarks that the researcher particularly wants to explore. The development and presentation of an argument may refer not only to what was talked about but the way it was talked about in the group setting. This became a significant aspect of the focus group research conducted by Gibson, Cameron, and Veno (1999, 29):

> The stories of success and hope that emerged when the discussion was shifted onto the terrain of community strengths and innovations were numerous. They came stumbling out in a disorganised manner suggesting that these stories were not readily nor often told. In the face of dom-

> > Bianka: The whole politics thing. They're all liars; they're all full of it. Craig: All the media carry on with is stuff like when they asked Hewson [a former politician] if [the] GST [Goods and Services Tax] would be applied to a birthday cake and they just blew that up. Who gave a shit?
>
> The respondents felt news and current affairs did not help them develop a political identity. They also expressed distrust in politicians who attempt to "persuade" them to choose a lesser evil. As Bianka points out: "They all change their minds when they get what they want. I mean what's the point?" What eventuates is a distrust of not only politicians but also the media that is supposed to decipher the positive and negative elements of each candidate's actions.

Note how the findings are reported only in terms of the focus groups and in tentative terms with the use of phrases such as "responses suggest," "respondents felt," and "[w]hat eventuates." Main themes that emerged from the focus groups are summarized by the authors, and quotes from participants are used to illustrate and elaborate these themes.

Source: Evans and Sternberg 1999, 105.

inant narratives of economic change perhaps such stories are positioned as less important or effective. It is clear that there is a lack of a language to talk about this understanding of community capacity; yet, as we will argue, this understanding has the potential to contribute to the ability of a region to deal effectively and innovatively with the consequences of social and economic change.

In research on peat mining in Ireland, Collier and Scott (2010) also comment on the significance of shifts in tenor of the discussion. They found that when presented with photomontage images of mined and restored landscapes "there was a palpable change in participant attitude" and that the focus groups became so animated the researchers had to ask participants to slow down so that individual comments could be heard (p. 310). Collier and Scott note that at this point the focus groups became a setting for social learning with participants being much more willing to put forward and discuss different ideas, and to ask the researchers for more information that might help them explore options (p. 311).

One important element in the process of writing up (or "writing-in," as Mansvelt and Berg call it in Chapter 19) is to find a balance between direct quotes and your summary and interpretation of the discussion. When too many quotes are included, the material can seem repetitive or chaotic. Too few quotes, on the other hand, can mean that the vitality of the interaction between participants is lost to the reader. Morgan (1997, 64) recommends that the researcher should aim to connect the reader and the original participants through "well-chosen" quotations (see Chapter 20).

Conclusion

Focus groups demand careful preparation on the part of the researcher. The selection and recruitment of participants, the composition, size, and number of groups, and the questions and topics to be explored are all key points to consider during the planning stage. Even the apparently mundane details of appropriateness of venue, provision of refreshments, and quality of audio equipment are critical to the success of focus groups. A well-prepared researcher also gives thought beforehand to the process of facilitation, including the points to cover in the introduction, the wording of key questions, topics, and phrases, the probes and prompts that might be useful in further exploring a theme or topic, and strategies for drawing out different points of view, keeping the discussion on track, and dealing with both the more talkative and quieter members of the group. As soon as possible after the focus group, start the process of analysis, beginning with transcribing the audio recordings, followed by reading and rereading the transcripts, summarizing main points, and identifying central themes.

Although they require careful planning beforehand and a great deal of reflection afterwards, focus groups are an exciting and invaluable research tool for geographers to use. Participants almost invariably enjoy interacting with each other, and the discussion can generate insights and understandings that are new to both participants and researchers. The interactive element makes focus groups ideally suited to exploring the nuances and complexities of people–place relationships, whether the research has a primarily data-gathering function or is more concerned with the collective practice of knowledge production.

Key Terms

anecdote
asynchronous groups
bulletin board
facilitator
margin coding
moderator
online focus groups
over-disclosure
probe
purposive sampling
recruitment
saturation
standardized questions
synchronous groups
synergistic effect
transcript
under-disclosure

Review Questions

1. What is meant by the "synergistic effect" of focus groups and why might this effect be important for geographers interested in qualitative research?
2. Focus groups are useful for both collecting knowledge and generating knowledge. Give some examples of focus group research in which collecting knowledge has been the main aim, and focus group research in which generating knowledge has been the main aim.
3. Why might focus groups be used in combination with quantitative research methods?
4. What are some key issues that researchers need to consider when planning focus group research, particularly in terms of the composition of focus groups; the size and number of groups; and the organization of questions and topics?
5. What are some strategies that researchers can use to ensure that focus groups run as smoothly as possible?
6. What are some characteristics of how focus group findings are reported?

Review Exercises

1. Find a research project from a recent issue of a geographical journal that you think could usefully have been conducted using focus groups. Why do you think focus groups would be appropriate? Discuss the participants you would select, the composition of the focus groups, the size and number of groups you would use, the questions you would ask or themes you would use, and strategies for recruiting participants.
2. The fictitious University of Pacifica is redesigning the "For Current Students" section of its web page. Your research company has been commissioned by the university to conduct a focus group study on what students think about the current design and what they think is important in the redesign. Devise a series of questions that you would use to canvass students' views. Select two students to co-facilitate the focus group, and select seven to eight students to participate in the focus group. Other students observe, noting how the content of the discussion flows, and the verbal and non-verbal interactions between participants and the facilitators. You could do this exercise around other topics. For example, building on Evans and Sternberg's (1999) research, discussed above, into the impact of television news and current affairs on young people's political participation and active citizenship, you could explore the media young people use to find out about news and current affairs and how this affects their view of politics.

Useful Resources

Barbour, R. 2007. *Doing Focus Groups*. Los Angeles: Sage.

Barnett, J. 2009. "Focus groups tips for beginners." TCALL Occasional Research Paper no. 1. Texas Center for Adult Literacy and Learning. http://www-tcall.tamu.edu/orp/orp1.htm#1.

Bryman, A. 2012. *Social Research Methods*. 4th edn. Oxford: Oxford University Press.

Morgan, D.L. 1997. *Focus Groups as Qualitative Research*. 2nd edn. Thousand Oaks, CA: Sage.

Rezabek, R. 2000. "Online focus groups: Electronic discussions for research." *Forum Qualitative Social Research* 1 (1). http://www.qualitative-research.org/fqs-texte/1-00/1-00rezabek-e.htm.

Stewart, D., P. Shamdasani, and D. Rook. 2007. *Focus Groups: Theory and Practice*. 2nd edn. Thousands Oaks, CA: Sage.

11 Historical Research and Archival Sources

Michael Roche

Chapter Overview

In the second edition of this volume I claimed, citing Driver (1988), that although human geographers have become more willing to incorporate a historical dimension into their work they have still tended to overlook historical approaches and archival sources. The interrogation of archival sources is fundamental to much historical geography research but it need not be restricted to historical geography and can be applied more widely across human geography. That this is not the case is at least partly because historical geographers have tended to treat archival research as an in-house craft and made only limited attempts until recently to provide more general advice and guidance to other geographers. I consider that human geographers more generally can benefit from having some familiarity with archival research even for research that is of a more fundamentally contemporary nature (McLennan and Prisen 2014). In addition, the now widespread use of digital cameras in the archive has serendipitously brought archival research towards the mainstream of human geography.

Introduction

After 30 years of working as a historical geographer, I still relish the opportunity to undertake archival research. There is a continuing sense of delving into the unknown, of engaging in academic detective work trying to understand inevitably fragmentary and partial surviving **records**, of striving to make sense of the sometimes highly partisan evidence you are scrutinizing. My own introduction to **archival research** was orchestrated only at the graduate level. To a large degree, I was able to learn by trial and error, following the tendency of historical geographers at the time to regard archival research as part of their craft, something to be acquired on the job rather than by reading about how to do it. Good archival scholarship was to be inferred from reading its products in the form of journal articles or books by leading historical geographers and from discussions with supervisors. In many ways, this was a laudable model, one that allowed me to develop my skills and understanding at my own pace, but geography students of today wishing to use archival sources can benefit from a more overt discussion of the fundamentals of archival research. This is also the case because archival research needs to engage with wider disciplinary theory and the research ethics that are also a part of historical inquiry. Even so, like other methods, archival skills can to some extent only be learned by doing archival research. Expertise improves with experience, and this is not easily reduced to a checklist of best practice.

What Is Archival Research?

> Archival scholarship at its best, it seems to me, is an ongoing, evolving interaction between the scholar and the voices of the past embedded in the documents. [Harris 2001, 332]

Archival sources are a subset of what historical geographers and historians refer to as **primary sources**. They include non-current records of government departments held in public **archives** but can be extended to include company records and private papers. As well as **documents**, handwritten and typed, these sources can embrace personal letters, diaries, logbooks, and minutes of meetings, as well as reports, plans, maps, and photographs. More recently, they have included records created in electronic format, which brings with it new challenges (Davison 2003). This chapter concentrates on official papers, including manuscript and typescript files.[1] Most of the comments are also applicable to company archives and private papers. With the target readership of this book in mind, the chapter concentrates largely on government archives, on the past century or so, and on the "New World." This focus simplifies the discussion, since much of this documentation is typewritten and the language of the more recent past is relatively easy to comprehend today. As a collection of unique, single documents, created contemporaneously with the events they discuss, the materials lodged in archival repositories provide a particular window on the geography of earlier times. Historical approaches applied to archival sources will not allow all of the research questions of human geography to be addressed; however, they do provide a means of answering questions about the recent as well as the more distant past that are not recoverable by the other techniques or from other sources available to human geographers and can provide a way of helping to "verify" other sources of information such as oral testimony and official statistics.

More often than not, the researcher will make use of public archives housed in a government agency charged with the preservation of non-current records. On other occasions, small regional collections, such as those associated with some museums, may be targeted. Sometimes, access to the records of private organizations may be sought. Miles Ogborn (2006) offers some rich insights into locating and scrutinizing archives from the local to the national, be they paper-based originals or online collections. Increasingly, the web provides the initial contact point with research and archival collections. Box 11.1 lists some major repositories and their "www" addresses. In addition there are various more specialized archives concerned with women (Mason and Zanish-Belcher 2007), with life and labour surveys in late nineteenth- and early twentieth-century London (http://booth.lse.ac.uk/), and making use of digital technology providing a somewhat wider interpretation of archives that I have adopted here. Examples of the latter are provided by the Tantramar marshlands of New Brunswick in Canada (http://www.mta.ca/marshland/) and the "Discovering Anzacs" site hosted by National Archives of Australia relating to Australian soldiers who served in World War One (http: http://discoveringanzacs.naa.gov.au/).

Identifying the Archives

Archives New Zealand:
http://www.archives.govt.nz/

Library and Archives Canada:
http://www.collectionscanada.ca/index-e.html

National Archives of Australia:
http://www.naa.gov.au

National Archives of Ireland:
http://www.nationalarchives.ie

National Archives of Scotland:
http://www.nrscotland.gov.uk/

National Archives of South Africa:
http://www.national.archives.gov.za

Public Record Office (England and Wales):
http://www.nationalarchives.gov.uk

US National Archives and Records Administration:
http://www.archives.gov

For Mayhew (2003), historical geography has a twofold significance for the discipline as a whole that lies beyond increasing the understanding of the geography of the past: to re-evaluate taken-for-granted concepts and to develop a comparative perspective so that as geographers we might more fully appreciate what is distinctive about today's world and how we understand it in disciplinary terms. Historico-geographical research based on archival research underpins both of these objectives.

Advice on Conducting Good Historical and Archival Research

> You may have to go to the data rather than having them come to you. [Ogborn 2006, 111]

So where to begin? Good archival research is difficult to reduce to a checklist of points. However, it is useful to explore sequentially the sorts of things you might need to do and the obstacles that you might encounter in undertaking archival research. At the start, like any other research project, your work ought to be informed by prior in-depth reading on current scholarship around the topic but accompanied by an openness about the ultimate direction of the research. This point is well made by the eminent Canadian historical geographer Cole Harris (Box 11.2).

Archival research begins before you arrive at the archive. You ought to be familiar with the existing **secondary literature** on your research topic before beginning any search for archival material. For instance, for a project involving state agencies, it is important to read any previously published institutional histories as well as the annual reports of that agency and to look at parliamentary debates (or their equivalent) before venturing into the archive. With larger official archives it may now only be a matter of registering as a user and then searching for and ordering files through an online system similar to that of a library catalogue. This may even be possible at distance and in advance of arriving at the archives. Even experienced archival researchers may wish to seek some further advice from the archivists on duty about their research. Do check in advance on the institution's website for registration requirements and ensure that you have the appropriate sorts of identification. For smaller regional archives it may be advisable to email or write ahead of time outlining your research topic.

Sometimes ingenuity is called for in that you may only be able to address research questions obliquely. The **archivists** can sometimes provide helpful, expert advice about record sets that you may not have considered useful. Typically, as a new user you will be given the opportunity to explain what you are researching and why. The archivist will tell you how the **finding aids** work and can offer

Approaching Archival Research

The first point to make about archival research is that it cannot be contained within a single methodology. Any sizable archive holds a vast array of material, and even if one's research questions are fairly specific, the chances are good that there will be far more potentially relevant documents than there will be time to examine them . . .

. . . [A]rchival research tends to gravitate towards one of two polar reactions—neither, I think particularly helpful. It is easy enough to be taken over by the archives, to attempt to read and record all their relevant information. In this way months and perhaps years go by, and eventually the investigator has a vast store of notes and, usually, rather weak ideas about what to do with them. A fraction of the archives have been transferred from one location to another, while the challenges of interpretation have been postponed . . .

In effect the archives have swallowed the researcher. At the other pole are those who come to the archives with the confidence that they know precisely what they want. They have conceptualized their research thoroughly in advance. They pretty much know how they will argue their case and what their theoretical position is. But they do need a few more data, which is why they return to the archives. As long as they cleave to their initial position, either they will find that data they need and leave fairly quickly or they will not find them and also leave. Fair enough for certain purposes. But they are imposing their preconceptions on the archives. They have solved the problem of archival research by, in effect, denying the complexity of the archives and the myriad voices from the past contained in their amorphous record (Harris 2001, 330–1).

suggestions about where to start looking. While their experience and expertise can often prove invaluable, it is important to remember that they may have limited time available to assist individual researchers.

A crucial difference between a library and an archive lies in the way that each organizes and stores material. Libraries typically catalogue books and journals by either the "Dewey Decimal" or the "Library of Congress" classification system that groups together all books on similar subjects. In contrast, public archivists seek to maintain the integrity of the record sets they obtain from government departments or other agencies in terms of preserving the place of specific files in the broader record set, maintaining the original ordering of documents in the file, providing storage conditions that will ensure the long-term survival of the records, and making them available to the public. Archivists place great emphasis on the **provenance** of the **files**; the actual order of the material within the files in itself tells the researcher something about the situation that prevailed when the file was being created.[2] Thus, whereas in a library you can refer to a catalogue to find a book on a particular subject on an open shelf, in an archive basic paper-based finding aids take the form of sequential **series lists** of all the files held by particular agencies. These lists itemize all the files created by an organization using the original description system (usually numerical but sometimes alpha-numerical).

In most archives, electronic searching of the collections is now possible. This means that you can search for specific items in the same way, superficially at least, that you would use a library catalogue. Some have more advanced search systems also allow you to look for specific files by number or for selected file series. This can be useful if you need to look systematically through a batch of material where the individual file names may not allow files to be easily recovered. Do not forget about provenance. What survives in the file is likely to be only a fragment, and it may be quite partial in terms of providing any insights about the past. With small archives, you may have only the series list of files to guide you. Inevitably, you will find that some of the originating records staff have been more thorough and less idiosyncratic than others in managing their filing systems. The name of a file may not always be a clear guide to its contents, material may have been misfiled, and some files may have been lost or destroyed. For instance, I recently found that files marked "railways accommodation" had nothing to do with railway department housing, the topic I was working on, but actually in a technical sense referred to the number of passenger carriages and freight wagons that could be "accommodated" at the railway station yards. Similarly, a file (mis)labelled as "hoses" was actually about houses (here the file number provided the clue).

In some national collections, precious and fragile originals may have be made available on microfilm but more recently are often electronically scanned and made available online (e.g., Hackel and Reid 2007). As Summerby-Murray (2011, 117) has noted, "an unexpected side effect of digitalization" has been "a dramatic improvement of legibility" as well as making items more widely and conveniently available to researchers.

You may also find that there is restricted access to some files. Personnel files typically fall within this category. The period during which restricted access applies varies from one country to another, but 30 years after the closing of the file is typical. In some instances a lesser degree of restriction applies, and permission to look at files may be granted by a senior archivist, government official, or someone associated with the organization that created them. A formal written request outlining your research project may result in the granting of access; however, some conditions may be attached—for example, you may be permitted to read only a particular portion of the material while the remainder of the file remains physically sealed. In other cases, the researcher has no choice but to wait patiently until the material is released.

Photocopying material is usually possible, but it can be costly, and you may have to pay in advance. Some material may be deemed too fragile or, if bound, too difficult to photocopy. It is therefore advisable for you to find out what the policy is beforehand. Plans, maps, and charts larger than A3 size can be copied by means other than photocopying, but this is sometimes quite expensive. However, it may be the only means of obtaining a copy of an essential document. Be prepared to only be able to use a pencil for note-taking—the so-called **"pencil-only" rule** intended to ensure file materials are not damaged by a careless researcher writing with a pen. Most archives today permit digital copying and possibly a file is now more at risk from the researcher folding or crushing papers under their laptop or tablet. But you may still on occasion find yourself wearing white cotton gloves and wielding a pencil, so be ready for this possibility.

Where archives permit researchers to make their own digital copies of documents, various conditions will apply, including registering your camera and completing associated documentation and agreeing not to use flashes or tripods or not to fold documents. The use of digital images also raises new issues regarding labelling and storage if you are to make effective use of such materials (Box 11.3).

When the archivist gives you the file to work on, you will find in most cases that new items are on top of the older material, particularly if the material is secured by paperclip. You will probably need to work from back to front. Will the material answer any of your research questions? It may be immediately obvious that the material is relevant to your inquiry, or it may appear only tangentially relevant or even irrelevant. Sometimes it is difficult to make a judgment at first glance, and you may have to recall material you examined previously but whose significance you did not appreciate at the time. Alan Baker, a British historical geographer, has offered some guiding thoughts on evaluating primary sources, including archival materials (Box 11.4).

Baker's words seem to me to be crucial for those using archives as qualitative sources in human geography. It is essential to understand as fully as possible the original purpose of the document, who created it, what position they held, and how and when it was made. Some generic questions to pose when assessing documentary sources are laid out in Box 11.5.

The questions raised in Box 11.5 provide a useful start, although I would make three qualifying points. First, it is possible to extend "document" to include maps and plans (see Harley 1992). Second, this approach tends to privilege the ideas behind actions. That is to say, the past is being understood in idealist terms whereby the thought behind the action is regarded as providing the understanding

Digital Images

It is now common for researchers to make their own digital copies of material. This has obvious advantages in terms of the ability to make images of, for instance, bound volumes that would not have been photocopied because of concerns over preserving their binding. Digital images are also less expensive than photocopying, and they do not involve a lengthy delay in obtaining them, an important point for time-pressed students who cannot wait for an archival turnaround time of several weeks. However, I would be mindful of Harris's (2001) comment about transferring the archive from one location to another. Ease of copying in itself can create other difficulties if reference details are not kept meticulously. As a checklist, resting on painful personal experience, I would suggest the following points:

1. Ensure the memory card in the camera is clear and the battery charged (take spare batteries).
2. When copying lengthy documents, be wary of making blurred images and of missing pages.
3. Recognize that you may require some maps and images to be reproduced with greater clarity than you can obtain with your hand-held camera, and be prepared to pay for photocopier or high-quality camera images.
4. Ensure that you have a reliable system for linking the digital image to the source file. (My low-tech approach to this has been to include a slip of paper with the file details on it alongside the page when I take the photograph so that I have a visual reference on each digital image.)
5. Store the digital images so that they can be located and retrieved easily.
6. Post-archive editing of the digital images may be necessary to improve readability but in so doing take care not to crop off any key details or the file reference information.

Assessing Evidence in Historical Geography

No source should be taken at face value: all sources must be evaluated critically and contextually. The history and geography of a source needs to be established before it can legitimately be utilized and incorporated into a study of historical geography. The historical sources we use were not compiled and constructed for our explicitly geographical purposes; they were more likely to have been prepared, for example, for the purposes of taxation and valuation, administration and control. We also have to understand not only the superficial characteristics of a specific source but also its underlying motivation, background and ideology of the person(s) who constructed it. In order to make the most effective and convincing use of a source we must be aware of its original purpose and context and thus its limitations and potential for our own project.

Source: Baker (1997, 235).

Questions to Ask of Documentary Sources

1. Can you establish the authenticity of the source—is it genuine? Are you looking at the original?
2. Can you establish the accuracy of the document—how close is it to the source of events or phenomena? How accurately was the information recorded? (Cross-check with other sources.)
3. What was the original purpose for collecting the information? How might it have influenced what information was collected?
4. How has the process of archiving the information imposed a classification and order upon historical events?

Source: After Black (2006).

necessary to interpret these events. Historical geography can be written legitimately from a viewpoint other than that of contemporary observers (Baker 1997). Third, the documents themselves cannot be read in isolation but must be understood in their wider context, and even then any conclusions will be provisional rather than absolute.

To some extent, all archival researchers develop individualized approaches to note-taking from archival materials. However, there are two basic and contrasting strategies. The first involves collecting material by topic, noting specific details and suitably referenced quotations. Classically, historical researchers used large index cards for this purpose, although most now use computers to organize their notes. New topics can be noted as more files are read and new research questions formulated. The alternative approach is to record chronologically any pertinent information from each file and then subsequently identify themes that emerge across the files. Both strategies have advantages and disadvantages. The former depends on identifying key topics at the beginning of the project within which to collect information. Such an approach still allows you to add new topics or identify dead ends and see how themes merge or diverge. My personal view is that while this approach means that many diverse sources are brought together, it can blur a researcher's capacity to make good inductive judgments when in the archive. The latter method is more sensitive to the provenance of files and can give a clearer sense of the role of particular officials or departments. It does, however, involve a degree of double-handling in that evidence that has already been collected by the researcher needs to be reorganized after each visit to the archives and perhaps annotated further. It is important to follow up other research questions that may emerge from this re-sorting process. The latter approach is one that I have used over many years. It suits me, and as a full-time academic, I can incorporate it into my way of working. But I would acknowledge that it probably works best when one is working in a familiar area, even if the specific contents of the files are unknown. Students with limited time for archival research may prefer to adopt the first strategy and will probably be using computers, particularly when they have the keyboard skills (speed and accuracy) that I lack!

After you have located and extracted archival evidence, it must be adequately cited in the written products of the research. The first step is to carefully record the specific document description and file reference. For example, the personnel file of L.M. Ellis, a Canadian who was first Director of Forests in New Zealand (1920 to 1928), is located among the Forest Service files at Archives New Zealand in Wellington. The specific item reference is R21098142, the agency code is ADS Q and Series 185/38, the accession is W607, and the item is 4/4i. This accuracy is crucial, especially if you have to complete a written request for the item. Fortunately the Ellis file can be identified and ordered online more or less at the click of a button, though locating it illustrates some of the other challenges of archival work. Ellis's given name was Leon but he never used it and searching for him under this heading throws up two unrelated files. To further complicate matters, Ellis preferred the spelling MacIntosh, but on his official file it was recorded as McIntosh—thus MacIntosh Ellis throws up his only original job application and a draft report, but not the personnel file. Be prepared to think around a problem rather than immediately accepting that the material you are looking for is not held. The crucial thing is to record the details carefully. This is important for two reasons. First, it enables you as a researcher to keep track of where you found specific information. Second, it enables a subsequent researcher to relocate the material. The idea is simple enough, but given the nature of archival material, it is somewhat more exacting than, for example, the standard bibliographic requirements of author, date, title, and publisher/place for a book in the reference list of a thesis. Citing archival materials correctly can also pose problems in that human geography has tended to adopt versions of in-text citation systems, such as Harvard (see Hay 2012). Most archival sources sit uncomfortably within this framework and are generally better referenced in footnotes or endnotes typically used by historians. Students undertaking archival research may need to negotiate a variation from human geographies preference for social science–oriented formats for referencing.

The archive does not constitute the only source for historical research. For instance, newspapers, private papers, and unpublished memoirs may provide valuable material for cross-referencing with the archival record. As well, once archival work is completed, the researcher may need to follow up on unfamiliar key actors by checking old editions of *Who's Who* or newspaper obituaries, as well as on unfamiliar organizations or period issues; here contemporary newspaper accounts can be invaluable (and some are now available electronically, see for instance Trove for Australia and Paperspast for New Zealand). This "post-archive" work can of course help to shape and inform the purpose of subsequent trips to the archive.

Moreover, files are not the end point of research, as US cultural geographer Carl Sauer reminded historical geographers nearly 70 years ago:

> Let no one consider that the historical geographer can be content with what is found in archive and library. It calls, in addition, for exacting fieldwork. One of the first steps is the ability to read the documents in the field for instance of an account of an area written long ago and compare the places and their activities with the present, seeing where the habitations were and the lines of communication ran, where the forests and the field stood, gradually getting a picture of the former cultural landscape behind the present one. [Sauer 1941, 13]

Although Sauer's words may indicate nostalgia for a pioneering rural past while your focus could just as easily be urban and social present-day focused and not at all wistful, his challenge remains pertinent. It is recast in today's terms in Keighren's (2012) case study of archives-based fieldwork that formed part of an undergraduate field excursion to New York, which explored its "moral politics" during the early twentieth century. The archival case study enabled the students to better contextualize the city's current social geography and in so doing re-forges the link between archive and fieldwork long ago discussed by Sauer (1941).

Challenges of Archival Research

There are two types of challenges facing researchers working with archives. The first is intellectual and the second technical. When dealing, for instance, with the file materials contained in an official government archive, it is important to bear in mind the sorts of power relations inherent in the surviving materials. This is rather more than just acknowledging that the surviving files are fragmented and partial. The records are those created by politicians and officials. They reflect the outlooks and understandings of the dominant groups in the national context at the time they were created. Duncan (1999) writes of these concerns in terms of complicity stemming from use of the "colonial archive." For much of the nineteenth century and well into the twentieth century, these records were created largely by men in the upper echelons of society, and in states such as Australia, Canada, and New Zealand, they are predominantly the records of colonizing British settlers. Summarizing the contents of files from the archives merely reproduces these uneven power relations rather than interpreting them. The records of nonofficial, community, or even sporting groups may provide a way into understanding the concerns and aspirations of those who had no position in the public political sphere. In the same way, oral histories from the recent past may provide insights into gendered and minority concerns. Furthermore, an awareness of the power relations within the archival material may allow the researcher to reinterpret surviving materials. For instance, what I once mapped as examples of illegal felling of forest in New Zealand in the 1870s I would now be inclined to understand as resistance on the part of Māori forest owners to the imposition of authority by the Crown and as the flouting of government regulations by European timber-cutters who had limited alternative means of supporting themselves (see Chapters 2 and 4 for further discussion of power relations in qualitative research).

The most fundamental technical difficulty relates to the ability to actually read the documents retrieved in the archive. During the first half of the nineteenth century, many official documents were handwritten in **copperplate script**. This script looks elegant, but it can take some time for novices to learn to read it proficiently, a situation that may be exacerbated when officials wrote both across and along a page in order to save paper. Perseverance will pay off. Archives from the latter part of the nineteenth century are commonly written in a **modern hand**. They are generally readable with a bit of effort. However, original manuscripts concerned, for example, with the early settlement of North America before 1700 may be written in **secretary hand**. This was the script of professional scribes of the time, and it is difficult to read without specialized instruction and its translation requires additional palaeographical skills beyond the scope of this chapter. In any case, all kinds of handwritten documents made in the past tend to be difficult to decipher, especially when the investigator is trying to read a faint letterbook copy of the original. For instance, in the mid-nineteenth century, the "long s" written much like an "f" was frequently used in official correspondence. Words with a double "s"—"lesson," for instance—are rendered as what looks to us like "lesfon." Not only do spellings change as you move back in time, but so does the very construction of the English language.[3] This makes it more of a challenge to understand the world view of these earlier times. From around the 1880s, typewritten material becomes more common in government files, but important marginal annotations will be handwritten and often cryptic in meaning. These annotated comments are particularly important for the insight they can give into discussion within an organization about the issue to which the larger document relates.

A fundamental challenge can be posed by difficult-to-read handwriting; where the document was particularly important, I have on occasions been reduced to identifying how each letter of the alphabetic was written and then transcribing the document letter by letter. Anyone contemplating the use of archive material as a source for qualitative research in human geography must be prepared to be patient and resourceful; using documentary evidence is rarely easy and may require a considerable deal of time. The decision about which sources to start work on should be carefully considered. The units of land area and currency may also be different from those in use today (for example, acres rather than hectares). This raises the issue of whether to convert every measurement to the current system or to give a general conversion factor and use the units of the period (generally, I prefer the latter). Some facility with the original units is useful. Appreciating that there are 640 acres to a square mile makes it possible to recognize, for example, that the apparently precise data on the forest areas in Otago Province in New Zealand in 1867 are actually only estimates to the nearest quarter square mile, or 160 acres. You may also need to understand more specialized measures, depending on your field of research. For example, throughout much of the British Empire in the twentieth century, quantities of sawn timber are often given in superficial feet (colloquially referred to as "superfoot")—that is, 12 inches by 12 inches by 12 inches (30 cm by 30 cm by 30 cm), but in North America the equivalent term "board foot" was used.

If you are working through a file and time runs short, you may find yourself copying whole documents that you think could be of use because you do not have the time to read them carefully and decide whether they are. In the end, you find yourself with page after page of material that may subsequently prove to be marginal to the research. This is particularly important in that when you return to your research material, it is too easy for the digital copy to overshadow your handwritten notes so that you again end up with the situation Harris refers to as the archive "swallowing the researcher" (see Box 11.2).

Mistakes in interpretation can and do occur. I once mistook the numbered applications for the position of director of forests in New Zealand for the rankings of candidates. The result was an apparently nonsensical list of candidates. On closer subsequent investigation of the date stamps showing the receipt of the applications, it became clear that the numbers related only to the order in which they had been received. Retrospectively, I can draw three points from this episode. First, scrutinize documents carefully. In this case, the answer was there in the documents, but I did not see it the first time around. Second, if you are uncertain about what the documents indicate, acknowledge this, and do not make too definite a claim regarding the surviving evidence. Third, by "learning the ropes"

BOX 11.6 Changing Archival Practice: A Cautionary Tale

The first decade or so of my archival work was characterized by assiduous note-taking from files, even including making simple sketches of key portions of maps and plans. This detailed transcription of the primary material and incidental notes about interpretations and other lines of inquiry certainly helped imprint the contents on my memory. On a few occasions I ordered photocopies of some important maps. At the time this was both expensive and time-consuming—some weeks would elapse before I received the material.

The second decade revealed some changes as, supported by some research funding but facing greater time pressures and restricted time for out-of-town archival work, I began to order photocopies of larger amounts of material, although still maintaining a fairly full set of written notes and comments on the original material. This behaviour was rationalized on the grounds of photocopying being cheaper than a return airline fare. It was sometimes advantageous when a portion of document not originally thought to be significant proved so and was available in my office. At times I must have come close to fulfilling Cole Harris's analogy of being swallowed by the archive.

In recent years, particularly once digital cameras were permitted in the archive, my practice has further changed, and not necessarily for the better.

as an undergraduate or graduate student, you can avoid some of the more obvious pitfalls of interpretation before you have anything published.

Ethics and Archives

It is all too easy for archival researchers to dismiss ethical issues as something relevant only to geographers working on present-day topics using other qualitative or quantitative methodologies. Actually, archival researchers also have ethical obligations, accentuated by the fact that the individuals who created—or are the subjects of—the records in question are in all likelihood now deceased and unable to represent themselves. Other ethical dimensions of archival research have less to do with safety, harm, and risk to the researcher or research participants and more to do with retaining the integrity of the material contained in the files and the preservation of the archives themselves. After all, those using archives should regard access to the material as a privilege. Historical records are precious and often irreplaceable. All researchers are under an obligation to

I find that I am now tending to digitally copy much of the contents of files that look as if they may be of use. This has some advantages; many original plans and maps are coloured and this can now be captured on the camera and help with the interpretation, although as Rekrut (2011, 152–4) notes, digital copying can also result in the loss of some information. By way of compensation I have tended to read the file more fully before copying anything in it. I have also kept a log of items digitally copied and write a commentary on the nature of the material in the file, what is of interest and why (especially if this is not necessarily reflected in its title). Because a large part of my research collection is paper based, I find I am caught with a hybrid system that risks assuming the worst features of both. This is not something the "digital natives" of today will face. Playing with Harris's phrase, the danger is now not some my much being "swallowed" by the archive but choking on it.

That said it is not all bad news; various digitization projects are making primary source material available online so that researchers are freed from being physically in the archive. While this is disadvantageous in terms of the loss of "mental space" for focused inquiry without interruption, it is advantageous in that it helps relocate historico-geographical research to a position alongside and much closer to a range of other qualitative approaches used in human geography.

look after archival material and to ensure that it is preserved in good order for any subsequent scholars.

The US National Council of Public History identifies three guidelines for using archive material.[4] We can reasonably substitute "geographer" for "historian" in each of these guidelines.

1. Historians work for the preservation, care and accessibility of the historical record. The unity and integrity of historical record collections are the basis of interpreting the past.
2. Historians owe to their sources accurate reportage of all information relevant to the subject at hand.
3. Historians favour free and open access to all archival collections (National Council on Public History 2003).

Another situation in which ethical issues may arise happens when files that contain classified or otherwise restricted material are issued to you by mistake. While it may be tempting to capitalize on an archivist's error in issuing a file before a time embargo or other restriction has elapsed, in the longer term this is counter-productive; it is equivalent to an unsanctioned questionnaire or an interview in which the participant does not know the true purpose of the research. Such behaviour can result in tighter lending conditions being imposed on all subsequent archival users.

Public archives typically specify conditions to which users must agree to adhere when they sign in or request a reader's card. Not all primary documents are in public archives, however, and having access to such documents can present practical and ethical issues. Finer (2000) recounts an episode in which, after she initially received unlimited access to the records of a prominent Italian social reformer, she learned partway through the project that new conditions governing access, the scope of the research, and its objectives were being imposed (Box 11.7).

Presenting the Results of Archival Research

There is no single correct way of presenting the results of archival research. The theoretical foundation of the research project, the sorts of empirical information retrieved, and the writing style of the researcher all shape how the research project or thesis is expressed.[5] Typically, however, archival researchers will make use of direct quotations from key documents to demonstrate their case. They will also be mindful of the actions of key actors within organizations (and sometimes the importance of the role of obscure bureaucrats as well) in shaping decisions and policy that may have had far-reaching geographical significance. They also make

Researching in a Private Archive

For researchers using public archives, the protocols are fairly well-established and reinforced in documentation that is part of the user registration process. On occasions, researchers will have access to private papers or records of small organizations. On the basis of a particular research project, Finer (2000) puts forward four "negotiations" that ought to be undertaken on those occasions to ensure the smooth running of the project. They are:

1. To insist on and ideally participate in the drafting of a detailed written agreement regarding precisely what is to be attempted in the research and to what end.
2. To draft a timetable agreed to in advance by staff members who are in a position to affect access to records or to other facilities such as photocopying.
3. To reach agreement in advance on the handling of sensitive material and the extent to which and on what terms it is to be cited.
4. To reach agreement on matters of faith/ideology—that is, the extent to which it is or is not considered necessary for the researcher to be of the same persuasion as the person(s) being researched.

While Finer's project had a biographical dimension to it rather than much in the way of human geography research, her points have a general utility.

use of case study material to illustrate points. On occasions, good use may be made of cartographic or pictorial material.

Adept researchers are often able to move easily from specific points of detail to sketch a much larger picture and to relate it to what is known about related topics. I would recommend critical perusal of recent issues of the *Journal of Historical Geography* and *Historical Geography* and books by recognized figures in the field (e.g., Morin, 2008; Withers, 2010). However, it is not just a matter of identifying key quotations but rather of building an argument. This obliges you to select ideas in a logical way from the pre-existing literature and then to use them to provide an informed discussion based on what you have found in the archives. The desirable end point, however, is to be in command of the source material. Rather than merely reproducing a chronicle of part of what is contained in the archive, strive to make your writing a synthesis of specific detail and informed interpretation.

Conclusion

Although it has much to offer human geography in general, archival research has tended to be neglected by other than historical geographers. As a research method, historical research using archival sources:

- calls for creative thinking in identifying source materials relevant to your research problem;
- needs patience, precision, and critical reflection in collecting and evaluating material;
- requires a sense of historico-geographical imagination in interpreting source material whereby theorization does not outstrip the evidence;
- is partial and requires that you relate archival material to other contemporary sources of a textual and pictorial sort that may be held in other collections;
- asks researchers to continually negotiate between the theoretical and the empirical.

Archival work can be time-consuming and, superficially at least, frustrating in that the information retrieved may offer only partial answers, particularly when you find yourself under time pressure to complete a research project. Archival work done properly requires patience. Rarely will the surviving archival material provide "full" answers to the questions you pose. In the case of public archives, the surviving material typically says more about politics, economics, the concerns of elites, and men than it does about social and private spaces, women, and minority groups. It is, however, still possible to use these records to recreate something about the lives of ordinary people. But surrendering to the temptation to merely summarize the content of files, a trap into which inexperienced archival researchers can fall, is another way of being—as Harris terms it—"swallowed by the archive."

It is all too easy in discussing archival research to create the impression that there is no room for novices, when in fact more human geographers need to be encouraged to incorporate archival work into their research programs. I would simply describe archival research as somewhat akin to confidently accepting the challenge of working on a jigsaw puzzle even though you can be reasonably certain that pieces are missing and that the box cover with the picture of the completed puzzle will never be found. Good archival research can be extremely satisfying, both in learning the skills to conduct it and in the presentation of results.

Key Terms

archival research
archives
archivist
copperplate script
document
files
finding aids
modern hand
"pencil-only" rule
primary sources
provenance
records
secondary literature
secretary hand
series lists

Review Questions

1. What sorts of research questions can be addressed using archival sources?
2. What are the problems of a researcher being, as Harris terms it, "swallowed by the archive"?
3. What is meant by provenance, and why it is important to archival researchers?
4. What steps are required in assessing historical evidence?
5. What approach would you adopt to organize and store digital images from archival collections?

Review Exercise

The one-hundredth anniversary of the beginning of World War One has been marked by various commemorations and national observances. There are also signs of renewed scholarly interest in the war and its aftermath. For Australia and New Zealand, the Gallipoli campaign of 1915 is associated with foundational nationhood myths of both countries. Accordingly, this exercise asks you to log on to several archives to explore some projects on the war and to access selected archival material available online. The intent of the exercise is to give you some experience at locating files, in reading period documents, and in searching for information therein.

National Archives of Australia has created a "Discovering Anzacs" website (see http://discoveringanzacs.naa.gov.au/). ANZAC is the acronym for the Australia and New Zealand Army Corps formed for the Gallipoli campaign. The "Discovering Anzacs" site is especially sensitive to geography in that individuals can be searched for by place of birth and of enlistment, which makes it useful for understanding the demographic impact of the war on particular localities. Choose a location in a state of your choice and select from the list of locally born or enlisted or search by name and then click on their service record—for instance James Henry Berry who enlisted at Colac in Victoria in

1915 and was killed in France in 1917. The nationality, age, occupational, and biometric information can help you build up a picture of the enlistees that can complements the bare statistics of numbers enlisted.

The postwar experiences of some of these men can be traced through the NSW State Records "A Land Fit for Heroes?"" site (http://soldiersettlement.records.nsw.gov.au/) that includes 189 case studies drawn from the 9000 soldier settlers in the state. The case studies provide some harrowing accounts of attempts to farm land during the 1920s and 1930s. It is possible to match up soldier enlistment records with postwar settler records; see, for example, Percy Bailey who enlisted in Condobolin, NSW and later farmed in the same area.

Archives New Zealand has begun to digitize WWI service records where they can be accessed on the archive website (http://www.archway.archives.govt.nz/). By working through a selection of service records you can build up a picture of the ages and occupations of the men. There were volunteers throughout the war but conscription was introduced in 1916. By examining their service records you can gain a sense of the mobility of their army careers. This may involve some patience both in interpreting the largely handwritten records and in decoding the military abbreviations. Individuals who may be searched for by name in the simple search function on Archway. For the purposes of this exercise, search John Smith Gandell, Hoana Mete, Peter Prendergast, and George John Smith. (N.B. This group is not intended to be representative; rather the point is to give you experience in reading and extracting information from the service records.)

Useful Resources

Black, I. 2006. "Analysing historical and archival sources." In N. Clifford and G. Valentine, eds, *Key Methods in Geography*, 477–500. London: Sage.

Keighren, I. M. 2012. "Fieldwork in the Archive." In R. Phillips and J. Johns, eds, *Fieldwork for Human Geography*, 138–40. London: Sage.

McLennan, S., and G. Prinsen. 2014. "Something old, something new: Research using archives, texts, and virtual data." In R. Scheyvens, ed., *Development Fieldwork: A Practical Guide*, 81–100. London: Sage

Rekrut, A. 2011. "Connected constructions, constructing connections, materiality of archival records as historical evidence," In K. Gray and C. Verduyn, eds., *Archival Narratives for Canada Re-telling Stories in a Changing Landscape*, 135–157. Halifax and Winnipeg: Fernwood.

Ventresca, M., and J. Mohr. 2005. "Archival research methods." In J. Baum, ed., *The Blackwell Companion to Organizations*. Blackwell Reference Online. http://www.blackwellreference.com/subscriber/tocnode?id=g9780631216940_chunk_g978063121694040.

Notes

1. Much of this chapter also relates to photographs, but this sub-field has a literature of its own—for example, Schwartz and Ryan (2003) and Quanchi (2006).
2. A discussion of the behind-the-scenes work of the archivist in appraising, arranging, and describing records is provided by Harvey (2006).
3. Indeed, you may be dealing with records in another language. For instance, there is good deal of correspondence in Te Reo Māori in Archives New Zealand, much but not all of which is accompanied by an English version prepared by official translators.
4. These guidelines are expanded on in the American Historical Association's "Statement of standard of professional conduct," which is available online at http://www.historians.org/PUBS/Free/ProfessionalStandards.cfm.
5. Given that this volume focuses on qualitative methods, I have omitted discussion of how a researcher might extract and present in tabulated form quantitative information derived from archival sources.

Using Questionnaires in Qualitative Human Geography

Pauline M. McGuirk and Phillip O'Neill

Chapter Overview

This chapter deals with questionnaires, an information-gathering technique used frequently in mixed-method research that draws on quantitative and qualitative data sources and analysis. We begin with a discussion of key issues in the design and conduct of questionnaires. We then explore the strengths and weaknesses for qualitative research of various question formats and questionnaire distribution and collection techniques, including online techniques. Finally, we consider some of the challenges of analyzing qualitative responses in questionnaires, and we close with a discussion of the limitations of using questionnaires in qualitative research.

Introduction

Qualitative research seeks to understand the ways people experience events, places, and processes differently as part of a fluid reality, a reality constructed through multiple interpretations and filtered through multiple frames of reference and systems of meaning-making. Rather than trying to measure and quantify aspects of a singular social reality, qualitative research draws on methods aimed at recognizing "the complexity of everyday life, the nuances of meaning-making in an ever-changing world and the multitude of influences that shape human lived experiences" (DeLyser et al. 2010, 6). Within this epistemological framework, how can questionnaires contribute to the methodological repertoire of qualitative human geography? This chapter explores the possibilities.

Commonly in human geography, questionnaires pose standardized, formally structured questions to a group of individuals, often presumed to be a **sample** of a broader **population** (see Chapter 6). Questionnaires are useful for gathering original data about people, their behaviour, experiences and social interactions, attitudes and opinions, and awareness of events (McLafferty 2010; Parfitt 2005). They usually involve the collection of quantitative *and* qualitative data. Since such **mixed-method** questionnaires first appeared with the rise of behavioural geography in the 1970s (Gold 1980), they have been used increasingly to gather data in relation to complex matters like the environment, social identity, transport and travel, quality of life and community, work, and social networks.

While there are limitations to the qualitative data that questionnaires are capable of gathering, they have numerous strengths. First, they can provide insights into social trends, processes, values, attitudes, and interpretations. Second, they are one of the more practical research tools in that they can be cost-effective, enabling extensive research over a large or geographically dispersed population. This is particularly the case for questionnaire surveys conducted online where printing and distribution costs can be minimized (Sue and Ritter 2012). Third, they are extremely flexible. They can be combined effectively with complementary, more intensive forms of qualitative research, such as interviews and focus groups, to provide more in-depth perspectives on social process and context. For instance, McGuirk and Dowling's (2011) investigation of the planning and development of masterplanned estates and the everyday lives of residents combined key informant interviews with planners and developers, questionnaires with local residents, and follow-up in-depth interviews with volunteers who had participated in the questionnaire. Data from the questionnaire provided a framework for the in-depth interviews, allowing key themes, concepts, and meanings to be teased out and developed (see Mee 2007, Askew and McGuirk 2004, for similar examples). In this mixed-method format particularly, questionnaires can be both a powerful and a practical research method. Comparatively, Beckett and Clegg (2007) report on the success of qualitative research into women's experiences of lesbian identity using only postal questionnaires to gather rich accounts from respondents. This process allowed respondents the privacy and time to consider and develop their responses to sensitive questions. The questionnaire as a research instrument, then, seems to have nurtured rather than constrained the data collection exercise.

Questionnaire Design and Format

While each questionnaire is unique, there are common principles of good design and implementation. Producing a well-designed questionnaire for qualitative research involves a great deal of thought and preparation, effective organizational strategies, and critical review and reflection, as an array of literature suggests (for example, de Vaus 2014; Dillman 2007; Fowler 2002; Gillham 2000; Lumsden 2005; see also the relevant chapters in Babbie 2013; Bryman 2012; Clifford and Valentine 2003; Flowerdew and Martin 2005; Hoggart, Lees, and Davies 2002; and Sarantakos 2012). The design stage is where a great deal of researcher skill is vested, and it is a critical stage in ensuring the worth of the data collected.

Notwithstanding the quality of the questionnaire devised, we are beholden as researchers to ensure that we have sufficient reason to call on the time and energy of the research participants. The desire to generate our "own" data on our research topic is insufficient justification (Hoggart, Lees, and Davies 2002). As with any study, and as discussed in many of the chapters of this book, the decision to go ahead with a questionnaire needs to be based on careful reflection on detailed research objectives, consideration of existing and alternative information sources, and appropriate ethical contemplation that is attuned to the particular cultural context of the research.

The content of a questionnaire must relate to the broader research question as well as to your critical examination and understanding of relevant processes, concepts, and relationships. As a researcher, you need to familiarize yourself with relevant local and international work on your research topic. This ensures clarity of research objectives and will help you to identify an appropriate participant group and relevant key questions. You need to be clear on the intended purpose of each question, who will answer it, and how you intend to analyze responses. You also need to be mindful of the limits to what people are willing to disclose, being aware that these limits will vary across different social and cultural groups in different contexts. Public housing tenants, for instance, might be wary about offering candid opinions about their housing authority. Respondents might be cautious about what they are willing to disclose in questionnaires administered via email because of the loss of anonymity that occurs when email addresses can be matched with responses (Van Selm and Jankowski 2006). Every question, then, needs to be carefully considered with regard to context and have a clear role and purpose appropriate to the social and cultural norms and expectations of the participant group (Madge 2007).

Begin by drawing up a list of topics that you seek to investigate. Sarantakos (2012) describes the process of developing questions for a questionnaire as a process of translating these research topics into variables, variables into indicators, and indicators into questions. Identify the key concepts being investigated, and work out the various dimensions of these concepts that should be addressed. Then identify indicators of the dimensions, and use them to help you formulate specific questions. Doing this will ensure that each question relates to one or more aspects of the research and that every question has a purpose. De Vaus (2014) suggests that it is helpful to think about four distinct types of question content:

1. *attributes:* Attribute questions aim at establishing respondents' characteristics (for example, age or income bracket, dwelling occupancy status, citizenship status).
2. *behaviour:* Behaviour questions aim at discovering what people do (for example, recreation habits, extent of public transport use, food consumption habits).
3. *attitudes:* Questions about attitudes seek to discover what people think is desirable or undesirable (for example, judgment on integrating social

housing with owner-occupied housing, willingness to pay higher taxes to fund enhanced social welfare services).

4. *beliefs:* Questions about beliefs aim at establishing what people believe to be true or false or preferred (for example, beliefs on the importance of environmental protection, beliefs on the desirability of social equity).

A guiding principle for question types, however, is to ensure that your target participant group will understand the questions and has the knowledge to answer them (Babbie, 2013). As is the case in newsprint journalism, it is recommended that unless you are targeting a specialized and homogenized group, you phrase questions to accommodate a reading age of approximately 11 years (Lumsden 2005). Rather than "dumbing down" your questionnaire, this tactic helps with clarity and direction. It also encourages respondents to answer the questions: for instance, a complex question asking whether government planning policies contribute to local coastal degradation may lead respondents to abandon the questionnaire.

Apart from the typology of question content, there is a range of question formats from which to draw. We commonly make a distinction between closed and open questions, each of which offers strengths and weaknesses and poses different challenges depending on the mode through which the questionnaire is being administered (e.g., mail, face-to-face, email). **Closed questions** may seek quantitative information about respondent attributes (for example, level of educational attainment) or behaviour (for example, how often and where respondents buy groceries). You should provide simple instructions on how to answer closed questions (e.g., how many responses the respondent can tick). Some examples are set out in Box 12.1. Closed questions can ask respondents to select categories, rank items as an indicative measure of attitudes or opinions, or select a point on a scale as indicative of the intensity with which an attitude or opinion is held (see Sarantakos 2012, chapter 11). A major benefit of closed questions is that the responses are easily coded and analyzed, a bonus when interpreting a large number of questionnaires. Indeed, for web-based questionnaires, a data file can be assembled automatically as respondents type in their answers. Closed questions are demanding to design, however, since they require researchers to have a clear understanding of what the range of answers to a question might be. Respondents' answers are limited to the range of categories designed by the researcher, and this can be a constraint. It has also been found that when respondents are asked to "tick all appropriate categories" on a list (see the category list question in Box 12.1), they can turn to **satisficing behaviour**; that is, they keep reading (and ticking) until they feel they have provided a satisfactory answer and then stop. Relatedly, a significant limitation of closed questions is that they rest on the assumption that words, categories, and concepts carry the same meaning for all respondents, which is not always the case. For example, how a respondent answers the question "How often have you been a victim of crime in the past two years?" will depend on what the respondent sees as a crime (de Vaus 2014). It is worthwhile to be aware, too, that the ways particular questions are posed or how they relate to preceding questions can influence respondents' answers. Babbie (2013) shows that greater support in questionnaire surveys is indicated habitually for the phrase "assistance to the poor" rather than "welfare" and for "halting rising crime rate" rather than "law enforcement." A further criticism of closed questions is that the loss of spontaneity in respondents' answers and the removal of the possibility of "interesting replies that are not covered by the fixed answers" (Bryman 2012, 250). This limitation might be overcome by offering an answer option such as "other (please specify)" or by using **combination questions** that request some comment on the option chosen in a closed question (see Box 12.1).

In general, **open questions** have greater potential to yield the in-depth responses that match an aspiration of qualitative research: to understand how meaning is attached to process and practice. Open questions offer less structured response options than closed questions, inviting respondents to recount understandings, experiences, and opinions in their own style. Rather than offering alternative answers, which restrict responses, open questions provide space (and time) for free-form responses. Open questions also give voice to respondents and allow them to question the terms and structure of the questionnaire itself, demonstrate an alternative interpretation, and add qualifications and justifications. This capacity acknowledges the **co-constitution of knowledge** by researcher and research participant (Beckett and Clegg 2007). For instance, Mee (2007) used open questions in her questionnaire-based research exploring public housing tenants' experiences of home in medium-density unit dwellings in Newcastle, Australia. Despite normative assertions that link ideas of home to home ownership and detached housing, respondents used the open questions to describe their rented apartment homes as "heaven," "a blessing," and as "wonderful" and "beautiful." Open questions, then, are capable of yielding valuable insights, many of them unanticipated, and they can open intriguing lines of intensive inquiry in scenarios where extensive research is the main focus or where a more intensive person-to-person approach is not possible (Cloke et al. 2004). Such scope, however, means that open questions can be effort-intensive for respondents to answer and time-consuming to code (Bryman 2012). An open format can also throw up responses that lack consistency and comparability. Certainly, respondents answer them in terms that match their interpretations. So open questions and the responses they yield are certainly more challenging to analyze than are their more easily coded closed counterparts (see Chapter 18). But welcome to "the rich yet ambiguous and messy world of doing qualitative research" (Crang 2005b, 231)!

BOX 12.1 Types of Questionnaire Questions

Closed questions

Attribute information

How often do you shop at this shopping mall? (please tick the appropriate box)

- Less than once a week ☐
- Once a week ☐
- Twice a week ☐
- More than twice a week ☐

Category list

What was the main reason you chose to live in this neighbourhood? (please tick the appropriate box)

- Proximity to work ☐
- Proximity to family and friends ☐
- Proximity to schools or educational facilities ☐
- Proximity to shopping centre ☐
- Proximity to recreational opportunities ☐
- Environment ☐
- Housing costs ☐
- Good place to raise children ☐
- Pleasant atmosphere of neighbourhood ☐
- Other (please specify)

Rating

Please rank the reasons for buying your current house (please rank all relevant categories from 1 [most important] to 6 [least important]).

- Price ☐
- Location ☐
- Size ☐
- Proximity to job/family ☐
- Investment ☐
- Children's education ☐

Scaling

Please indicate how strongly you agree/disagree with the following statement (please tick the appropriate box):

Having a mix of social groups in a neighbourhood is a positive feature.

- Strongly disagree ☐
- Disagree ☐
- Neutral ☐
- Agree ☐
- Strongly agree ☐

Grid/matrix question

Think back to when you first got involved in environmental activism. What initially inspired you to get involved? (please tick the appropriate box for *each* reason)

	Very influential	**Fairly influential**	**Not very influential**	**Not influential**
Spirituality/ religious beliefs				
Fear/anxiety about ecological crisis				
Desire to change the world				
Nature/ecology experiences and care for the environment				
Political analysis				
Commitment to justice				
Felt like you could make a difference				
Influential person (please specify)				
Influential book/ film (please specify)				
Key event (please specify)				
Contact with an organization, campaign, or issue (please specify)				
Outreach activities by an organization (please specify)				
Wanted to meet new people				
Want to learn new skills				
Sense of personal responsibility				
Other (please specify)				

Combination question

Have changes in the neighbourhood made this a better or worse place for you to live? (please tick the appropriate box)

Changes have made the neighbourhood better ❑
Changes have not made the neighbourhood better or worse ❑
Changes have made the neighbourhood worse ❑

____

____

____

Open questions

What have been the biggest changes to the neighbourhood since you moved in?

____

____

____

____

____

____

____

What, if any, are the advantages for civic action groups of using the Internet, email, and cellular phones?

____

____

____

____

____

____

____

Please describe any problem(s) you encounter using public transport.

____

____

____

____

____

____

____

In summary, using open questions makes it possible to pose complex questions that can reveal people's experiences, understandings, and interpretations of social processes and circumstances. as well as their reactions to them. Closed questions are not capable of such in-depth explorations. Answers to open questions can also tell us a good deal about how wider processes operate in particular settings. Thus they point to the need for caution when imagining general processes, operating universally, and highlight the way wider processes are shaped in and by the messy contexts of everyday life. Thus, they enable research that addresses the two fundamental questions that Sayer (2010) poses for qualitative research: what are individuals' particular experiences of places and events? And how are social structures constructed, maintained, or resisted? (see Chapter 1). Beyond choice of question content and type, getting the wording, sequence, and format of a questionnaire right is fundamental to its success. Guidance on these is given in Box 12.2 with discussion revolving around clarity, simplicity, and logic. In question wording, you need to be sure that questions are sufficiently precise and unambiguous to ensure that the intent of your question is clear and well communicated. It is advisable to be familiar with the vernacular of the

Guidelines for Designing Questionnaires

- Ensure questions are relevant, querying the issues, practices, and understandings you are investigating.
- Keep the wording concise (about 20 words maximum), simple, and appropriate to the targeted group's vernacular.
- Ensure that questions and instructions are easily distinguishable in format and font.
- Avoid double-barrelled questions (for example, "Do you agree that the Department of Housing should cease building public housing estates and pursue a social mix policy?").
- Avoid confusing wording (for example, "Why would you rather not use public transport?"), and be alert to alternative uses of words (for example, for some people "dinner" implies an evening meal while for others it implies a cooked meal, even if eaten at midday).
- Avoid leading questions (for example, "Why do you think recycling is crucial to the health of future generations?"), and avoid loaded words (for example, "democratic," "free," "natural," "modern").

participant group. In online contexts, this may include becoming familiar with the jargon, abbreviations, and grammatical rules commonly used within the online community being approached (for instance, the language styles of specific blogger or social media groups) (Madge 2007). Remember that the language of a questionnaire is not just textual. Graphical and numerical modes might also be present. These modes work together to affect respondents' perception of the survey and are perceived in ways that are influenced by cultural context (Lumsden 2005). The web's capacity for global reach also means that online questionnaires may target international participants, not all of whom communicate expertly in English. There are software programs that allow the researcher to convert a questionnaire written in English into other languages (see http://www.objectplanet.com/opinio/howto/translation.html) as well as commercial providers (e.g., QuestionPro; http://www.questionpro.com/features/multi-lingual.html). Beyond issues of logic, clarity, and comprehension, questions should avoid threats or challenges to respondents' cultural, ethnic, or religious beliefs, which may arise from a researcher's insensitivity, ignorance, or lack of preparation, even in the absence of overt prejudice. The need for concern about respondents'

- Avoid questions that are likely to raise as many questions as they answer (for example, "Are you in favour of regional sustainability?" raises questions of what sustainability means, how a region is defined, and how different dimensions of sustainability might be prioritized).
- Order questions in a coherent and logical sequence.
- Ensure the questionnaire takes no more time to complete than participants are willing to spend. This will depend on the questionnaire context (for example, whether it is conducted by telephone, face-to-face, or online). Generally, 20 to 30 minutes will be the maximum, although longer times (45 minutes) can be sustained if the combination of context and research topic is appropriate.
- Ensure an uncluttered layout with plentiful space for written responses to open questions.
- Use continuity statements to link questionnaire sections (for example, "The next section deals with community members' responses to perceived threats to their neighbourhood.").
- Begin with simple questions, and place complex, reflexive questions or those dealing with personal information or sensitive or threatening topics later in the questionnaire.

cultural safety (Matthews et al. 1998, 316) is part of the researcher's broader ethical obligations.

The flow and sequence of the questionnaire are fundamental to respondents' understanding of the purpose of the research and to sustaining their willingness to offer careful responses and, indeed, to completing the questionnaire to its conclusion. Grouping questions into related questions connected by introductory statements will help here. In general, open-ended questions are better placed towards the end of a questionnaire, by which time respondents are aware of the questionnaire's thrust and may be more inclined to offer fluid and considered responses. In terms of layout, aim for an uncluttered design that is easy and clear to follow. Where you use closed questions, aligning or justifying the space in which the answer should be provided will contribute to clarity and simplify coding. With open-ended questions, particularly in hard copy, you need to be conscious of the need to leave enough space for respondents to answer without leaving so much as to discourage them from offering a response altogether.

All of these questionnaire design principles need to be observed regardless of how the questionnaire is being distributed: whether by mail, face-to-face, by telephone, by email, or online. However, there are additional design factors that are important to consider when using an online environment (Dillman 2007). Web-based questionnaire delivery makes it possible to incorporate novel features such as split screens, drop-down boxes, images, and sound tracks, although some of these features require that respondents have powerful computers, particular software, and ample download time. You need to consider whether the participant group has the ability and the capacity to receive and respond to the questionnaire and its mode of delivery. Web surveys with advanced multimedia features, for example, have high bandwidth requirements (Vehovar and Manfreda, 2008). You also need to remember that online questionnaires require respondents to think about how to respond to the questionnaire while simultaneously thinking about technical options, a matter that is particularly important if your target participant group is less computer literate. Keeping things simple and limiting the number of actions a respondent has to undertake is sensible. Finally, you need to take account of whether you will administer your questionnaire solely online or through other modes as well, in which case you need to be mindful of how questions will be posed in those other modes. Box 12.3 outlines additional key principles for the design of online questionnaires (adapted from Dillman and Bowker 2001).

Finally, whether developing a conventional or online questionnaire, you should include a cover or introductory letter or email. Box 12.4 offers examples. The letter or email needs to provide general information about the purpose of the questionnaire as well as information about confidentiality, how the respondent has been selected, how long the questionnaire will take to complete, and, when relevant, instructions on how and when to return the questionnaire.

Guidelines for Designing Online Questionnaires

- Introduce web questionnaires with a welcome screen page providing basic instructions and information and encouraging completion.
- Ensure the first question is interesting to respondents, easily answered, and fully visible.
- Use conventional formats for questions, similar to those normally used on self-administered paper questionnaires.
- Provide clear instructions including technical advice on how to respond to each question, and position them at the points where they are needed.
- Limit the length of the questionnaire. The typical length of a paper questionnaire may seem excessively long when completed on a website where a typical print page can take up several screen pages.
- Keep the layout, colour, and graphics simple to aid navigational flow and readability and ensure the format is maintained across different browsers and screen set-ups.
- Allow respondents to move to the next question without having to answer a prior question.
- Allow respondents to scroll from question to question without having to change screen pages.
- If the number of answer choices exceeds what can be viewed in a single column on one screen, display choices as a double bank.
- Include advice that indicates how much of the questionnaire the respondent has completed.
- Close with a thank-you screen page.

Sampling

Before administering a questionnaire, you will need to make a decision about the target audience, or sample. In quantitative research, questionnaires are used commonly to generate claims about the characteristics, behaviour, or opinions of a group of people ("the population") based on data collected from a sample of that population. The population might be, for example, tenants in public housing, the residents of a given local government area, or people living with HIV/AIDS. The sample—a subset of the population—is selected to be representative of the population such that the mathematical probability that the characteristics of the sample

Examples of Invitations to Participate in Questionnaire Studies

Sample cover letter

School of Geography
Geography Building
East Valley University
Kingsland 9222
Telephone: (04) 89889778
Facsimile: (04) 89889779
Email: E.saunders@evu.edu.ca

High-Density Residential Living in Port Andrew, East Valley

I am Edith Saunders, a research student with the School of Geography at the East Valley University. As part of my research on high-density residential environments in East Valley, I am investigating how people understand and create feelings of home in high-density neighbourhoods. The research is being conducted in collaboration with East Valley Council and is aimed at informing its policy and planning decision-making. The work is focused on the Port Andrew area, and you have been selected to receive this questionnaire as a local resident.

The questionnaire asks about the ways you understand and use your home and the ways you interact with your local neighbourhood spaces and services. The questionnaire will take approximately 30 minutes to complete, and completion is voluntary. The questions ask primarily about your experiences and opinions. There are no right or wrong answers. All answers will be treated confidentially and anonymously, and individuals will not be identifiable in the reporting of the research.

It would be appreciated if you could complete the questionnaire at your earliest convenience and no later than July 30. Please return the completed questionnaire in the reply-paid envelope provided. Return of the questionnaire will be considered as your consent to participate in the survey.

Your participation is greatly appreciated. Your opinions are important in helping to build understanding of high-density residential living and how it can be supported through local government planning and provision of neighbourhood spaces and services.

Questions about this research can be directed to me at the address provided.

Thank you in advance for your participation.

Yours faithfully,
Edith Saunders

The university requires that all participants be informed that if they have any complaints concerning the manner in which a research project is conducted, it may be given to the researcher or, if an independent person is preferred, to the university's Human Research Ethics Officer, Research Unit, East Valley University, Postcode OG9222, telephone (04) 8988 1234.

Sample email invitation to participate in an online questionnaire

From: kanchana.phonsavat@evu.edu.ca
To: [email address]

Subject: Survey on high-density residential college living

Dear Student,

I am a research student with the School of Geography at East Valley University (EVU). As part of my research, I am investigating how students understand and create feelings of home in high-density residential college environments. The research is being conducted in collaboration with EVU and East Valley Council. You have been selected to receive this invitation to participate as a student resident of one of EVU's residential colleges.

We are interested in the ways you understand and use your college accommodation and the ways you interact with your local neighbourhood spaces and services. The questionnaire will take approximately 30 minutes to complete and is completely voluntary and confidential. The data will be used to evaluate university and council policies and their support of high-density residential environments.

To complete the questionnaire, please click on the following link:

http://www.newurbanliving.evu.org.ca/surveys.html

It would be great if you could complete the questionnaire in the next two weeks. If you have any questions or need help, please email me at kanchana.phonsavat@evu.edu.ca.

Thank you in advance for your participation.
Kanchana Phonsavat

The university requires that all participants be informed that if they have any complaints concerning the manner in which a research project is conducted, it may be given to the researcher or, if an independent person is preferred, to the university's Human Research Ethics Officer, Research Unit, East Valley University, Postcode OG9222, telephone (04) 8988 1234.

are reproduced in the broader population can be calculated (May 2011). In such cases, a list of the relevant characteristics of the population, the **sampling frame**, is required so that a sample can be constructed. A sampling frame might be, for example, the tenant list of a given public housing authority, a local electoral register, or health register of all people in a given geographical area receiving treatment for diabetes. The rules surrounding sampling are drawn from the central limit theorem used to sustain statistical claims to representativeness, generalizability, and replicability (see McLafferty 2010; Parfitt 2005).

On the other hand, questionnaires used in qualitative research are usually used as a part of mixed-method research aimed at establishing trends, patterns, or themes in experiences, behaviours, and understandings. Important to the analysis, then, is uncovering the influence of a *specific context*, rather than making generalizable claims about whole populations (Herbert 2012). A more appropriate sampling technique for qualitative research is non-**probability sampling** where generalization about a broader population is neither possible nor desirable. Sampling frames may not, in any case, be available. Some web surveys, for instance, involve self-selection by respondents where anyone who agrees to complete the questionnaire can be included in the sample. For example, Tomsen and Markwell's (2007) research into the perception and experience of safety at Australian gay and lesbian events included an online questionnaire. Respondents were invited to complete the questionnaire through targeted advertising in the gay and lesbian press, a media release, radio interviews, and providing information to 25 online chat groups and email lists. A total of 332 people from across the country participated in the questionnaire. Specifically, **purposive sampling** (see Chapter 6) is commonly used where the invitation to participate is made according to some common characteristic, be it a social category (for example, male single parents), a behaviour (for example, women who use public transport), or an experience (for example, victims of crime). There are no specific rules for this type of sampling. Rather, the determinants of the appropriate sample and sample size are related to the scope, nature, and intent of the research and to the expectations of your research communities.

As in all research, these considerations are overlain by resource constraints (time and money). Nonetheless, a lack of hard-and-fast rules and a need for pragmatism do not imply the absence of a systematic approach—quite the opposite. Complex and reflexive decisions need to be made about how to approach sampling. For instance, in research on what motivates "sea-changers" to abandon city life and relocate to regional coastal areas, researchers would need to take into account whether they should seek respondents in all age groups, all household types, and all income categories. Research on people living with HIV/AIDS would need to take into account whether the researchers should target, say, early-stage individuals only, both biological sexes, people of any sexual orientation, only individuals infected from a particular source, and so on. Each decision is liable to have ramifications for how sample recruitment proceeds and what mode of questionnaire distribution is suitable. Questionnaires administered online, for example, may be well suited to research on factors shaping environmental advocacy where the target respondents are likely to have web skills and access to computers as part of their work. By comparison, this mode of distribution may be poorly suited to research on perceptions of cultural displacement among low-income populations in gentrifying areas. These cases illustrate the importance of research scope, purpose, and intent in shaping the sampling approach and in determining appropriate sample size. Bryman (2012) provides details of various types of purposive sampling, along with a discussion of sample size, and Chapters 6 and 7 in this book provide an extended treatment of further questions regarding selecting cases and participants. In the end, decisions about samples are shaped by compromises between cost, need for targeting, the nature of the research, and the limits of possibility.

Pre-Testing

You must try out a questionnaire before it is distributed. **Pre-testing** is piloting or road-testing a questionnaire with a sub-sample of your target population to assess the merits of its design, its appropriateness to the audience, and whether it does in fact achieve your aims. For web-based questionnaires, rigorous testing of the questionnaire on a range of platforms and browsers should be undertaken to identify and weed out potential technical problems. In web-based contexts, technical bugs are very likely to result in respondents abandoning the questionnaire. Getting feedback from those with extensive questionnaire-design experience and from those who might use the data generated (for instance, in the example in Box 12.4, a local authority and a university) will allow possible problems to be identified and improvements made. Scheduling a pre-testing stage provides the opportunity for post-test revisions that might dramatically increase the questionnaire's effectiveness.

Both individual items and the overall performance of the questionnaire need attention at this stage. Are instructions and questions easily understood? Would any of them benefit from the addition of written prompts? Do respondents interpret questions as intended? Do any questions seem to make respondents uncomfortable? Discomfort and sensitivity (perhaps the question is considered too intrusive) might be indicated by respondents skipping or refusing to answer a question or section. Alternatively, such outcomes could mean that respondents do not understand the question or do not have the knowledge or experience to answer it. Consider too how respondents react to the order of the questions. Does it seem to them that the questions flow logically and intuitively? Are there parts where the questionnaire seems to drag or become repetitive? Technical aspects can also be tested: Is there enough space for respondents to answer open questions? How long will the questionnaire take to complete? Do the data being generated present particular problems for analysis? If you plan to conduct the questionnaire face to face with respondents, the pre-test stage can also be a useful exercise in training and confidence-building.

Modes of Questionnaire Distribution

Consideration of the mode of questionnaire distribution should be one of the earliest stages of your questionnaire design. This has implications for design, layout, question type, and sample selection. The main distribution modes are mail, face-to-face, telephone, and the Internet-mediated modes of email and the World Wide Web. Each mode has distinctive strengths and weaknesses, and our choice depends on the research topic, type of questions, and resource constraints. The best choice is the one most appropriate to the research context and target participant group, while the success of any particular mode is dependent on a design appropriate to context and participant group. So the question is: what should researchers interested in qualitative research be aware of to guide them in the choice of mode?

Mailed questionnaires have clear advantages of cost and targeted coverage. They can be distributed to large samples over large areas (for example, an entire country or province) at a relatively low cost. The anonymity they provide may be a significant advantage when sensitive topics are being researched—for example, those dealing with socially disapproved attitudes or behaviours, such as racism or transgressive sexual behaviour, or topics involving personal harm, such as experience of unemployment or crime. Respondents may also feel more able to take time to consider their responses if unimpeded by the presence of an interviewer. Clearly, too, the absence of an interviewer means responses cannot be shaped by how an interviewer poses a question, interacts with the respondent, or interprets cues in the conversation in culturally specific ways.

Nonetheless, mailed questionnaires are generally the most limited of the three modes in terms of questionnaire length and complexity. The scope for complex open questions is particularly limited by the need for questions to be self-explanatory and brief, and this may be a significant consideration for qualitatively oriented research. Once the questionnaire is sent out, there is little control over who completes it or, indeed, over how it is completed; respondents may choose to restrict themselves to brief, unreflective, or patterned responses. A response to the question "what do you value about living in this community?" might yield a response of several paragraphs from one respondent and the comment "friends and neighbours" from another. There is no opportunity to clarify questions or probe answers. Nor is there control over the pattern and rate of response. Some parts of the target participant group may respond at a higher rate than others. It is common, for instance, for mailed questionnaires to achieve significantly higher response rates in wealthy neighbourhoods than in less socially advantaged neighbourhoods. Finally, mailed questionnaires can be subject to low response rates unless respondents are highly motivated to participate. Response rates of 30 to 40 per cent are considered good (Cloke et al. 2004), although effective follow-up steps can increase a rate somewhat (May 2011).

Distributing questionnaires electronically is a recent variation on mail distribution and brings new potential for innovation and experimentation (Babbie 2013, 284). There are three main means of electronic distribution: (1) sending the entire questionnaire to respondents as an email attachment, (2) posting or emailing respondents an introductory letter with a hyperlink to a web-based questionnaire, and (3) distributing a general request for respondents (for example, via an online newsgroup) to complete a web-based questionnaire. You might also use a mix of these distribution strategies (Bryman 2012, 672). A major benefit of electronic distribution is that it "compresses" physical distance and expands enormously the reach of the questionnaire. Groups can be reached that are difficult to contact with paper questionnaires. These could include, for example, people with restricted mobility who might find it easier to respond online than to mail a completed questionnaire. Furthermore, people practising covert or illegal behaviours—for example, graffitists or drug users—may be more easily recruited through the Internet. The Internet is also a powerful way of gaining access to self-organized groups—for example, those with common interests, lifestyles, or experiences organized into chat-rooms, newsgroups, and online forums. For example Banaji and Buckingham's (2010) study on Internet activism and young people sought out specific activist websites and conducted a questionnaire with 3000 users. Mailing lists or online newsgroups can be used for circulating the questionnaire or inviting participants to complete an online questionnaire. However, some groups are sensitive to the intrusion of researchers via mailing lists and newsgroups (Chen, Hall, and Johns 2004). Many discussion groups state their privacy policy when you join, so researchers should check the welcome message of public discussion lists for guidelines before using them to recruit potential participants (Madge 2007).

Regardless of the specific means of electronic distribution used, the recruitment of participants will be affected by the age, class, and gender biases that shape computer use, email, and online patronage (see Gibson 2003). For instance, online delivery of a questionnaire investigating the leisure habits of elderly people is likely to confront participation problems, given that elderly people are less likely to complete online surveys due to lack of Internet access. Low-income groups would similarly have restricted access (Babbie 2013, 283).

Other benefits of electronic distribution include cost savings and efficiencies. Electronic dissemination enables the use of attractive formats and colour images without associated printing costs, although you should avoid overloading online questionnaires with cluttered design features or complex graphics that require excessive download time. Electronic distribution opens up opportunities for flexibility in question design, for more complex questions, for incorporating adaptive questions with encoded skip patterns (thus removing the need for complex instructions and filter questions), and for increasing the potential to generate rich and accurate qualitative data with fewer unanswered questions (Bryman 2012). Researchers who have deployed electronic distribution report lower response rates than conventionally distributed questionnaires; although rates can be comparable when pre-notification and follow-up emails are used (Fan and Yan 2010). Online respondents characteristically submit lengthy commentaries on open questions (Van Selm and Jankowski 2006), a plus for qualitative research. Apart from saving on print and postage costs, the electronic collection of data offers the major advantage over paper questionnaires of eliminating the need for a separate labour-intensive phase for data entry and coding of closed questions (Van Selm and Jankowski 2006).

Mailed and online questionnaires do, however, present a particular set of challenges surrounding hidden costs, ethical issues, and technical capacities and failures. The cost and labour savings of avoiding coding and data entry through electronic data capture can be offset by the costs of design and programming (Hewson et al. 2003). To run a web-based questionnaire, you need to be proficient in producing HTML documents, to use survey construction software packages, which can be costly, or to use the commercial services of a web survey host (see Sue and Ritter 2012). Costs can vary significantly. When it comes to ethical issues, obtaining informed consent, and managing privacy and confidentiality all present challenges (Vehovar and Manfreda 2008). It can be difficult to obtain adequate online informed consent. In terms of privacy, the identity of web-based questionnaire respondents can be protected if they withhold their names, although technically adept researchers can collect data about web-based participants using, for example, user log files or Java Applets (Lumsden 2005; Bryman 2012). Anonymity cannot be provided to email questionnaire respondents when the returned questionnaire attaches an email address. Responses stored on computer files, and online, can be accessible to hackers, and this may be a particularly important concern if the study being conducted involves sensitive and personal data. Using encryption to increase the security of data during transfer and storage and backing up and storing data in a secure off-line location are advisable.

Qualitative research is often very effective if questionnaires are administered face to face, although this is a costly option. The major benefits of this mode flow from the fact that an interviewer's presence allows complex questions to be asked (see Chapter 8). As well, an interviewer can take note of the context of the interview and of respondents' non-verbal gestures, all of which add depth to the data collected (Cloke et al. 2004; May 2011). As an interviewer, you can motivate respondents to participate and to provide considered, informative responses. Moreover, people are generally more likely to offer long responses orally than in writing. However, as Beckett and Clegg's (2007) work on lesbian identity suggests, this outcome is context dependent. Perhaps more crucially, face-to-face questionnaires give an interviewer the opportunity to clarify questions and probe vague responses (see Chapters 8 and 9 for related discussions). For example, adding probes like "why is that exactly?," "in what ways?," or "anything else?" can elicit reflection on an opinion or attitude. Long questionnaires can also be completed because direct contact with an interviewer can enhance engagement. The ability to pose complex questions and elicit more in-depth and engaged responses is a major benefit for qualitative research. Moreover, this high level of engagement can also secure high response rates with a minimal number of nil responses and "don't know" answers (Babbie 2013). However, the level of interviewer skill and reflexivity required to secure optimal outcomes should not be underestimated.

As Kevin Dunn discusses more fully in Chapter 8, the presence of an interviewer can be a powerful means of collecting high-quality data, but it introduces limitations as well. Interviewer/respondent interaction can produce "interviewer effects" that shape the responses offered. People filter their answers through a sense of social expectation, especially when interviewed face to face (Lee 2000). They may censor or tailor their answers according to perceived social desirability. That is, they may avoid revealing socially disapproved behaviours or beliefs (such as racism or climate change skepticism) or revealing negative experiences (for example, unemployment). Beckett and Clegg (2007) chose postal questionnaires specifically to ensure the *absence* of an interviewer. Their argument was that participants should be allowed to recount their stories in their own terms, without any identification with the researchers' associations with particular geographical spaces or social and cultural attributes and without fear of judgment by the researcher. When interviewers are used, one means of dealing with respondents' self-censoring is to incorporate a self-administered section in the questionnaire or to reassure respondents through guarantees of anonymity. Moreover, the interviewer's presence (as an embodied subject with class, gender, and ethnic characteristics) can also affect the nature of responses given. For instance, Bryman (2012) suggests that the gender, ethnicity, and social background of the interviewer can introduce significant variations. So while distinct benefits arise from using face-to-face distribution, there are drawbacks. Perhaps the most limiting is the practical consideration of cost. Interviewer-administered questionnaires are expensive and time-consuming and tend to be restrictive both spatially and with respect to population coverage. However, as we suggested before, this factor may not be a significant drawback if a particular, localized participant group is targeted.

While the opportunities for personal interchange are more restricted in telephone than in face-to-face questionnaires, the telephone mode still offers the possibility of dialogue between researcher and respondent and can provide some of the benefits of an actual face-to-face interview but with a level of anonymity that may limit problematic interviewer effects. Conducting questionnaires over the phone may encourage respondent participation because it may be seen as less threatening than opening the door to a stranger wanting to administer a questionnaire. However, telephone delivery constrains the scope for lengthy questionnaires, with about 30 minutes being the maximum time respondents are willing to participate (de Vaus 2014). Furthermore, because the mode relies on a respondent's memory, the question format must be kept simple and the number of response categories in closed questions needs to be limited. However, the advent of **computer-assisted telephone interviewing** (CATI) and **voice capture** technology is significantly enhancing telephone questionnaires (see Babbie 2013, 281) and extending their potential. Moreover, they can be administered with great convenience and at relatively low cost.

Telephone questionnaires may rely on a telephone directory as a sampling frame, and this can introduce class, age, and gender biases among respondents as well as ruling out people whose numbers are not listed. Moreover, as cellular phone use increases, landline directories are becoming less useful as a sampling frame. If telephone numbers are available for a selected group of people, this may not pose a problem. Historically, telephone surveys have had good response rates. However, growing public annoyance with unsolicited marketing calls means approaches by telephone face rejection or screening by answering machines (Guthrie 2010; Dillman et al. 2009).

Maximizing Questionnaire Response Rates

Questionnaire response rates are shaped by the research topic, the nature of the sample, and the quality and appropriateness of questionnaire design as much as by the mode of distribution. In any case, questionnaire response rates tend to be higher when using a purposive sample—as is common in qualitative research—wherein interest in the research topic may be strong. There is good evidence that response rates for online questionnaires are stronger if the questionnaire is relatively brief, taking no longer than 20 minutes to complete; is not complex to complete; is simple in design; and does not require participants to identify themselves (Lumsden 2005). Regardless of the mode of distribution, response rates can be improved by undertaking a series of strategies before questionnaire distribution and as follow-up (Dillman 2007; Bryman 2012). Box 12.5 summarizes the strategies that enhance questionnaire response rates according to the different modes of distribution.

BOX 12.5

Strategies for Maximizing Response Rates

Strategy	Face-to-Face	Telephone	Mail	Online
Ensure mode of distribution is appropriate to the targeted population and research topic.	√	√	√	√
Send notification letter letter (or email pre-notification) introducing the research and alerting to the questionnaire's arrival (or posting online).		√		√
Place newspaper or online advertisement in: local community newspaper/ magazines or online chat rooms/newsgroups introducing the research and alerting to the conduct of the questionnaire.	√	√	√	√
Ensure questionnaire is concise.	√	√	√	√
Ensure appropriate location of approach.	√			
Ensure appropriate time of approach.	√	√		
Vary time if no contacy is made initially.	√	√		
Pre-arrange time/ location for conduct of questionnaire, if appropriate.	√	√		
Print questionnaire on coloured paper to distinguish it from introductory material or other mail.			√	
Send follow-up postcard/ /email thanking early respondents and reminding others (about one week after initial receipt).			√	√

continued

Strategy	Face-to-Face	Telephone	Mail	Online
Ensure reply-paid envelope is included in mail-out.			√	
Send follow-up letter/ email and additional copy of questionnaire (two to three weeks after initial receipt).			√	√
Avoid abrasive manner.	√	√		
Dress appropriately to the target population.	√			

Analyzing Questionnaire Data

Analyzing questionnaires used in mixed-method research that blends qualitative and quantitative data requires an approach that distinguishes between closed questions in which responses are provided in an easily quantified format and open questions that seek qualitative responses. Quantitative data arises primarily from closed questions that provide counts of categorical data (for example, age and income bands, frequency of behaviour) or measures of attitudinal or opinion data (see Box 12.1 for examples). Questions such as these are relatively easy to code numerically and analyze for patterns of response and relationships between the variables that the questions have interrogated (May 2011). Indeed, as noted earlier, response categories can be pre-coded on the questionnaire, simplifying matters even further (see de Vaus 2014 for more detail), while data can be collected readily and easily collated within the electronic environment. The analysis of qualitative responses is more complex. The power of qualitative data lies in its uncovering of a respondent's understandings and interpretations of the social world, and these data, in turn, are interpreted by the researcher to reveal the understandings of structures and processes that shape respondents' thought and action (for elaboration, see Crang 2005b). Chapters 18 and 19 discuss the techniques and challenges of coding and analyzing qualitative data in detail. Nonetheless, it is worth raising some important points specific to analyzing qualitative data arising from questionnaires.

In qualitative responses, the important data often lie in the detailed explanations and precise wording of respondents' answers. For qualitative research, then, it is best to go beyond classifying qualitative responses into simple descriptive categories so as to confine reporting to quantitative dimensions, stating, for example, that "49 per cent of respondents had positive opinions about their neighbourhood." There are two problems here. First, such reporting may well be statistically misleading given they might have been derived from a relatively small purposive sample and could be used incorrectly to frame generalizations. Second, this approach involves "closing" open questions so that the richness of how respondents constructed, in this example, their positive understandings and experiences of their locality, is lost. Certainly, classifying qualitative responses into descriptive categories allows us to simplify, summarize, compare, and aggregate data. Yet, in so doing, we should be careful not to forfeit the nuance and complexity of the original text that was collected as a qualitative exercise to help our understanding of the meanings and operations of social structures and processes and people's interpretations and behaviour in relation to them. Analysis that is more attuned to the thrust of qualitative research will analyze questionnaire data gained by sifting and sorting to identify key themes and dimensions as well as the broader concepts that might underlie them (see the discussion of coding in Chapter 18). Reporting findings in these terms is much more meaningful than falling back on awkward attempts at quantification.

Further, in analyzing qualitative responses, we need to be aware that qualitative research makes no assumption that respondents share a common definition of the phenomenon under investigation (be that quality of neighbourhood, experience of crime, understanding of health and illness, and so on). Rather, it assumes that variable and multiple understandings coexist in a given social context. We need to incorporate this awareness into how we make sense of respondents' answers. Indeed, one of the strengths of using questionnaires in qualitative research is their ability to identify variability in understanding and interpretation across a selected participant group, providing the groundwork for further investigation through additional and complementary methods such as in-depth interviews.

Finally, keep in mind that qualitative data analysis is sometimes referred to as more of an art than a science (Babbie 2013) in that it is not reducible to a set of neat techniques. Although useful procedures can be followed (see Chapter 18), they may need to be customized to the distinctive concerns and structure of each questionnaire and the particular balance of quantitative and qualitative data it gathers. For this reason, and others, at all stages of the process of analysis we need to be mindful of engaging in critical reflexivity, especially when considering how our own frames of reference and personal positions shape the ways in which we proceed with analysis (see Chapters 2 and 19).

Conclusion

In seeking qualitative data, questionnaires aim not just at determining attitudes and opinions but at identifying and classifying the logic of different sets of responses, at seeking patterns or commonality or divergence in responses, and at exploring how they relate to concepts, structures, and processes that shape social life. This is no easy undertaking, and questionnaires struggle with the tensions of seeking explanation while being generally limited in their form and format to obtaining concise accounts.

Hoggart, Lees, and Davies (2002) argue that the necessarily limited complexity and length of questionnaires prevent them from being used to explain action (since this requires us to understand people's intentions), the significance of action, and the connections between acts. Compared with the depth of information developed through more intensive research methods such as in-depth interviews, focus groups, or participant observation, questionnaires may provide only superficial coverage. Nonetheless, they go some way in the explanation in that they are useful for identifying regularities and differences and highlighting incidents and trends (see de Vaus 2014 for an extended critique). Indeed, as Beckett and Clegg's (2007) work shows, in some contexts they can enable the collection of full and frank, thoughtful, and detailed accounts in ways that more intensive methods involving interviews and interviewers' presence may inhibit.

There are ways of constructing and delivering effective questionnaires that are largely qualitative in their aspirations, being mindful of the possibility of acquiring deep analytical understandings of social behaviours through careful collection of textual materials. Certainly, the interview, through its record of close dialogue between researcher and respondent, provides a particularly powerful way of uncovering narratives that reveal the motivations and meanings surrounding human interactions (see Chapter 8), and questionnaires can only ever move incompletely in this direction. However, by not requiring close and prolonged engagement with the research subject, the questionnaire offers opportunities to reach a wider range and greater number of respondents, in particular through online applications, and to collect data on people's lived experiences. This extensiveness and diversity makes questionnaires an important, contemporary qualitative research tool.

Key Terms

closed questions
co-constitution of knowledge
combination questions
computer-assisted telephone interviewing (CATI)
cultural safety
mixed-method research
open questions
population
pre-testing

probability sampling
purposive sampling
sample
sampling frame
satisficing behaviour
voice capture

Review Questions

1. Why are open questions more suited to qualitative research than closed questions?
2. Why is the choice of the mode of questionnaire distribution specific to the nature of the sample and the nature of the research topic?
3. Why should we avoid "closing" open question responses for the purpose of reporting findings?
4. What are the limitations of the use of questionnaires for qualitative research?
5. What are the particular benefits of administering questionnaires online?

Review Exercise

Sydney, Australia, continues to experience population growth. As a counter to its sprawling suburbs, the city is building a new high-rise urban community on old industrial land a few kilometres from the central business district. The new area is called Green Square. By 2030 it is expected to house 40,000 new residents and 22,000 new workers. Sydney's mayor, Clover Moore, says Green Square, " . . . is fast becoming a really great place to work, live, and play."

Part A

Imagine it is now 2030 and the residents and workers are in place. Your task is to draw up a table that guides a research topic called "An investigation of positive social relations in Green Square." The table should show, first, five variables that could be assumed to underpin positive social relations. Consider variables such as maintenance and care of the built environment, neighbourhood friendship networks, attitude to strangers, use of public space, vitality of social enterprises, viability of commercial recreational services (e.g. cafés). Then, second, your table should suggest two or three indicators for each of these variables. For example, indicators of a positive attitude to strangers might be demonstration of accepting gestures (say a smile or a nod), assistance rendered to someone unknown (say helping lift a baby carriage across a curb), or a feeling of calm in the presence of unknown shabbily dressed people in a dark street.

Part B

1. Select one of the variables from your table. Write an open question for each of the indicators you have nominated for this variable.
2. Select one of the other variables. Write a closed question for each of the indicators you have nominated for this variable.

Part C

Outline the methods you would use to analyze data collected from the questions you devised for Part B above.

Useful Resources

Babbie, E. 2013. *The Practice of Social Research*. 13th edn. Belmont, CA: Wadsworth. See Chapters 9 and 13.

Bryman, A. 2012. *Social Research Methods*. 4th edn. Oxford: Oxford University Press.

Cloke, P., et al. 2004. *Practising Human Geography*. London: Sage. See Chapter 5.

de Vaus, D.A. 2014. *Surveys in Social Research*. 6th edn. Sydney: Allen and Unwin. See Chapters 7 and 8.

Duke University's Initiative on Survey Methodology. http://dism.ssri.duke.edu/survey_mode.php. An interdisciplinary initiative on survey methodology containing extensive tips and resources on survey research methods.

Fielding, N. G., R.M. Lee, and N. Grant. 2008. *The Sage Handbook of Online Research Methods*. London: Sage. See Chapters 10–13.

Hoggart, K., L. Lees, and A. Davies. 2002. *Researching Human Geography*. London: Arnold. See Chapter 5.

Parfitt, J. 2005. "Questionnaire design and sampling". In R. Flowerdew and D. Martin, eds, *Method in Human Geography: A Guide for Students Doing a Research Project*. Harlow: Pearson/Prentice Hall. See pp. 78–109.

Sarantakos, S. 2012. *Social Research*. 4th edn. New York: Palgrave Macmillan. See Chapter 11.

———. 2012. "Social research 4e". Palgrave Macmillan. http://www.palgrave.com/sociology/sarantakos4e/workbook/. This is a companion website for Sarantakos's book *Social Research*. It offers a workbook on questionnaire surveys.

Sue, V., and L. Ritter. 2012. *Conducting Online Surveys*. 2nd edn. London: Sage.

SurveyMonkey.com. 2013. "SurveyMonkey.com—create surveys." http://www.surveymonkey.com. This is a commercially available web-based interface for creating and publishing custom web surveys and then viewing the results graphically in real time.

SurveyMonkey—The Monkey Team. n.d.. "Smart survey design." s3.amazonaws.com/SurveyMonkeyFiles/SmartSurvey.pdf. SurveyMonkey's guide to effective design and question-writing for online questionnaires.

13 Visual Methodology

Jim Craine and Colin Gardner

Chapter Overview

Our world of today is a landscape of visual representations. The study of that landscape has captured the imagination of geographers. We use this chapter to introduce a selection of visual methods, ranging from discourse analysis to geovisualization to the use of the virtual in visual methods to better provide an understanding of how geographic knowledge is conveyed visually and discuss the methods used to explore this knowledge. We begin with a short explanation of various forms of visual analysis, including psychoanalysis, semiotics, and discourse analysis. We then move on to digital and virtual visual representations and how new methods of engagement are evolving, including geovisualization. As geographers, we are now, more than ever, exploring immersive virtual environments in the course of our research. The technologies have moved from basic telemantics (the synthesis of telephony and digital computation) to new virtual technologies. We now have the methods that can help us uncover the importance of understanding the spaces of representation inherent in film, video, television, advertising, landscapes, and other visual representations, thereby providing the opportunity to work towards the advancement of this exciting field of geography.

Visual Methods in Geography

Geography has always been a visual discipline: Rose (2003b, 219) succinctly states, "Critical geographers need to explore the visualities of the discipline more carefully" while Crang (2003a, 238) says, "The role of vision in constituting geographic knowledge is fertile terrain." However, instead of visuality as a defining property of the traditional geographic object (a "map" for instance), it is the practices of *looking* invested in any object that constitute the object's geography: its historicity, its social anchoring, and its openness to the analysis of its visual components. It is, therefore, the act of seeing, what Foucault (1975, ix) calls the look of knowing, and not the materiality of the object seen, that decides whether the object can be considered from the perspective of **geovisualization**. New methodologies that address the current engagement between geography and visual representations have increased our understanding of the symbolic and ideological meanings that form these representations: i.e., what they analyze. Our geographic knowledge is no longer simply limited to cognition—it is *performed* in the same ways of looking that it describes, analyzes, and critiques. Visual methodologies now have the capacity to direct and colour our gaze, thereby making visible those aspects of geographic objects that otherwise remain invisible (Foucault 1975,15). Of course, the converse is also true: geographic knowledge can also determine what other aspects and what other objects remain invisible. Thus, following Hooper-Greenhill (2000), visual methodology can work towards a social theory of visuality, focusing on questions of what is made visible, who sees what, and how seeing, knowing, and power are interrelated. We can also now extend our methodologies to virtual and digital environments and understand how those formats can be used to examine the act of seeing as a product of the tensions between external images and internal thought and cognitive processes.

Visualization is an act of interpreting, and thus interpretation can influence ways of seeing—it can produce many possibilities of visual knowledge. Consequently we can then use visual methods to suggest alternative orderings of knowledge. Visual methods can take cultural representations that are presented as natural, universal, or true and analyze them so that alternative narratives, based on geography, become visible. We can explore and explain, for example, the bond between visual culture and nationalism, racism, or gender relations. We can also gain some understanding of how dominant classes set themselves up as orthodoxies to recognize and follow, allowing us to understand the political interests underlying the production of these cultural representations by using visual methods to study and **decode** their apparent transparency. Visual methodology also promotes the subjective look of the viewer/consumer—understanding that visual production comes first, followed by the perception it guides. We can thus be exposed to the interconnections between public and especially private meanings and the latter's uses in memories, family histories, and the ingrained visual tools that help to define them. We can also expose the interconnections between the physical and its digital virtual counterpart by, in essence, becoming virtual. These characteristics can instill emotional comfort or distancing, confinement, intimacy, or threat, but also, as a cognitive mode of understanding, they can provide a "scientific" method for grasping the complexities of the postwar world. Finally, and most basically, the intertextual relationships between cultural representations (that we can consider as objects or texts), and our different participating senses—the *affectivity* of these representations in other words—work extremely well with our new forms of visual methodologies and analysis.

The cultural theorist Raymond Williams (1993, 6) once said, "Every human society has its own shape, its own purpose, its own meanings. Every human society expresses these, in institutions, and in arts and learning." Williams often talked about painting, a form of visual representation, and the study of these forms of cultural representations has resulted in new forms of visual methodologies based on the spatial aspects contained therein. The use of visual representations as a pedagogic tool in geography can take many different forms—film, advertising, painting, video, cartography all come to mind—and the study of how these representations are created and the meanings they contain can offer great insights into the way individuals and social institutions interact and how individual and societal identity is constructed. Visual imagery, be it analogue or digital, real or virtual, serves as mimetic devices that provide representations of real-world places and people; they can provide sites that permit the exploration of any number of social issues, not to mention their purpose as spaces of resistance and contestation, or conversely, oppression and compliance. Perhaps foremost, visual representations can be thought of as geographical images that are *landscapes*—spaces that offer a site of engagement into the binary of ideological formations and social contestation. Visual methodology employs these representations to uncover patterns and relationships constructed and located within their landscapes.

Since the publication of Sauer's *The Morphology of Landscape* in 1925, cultural geographers have recognized landscape as a central concept and subject, particularly how landscapes reflect and symbolize the activities and cultural ideas of a place. Visual methodologies provide a way to engage landscape representations as metaphors that can be interpreted and analyzed in order to uncover how these images work to create distinct social and cultural space. That is, the representational *image* of a landscape can be interpreted to show the socio-cultural and political processes that shape actual physical landscapes. According to Rose (2001, 14–15), critical visual methodology embodies three important tenets:

1. Take images seriously: images are more than just reflections of their social context; visual representations have their own effects and can be used by many people for many different reasons.
2. Think about the social conditions and effects of visual objects: visual representations both depend on and produce social inclusions and exclusions, and the way these images articulate meanings needs to be carefully understood.
3. Consider your own way of looking at images: ways of *seeing* are historically, geographically, culturally, and socially specific; how we look at these images is not natural or innocent and is always constructed through various practices, technologies, and knowledges.

Rose's guidelines offer a unique way of interpreting visual texts, providing the foundation for further examination through other qualitative methodologies that can be applied to visual interpretation such as **psychoanalysis**, the range of theories that deal with human subjectivity, sexuality, and the unconscious; semiology, how images make meanings whereby "something" stands for "something else"; **discourse analysis**, the analysis of, in this case, visual representations and how and under what circumstances they are created; and newer critical visual methodologies that explore virtual and digital representations of landscape using methods brought into geography from media and communications studies.

Psychoanalysis and Semiotics

Let us first discuss how psychoanalysis and **semiotics** (discussed further in Chapter 14) are used as visual methodologies. As a critical visual methodology, psychoanalysis is concerned mostly with subjectivity, sexuality, and the unconscious (Rose 2007). Sexual difference is the key focus and is understood as relational, structured between the male and female objects within the representation and the members of the representation's audience. The relationships are articulated by different looks, and psychoanalytical analysis interprets how these looks are structured. Using cinema as an example, Rose further explains how film manipulates the visual, the spatial, and the temporal in an attempt to structure "looking," which in turn affects the gaze of the spectator and also maintains that there are important visual components to geography that should be further explored (2003b).

In general (but by no means exclusively), the psychoanalytical analysis of visual representations centres on how the unconscious of patriarchal society has structured the visual form. For example, Mulvey (1992) notes the significance of space as a site of gender differences: a home or homestead as a signifier of stable space, femininity, and family, the scene of domestic space that is the location of narrative events. Opposed to this is the outside, the masculine space of adventure, action, and movement. A classic film example of this binary opposition is John Ford's *The Searchers* (1956), where John Wayne's Ethan Edwards represents the nomadic, masculinist world of violence and racism in his attempts to track down his niece, Debbie (Natalie Wood), kidnapped years earlier during a Comanche raid on the family homestead. At film's end, Debbie is rescued and unity is seemingly restored but Ethan's violent racism is such that he is unable to be rehabilitated into the restored domestic (read: feminine) family unit. Ford expresses this combination of alienation and ostracism in a touchingly brilliant final shot from inside the darkness of the house whereby Ethan, framed in the doorway, is silhouetted against the unforgiving wilderness of the landscape—the site and location of his patriarchal bigotry. As the Sons of the Pioneers sing the film's title song on the soundtrack, the door closes on Ethan as he turns and walks away from the camera, excluding him spatially from both his relatives and the film's audience.

It is through these codes that meanings are constructed not only through visual images themselves but also through their ability to control the dimensions of time and space through their respective modes of production. How the viewer interacts with the images becomes an important aspect of any interpretation. The viewer becomes, in the act of watching, an integral part of a spatial environment in which all aspects of that space combine to give the spectator the illusion of looking in on a private world. As in the case of *The Searchers*, the content found in the representations takes on an ideological significance demanded by the dominant patriarchal order and allows for the cultural production of a landscape that contributes to a more general cultural construction of identities (in this case sedentary/nomad, inside/outside, female/male). In its application to visual methodology, psychoanalytical interpretations of these very different landscapes provide insights into how the narrative spaces of the image are representations of an actual or imagined environment viewed by a spectator.

Crang (2003a) links visuality to the goal of obtaining geographical knowledge and also describes semiotics as the study of signs and the construction of meaning; ways in which words, things, pictures, and actions are constructed as signs that convey meanings in particular times and places. Crang further restates the views of most semioticians that meanings are relational rather than fixed in that signs derive their meaning from other signs and from the wider system of signs, and not just from their actual form or content. This view validates McLuhan's (1964) concept of **reified** sensory ratios in which our way of understanding the world is related to how, and to what degree, our senses (primarily visual and tactile) derive meaning from what they perceive. This theory is later discussed by Williams (1977), who saw visual media as distinct material social practices that embody the social conditions and technologies of the culture using the medium to produce a representation of that culture. Saussure (1959) explains the concept of the sign, the basic unit of language and its two parts, the signified (the idea or meaning) and the signifier (the acoustic or visual image), stressing that there is an arbitrary relationship between the two. Thus the word/sound "cat" would signify different things in different contexts (a household pet, a lion or tiger in a zoo, a hipster in a jazz club, an earth mover on a building site). Semiotic analysis uses the term *codes* to discuss the systems of meanings found within the signs of a text, thereby allowing the users to gain a better understanding of the ideologies at work in the production of the text. Different languages use different words for the same signified (e.g. the French *chat* for "cat") and the same signifier can have different meanings, a view also advocated by Metz (1974) who additionally discusses the use of psychoanalytic theory in the semiotic exploration of cinema (1977). In film, for example, a knife can be a benign household object but also a murder weapon.

More famously, because of Hitchcock, audiences can never think of birds, showers, ropes, or ovens without also associating them with death and/or mayhem. Aitken (1996) also sees semiotics as a visual methodology that can be used to uncover the meaning of texts by identifying what the signs are and how they function. The complex patterns of associations, or codes, are common to a particular society at a particular time and they affect how signs are interpreted. Aitken states that to be socialized into a culture means to be taught these codes that semiotics can be used to decipher, a theory offered earlier by Bruno (1987) in her discussion of the film *Blade Runner* and how a meaning of that film can be found in the relationship between signifiers and the human capacity to interpret them.

As a methodology, Rose (2001, 91) suggests a series of steps through which a semiotic analysis can be accomplished:

1. decide what the signs are
2. decide what they signify "in themselves"
3. think about how they relate to other signs both within the image and in other images
4. explore their connections to wider systems of meaning
5. return to the signs via their codes to explore the precise articulation of ideology

Thus, in terms of visual methodology, semiotic analysis of a visual artifact focuses on the image(s) contained within the representation, using the sign to explore social conditions and social effects coded into the artifact.

Discourse Analysis

An easily accessible and dependable visual methodology is discourse analysis (discussed in more detail yet accessibly in Chapter 14), a method of geographical insight based on the writings of Michel Foucault that addresses the absences in other analytic frameworks because it looks specifically at the social consequences of difference through power in tangent with the construction of identity (the primary focus of feminism and psychoanalysis). Discourse refers to groups of statements and practices that not only structure the way we think about things but also allow us to make specific assertions of knowledge in the first place. In this way it is possible to speak of a geographical discourse that refers to the special language of the discipline and how it is practised, with its focus on spatial relations, uneven developments, and the power of places. The discourse produces subjects but it also produces technologies like maps and GIS that enable particular ways to visualize the world. Discursive practices—the ways meanings are connected through representations, texts, and behaviours—are also a form of disciplining and so discourses are also about power and knowledge production (Aitken and Craine 2009).

Before beginning a discourse analysis, one must first read through and around the context of the image. It is important to know something about all aspects of the image that one intends to analyze. For example, literature on Sir Edwin Landseer's paintings enables researchers to contextualize those works within the concept of imperialism. The writings of art critics can give insight into how Landseer or Bierstadt or other landscape painters and photographers portrayed their subjects through use of light and the placement of artifacts. This additional research provides knowledge about the artists and their connections with the cultural, political, and social structures of their times. To avoid what Rose (2001, 29) calls "analytical incoherence," there is a need to situate oneself in the debates of visual culture and embrace the "modalities" that are most important for interpreting the chosen image in question. In the example of the represented landscape of the American West, the focus is on a dialectic that elaborated imperial domination. It is important to understand the theoretical position that the researcher brings to any analysis and that the methods chosen carry with them "baggage" that is also part of your analysis. There are many different ways of understanding visual imagery; different theoretical positions have different methodological implications, and one must be clear on that standpoint before embarking on a piece of research. That said, a theoretical standpoint often brings clarity to the methods and the image under consideration.

One example that can be easily grasped by anyone new to visual methodology is the analysis of a particular artist or specific work of art. In *The Iconography of Landscape* (1988), Dennis Cosgrove and Stephen Daniels peer into the relationship between art and geography and how that intertwining can be used to perpetuate dominant ideologies. For example, Trevor Pringle (1988) describes the appropriation of Scottish history and geography through the commissioned landscape paintings of Landseer. By positioning Queen Victoria and members of her court in deliberate ways and painting them with particular demeanours, Landseer represents the loyal Scot nobility and a royal appropriation of the Highlands. Landseer's visual texts tell a story of loyalty and subjugation at a time when the English crown needed to consolidate its territorial possessions for imperialist expansion.

Using representations of the United States, another example of how discourse analysis looks at different forms of imperialist expansion is Manifest Destiny.[1] This allows us to see how representations of the American West can elaborate ways to get through and below visual images. The representations provide sites that can embody many different scales of spatial relations and meanings and are thus capable of showing and revealing the complex nature of landscape. These geographical representations are viewed as cultural images that represent, structure, and symbolize our surroundings—the images become *places* that can be analyzed to better understand lived experience. The images are important cultural signifiers that, when interpreted, reveal social attitudes and material processes where ideologies are transformed into concrete material forms.

The American West is certainly a representational landscape that contains any number of systems of meaning. Here, like Queen Victoria's appropriation of the Scottish Highlands but within the spatial context of the open spaces of the American West, is a landscape waiting to be occupied by the imperialist forces of manifest destiny. The *See Your West* advertising program used by Standard Oil of California beginning in 1938 provides an inventive way to apply discourse analysis methodology. The advertising was in the form of a series of collectable high-quality photographic prints of—as the collectible album stated—"scenic views of the West prepared for your enjoyment: to recall to memory certain favorite spots you visited in the past, and to help you visualize the beauties of those regions you have yet to see." Like Albert Bierstadt's nineteenth-century images of Manifest Destiny, the Standard Oil series linked images and ideas with specific places in the western United States, defining the meanings of Western landscapes and then reifying those definitions by inscribing them onto landscapes and promoting automobile tourism as a way to populate the West.

In terms of visual methodology, this example of advertising from the post-World War II years provide a particularly interesting time to observe the reverberations of Manifest Destiny in US culture. A deluge of tourism promotions encouraged the nation to travel. Standard Oil's *See Your West* campaign consisted of 25 9" ×12" Kodachrome prints distributed free of charge to motorists at Standard and Chevron Gas stations and garages and other distribution sites in 15 western states, British Columbia, and Hawaii. Each year a new set of 25 prints was distributed. Information about each location was presented in accompanying text. The use of famous photographers (e.g., Ansel Adams) and writers (e.g., Ernie Pyle, Erle Stanley Gardner) increased the desirability of these prints, enabling Standard Oil, in their effort to sell more products to our new and highly mobile society, to take advantage of the postwar travel boom and the burgeoning westward movement of the population. By the end of the 1955 program, more than 130 million full-colour prints, featuring over 200 different scenes, had been distributed to customers.

Like Bierstadt's nineteenth-century images of Manifest Destiny, *See Your West* depicted the wilderness in pristine and virginal (note the specific gendering of the landscape as something to be either preserved or conquered) form, offering on one level a familiarly romantic, nostalgic vision of nature. It did so with great specificity, singling out places to go and objects to admire, even angles from which to view them. In this way it reified those places and objects. *See Your West* does not digress into permutations; it reveals *the* West and, like the nineteenth-century images of Manifest Destiny, *See Your West* conveys a tone of palpable pride as Americans encounter the West. As an advertising campaign, *See Your West* attends mainly to the image of Standard Oil. The company is the "expanding and technologically advanced" object of admiration this time around. Only a great corporation could marshall high-calibre talent to "conquer" the West with an annually replenished series of photos and written descriptions that capture the region's essence in handsome, collectible form—*and* provide the gasoline for Americans to motor to the region en masse themselves.

Discourse analysis shows us that perhaps the most interesting sense in which landscape becomes artifice lies in the medium of representation itself. *See Your West* consists of photographs and text, not oil paintings like nineteenth-century images of Manifest Destiny. Photographs, through their **indexicality**, wield great potential to reify their subjects. This power derives, writes Steven Hoelscher, from the medium's "factuality" as an apparently transparent window into its subject. This transparency renders photography "a superb vehicle for cultural mythology": "As it presents a thoroughly convincing illusion of factuality while providing a repertoire of techniques that enable a photographer to make a powerful statement, the medium is uniquely positioned to naturalize cultural constructions" (Hoelscher 1998, 551).

Hence the 25 photographs in each annual *See Your West* series present a more "objective" visual record of the West than a gallery's worth of nineteenth-century landscapes could have. Moreover, the collectability of *See Your West* prints unleashes their reifying power on an imperial scale. To visit each of the sites featured in *See Your West*, as the series invites one to aspire to do, one would have to conquer thousands of miles of highway. By celebrating the marriage of automobile and camera in the postwar travel boom (as opposed to the twinning of the railroad and painting in the Bierstadt era), the Standard Oil series embraced and contributed to the democratization of the technological conquest of the West by reflexively engaging in the project of making and reaffirming the West through photography.

It is important for the practitioners of visual methodology to recognize that images, no matter their format—painting, photography, film, and advertising—are more than just passive visual images; they are symbolic representations of a dominant order and organization, and, when subjected to a visual analysis, we see that their production and consumption are socially mediated metaphors and systems of metaphors about landscapes of power. By utilizing a methodology like discourse analysis, the various types of visual images can help geographers understand how particular spaces are produced by manipulating images of the landscape and the built environment. To the geographer, the appeal of using visual methodologies is to uncover the ability of all types of visual images to use the physical world as a metaphorical space for the representation of the spatial relationships that exist within the environment (Dittmer's 2005 discussion of Captain America comic books and the religious tracts of Jack Chick [2006] are also interesting applications of discourse analysis to pop culture visual representations).

Exploring the Virtual

An interesting avenue for geographers using visual methodology is the work of French media critic Pierre Levy, who sees our *virtual* engagement with digital space as a method to analyze the datasets located therein, thus giving us a better understanding of the *geographica*l information contained within these datasets. Levy (1998, 131) refers to semiotics as "mutant hordes of representations, images, signs, and messages of all shapes and kinds (aural, visual, tactile, proprioceptive, diagrammatic) that people the space of connections" but, nonetheless, also move semiotic analysis into a virtual dimension (1998, 63):

> I am no longer interested in what an unknown author thought, but ask that the text make *me* think, here and now. The virtuality of the text nourishes my actual intelligence.

And, further, Levy (1997, 49) applies virtualized semiotics to digital media:

> Digital hypertext automates and materializes the operation of reading, and considerably magnifies its scope. Always in the process of reorganization, it offers a reservoir, a dynamic matrix through which a navigator or user can create individual text based on the needs of the moment. Databases, expert systems, spreadsheets, hyperdocuments, interactive simulations, and other virtual worlds are so many potential texts, images, sounds or even tactile qualities, which individual situations actualize in a thousand different ways. Thus digitization reestablishes sensibility within the context of somatic technologies while preserving the media's power of recording and distribution.

Thus, in the context of visual methodology, the significance of the virtual environment and its importance to how humans in today's world interact with digital datasets requires a greater understanding. Virtual space is often closely connected to non-spatial concepts like imagination, dreams, memory, religion, etc. By affectively engaging digital virtual worlds—by gaining a deeper understanding of how the virtual works in tandem with geospatial information—geographers can create a space in which the virtual worlds are transformed from conceptual spaces to actual places through the use of geovisualization methodology. Thus, a virtual world emerges from a technical system that allows a user to interact synchronously. Since we all have experience from being in the physical world, the process of understanding a virtual world is more about trying to map our understanding of the physical world onto the virtual world, but at the same time we are always considering new opportunities in the form of aspects of being that are *unlike* rather than *like* the physical world.

Some easily accessible examples to consider are DVDs and Blu-ray discs. Their digital datasets are certainly symbolic or semiotic operators and complexes of sign functions; they are also extensions of the senses, calibrations of McLuhan's (1994/1964) reified sensory ratios, that are embedded in the practice, experience, and technologies of digital virtualization and can be made accessible via geovisualization methods. Media geographers know that visual representations are constructed in historically and geographically specific ways that overly generalized psychoanalytic theory is unable to shed light on. Geovisualization methods explore how the specific meanings, and the construction and reproduction of meanings, that define landscape representations and the physical appropriation and use of that digital space are encoded in these spaces. More recent forms of visual methodology focus on the affective properties of cultural artifacts (see Craine, Curti, and Aitken's 2013 discussion of the film *Videodrome* for a more extensive analysis of affectivity and its place in visual methodology). The interplay of social constructions upon the landscape through space and time is a common thematic element, and these representations can provide a site for geographers to uncover and define meanings and identities. Levy's study of the virtual provides geographers with uniquely valuable understandings and knowledge of the geospatial environment. Thus *geographical* methods can be directly applied to the visual landscapes, spaces, and places without having to resort to incorporating a methodology (like film criticism) that relegates geography to secondary status.

Geovisualization as Method

As discussed, in their application of visual methods, geographers have often used psychoanalytical and semiotic methods. The visual methods of our digital environment are, however, still evolving. We now use methods to fully engage the virtual and our new definitions of movement and spectatorship. To come almost full circle, engagement with any of the recent visual theories and methodologies can be applied to a well-understood, long-standing qualitative methodology. Visual methodology includes the interpretation of texts such as film, maps, or photographic images as a way to gain a subjective understanding of the cultural codes contained therein. No matter the method, at its most basic, we understand that textual representations hold descriptions of the lived experiences associated with everyday life, and interpreting these texts can provide insight into the social construction of space and place.

Geovisualization is often associated with GIS cartographic applications and the analysis and display of quantitative data. However, the basic principles of geovisualization as a visual method are equally applicable to uncovering and revealing geographic unknowns within the context of qualitative geography, particularly virtual and digital data. Geovisualization, as defined by Buckley et al. (2000), focuses on visualization and its application to all stages of problem solving in geographical analysis. Dibiase (1992, 202) emphasizes the importance of graphic representation in the realm of visual thinking by stating, "Photographs and imagery, whose spatial dimensions correspond with those of the physical object being depicted, are more realistic than graphs, whose spatial dimensions represent nonspatial quantitative data or diagrams in which spatial relations are topological." Thus geovisualization methods can be used to determine and understand how social order is constructed. The combination of geovisualization theory and conceptual thought with other methods such as discourse analysis or textual analysis can create a new and valid research process. The methodology can then be used to further explore spatial data and generate new geographic knowledge by expanding the domain of qualitative methodology.

As an example, take the previous discussion of the Landseer paintings or the Standard Oil *See Your West* advertising campaign. Using geovisualization, the researcher perceives the data set (the paintings or the photographs) as derived from the body of work of a particular entity (Landseer or Standard Oil). Alternatively, the researcher can view both as a group of similarly themed representations or as simply just a single painting or even a single photograph. All the components of the image (location, colour, composition) can be used as texts and can help inform geographical interpretations of the cultural landscape. This makes the researcher aware of the role images play in the organization of social, economic, and political spaces and how a detailed analysis can uncover new ways of thinking about space, place, landscape, and identity.

Geovisualization as a method can be a very valuable tool to the visual researcher because it is, initially, quite easy to use and, even at the beginning stage, is capable of providing great insight. As the researcher explores its greater capabilities, its utility is increased and a greater level of depth and understanding attained. In most cases, other forms of qualitative analysis, such as a semiotic or a structural analysis (like discourse analysis or **Actor Network Theory** [ANT]), that allows the researcher to determine the relationships between theory and data, can also be utilized (see Aitken and Craine 2014 and Craine and Aitken 2005 for more specific application of geovisualization methods).

Conclusion

At this point in the evolution of geovisual methodology, geographers must still challenge the current visual methods; we must continue to explore and dig deeper into all of the forms of visual representations produced by our technogenetic society. Geographers must understand that visual methods are not a domain of our discipline only and should consider the advantages that can be found in being interdisciplinary. Media theory and communication studies both have much to offer and are ideally suited to supplement the development of visual methodologies that are grounded in geography. While methods like discourse analysis and geovisualization and even semiotics are useful, geographers should draw upon visual theories used in cognitive science, and media and communication studies. New methods can provide geographers access to geospatial information and knowledge in a manner that takes advantage of our unique cartographic and visualization skills. This takes geographical knowledge into new realms that provide an understanding of exactly how the new forms of digital and virtual data can be studied in more depth. Geographers can now more readily take advantage of the information contained within representations in not only the traditional analogue sense but also in their digital formats, and this, in turn, will lead us down new paths of exploration. Thus, in the end, visual methodology gives geographers the opportunity to better engage visual representations and thus gain a deeper appreciation and wider comprehension and understanding of spatial information.

Key Terms

Actor Network Theory [ANT]
decode
discourse analysis
geovisualization
GIS
indexicality
psychoanalysis
reification
semiotics

Review Questions

1. In what ways can geography be understood to be a visual discipline? What are some of the visual "texts" that geographers can use in their interpretations of social and physical worlds?
2. Of what significance to geography is the arbitrary relationship within signs between the signified (the idea or meaning) and signifier (the acoustic or visual image)?
3. What are discursive practices and how are they a form of disciplining?
4. What are some of the ways in which geographical methods can be applied directly to visual landscapes, spaces, and places?

Review Exercises

1. Select a visual representation of a landscape. Using a semiotic approach, analyze the image. Describe what is signified and what functions as the signifier. What are the "signs" inscribed in the representation? How might your analysis change if you were to use a psychoanalytic method?

2. Using the Web or your library, find a landscape painting of the American West by Albert Bierstadt. Apply basic discourse analysis to the interpretation of this painting. Under what circumstances was it painted? What ideologies are contained therein? How does the painting work to reify positions of power? Who benefits and who is marginalized by the painting? Use as many sources as possible to compile a more comprehensive analysis.
3. Using a specific film in the DVD or Blu-ray format—for example, David Fincher's *Seven*—analyze the various spatial representations contained therein as a means of understanding how a digital dataset can create space and place. How does the disc set up connections between different environments (i.e., audio commentaries, deleted scenes, the film itself, the menu)? How do these different representations of the same *place* compare? How does the disc allow the audience to critically engage the digital/virtual representation?

Useful Resources

Bartram, R. 2010. "Geography and the interpretation of visual imagery." In N. Clifford, S. French, and G. Valentine, eds, *Key Methods in Geography*, 2nd edn, 131–140. Sage, London.

Craine, J. and S. Aitken. 2005. "Visual methodologies: What you see is not always what you get." In R. Flowerdew and D. Martin, eds, *Methods in Human Geography*, 2nd edn, 250–269. Harlow: Longman.

Crang, M. 2009. "Visual methods and methodologies." In D. Delyser, S. Herbert, S. Aitken, M. Crang, and L. McDowell, eds, *The SAGE Handbook of Qualitative Geography*. 208–225/ London: Sage.

Rose, G. 2007. *Visual Methodologies*. 2nd edn. Thousand Oaks, CA: Sage.

Note

1. The widely held nineteenth-century belief in the United States that American settlers were destined to expand throughout the continent because of their special virtues and because it was their irresistible duty to do so.

Doing Foucauldian Discourse Analysis—Revealing Social Realities

Gordon Waitt

Chapter Overview

Foucauldian **discourse analysis** is a well-established interpretive approach in geography to identify the sets of ideas, or discourses, used to make sense of the world within particular social and temporal contexts. This interpretive approach is underpinned by the challenging, yet insightful ideas of the French philosopher Michel Foucault. Following Foucault, discourse is a mediating lens that brings the world into focus by enabling people to differentiate the validity of statements about the world(s). The goals of this chapter are twofold. The first is to outline why Foucauldian discourse analysis is a fundamental component of geographers' methodological repertoire. The second goal is to provide a methodological template. The chapter begins by introducing Foucault's concept of discourse. Foucault's interest in discourse was to explain how some lines of thinking/being/doing are generally accepted as "true," while other possibilities are marginalized or dismissed. Foucauldian notions of discourse hold that discourse constructs knowledge through the production of categories of knowledge; thereby governing what it is possible to talk about and what is taken as common sense. For Foucault, discourse simultaneously produces and reproduces knowledge and power (**power/knowledge**) through what it is possible to think/be/do. Hence, discourse analysis offers insights into how particular knowledge of the world becomes common sense and dominant, while simultaneously silencing different interpretations. Next the chapter outlines a methodological template for the conduct of discourse analysis. Examples are considered to illustrate the benefits of discourse analysis for geographical research, particularly projects committed to addressing uneven social relationships and environmental injustice.

Introducing Discourse Analysis

What is discourse? This chapter relies upon Foucault's concept of discourse. Thus, it is important to briefly consider what Foucault understood by this term. Foucault's definitions of the notion of discourse cohere around the cultural production and circulation of knowledge. Foucault does not provide us with a clear-cut dictionary-like definition of discourse. Instead, Foucault employs the concept of discourse to critically explore what, and how, we know about the world about us. In Foucault's work at least three overlain explanations of the concept of discourse can be identified:

1. all meaningful statements or **texts** that have effects on the world;
2. a group of statements that appear to have a common theme that provides them with a unified effect;
3. the rules and structures that underpin and govern the unified, coherent, and forceful statements that are produced.

Crucially, remain mindful that Foucault's (1972) use of "discourse" is different from conventional linguistic definitions, which understand discourse as passages of connected writing or speech. Foucault is interested in how particular knowledge systems convince people about what exists in the world (meanings/representations) as well as shaping what they say (think), do (practices), experience (emotions), and become (subject). Given Foucault's interest in the role of knowledge production in making and remaking the co-constitutive relationships between people and place, his conception of discourse is inherently geographical.

The critical insights that Foucault sought through the concept of discourse may be found in the process of power/knowledge production. Discourse, as it is theorized by Foucault, allows us to critically explore what everyone would like to believe as true about people, animals, plants, things, events, and places. Such an approach is labelled **constructionist**—an approach that demands asking questions about the ways in which distinct social "realities" or categories become normative ways to think/be/do. The aim of discourse analysis is to investigate why particular lines of argument (constructed in and as discourse) have become taken as truths, while others are dismissed. For example, why in Western societies are scientists widely understood as the source of factual knowledge? Alternatively, why in Western societies are tropical islands often categorized as earthly paradises, or national parks as place for nature? Likewise, why are certain plants categorized as "native" while others are labelled as "weeds" or "invasive"? Simply put, a Foucauldian discourse analysis seeks to uncover the cultural and social mechanisms that maintain or rupture structures or rules of validity over statements about the world. The productive effect of power/knowledge between actors is always uncertain; on the one hand it may reinforce social norms, but on the other hand it may mobilize possibilities for change. In political terms, discourse analysis allows geographers to expose how inequalities and injustices are sustained by the resilience of certain underlying normative categories.

At the start, please note that there are many different types of discourse analysis. A website entitled "How to Use DAOL" hosted at Sheffield Hallam University,

UK, provides a helpful discussion on the range of different discourse analysis techniques and their application in various disciplines (http://extra.shu.ac.uk/daol/howto). Furthermore, remain mindful that Foucauldian discourse analysis is distinct from other forms of qualitative analysis including **semiology** (which considers qualitative source materials in terms of inherent meanings) and **manifest content analysis** (which quantifies the number of occurrences of particular themes or words). So, how to begin conducting a Foucauldian discourse analysis?

Interestingly, Foucault struggled with writing "how to" conduct discourse analysis. He feared that a methodological template would become too formulaic and reductionist. This absence is perhaps what has often led others to describe Foucault's methodological statements as "vague" (Barrett 1991, 127). Several qualitative method handbooks in the social sciences are equally hesitant to give formal guidelines (Phillips and Hardy 2002; Potter 1996). Some scholars argue that guidelines undermine the potential of discourse analysis as a "craft skill" (Potter 1996, 140); by limiting the researcher's ability "to customise" (Phillips and Hardy 2002, 78); or by inhibiting demands for "rigorous scholarship" (Gill 1996, 144) and "human intellect" (Duncan 1987, 473). For these scholars, the maxim is "learning by doing." Through a combination of scholarly passion and practice, discourse analysis is typically held to become intuitive. Thus, the methodology is often left implicit rather than made explicit. For scholars who defend Foucauldian discourse analysis as an "art," any methodological template would be understood as too systematic, mechanical, and formulaic (Burman and Parker 1993). In practical terms, such counsel is not especially helpful for those seeking advice on how to do discourse analysis! To help grapple with the methodological implications of doing a Foucauldian discourse analysis, this chapter draws on important lessons offered by geographer Gillian Rose (1996; 2001) and linguist Norman Fairclough (2003).

Doing Foucauldian Discourse Analysis

Initially, Foucauldian discourse analysis may not be intuitive. Hence, to help novices conduct a discourse analysis, Rose (2001) provides seven stages through which the technique moves (see Box 14.1). These axioms are not intended to stifle your own interpretations. Instead, they are offered as starting points in a conversation for your own thoughtful analysis.

In what follows, each stage is reviewed, using mostly examples from tourism geographies and geographies of sexuality. This choice of examples is not accidental. They illustrate a larger effort of the application of discourse analysis in geography to challenge what can be said, what is worthy of study, and what is regarded as factual. For example, until the 1990s, research on tourism and sexuality was often shunned in the discipline as unworthy of geographical analysis (Binnie and Valentine 1999). There is nothing intrinsic to sexuality and tourism that makes

Conversations to Join When Doing Foucauldian Discourse Analysis

1. Choice of source materials or texts
2. Suspend pre-existing categories: become reflexive
3. Familiarization: absorbing yourself in and thinking critically about the social context of your texts
4. Coding: once for organization and again for interpretation
5. Power, knowledge, and persuasion: investigate your texts for effects of "truth"
6. Rupture and resilience: take notice of inconsistencies within your texts
7. Silence: **silence** as discourse and discourses that silence

Source: Adapted from Rose (2001, 158).

them more or less relevant for geographers to study than, say, ethnicity or car manufacturing. Instead, dominant sets of ideas (discourses) in geography eschewed the topics of sexuality and tourism. The very idea of spaces and bodies being gendered and sexed, let alone mutually constituted through space, was unthinkable. Equally, tourism involved people having fun and so within a masculinist sets of ideas that framed academic geography until the 1990s was positioned as having little significance, irrespective of its impacts. Within this schema, professional geographers would obtain greater academic kudos and authority by quantitatively analyzing the economic geographies of men at work in the industries of steel, textiles, coal, and automobiles, spatialized by the **metaphor** of "industrial heartlands."

Choice of Source Materials or Texts

What kinds of source materials or texts are required for discourse analysis? Source materials or texts may include advertisements, brochures, maps, novels, statistics, memoranda, official reports, interview transcripts, diaries, paintings, sketches, postcards, photographs, and the spoken word. Choosing source materials for discourse analysis will in part be informed by the research goals. Here four examples will be helpful to illustrate the diversity of applications of discourse analysis and relevant source materials. First, John Urry (2002) investigated the role of the tourism industry in the transformation of people and places into tourist destinations. Urry envisaged the tourist experience as an exercise in Foucauldian "gazing." Urry selected brochures and guidebooks of the tourism industry to illustrate how tourist sights/sites are constructed by drawing on different sets of ideas—myth, science, cultural histories—that enable those sites/sights to be fashioned as extraordinary to be then "consumed" by visitors. Drawing on the example of Niagara Falls, he argued that because of the vast number of tourism industry images instructing visitors on how to "consume" or "gaze" at the Falls as extraordinary/spectacle, it is no longer possible for tourists to "see" the Falls themselves. Second, Kevin Markwell and Gordon Waitt (2009) explored the role that pride festivals play in processes of social change surrounding dominant ideas of heterosexuality in Australia. Source materials for this work included articles published in the lesbian and gay media. Third, Jessica Carrol and John Connell (2000), in analysis of the relationship between music, politics, and place, conducted a discourse analysis of the song lyrics of the inner-Sydney-based band, The Whitlams. A discourse analysis of the song lyrics of this band demonstrated shared understandings of an inner-city life being allegedly lost through the market-led process of gentrification. At the same time, the song lyrics draw on previously circulating discourses that fashion a geography of Sydney-suburbs upon essentialized differences between people. Finally, Andrew Gorman-Murray (2007) employed discourse analysis to examine how people who understand themselves as lesbian or gay challenge taken-for-granted ideas around heterosexuality that help stabilize notions of the Australian home. Source materials for this work were transcribed interviews conducted with participants in their homes. As these examples illustrate, discourses are expressed through a wide variety of source materials.

Different categories of source materials or texts can be further subdivided into **genres**. For example, there are many different genres of movies, including YouTube, home movies, and commercial films. The latter can be subdivided further into comedy, art-house, science fiction, horror, and romance. Similarly, there are different genres of written texts, including travel diaries, solicited diaries, biographies, autobiographies, fiction, travel writing, government papers, newspapers, scientific reports, and so on. Being mindful of the genre of text is important, because the producer of each genre or category of text (e.g., journalist, traveller, academic, medical practitioner) is addressing a particular **audience** and may have to conform to a particular style of writing. Each source will be produced, circulated, and displayed by means of a particular technology (such as printing, painting, photography, handwriting, and email).

Your research project may dictate the collection of one particular genre of text—say, solicited diaries, semi-structured interview transcripts, tourist maps, music album covers, tourist brochures, holiday photographs, or picture postcards. In each case, to inform the interpretation, you will need to remain alert to the social circumstance of production, text, and author in question. If you are using transcripts from interviews or focus groups, such critical reflection is an integral part of the method (see Chapters 8 and 10). Questions to help guide you through this process of critical reflections are outlined later in Box 14.3 on authorship, Box 14.4 on social circumstance of production, and Box 14.5 on audience.

Alternatively, your project may require that you to establish a research archive by collecting and interpreting several different types of texts, including photographs, advertisements, letters, newspaper articles, and so on. A research archive requires you to conduct a good deal of background research to answer the questions outlined in Boxes 14.3, 14.4, and 14.5. Consider, for example, some work by Lynda Johnston (2005). She examined the practice of suntanning on the beaches of Aotearoa/New Zealand. Her analysis drew on semi-structured interviews and young women's magazines (*She*, *Dolly*, *New Idea*, and *Cosmopolitan*) that advertise suntanning products. Her analysis required working across these source materials to explore how people navigate the dominant discourse that shapes the practice of changing the colour of one's skin—suntanning. This work highlights how the practice of suntanning is fashioned by Western ideas of femininity associated with leisure, youth, and beauty and contests sets of ideas that locate the skin as a "true" biological marker of the body's gender and race.

Intertextuality is the term often employed to describe the assumption that meanings, like those informing suntanning practices, are produced as a series of relationships between texts, rather than residing within the text itself. Robyn Longhurst's (2005) analysis of pregnant bodies in Hamilton, Aotearoa/New Zealand, is another excellent example of the application of intertexuality. To explore how pregnant women use clothing to constitute their subjectivity and social meaning in different places, her discourse analysis relied on a research archive comprising interview transcripts, maternity wear brochures, advertisements for maternity wear, Internet sites marketing maternity wear, newspaper clippings, and magazine photographs and reports (*New Zealand Women's Weekly*, *Woman's Day*, and *New Idea*). Intertextuality acknowledges that meanings are co-created with an active audience.

Your initial problem will most likely be having too many potential source materials. Identifying initial sources with which to begin your analysis is an essential research task. What, then, might inform your choice of texts? The following questions are designed to help you to identify your preliminary or opening texts:

- *Which source materials to sample?* In making decisions about selecting or excluding source materials, student researchers are often advised to select "rich" or "in-depth" texts. For example, Tonkiss (1998, 253) explains, "What matters is the richness of textual detail." Attention to richness of source materials allows the researcher to interpret the effects of discourse in normalizing understandings. But what qualities make a source material qualitatively rich? For example, why are source materials comprised of lists of numbers not considered qualitatively rich? Quantitative data may provide insights into the strength of an attitude, or the number of occurrences of particular events/practices. Yet, quantitative data provides no clues to the multiple sets of ideas that a person may draw upon to inform choices or their understandings of people, place, events, plants, or technology. Likewise, for an interview transcript to be categorized as a "rich" text, it requires more than "yes" or "no" answers. A qualitatively rich interview transcript may therefore require longer-term involvement and consideration of particular techniques (see Riley and Harvey 2007 for further discussion on how to enhance the richness of interviews).

- *How to select source materials?* A purposeful sampling strategy (see Chapter 6) may often be deployed initially to help identify source materials that are likely to provide insights. Effective purposeful sampling requires early and ongoing secondary research. Yet, because the research process is a continuing building process, you are likely to find that as the project unfolds, other, unexpected source materials may become relevant and important to its success. As a reflexive researcher, you must remain open to such unexpected possibilities. Justifying why a particular source material is meaningful to your project is an integral step in establishing **rigour**. Such justification is important to writing transparent research that is open to scrutiny by both the participant and interpretive communities. (See Chapter 6 for additional discussion.)
- *How many source materials?* As outlined in other chapters of this book, there are no rules for sample size in the context of qualitative research. Whereas in inferential statistics sample size can be prescribed by demands of representativeness, in discourse analysis the number of source materials depends on what will be useful and what is *meaningful* in the context of your project. Background research may lead you to the view that the sample size may be as small as one or extend into hundreds of texts. Again, explaining why you selected a particular number of texts for inclusion in your project becomes an integral step in establishing qualitative rigour (see Chapter 6).

What, then, might be a meaningful choice of source materials for investigating the following research question: Is Sydney, Australia, the "gay capital of the South Pacific"? What might your initial starting point be? What sort of background research of related works might be of assistance? Who might you wish to interview? What other source materials might you wish to consult to create a research archive? (See Box 14.2.)

Sydney as the "Gay Capital of the South Pacific": Choosing Texts for Discourse Analysis

Imagine that the goal of your project is to examine whether Sydney, Australia, is the "gay capital of the South Pacific." First: which initial source materials to select? In the first instance, you may find census statistics of same-sex couples in Australia. These statistics will tell you about the number and distribution of same-sex couples in Australia. They may provide a helpful starting point. Statistics can tell us if same-sex couples are more concentrated in Sydney than elsewhere (Gorman-Murray et al. 2010). But the numbers alone tell us nothing about the role of Sydney in the lives of people claiming a gay identity. Instead, interviews may offer a more suitable source material. However, this raises question of recruitment. Some background research may help in this process. In this case, helpful starting points may be the work of Johnston (2001), Markwell (2002), Waitt and Markwell (2006) and Waitt and Stapel (2011). The representation of Sydney as the "gay capital of the South Pacific" often relies on narratives that frame Oxford Street, Darlinghurst, Sydney, as a "gay homeland," reaffirmed each year through the annual Sydney Mardi Gras as a celebration of sexualities. Reading this research may alert you to the paradoxical understandings of Sydney as a "gay capital"; as site of both belonging and alienation. Hence you may wish to focus your research on people living their lives as straight, queer, gay, lesbian, bisexual, or transgender either within Sydney or elsewhere. Other source materials that might be helpful in this project are newspaper clippings, guidebooks, travel magazines, tourism advertisements for Sydney, and travel stories from the Internet.

Suspend Pre-Existing Categories: Become Reflexive

Foucault (1972) regarded the starting point of discourse analysis as reading, listening, or looking at your texts with "fresh" eyes and ears. The imperative of shelving preconceptions is underpinned by the objective of discourse analysis—to disclose the created "naturalness" of constructed categories, subjectivities, particularities, accountability, and responsibility. Foucault (1972, 25) pointed out that all preconceptions

> must be held in suspense. They must not be rejected definitively, of course, but the tranquility with which they are accepted must be disturbed; we

> must show that they do not come about by themselves, but are always the result of a construction the rules of which must be known and the justifications of which must be scrutinized.

Foucault acknowledges that this request to defer pre-existing categories is an impossible task, given all knowledge is socially constituted. There is no independent position from which to suspend pre-existing knowledge. Instead, he calls for researchers to become self-critically aware of the ideas that inform their understandings of a particular topic. One strategy by which to implement Foucault's call is to deploy the techniques of reflexivity outlined by Robyn Dowling earlier in this volume (see Chapter 2). As a critically reflexive researcher, a key starting point is to acknowledge why you selected a particular research topic (context) and your initial ideas about the topic (partiality). The term **positionality statement** refers to a researcher explicitly "locating" themselves in the context of the project, and may include their lived experiences (**embodied knowledge**). In acknowledging your positionality, remain alert to Gillian Rose's (1997) warning that it may be impossible to ever fully locate oneself in research. The ideas we carry into the work may change while conducting the research. Hence, a vital part of a positionality statement is noting and reflecting upon these changes as the project unfolds. One crucial question to ask is: how has your self-understanding changed conducting the research? Finally, as discussed by Elaine Stratford and Matt Bradshaw (see Chapter 6), keep careful and transparent documentation of your interpretation process. Document your search for regularities and patterns and the ways you developed "hunches." The strategies outlined in the following section will help prevent you from imposing your own taken-for-granted understandings upon the source materials.

Familiarization: Absorbing Yourself in and Thinking Critically about the Social Context of Your Source Materials

Familiarization with your source materials is essential. A helpful starting point is to begin thinking critically about their "social production," becoming alert to their authorship, technology, and intended audience. These social dimensions of sources are a good starting point for a critical interpretation because discourses operate as a process, restricting not only what can be said about the world but also who can speak with authority (Wood and Kroger 2000). Foucault understood discourses to be grounded within social networks in which groups are empowered and disempowered relative to one another. He saw discourse as subtle forms of social control and power. One effect of discourse is the privileging of relatively powerful social groups. That is, particular voices and technologies are favoured over others, often counted as sources of "truthful" or "factual" knowledge, while other voices may be excluded and silenced, perhaps by becoming positioned as untrustworthy, anecdotal, hearsay, or folklore. Take, for example, the distinction in Australian national parks authorities between Indigenous and "scientific" approaches to environmental management. Until relatively recently, "objective" scientific knowledge was valued by most national park authorities at the expense of Aboriginal Australian knowledge. Scientific methods were equated with research, objectivity, and environmental management solutions. Oral Aboriginal Australian knowledge was constituted as lacking and was associated with negatively valued concepts, including the irrational and subjective (Lawrence and Adams 2005).

Foucault's ideas on how knowledge about a topic is constituted and maintained within social networks have implications for doing discourse analysis. The key point is that all texts are the outcome of an uneven power-laden process, fashioned within a particular social context. Hence, an integral part of the familiarization process is to conduct background research to help anchor your texts within a particular historical and geographical context. Boxes 14.3, 14.4, and 14.5 provide a series of questions to help you investigate all texts as expressions of knowledge production and a subtle form of social power that constitutes particular social realities.

First, consider authorship as the outcome of a highly social process (Box 14.3). Regardless of whether your source material is a transcript, photograph, painting, song, or novel, you must investigate the author's relationship with the intended audience. Hence, you must reflect carefully on what social dynamics are carried into the production of the source material and may operate as a subtle form of social control. Box 14.3 provides a starting point to interpreting the relationship between discourse, knowledge, and power. *Who* produced your selected materials? Give particular attention to how the author is positioned within historically and geographically specific understandings of gender, class, sexuality, and ethnicity. Remember that, like the construction of sexuality, gender, class, and ethnicity are not "natural" categories but are instead constituted through discourse. *When and where* was your selected source material produced? *Why* was it produced? Are the social identities of the author or audience portrayed in your selected sources; that is, does the text help to naturalize belonging to a particular group in society—like a family snapshot?

Second, ask more specific questions about the social conventions linked to producing the source material (Box 14.4). The key point is that you must examine *how* your source materials were produced—in particular, the implications of particular technologies of production. More subtle questions then follow, including thinking carefully about how different categories of material have their own social histories and geographies.

Garlick (2002), Markwell (1997), and Nicholson (2002) provide helpful discussions of how the technology underpinning film and photograph production can be thought about as a way of constituting understanding of the world as factual knowledge. In Western societies, photography is commonly regarded as a science. Light reflected from the photograph object is etched into chemicals. The result possesses the objectivity of a science-based technology. Hence, photographs on Facebook or picture postcards or in brochures, advertisements, or family photograph albums appear to be a "true" or "accurate" representation of the "real" world, involving no subjective intervention. Forgotten are the highly social

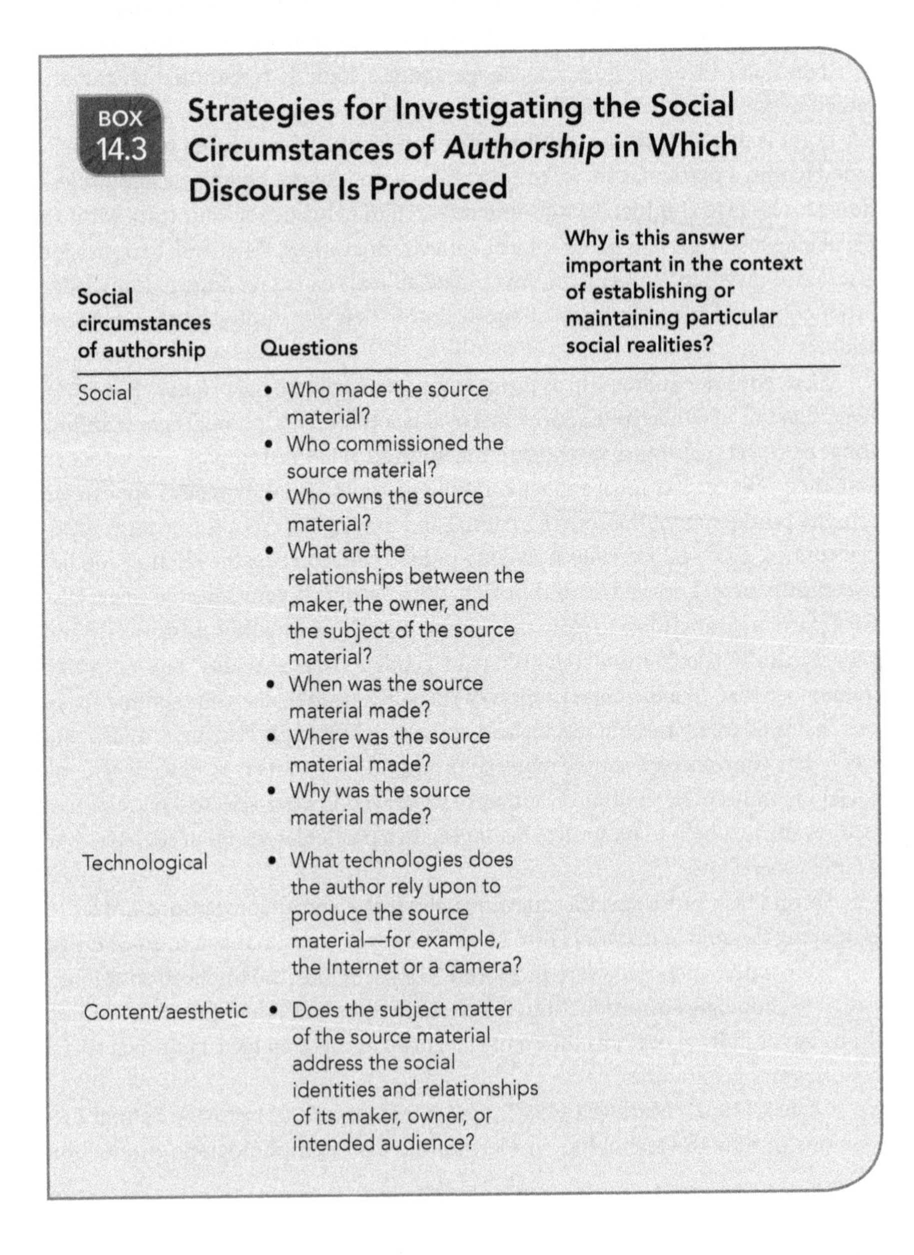

BOX 14.3

Strategies for Investigating the Social Circumstances of *Authorship* in Which Discourse Is Produced

Social circumstances of authorship	Questions	Why is this answer important in the context of establishing or maintaining particular social realities?
Social	• Who made the source material? • Who commissioned the source material? • Who owns the source material? • What are the relationships between the maker, the owner, and the subject of the source material? • When was the source material made? • Where was the source material made? • Why was the source material made?	
Technological	• What technologies does the author rely upon to produce the source material—for example, the Internet or a camera?	
Content/aesthetic	• Does the subject matter of the source material address the social identities and relationships of its maker, owner, or intended audience?	

Strategies for Investigating the Social Circumstances of the *Source Material* in Which Discourse Is Produced

Production and circulation of a source material	Questions	Why is this answer important in the context of establishing or maintaining particular social realities?
Social	• Does your selected source material re-circulate texts found elsewhere? For example, can the same source materials be found elsewhere—say, on the Internet, in tourist brochures, or on postcards?	
Technological	• How has technology affected the production of the source material? For example, what use is made of print, colour enhancement, photography, digital technologies (airbrushing)?	
Content/aesthetic	• What is the subject matter? • What is, or are, the genre(s) of your selected source material? For example, if a photograph, is it a family snapshot or advertisement? Alternatively, if your source material is in an electronic form, is it, for example, a television soap opera? • Is your source material one of a series? • What are the conventional characteristics of your selected genre of source material? • Is your selected source material contradictory, critical, or in some way different from those circulated elsewhere? • What is the "vantage point" of the viewer/reader in relationship to the source material? That is, how is the viewer/reader positioned in relationship to the source material?	

decisions of production, including choice of subject, lighting, colouring, cutting, cropping, etching, and cloning. When photographs are understood as a reflection of the "real" world, they actually conceal the opportunities for modification by the photographer for commercial or other purposes. Following Foucault, unpacking

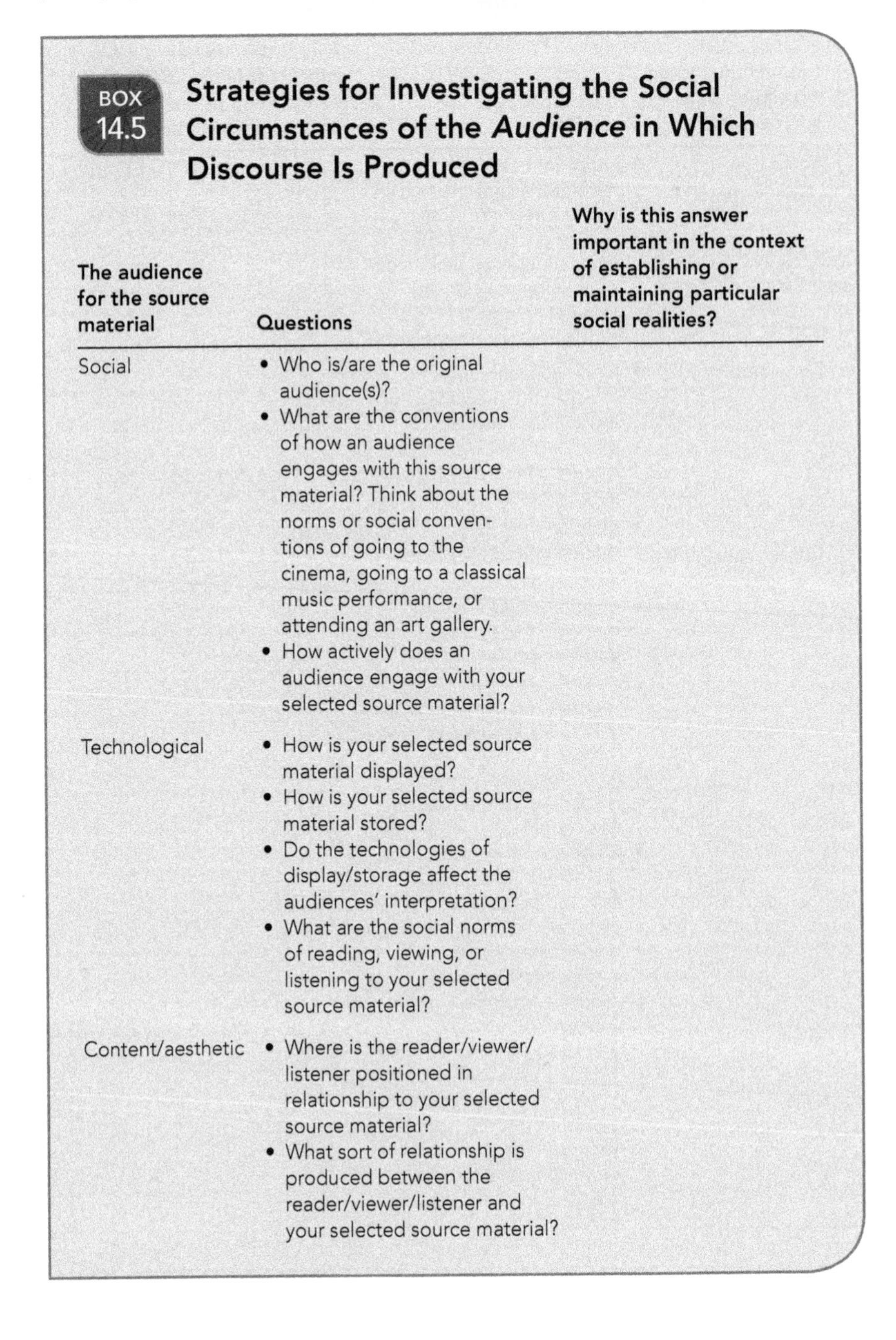

BOX 14.5

Strategies for Investigating the Social Circumstances of the *Audience* in Which Discourse Is Produced

The audience for the source material	Questions	Why is this answer important in the context of establishing or maintaining particular social realities?
Social	• Who is/are the original audience(s)? • What are the conventions of how an audience engages with this source material? Think about the norms or social conventions of going to the cinema, going to a classical music performance, or attending an art gallery. • How actively does an audience engage with your selected source material?	
Technological	• How is your selected source material displayed? • How is your selected source material stored? • Do the technologies of display/storage affect the audiences' interpretation? • What are the social norms of reading, viewing, or listening to your selected source material?	
Content/aesthetic	• Where is the reader/viewer/listener positioned in relationship to your selected source material? • What sort of relationship is produced between the reader/viewer/listener and your selected source material?	

the set of ideas that sustain photography mask the making of particular social realities. In political terms, discourses of photography may help to naturalize particular understandings of the world as "true." For example, Waitt (1997) and Waitt and Head (2002) discuss the deployment of different genres of photography by the Australian tourism industry, including postcards and advertisements to mobilize certain racialized truths about the outback as "wilderness." Photography is integral to "circulating" and "maintaining" particular social realities or "truths" about places, including holiday destinations.

Alternatively, Lisa Law (2000) provides an interesting discussion of how map-making helps to maintain particular sets of ideas. In the context of the Health Department of Cebu, the Philippines, she notes how the making of the commercial sex industry mobilized certain gendered and sexualized "truths" of "good" and "bad" women. She noted how the appearance of the so-called "red-light district" as an empirical "reality" on the map made apparent the Catholic-inspired moral distinction between marital and commercial sex. For us, the key point is that the technology of mapping is a power-laden process. In this case, mapping technology deployed within the Cebu Health Department may have helped convince many health officials that all women living in Cebu could be categorized along a simplistic spatialized division of good/bad. An integral part of discourse analysis is remaining alert to the role that different technologies play in **strategies of conviction** by producing or mobilizing certain "truths."

Finally, investigate audiences along the strategies outlined in Box 14.5. Audiences are not pre-given but are again thought of as the outcome of a highly social process. How a source material is produced is in part dependent on its intended audience. In other words, an author will draw on particular discourses, mindful of the needs, demands, and fantasies of the intended audience. Hence, audiences can be conceptualized as co-authors of a source material.

For example, both Crang (1997) and Markwell (1997) illustrate the active role of prospective audiences as co-authors of vacation snapshots. They demonstrate how this genre of photograph is always about framing, making, and circulating social realities for a particular audience. Vacation snapshots are the outcome of a highly selective social process influencing the choice of subjects to photograph. At one level, this choice reveals how a person is set within, or challenges, particular common-sense understandings of, for example, tourism discourses. Making vacation snapshots is not value-free but reflects particular ideologies and sets of ideas about appropriate leisure practices for a specific place, produced and circulated in tourist brochures, guidebooks, and Internet sites. At another level, each photographer frames a place in a particular way to meet the needs of audience, which could be themselves, family, friends, or strangers. The vital point, as Crang (1997, 362) emphasizes, is that, "It is not a case of pictures showing what is 'out there' . . . but rather how objects are made to appear for us." How vacation snapshots are framed is an example of how individuals make sense of and then, often with a particular audience in mind, communicate particular versions as "truths" about places which are made "real" by the photograph.

Of course, Foucault did not envisage audiences as passive recipients of meanings of source materials. As emphasized earlier, interpretation of source materials is always a socially, spatially, and temporally contingent process. Kathleen Mee and Robyn Dowling (2003) illustrate how film reviewers—one audience for an Australian film entitled *Idiot Box*—maintained conventional understandings of life for young men in Australian suburbia despite the alternative meanings intended by the producers. Filmed in the western suburbs of Sydney, *Idiot Box* is a film about a fictional bank robbery. The planning and implementation of the unsuccessful robbery by the key characters, Mick and Kev, was—according to the writer-director David Caesar—intended to challenge common-sense assumptions among cinema audiences about young, unemployed, suburban, working-class men as "yobbos." Yet, as Mee and Dowling (2003) illustrate, the meanings of *Idiot Box* for film reviewers were shaped by previously circulating discourses presenting young working-class men in the western suburbs of Sydney as "rebels" and suburbia as "boring." In this way, film reviewers sustained one set of understandings about the socio-spatial identities of western Sydney as "truth." Two key points can be drawn from the *Idiot Box* example. First, the meanings of a text are never singular or uni-directional. Each audience segment will bring different meanings and power dynamics linked to their personal affiliations. Second, critical reflexivity may help you identify the ideas you bring to a project and help prevent you from imposing your own taken-for-granted understandings on to the source materials (see Chapter 2).

Coding: Once for Organization and Again for Interpretation

Coding is a process by which researchers structure and interpret qualitative data. When doing discourse analysis, coding serves two primary functions: organization and analysis of source materials. Hence, one possible way to start coding is to draw on two different types of codes: descriptive and analytical. In Chapter 18, Meghan Cope provides a helpful discussion of the difference between these two coding structures. Descriptive codes of "who," "where," "when," "how," and "why" offer a starting point to organize your data. From this starting point, you will need to pay closer attention to the "conditions," "interactions between actors," "personal reflections on experiences," "actions," and "outcomes." Mechanistic application of these headings should be avoided. For example, Waitt and Warren (2008) employed four category labels to interpret the transcripts of surfers:

1. context (for example, in an analysis of diaries kept by surfers, *where*, *when*, and *who* participated in surfing);

2. practices (the *events* [for example, what happened while surfing], *interconnections* [for example, who influenced the surfers' style of surfing and how they interact with other surfers and beach-goers], and *actions* [for example, what type of manoeuvres they performed while surfing]);
3. attitudes (for example, statements of judgment about other surfers, beach-goers, the ocean, or sharks);
4. experiences/emotions (for example, statements of emotions about surfing or interactions with other surfers or beach-goers).

For each category label, start a list of codes. However, remember that coding is an iterative process. Hence, your initial descriptive codes may change. In some cases, you will need to divide initial codes into finer detail. In other cases, you will have to amalgamate descriptive codes into broader categories. Following each "reading" of your source material, your coding structure will become more refined.

Alternatively, manifest content analysis may be deployed as the starting point for discourse analysis. Manifest content analysis is essentially a quantitative descriptive coding technique. Following this approach, key themes emerge from quantifying the instances of particular words, cluster of words, or lines devoted to a particular statement (Lutz and Collins 1993). Thomas McFarlane and Iain Hay (2003) provide a helpful example that illustrates the application of content analysis. In the context of the World Trade Organization (WTO) ministerial conference in Seattle during December 1999, they coded the word clusters by quantifying the number of lines used to describe the protests in articles that appeared in Australia's only national newspaper, *The Australian*. They demonstrated how articles in *The Australian* effectively demonized and marginalized anti-WTO protestors. The free software Wordle One is one useful way to generate graphic "word clouds" or "clusters" from text (http://www.wordle.net). Those words that appear most frequently in your text will appear most prominent in the Wordle cluster. Remember, when doing discourse analysis, descriptive codes or content analysis are envisaged only as an initial starting point. The frequency of words may help identify shared understandings, but provide little insights into the particular sets of ideas that maintain statements as taken-for-granted or "true."

Coding for a Foucauldian discourse analysis also involves devising a list of analytical codes. Analytical codes involve some form of abstraction or reduction. Hence, analytical codes may be envisaged as interpretative themes rather than descriptive labels. In discourse analysis, analytical codes typically provide insights into why an individual or collective holds particular sets of ideas by which they make sense of places, themselves, and others. Take for example Goss's (1993) discourse analysis of the Hawai'i Visitors Bureau's (HVB) portrayal of the Hawai'ian Islands for the North American tourism market. For his analysis of the

34 advertisements commissioned by HVB and published between 1972 and 1992, he initially coded the recurring images and words employed to portray the island, including Indigenous Hawai'ians, plants, volcanoes, location, and climate. This coding then suggested five analytical themes deployed by the HVB to "invent," or pitch, the Hawai'ian Islands outside the "normal" construction of North American geography: earthly paradise, marginality, liminality, femininity, and *aloha*. Goss went on to interpret each of these themes. He argued that the tourism industry reproduced long-standing colonial knowledge about Hawai'i. To provoke travel desire among potential tourists, the advertisers' preferred image of Hawai'i remained with the socially constructed fictional colonial knowledge of a sexually permissive black female sexuality and verdant tropical nature. Hawai'i is portrayed as a timeless location, a portal to the past, inhabited by people who are "naturally" friendly. In this (socially constructed) timeless and luxuriant tropical island paradise, visitors are promised possibilities to become themselves, free from the repressive regulations of mainland North America. The lesson for doing discourse analysis is that coding must be implemented as a twofold process, once for organization and again for interpretation.

Power, Knowledge, and Persuasion: Investigate Your Source Materials for "Effects of Truth"

Persuasion entails establishing and maintaining sets of ideas, practices, and attitudes as both common sense and legitimate. Foucault conceptualizes persuasion as a form of disciplinary power that operates through knowledge. He positions the mutually interdependent relationship between power and knowledge as indistinguishable, arguing that "Truth isn't outside power. . . . Truth is a thing of this world; it is produced only by virtue of multiple forms of constraint. And it induces regular effects of power" (1980, 141). Hence, questions about the "truth" of knowledge are fruitless, for truth is unattainable. Instead, Foucault focuses on how particular knowledge is sustained as "truth" (truth effects). According to Foucault, the mutual relationship between power and knowledge is underpinned by **discursive structures**.

Discursive structures are the relatively rule-bound sets of statements that impose limits on what gives meaning to concepts, objects, places, plants, and animals (Phillips and Jørgensen 2002). Foucault uses this term to refer to sets of ideas that typically inform dominant or common-sense understandings of and interconnections among people, places, plants, animals, and things. In Western thought, for example, scientific knowledge is often understood as the most appropriate way of thinking about what exists in the world. Western rational thought is characterized by a set of binary rules underpinned by Descartes's mind/body separation. Today, this dualistic thinking is still evident in a whole series of hierarchically valued dichotomies, including rationality/irrationality, man/woman, mind/body, straight/gay, masculine/feminine, and humanity/nature. Hence, while Foucault understands discourses to be inherently unstable, discursive structures are understood to "fix" ideas of the world within particular social groups at specific historical and spatial junctures. In sum, for both individuals and collectives, discursive structures establish limits to, or operate as constraints on, the possible ways of being and becoming in the world by establishing normative meanings, attitudes, and practices. Simply put, discursive structures are a subtle form of social power that fix, give apparent unity to, constrain, and/or naturalize as common sense particular ideas, attitudes, and practices. Foucault refers to this form of social control as the "effects of truth."

Take, for example, the concept of sexuality. There are many different discourses about sexuality in the world. However, within a particular time and place, a specific set of ideas will come to define socially acceptable practices of sexuality. For example, in nineteenth-century Europe, North America, and Russia, the privileged knowledge about sexuality was inspired by religious, scientific, and medical institutional narratives. Among most scientists of this time, sexual difference was understood by theories of gender inversion. Homosexuals were mistakenly constituted as effeminate men. When combined with a lethal mix of fundamental Christian morals, same-sex-attracted men were "invented" through rhetorical strategies that grounded meaning in binary oppositions to heterosexual men as the "natural" and "healthy" sex. Hence, the cultural identity of the homosexual man was constituted as lacking. Homosexuality became associated with negatively valued concepts: the primitive, the irrational, the feminine, the diseased, and the sinful. In this way, it seemed self-evident that the homosexual was only worthy of study in efforts to find a medical cure. Simultaneously, the knowledge of science made same-sex-attracted women "invisible" by constituting sex as a penetrative act. Thus, social limits were placed on understandings about same-sex-attracted women and men. Indeed, the casting of homosexuality in the nineteenth century with so many negatively valued concepts often resulted in laws criminalizing sodomy in the West.

This example illustrates a number of key implications for doing discourse analysis. First, persuasion, or the effect of truth, is a power-laden process through which particular knowledge is deployed by institutions as a mechanism of social control. In this case, social power operates through the way that male-female distinctions came to define the appropriate and dominant conception of gender. Drawing on scientific knowledge, relatively powerful groups in society were able to naturalize meanings, attitudes, and practices towards another social group constituted as diseased and mad. Hence, when doing discourse analysis, it is imperative that you remain alert to institutional dynamics and the social context of a source material. Second, while discourses are always inherently unstable, multiple, and contradictory, discursive structures operate to give fixity, bringing a common sense order to the world. Particular sets of ideas become accepted and repeated by most people as "common sense," unproblematic, unquestionable, and apparently "natural." Hence, an essential part of doing discourse analysis is becoming aware of the ways in which particular kinds of knowledge become understood as valid, legitimate, trustworthy, or authoritative. The way that knowledge becomes understood as appropriate is not restricted to the use of particular technologies in the production of texts (for example, computers, maps, and photographs) but also encompasses the way that sets of ideas are legitimized by the subtle deployment of different knowledge-making practices (statistics, medicine, policies, anecdotes) or categories of spokesperson (politician, scientist, academic, lawyer, judge, priest, eyewitness).

Resilience and Rupture: Take Note of Inconsistencies within Your Sources

While one rationale for doing discourse analysis is to identify the limits on how a particular social group talks and behaves (that is, the discursive structures), another is to explore inconsistencies within your source materials. On the one hand, taken-for-granted sets of ideas about who and what exists in the world help to impose bounds beyond which it is often very hard to reason and behave. When particular relationships become understood as common sense, they set limits to the cultural know-how of a particular social group. As discussed earlier, Foucault understood dominant or common-sense understandings as discursive structures (see Phillips and Jørgensen 2002). On the other hand, while discursive structures may appear eternal, fixed, and natural, because they are embedded within different social networks they are fragile and continually ruptured. Hence, there are always possibilities for meanings, attitudes, and practices to change or be challenged. Therefore, an essential part of doing discourse analysis is to be alert to possible contradictions and ambiguities in texts.

For example, take again Gorman-Murray's (2007) research on the Australian home. He demonstrated the resilience of social norms within Australian building design and government policy that shape and reshape the detached suburban dwelling naturalized as the heterosexual nuclear family form. Consequently the parental home may simultaneously be, for some lesbians and gay men, a site of belonging and alienation as they reconcile their sexuality. Gorman-Murray illustrated how the homemaking practice of people living their adult lives as lesbians and gay men in suburbia can rupture normative heterosexual domestic ideals. Inconsistencies become evident. While the domestic realm is often understood as a site of affirmation of self and family, not all people living in suburban homes are involved in reproducing heterosexuality and nuclear families through their home-making practices. Consequently, domestic spaces may affirm, legitimize, and nourish sexual difference. To generalize, doing discourse analysis involves remaining alert to contradictions and ambiguities within your texts.

Lesley Head and Pat Muir's (2006) work on suburban gardens in Wollongong, New South Wales, also illustrates the concept of resilience and rupture. Interviews with suburban gardeners provided insights into their understandings, attitudes, and practices towards nature. Head and Muir discuss the resilience among many European-Australian gardeners of the idea of cities as places devoid of nature. According to some of the gardeners, nature is only found beyond the spatial limits of the city in, say, national parks or the Australian "bush." Yet others talked about cultivating nature in their backyards, thereby rupturing conventional ideas about where nature is found and repositioning humans as nurturing rather than destroying nature. Remaining alert to such ambiguities and contradictions is a fundamental challenge of doing discourse analysis.

Silences: Silence as Discourse and Discourses That Silence

Finally, becoming attuned to silences in your texts is as important as being aware of what is present. Gillian Rose (2001, 157) reminds us that "silences are as productive as explicit naming; invisibility can have just as powerful effects as visibility." Similarly, Elizabeth Edwards (2003), while discussing her approach to discourse analysis as "dense context," draws attention to the importance of silences:

> "Dense context" is not necessarily linked to the reality effect of the photograph in a direct way, indeed to the extent that it is not necessarily *apparent* what the photograph is "of." Often it is what photographs are *not* "of" in forensic terms which is suggestive of a counterpoint. [Edwards 2003, 262–3, emphasis in original]

The key point is that identifying silences produced by texts is an integral part of discourse analysis. However, becoming alert to silences is always challenging. Clearly, to be able to interpret your texts for what is omitted requires conducting background research into the broader social context of your project and texts. Only then will you become aware of the existence of various social structures that inhibit what is present in your texts.

According to Foucault (1972), silences operate on at least two levels. First, silence as discourse is a reminder of how speakers' subjectivities are created within discourses. Who has the right to speak or is portrayed as an authority is itself constituted through discourse. Considering the social circumstances of authorship leads us to consider who is given the authority to speak. Silence may become particularly relevant when the voices within texts are considered in terms of the intersections of gender, age, class, ability, sexuality, and race. Being mindful of whose "voices" are silenced within your texts is an integral part of discourse analysis. For example, think about an advertisement for McDonald's restaurants. These advertisements often contain the "voices" of children. Silenced are the voices of parents who are resolving the conflicting sets of ideas about fast food and parenting. These are silences to which you can speak.

Second, Foucault's ideas alert us to how a **privileged discourse**—or dominant discourse—operates to silence different understandings of the world. Here, of vital importance are Foucault's arguments concerning the intersection between power, knowledge, and persuasion. According to Foucault, silence surrounding a particular topic is itself a mechanism of social power within established structures. For example, Sarah Holloway, Mark Jayne, and Gill Valentine (2008) demonstrate how policy debates, media attention, and epidemiological research about alcohol in recent years in England have centred on questions of drinking in public spaces, particularly by younger people participating in the inner-city nighttime economy of metropolitan centres. As they go onto argue: "Like many geographical imaginaries, this vision is a highly partial one and is as much of interest for what it excludes as that which it includes" (p. 533). The inner-city focus silenced questions about alcohol consumption in the domestic realm; that is a significant part of the English market. Hence, on the one hand, the focus on the nighttime economy served the interests of municipal, law-enforcement, and health authorities seeking to curb binge drinking. On the other hand, the focus of the night economy and the youthful binge drinker also served the interests of older alcohol consumers drinking in the domestic realm, supermarkets, and alcohol companies supplying the retail sector.

Another helpful example of a discourse that silences the voices and interests of some is the elaborate colonial Western fiction of the frontier. In the nineteenth century, frontier discourses among European politicians, scientists, priests, ministers, social scientists, and many settlers helped to naturalize and justify colonial settlement and nation making in North America (see Goss 1993; Turner 1920) and Australia (see McGregor 1994; Schaffer 1988; and Ward 1958). Frontier discourses relied upon maintaining Western understanding of colonized places as "uncivilized," "empty," "natural," "wild," and "timeless." In frontier places, European history was understood as yet to begin. As Deborah Bird Rose (1997) argues, settlers' discursive strategies towards the frontier erased the presence of Indigenous people or, at best, cast them as "primitive" people whose natural fate according to social Darwinism was extinction. In Australia, frontier discourses that portrayed Australia as a place waiting for history to begin became law, dictating that before the arrival of Europeans, the continent belonged to nobody (*terra nullius*). It was only in 1993, following the Australian High Court's Mabo decision, that the "truth" of *terra nullius* was overturned. No longer could Aboriginal Australians' knowledge be ignored in decisions over land ownership.

Nevertheless, Waitt and Head (2002) note how the colonial frontier discourse still enjoys widespread currency within the contemporary Australian tourism industry. Potential tourists are offered a portal to a timeless land. Imagined as timeless, the Kimberley—a northwest region of Australia—can then fulfill the specific market demands of primarily metropolitan Australian and international visitors, including: to discover the "real" Australia, imagined as the outback; to experience places imagined as wilderness; to gaze upon places portrayed as offering sublime "natural" beauty; or to explore the Kimberley as an adventure setting. However, the Miriuwung-Gajerrong's (the Indigenous people) knowledge constituting this place as "home"—named, known, and cared for over tens of thousands of years—is silenced by the colonial frontier discursive structures. Also silenced are the geographical knowledges constituted by Anglo-Celtic Australian pastoralists working on cattle stations. Revealing silences requires appropriate background research to identify who or what is missing from source materials.

Reflecting on Doing Foucauldian Discourse Analysis

The goal of Foucauldian discourse analysis is to reveal how particular ideas that help to forge social and spatial realities become understood as common sense. In political terms, discourse analysis allows insights into processes of social and environmental injustice. Conducting Foucauldian discourse analysis requires both understanding the concept of discourse and familiarity with a methodological template. Foucault's concept of discourse alerts us to the social constitution of all knowledge. Further, Foucault warns us that within many competing knowledge systems, particular sets of ideas emerge as dominant in the definition of appropriate forms of knowledge. According to Foucault, how certain knowledge is constituted as "truth" is not an accidental or haphazard process. Particular forms of knowledge become ascendant in ways that serve the interests of particular social groups. To unravel how, and why, requires careful tracing of ideas expressed in the authorship, technological production, and circulation of diverse texts including government reports, policy documents, paintings, diaries, photographs, and interviews. When conducting discourse analysis these texts become your source materials.

Foucault does not provide a formal set of guidelines for the analysis of discourse. Instead, geographers, along with other social scientists, have contributed to designing approaches to analysis (Box 14.1). Are there disadvantages to relying on a methodological template? One objection, perhaps, is that the effect of writing geography that employs these criteria is inevitably both selective and prescriptive. To implement the checklist criteria uncritically would mask their potential to both silence and demand particular responses. In other words, the checklist may wrongly become equated with objectivity and rationality at the expensive of subjectivity and creativity.

Nevertheless, with this caveat in mind, the checklist has at least four crucial advantages for the researcher new to Foucauldian discourse analysis. The first is to keep in mind the iterative relationship between the researcher and their work: the researcher shapes the analysis as much as the analysis shapes the researcher. The researcher must be conceived as integral to rather than separate from the discourse analysis. Therefore, reflexivity, or writing one's self into a project, is a vital part of discourse analysis (for a more detailed discussion of related ideas, see Chapter 19). The second is that the checklist encourages researchers to remain alert to a key aim of discourse analysis: discourse analysis is not about determining the "truth" or "falsity" of statements but instead seeks to understand the geographical and historical circumstances that privileged particular discourses that become fixed within discursive structures. Thus, in addition to selecting, familiarizing, and coding, discourse analysis requires background research into a source material's or text's social circumstances, including authorship, production, and circulation. Third, the checklist operates as a reminder that all knowledge production is caught up in power relationships. Certain ways of knowing the world are privileged while others are silenced. Discourse analysis involves being alert to different strategies of conviction deployed by authors to help persuade audiences that a particular form of knowledge is intrinsically better than others. Finally, it is crucial to understand that while discourses may manifest themselves in ways that bring order to social life as rules, maxims, common sense, or the norm, they are always unstable and may be ruptured. At the heart of discourse analysis is remaining alert to such instability, ambiguity, and inconsistency. Well-conducted and thoughtful Foucauldian discourse analysis enables insights into the resilience and rupture of multiple and sometimes conflicting discourses that produce and reproduce meaning of our always spatially located everyday lives.

Key Terms

audience
constructionist approach
discourse analysis
discursive structure
embodied knowledge
genre
intertextuality
manifest content analysis
metaphor
positionality statement
power/knowledge
privileged discourse
rigour
semiology
silence
strategy of conviction
text

Review Questions

1. Why should both background research and the researcher be regarded as integral parts of a Foucauldian approach to discourse analysis? To answer this question think about how discourse constructs knowledge that produces generally accepted ways of thinking/being/doing.
2. According to Foucault's notion of discourse, in what ways are the terms "real" and "truth" misleading?
3. What are the differences between descriptive and analytic codes? How do these different codes relate to one another in a Foucauldian approach to discourse analysis?
4. Why are there tensions surrounding presenting a checklist or methodological plan for "doing" Foucauldian discourse analysis?

Review Exercises

1. *Nature Talk*. Foucault argues that discourse governs what is possible to talk about through the production of categories of knowledge. This process is not accidental. At the same time, discourses reproduce power by governing what is generally accepted as true. To explore this concept think about your own "nature talk."

   First, make a list of all the things/places/people that you generally talk/think about as belonging to nature.

   Next, using your "nature list," explore the following questions:
   a. What is disallowed?
   b. What is normalized?
   c. Whose interests are being mobilized and served by this list?

   To further explore ideas associated with this exercise, see Castree, N. (2005) *Nature: The adventures of a concept*. London: Routledge.
2. *Sweaty Geographies*. Foucault argues that discourse defines categories of thinking or lines of argument that frame ways of thinking/being/doing as generally accepted truths, at the same time as other ways of thinking/being/doing are dismissed. To explore this concept, write down the places where you think it is both appropriate and inappropriate to sweat. Start with the places where sweating is appropriate. What sets of ideas makes it possible and or desirable to sweat here? Why is sweat normalized in these places? How is sweat understood in these contexts? How does the appearance of sweat in these locations help constitute a particular subject? What interests are being mobilized and served by thinking/talking about sweat in this context? Now turn to the places where you think sweating is inappropriate. What makes sweat a problem in these locations? Why is sweat disallowed? What identities become made impossible? Explore

the differences and similarities between class members' answers to these questions. What are the social norms governing "truths" about where sweaty bodies are welcomed or marginalized?

To further explore ideas associated with this exercise, see Waitt, G. (2013) "Bodies that sweat: the affective responses of young women in Wollongong, New South Wales, Australia," *Gender, Place and Culture* doi: 10.1080/0966369X.2013.802668

Useful Resources

Linda Graham, Queensland University of Technology, and Helen McLaren, University of South Australia provide helpful papers about how Foucauldian discourse analysis is used in education and feminist research. They are available at: http://eprints.qut.edu.au/archive/00002689/01/2689.pdf.; and http://www.unisa.edu.au/Documents/EASS/HRI/foucault-conference/mclaren.pdf

Helpful starting points for more comprehensive reviews of Foucault's concepts include Cousins and Houssain (1984), Hall (1997), Hook (2001), McNay (1994), Mills (1997), and Thiesmeyer (2003).

Alternative discussions of "doing" discourse analysis are provided by Dryzek (2005), Phillips and Jørgensen (2002), Rose (2001), Shurmer-Smith (2002), and Tonkiss (1998).

Placing Observation in the Research Toolkit

Robin A. Kearns

Chapter Overview

> . . . for a discipline often preoccupied with the visual, it seems . . . [geography] has not studied the practices of seeing rigorously enough. [Crang 1997, 371]

"Seeing is believing," as the saying goes. While visual observation is a key to many types of research, there is more to observation than simply seeing: it also involves touching, smelling, and hearing the environment and making implicit or explicit comparisons with previous experience. Further, seeing implies a vantage point, a place—both social and geographical—at which we position ourselves to observe and be part of the world (Jackson 1993). What we observe from this literal or metaphorical place is influenced by whether we are regarded by others as an "insider" (i.e., one who belongs), an "outsider" (i.e., one who does not belong and is "out of place"), or someone in between. The goal of this chapter is to reflect on the reasons why observation is fundamental to geographical research and to consider critically what it means to observe. I explore various positions the researcher can adopt vis-à-vis the people or materials that are observed—from the viewing of secondary materials such as photographs or video footage to participating in the life of a community. Two key contentions are:

1. that observation has tended to be an assumed, and consequently undervalued, practice in geographic research; and
2. that ultimately all observation is a form of participant observation.

In the chapter, I argue that observation has been taken for granted as something that occurs "naturally." Indeed, observation is commonly—and unfairly—regarded as "inherently easy" and "of limited value" (Fyfe 1992, 128). As a research approach, it has not been regarded as requiring the degree of attention granted to more technical aspects of our methodological repertoire as human geographers (e.g., questionnaire design, survey sampling). With critical reflection, however, observation can be transformed into a self-conscious, effective, and ethically sound practice.

Purposes of Observation

Among the definitions of **observation** in the *Oxford English Dictionary* is "accurate watching and noting of phenomena as they occur," implying that observation has an unconstrained quality. To regard observation as random or haphazard would be a mistake, however, for we never observe everything there is to be seen. Observation is the outcome of active choice rather than mere exposure. Our choice—whether conscious or unconscious—of first *what* to see and second *how* to see it means that we always have an active role in the observation process. Following Mike Crang, I argue for observation as a way of "taking part in the world, not just representing it" (Crang 1997, 360).

There is a range of purposes for observation in social scientific research, and they can be summarized in three words, which conveniently all begin with "c": counting, complementing, and contextualizing. The first purpose—*counting*—refers to an enumerative function for observation. For example, we might accumulate observations of pedestrians passing various points in a shopping mall or airport in order to establish daily rhythms of activity within these places. Research in the time-geographic tradition has used this approach to chart the ebb and flow of spatio-temporal activity (Neutens et al. 2011). Under this observational rationale, other elements of the immediate setting are (at least temporarily) ignored while the focal activity occurs. The resulting numerical data is then easily displayed graphically or analyzed statistically. This is an approach to observation that may be useful for establishing trends but is ultimately too reductionist and devoid of human experience for developing any comprehensive understanding of place.

A second purpose of observation is providing *complementary* evidence. The rationale here is to gather additional descriptive information before, during, or after other more structured forms of data collection. The intent is to gain added value from time "in the field" and to provide a descriptive complement to more controlled and formalized methods such as interviewing. Complementary observation might involve "hanging out" in a neighbourhood after completing a household survey and taking notes on the appearance of houses, the types of cars, and the upkeep of gardens. In other words, qualitative research requires the researcher to be engaged, to a greater or lesser extent, in the lives of participants—"to hear their stories, grasp their point of view, and understand their meanings" (Cobb and Forbes 2002, 197). While an interviewer is primarily bound by the questions to be asked, interviewers can also reflect on their own experience of being-in-the-research, as well as their understandings of participants' meanings, perceptions, and experiences (Ansell and van Blerk 2005; Browne 2003). I used this approach in research on housing problems in Auckland and Christchurch. Interviewers were instructed to observe the dwellings of those they interviewed. Their **field notes** added to (and often contrasted with) what people said about their own dwellings

(Kearns, Smith, and Abbott 1991). Such information complements the aggregated data gathered by more structured means and assists in interpreting the experience of place. It allows the "seeing" of multiple viewpoints including not only the participants' experience of place, but also the field itself and the place occupied by researchers within it.

The third purpose of observation might be called *contextual* understanding. Here the goal is to construct an in-depth interpretation of a particular time and place through direct experience. To achieve this understanding, the researcher immerses herself or himself in the socio-temporal context of interest and uses first-hand observations as the prime source of data. In this situation, the observer is very much a participant. An example from the "methodological toolbox" is the "go-along" interview in which the researcher and participant walk together, talking as they observe surroundings together (Kusenbach 2003).

These purposes for observation are not mutually exclusive, however. I will later show that one can approach observation with mixed purposes, seeking, for instance, to both enumerate and understand context during a period in the field. My examples will be mainly drawn from projects early in my career that deeply influenced my thinking as well as from the recent fieldwork of Auckland doctoral graduate Tara Coleman.

Types of Observation

Some social scientists identify two types of observation: *controlled* and *uncontrolled*. **Controlled observation** "is typified by clear and explicit decisions on what, how, and when to observe" (Frankfort-Nachmaias and Nachmaias 1992, 206). This style of observation is associated with natural science and its experimental approach to research, which has been imported into physical geography. Thus, a geographer might set up an electronic stream gauge to gather data at periodic intervals throughout a flood, resulting in a series of so-called "observations." However, such data are collected remotely without the aid of human senses. Indeed, the human eye may have only been involved in observing the gauge subsequent to the actual data collection. This recognition that controlled observations can be made through mechanical means adds weight to the common perception that such research is rigorous and easily replicated. However, controlled observation is also limiting in terms of the sensory and experiential input that is admissible as "findings." Human geographers are unlikely to employ such controlled methods of observation, except perhaps in the earlier example of pedestrian counts.

In two respects, controlled observation is necessarily limiting. First, there is an imposed focus on particular elements of the known world, and second, it is only *directly* observable aspects that are of interest. Thus, imputed characteristics of place and the feelings of residents may be out of range. Most of the observation conducted by contemporary social and cultural geographers could be described as **uncontrolled observation**. Such observation is certainly directed by goals and ethical considerations but is not controlled in the sense of being restricted to noting prescribed phenomena. Therefore, although observation refers literally to that which is seen, in social science it may involve more than just seeing. Most obviously, observation also includes listening, a critical aspect of field research. Effective listening has been described as the most important skill in interviewing (Seidman 2013) and can assist visual observation by both confirming the place of the researcher as a participant and by attuning oneself to **soundscapes** within social settings (S.J. Smith 1994). Indeed, in some contexts like a music festival, sounds are critical characteristics that arguably create the place through attracting a crowd (Kearns 2014).

One could argue that all research involves observation or at least comprises a series of observations. Thus, in a social setting, we can "observe" the population by employing questionnaires through which the researcher establishes the frequency of certain variables (for example, occupation, sex). The activity of conducting a questionnaire survey invariably places the researcher in the position of an "outsider," marked as "other" by purpose, if not by appearance and demeanour (for example, clothing, age, ethnicity, or the type of language used can accentuate "outsiderhood"). Identifying a sample and evaluating the responses to questions addresses the goal of generalizability but the cost is that only a subset of social phenomena is implicitly deemed to be of interest. These phenomena can too easily be isolated from their context, if unaccompanied by less directly observable values, intentions, and feelings. Qualitative approaches, on the other hand, allow the consideration of human experience and emotion, potentially suspending conventional concerns about researcher bias and recognizing instead the relationship between the researcher and the people and places he or she seeks to study.

A further distinction can be drawn between **primary observation** and **secondary observation**. The former activity would have us adopt the position of participants in and interpreters of human activity; as secondary observers we are interpreters of the observations of others. Examples of secondary observation are analyses of picture postcards available to tourists (e.g., Marwick 2001) or the photography of one "racialized" group by a representative of another (e.g., Alderman and Modlin 2013). In this chapter, rather than dwelling on the use of secondary materials I will rather focus on the ubiquitous nature of participant observation. The fact that participation is inescapable can be illustrated through the point that a solitary researcher observing images of place is also an active participant in the process. This is because the observer is co-creating meaning through bringing her or his own perspectives and life experiences to their analysis and interpretation. Further, **secondary data** is increasingly being generated *through* participatory methods. **Photovoice**, for instance, is an approach formulated within community health development in which participants are given cameras to record sites and objects of significance to them. Once printed, the photographs serve as catalysts for discussion, which in turn is recorded and interpreted (Wang and Burris 1997). Variations on this approach have been adopted by geographers for use with, for example, children (e.g., Mitchell, Kearns, and Collins 2007). The relevant point for our discussion is the importance of engaging with *participants* (not "respondents") and at least partially participating in their worlds. Only then can we meaningfully observe their photographs and engage in interpretive discussion. Similarly, but with more ethical complexity, video has been used as an observational complement to direct engagement in social settings such as cafés (Laurier and Philo 2006). The blurring of boundaries between primary and secondary data, with participation as the "glue," is illustrated in Tara Coleman's field notes, set out in Box 15.1 below.

Participant Observation

Participant observation has its roots in social anthropology (Sanjek 1990) and significantly, its profile among geographers was raised by Peter Jackson (1983), who had studied both disciplines. The approach has been adopted and adapted by geographers seeking to understand more fully the meanings of place and the contexts of everyday life. Examples include a number of classic studies within the humanistic tradition, such as David Ley's (1974) work on Monroe, a neighbourhood

Blurring Boundaries

Taking photographs with seniors and at their direction allowed their experiences, values, and meanings relating to the scene in their houses being photographed to be spontaneously shared and discussed. It was less about the actual photographs and more about engaging in an act with participants in order to observe their choices, feelings, actions. The act of photography was given centrality and enabled participants' experiences and meanings related to place, being aged, and well-being to be analyzed in the moment, as they unfolded. Thus, seniors' thoughts, feelings, and decisions with respect to home and community were explored as they occurred and were also reflected upon later many times during interpretation. I also used the photographs I printed to invite further conversations: formally, in the participants' journals, and informally, through giving photographs to participants for them to keep.

in inner-city Philadelphia and Graham Rowles's (1978) research on the ways older people experience place. Like many social geographers, these writers talked to "locals" in the course of their research, but it was the depth of their involvement in a community, their sustained contact with people, and their relatively unstructured social interactions that distinguished their work. While their contemporaries were mainly interested in perception or behaviour, these geographers were concerned with experience.

Developing a geography of everyday experience requires us to move beyond reliance on formalized interactions such as those occurring in interviews. As Mel Evans (1988, 203) remarks, "although an interview situation is still a social situation . . . it is a world apart from everyday life." Evans is suggesting that no matter how much we are able to put people at ease before and during an interview, its structured format often removes the researcher from the "flow" of everyday life in both time and space. In other words, an interview ordinarily has an anticipated length and occurs in a mutually agreeable place often set apart from other social interactions. In contrast, the goal of participant observation is to develop understanding through being part of the spontaneity of everyday interactions.

There is a consensus that, although definable, participant observation is difficult to describe systematically. Commentators have remarked on its "elusive nature" (Alder and Alder 1994) and its breadth (Bryman 1984). Students who are curious about the approach have commonly been referred to classic studies such as William Whyte's *Street Corner Society* (1943) for models. One explanation for this tendency to refer to examples rather than to offer step-by-step guidelines is that every participant observation situation is unique. Another reason offered by Evans (1988, 197) is that the success of the approach depends less on the strict application of rules and more upon "introspection on the part of the researcher with respect to his or her relationship to what is to be (and is being) researched." The extent to which a researcher is able to engage in this introspection will be determined in part by both disposition and discipline.

While introspection or reflection on what we see and experience is important, guidance in the act of observing is also needed. Jackson looks back to Kluckhohn (1940) for a concise description of participant observation as "conscious and systematic sharing, in so far as circumstances permit, in the life activities and, on occasion, in the interests . . . of a group of persons" (1983, 39). It is therefore the intentional character of observations that contrasts the activities of a participant observer with those of routine participants in daily life (Spradley 1980). To generalize, participant observation for a geographer involves strategically placing oneself in situations in which systematic understandings of place are most likely to arise.

One rationale for participant observation is recognition that, in a more structured setting, the mere presence of a researcher potentially alters the behaviour or dispositions of those being observed. This point is illustrated in a "Far Side" cartoon by Gary Larson. In the drawing, two men wearing pith helmets are approaching a thatched hut. The occupants, stereotypically adorned with head-dresses and bones through their noses, are exclaiming "Anthropologists! Anthropologists!" as they rush to hide their television set. The cartoon ironically suggests that scholarly explorers do not expect "primitives" to have technology (and that primitives might want to live up to this stereotype). In terms of research methods, the cartoon's message is that the undisguised entry of others into a social situation is bound to alter behaviour. The point for us as qualitative researchers is that conscious participation in the social processes being observed increases the potential for more "natural" interactions and responses to occur.

Use of the term "observation" in highly controlled scientific research might tempt us to think too easily in terms of a simple dichotomy between participant and non-participant. According to Atkinson and Hammersley, this is an unhelpful distinction "not least because it seems to imply that the non-participant observer plays no recognised role at all" (1984, 248). Indeed, as the "Far Side" cartoon succinctly illustrates, there is no such thing as a non-participant in a social situation: even those who believe they are present but not participating in a research context often unwittingly alter the research setting. To move beyond this false binary construct of participant–non-participant, Gold (1958) suggests that there is a range of four possible research roles:

1. **complete observer** (for example, a prisoner being observed through closed-circuit TV cameras)
2. **observer-as-participant** (for example, a newcomer to a sport being part of the crowd—see Latimer 1998);
3. **participant-as-observer** (for example, seeking to understand personally familiar places in a new light—see Kearns and Fagan 2014)
4. **complete participation** (for example, living in a rural settlement to understand meanings of sustainability).

While it is difficult to imagine how being a complete observer might be incorporated into geographical research, a potential example could be remote surveillance of a public square, perhaps with access to footage from closed-circuit television (CCTV). This technology allows the "remote" observation of activity. Behaviour may be less influenced by the observer "being seen" than it would be if the research were conducted in the presence of an embodied and visible observer. Frequently, however, the mere presence of cameras influences and moderates behaviour (Williams and Johnstone 2000). Further, the use of such footage for research purposes raises various ethical issues around consent and privacy. As the foregoing references indicate, work by geographers can be easily categorized according to the remaining three roles (i.e., not a "complete observer"). Each category represents a form of participant observation, and the difference is essentially a matter of the degree of participation involved. This division implies a continuum of involvement for the observing researcher from detachment to engagement. However, whatever one's positioning vis-à-vis "the observed," it is important to acknowledge that the act of observation is imbued with power dynamics.

Power, Knowledge, and Observation

Before considering the stages of observation in the field in greater detail, it is important to reflect on aspects of the process itself. As emphasized earlier, observation involves participating, both socially and spatially. We cannot observe directly without being present in a place, and our bodily presence brings with it personal characteristics that mark our identity such as "race," sex, and age. Belonging to a dominant group in society can mean that we carry with us the power dynamics linked to such an affiliation (Kearns and Dyck 2005). For instance, being a white adult male creates challenges for engaging with a group whose members do not share those characteristics, such as a new mothers' support group (Kearns et al., 1997). Similarly, embodied difference can be a talking point. Over the course her fieldwork, Tara Coleman found herself both limping from a hip injury, and then pregnant. These situations generated comment from her older participants and the resulting interactions resulted in a blurring of the boundaries of the research relationship, which in turn complicated power dynamics. In other words, our difference in terms of embodiment contributes to our (in)ability to be "insiders" and participants in the quest to understand place.

More subtle challenges may be generated by our level of education and affiliation with a university. In undertaking research, we can subtly carry institutional dynamics into our acts of observing. Social control can occur through the ways in which members of one group (the relatively powerful) are able to maintain watch over members of another (the relatively disempowered). Perhaps the most memorable image of this dynamic is Bentham's **panopticon**, a circular prison designed to maximize the ability of guards to see into cells and watch prisoners (Foucault 1977a). Prisoners do not know exactly when they are being observed but learn to act as if they are always being watched. Foucault sees this surveillance as a form of disciplinary power that is enforced through the layout of the built environment rather than through the exertion of force *per se.* The key to the resulting institutional dynamic is the knowledge on the part of prisoners that they can be observed at any time. This knowledge, according to Foucault, results in self-surveillance and self-discipline. The use of CCTV in prisons to maintain an ever-present eye on detainees reflects an ongoing manifestation of the principles of the panopticon.

Foucault's ideas have implications for observation as a research approach. The **surveillance** to which he refers is a very visual and disembodied form of observation. Gillian Rose (1993) has linked surveillance to the traditional geographical activity of fieldwork. To Rose and other feminists, geography has been an excessively observational discipline characterized by an implicit **masculine gaze**. The key point is that observation can be a power-laden process deployed within institutional practices. As we are based in and representative of academic institutions, it is imperative that we be aware of the ways in which others' behaviour may be modified by our presence (Dyck and Kearns 1995).

A challenge posed by feminist geographers and anthropologists is to see fieldwork as a gendered activity (Nast 1994; Rose 1993). Their argument is that in being participant observers, we unavoidably incorporate our gendered selves into the arena of observation. An extract from Tara Coleman's field notes illustrates how observation in human geography is far from a simple matter of uni-directional watching; rather, it involves interactions that are potentially laden with sexual politics. The resulting situations may shed light on the gendered constitution of the field site itself (Box 15.2).

The Gendered Field Site

T: Last time we talked, you mentioned that you no longer feel at home at the local bar where you used to play music. Could you tell me more about why you no longer feel at home there and whether you continue to go there at all?
P: Well, it's nice to be asked about myself by someone like you!
T: Oh, what do you mean?
P: To have the attention of a young woman, a pretty young woman. Actually, I like your small size. And so I want to talk to you. I don't want to talk to all the old farts up the road, they just make me feel old. That's why I don't go up there anymore. I want to feel young, that's how I still see myself. So, let me talk to a young woman like you and I will be happy to. Shall I show you my albums and you'll see what it used to be like around here . . .

> The interview then involved looking at albums that showed photos of him as a young fit man, as well as all his youthful musical achievements, and presenting himself as still young, attractive, and vital. He talked about all his sexual conquests and then directly propositioned me quite directly . . .

Stages of Participant Observation

Participant observation is far from haphazard. There are some commonly recognized stages through which the process moves, from choice of research site through to presentation of results. I will review these stages using case examples from my early research in the Hokianga district of Northland, New Zealand. In this work, my objective was to understand the influence that sense of place had in shaping the social meanings of the local health-care system. In the course of this research, I adopted three of the four roles identified by Gold (1958): observer-as-participant, participant-as-observer, and complete participant (see Kearns 1991b; 1997).

Choice of Setting

Choosing a setting for participant observation might be dictated by the goals of a larger research project, in which case the field site may be unfamiliar to the researcher. In other cases, one might focus on otherwise familiar neighbourhoods and attempt to see them in a new way in light of the experience of the social groups that are the research focus. For instance, in our "Kids in the City" study, we sought to understand the opportunities and constraints on play and independent mobility for children resident in medium- and high-density housing in central Auckland along with the influence of parenting discourses. We chose familiar settings, but tried to see places with "fresh eyes" through visiting—if possible with the members of the group of interest—and being attentive to the observations of participants (Carroll et al., 2015). In the case of more distant field locations, there would need to be a good deal of background research on the community in question as well as reconnaissance visits before a period of immersion into community life. In such an example, anthropologist Kathryn Scott spent a year in an isolated rural settlement almost four hours' drive north of Auckland in the quest to understand community sustainability (see Scott, Park, and Cocklin 1997).

It is likely that student researchers will choose field sites more akin to the first example; places that are at least partially familiar to them. But familiarity can bring pitfalls. There is a danger that the researcher may be over-familiar with the community, with the result that there is "too much participation at the expense of observation" (Evans 1988, 205). Indeed, doing fieldwork in one's own society can be just as challenging as it is in an unfamiliar one.

What, then, might be the ideal? Possibly the best balance to strive for in choosing a research setting is a mid-point between the "insider" and "outsider" statuses discussed earlier. The conventionally recommended stance is that of stranger, but this is not necessarily a position in which one simply does not belong. Rather, it is one in which the researcher's status is what Evans (1988) terms "marginal" in relation to the community. By "marginal," I take Evans to mean socially, and possibly spatially, on the edge of a community or group. (see Box 15.3).Thus, although a citizen of the same city, studying on a campus only a few kilometres away, MA student Annie van der Plas found herself a newcomer in her field area. This was in the east Auckland neighbourhood of Glen Innes, an impoverished public housing area in transition to newer housing and higher-income residents. Annie had chosen a "chalkboard" method to gauge community feeling towards such changes. She mounted a blackboard in two public places and returned each day to note what had been inscribed. In availing herself to opportunistic conversation, she participated in the research, but through her method she was, on balance, more observer than participant.

Access

Gaining entry to social settings is potentially a fundamental challenge. A crucial issue is identifying key individuals who can act as gatekeepers, facilitating opportunities to interact with others at the chosen research site. There are some settings into which one can simply walk and take on the role of participant. Investigating the way in which a public shopping mall is used is a good example. Because a mall

Reading Community: An Observer as Participant

The train ride offered some time to mull over the project and jot down thoughts, or other times it was a chance to completely let it go. . . . It became a chance to catch my breath after what were particularly draining visits to Glen Innes. The chalkboard method generated a range of feelings and emotions, perhaps more so than if I were carrying out an orthodox method like interviews. Through repeatedly spending time in a public space during the research process, I carried out actions that created attention and interest, often allowing me to speak with "locals" who felt strongly and passionately about Glen Innes. . . . I was already aware of hostile feelings from this community towards researchers "coming in" to take information, and I felt uncomfortable about this relationship, wanting to ensure that my own research was based more on a reciprocal relationship with Glen Innes and its people. I would arrive feeling nervous, and sometimes sick to my stomach about how the board would look, or what people may say to me. . . (van der Plas 2014, 96).

is commonly visited and at least perceived to be a public place, fewer permissions are needed to informally observe there than in more "private" settings such as health clinics. Nor is there a problem with "blending in," given the diversity of users. One can observe through simply participating in the mall's functions: shopping, resting on the seats provided, or using the food court. But gaining access is potentially more challenging when the place is smaller in scale or less public in character.

Gaining access may well be more straightforward if one has a known role. Even being "the visiting student" in a workplace may give one a role in a way that just being an anonymous visitor or stranger would not. However, there are often no convenient roles in hospitals or factories, so once ethical approval is gained (see the discussion in Chapter 2), perhaps there is good reason to resist being typecast into any role except that of outsider (see Kearns 1997). This ambiguous position worked for me in the health clinics of the Hokianga district. I could not legitimately pass as either a health professional or a patient, since local residents knew each other too well, so acting as a visitor reading the newspaper helped me to remain reasonably inconspicuous for a short period (see Box 15.4).

Field Relations

The role that the researcher adopts within an observed setting (for example, complete participant or participant-as-observer) will define the character of the relations she or he generates. The idea of impression management is critical here, for the impact you make on those encountered will determine, to a large extent, the

Gaining Access to a Field Site

My research in the Hokianga district of northern New Zealand began with a period of fieldwork during October 1988. My wife Pat had an opportunity to become a medical *locum tenens* in the district, and I was also keen to go north. Once in Hokianga, having a spouse with a temporary part in the health-care system gave me some legitimacy in seeking approval to undertake research at the hospital and community clinics. The medical director as gatekeeper to the health system knew who I was, and as I was already there, the formality of letters and telephone conversations could largely be circumvented. I hastily devised a research plan that involved spending time in the waiting areas of each clinic, where I observed social dynamics with the goal of understanding the social function of these places.

ease with which they will interact with you and incorporate you into their place. Embedded within the word *incorporate* is "corpus," the Latin word for body. My purpose in discussing incorporation is to stress the idea introduced earlier of the researcher's embodiment and to recognize that, as researchers, we take more than our intentions and notebooks into any situation: we also take our bodies. The way we clothe ourselves, for instance, can be a key marker of who we are or how we wish to be seen in the field. For example, while a doctoral student, I researched the inner-city experiences of psychiatric patients. I chose to wear older clothes to drop-in centres in order to minimize being regarded as yet another health professional or social worker intruding on patients' lives (Kearns 1987). Since I was a student at the time, it was easy for me to find suitable clothes for a drop-in centre! As an external element of appearance, clothes cannot erase difference when other markers (age, sex, ethnicity) are more potent. This point was evident in Coleman's (2007; 2013) research when, sitting in on a high school health class, she was attentive to her dress style, but was obviously not a student, and later, when taking part in a "care and craft" session at a seniors drop-in centre but clearly not of advanced years herself. While researchers like Coleman and I have been able to "blend into" field situations involving disempowered groups (e.g., school children, frail seniors, mental health patients), it is questionable whether a researcher could as easily assimilate if attempting to study more elite social relations of place (for example, the dynamics of a lawyers' convention or a restaurant frequented by politicians). At the most fundamental level, I would not have had the right clothes to wear. Indeed, it is generally easier to dress "down" than "up," and this perhaps explains in small part why participant observation is more often used in studies of people less powerful than researchers themselves. A further point is that our ability to relate to others in the field depends not just on appearance but also on the level and type of activity undertaken. Hence, passive activities such as sitting, watching, and conversing are more easily and regularly taken up by qualitative researchers than full participation in skilled occupations such as Crang's (1994) work on the dynamics of restaurant work.

A further point is that although concern for appearance and clothing is appropriate, it potentially reinforces geographers' fixations with the visual (see Rose 1993). Hester Parr (1998) describes how, while researching the geographical experiences of people with mental illness in Nottingham, she became conscious of non-visual aspects of her "otherness" such as smell: her perfumed deodorant served to set her apart from those she was attempting conversation with and inhibited easy interaction. Parr's reflection reminds us that senses such as smell add to the character of place (Porteous 1985) and merit consideration for the way they can mark as "other" the "bodiliness" of the researcher.

A related point influencing **field relations** is that codes of behaviour are attached to different settings. By way of example, Tara Coleman (2013) recounts a situation in her study of the meaning of aging in place for residents of Waiheke Island, Aotearoa/New Zealand. She was having trouble ending the research process with "Sam," a man in his early 70s. She found that meeting him at a café, rather than at his home, she was able to have more control over communicating the end of the research to him. That public setting meant she was less confronted by his attempts to keep their lives connected because he had less power in the neutral space and the codes of behaviour there were different. He didn't get as emotional as he had at home, he was more polite, and she was able to deliver her message more clearly. So, in the course of researching intimate spaces, like the home, using neutral spaces afforded a way to cope with the shifting boundaries of the field.

The lesson is that research relations may be enabled or constrained by the (often unspoken) ways in which social space is codified and regulated. Successful participant observation thus involves not only the (temporary) occupation of unfamiliar places but also the adoption of alternative ways of using time. Box 15.5 continues an account of Hokianga research that influenced my thinking, describing my use of time in community clinics.

Talking and Listening

We cannot blend in as researchers unless we participate in the social relations we are seeking to understand. Listening ought to precede talking so that we become attuned to what matters in a particular time, place, and social setting. There is a strong link between observation and silence. In times of bereavement, an assembly of people is said to "observe" a minute's silence. What might observing silence involve? According to Coleman's field notes (2013):

> paying attention to moments of silence frequently revealed participants' feelings and a sense of things beyond the questions I asked . . . It invited rich description since it allowed participants to take the lead and assisted me in deferring my own rationalisations.

The "how" and "where" of talking and listening are also crucially important. For research with children, for instance, adopting their level—both physically and in terms of style of language—may be the key to successful observation. Christina Ergler (2012) found that organizing games and playing with Auckland children was a useful precursor to asking them about their experiences about their neighbourhood. Similarly, Hannah Mitchell engaged with children within their "workplace" (the classroom) and "walk-place" (the space between home and school) (Mitchell, Kearns, and Collins 2007). Similarly, in engaging with seniors, Tara Coleman (2013) writes of routinely having morning tea and general conversation with participants, followed by helping them. This "performance" sometimes led to her being shown

Field Relations in an Observational Study

At some of the community clinics, I blended into the gathering of attendees, inconspicuously observing events under the guise of reading the newspaper. At others, however, I was clearly a Pakeha ("white" non-Indigenous) visitor, since all the others present were both "locals" and Māori and seemed to know each other well. It was immediately evident that going to the doctor involved more than medical interactions. Rather, the occasion frequently provided an opportunity for locals to tell stories and reflect not only on their own well-being but also on that of their families and friends. From this observational time in the clinic waiting areas, I noted that the most frequent conversation category was community concerns. Comments on the deleterious impacts of the restructuring of public services were frequently offered. Residents adopted a relaxed approach to clinic attendance and their use of waiting areas. I observed some arriving well before their appointment time (occasionally in the company of others who had no intention of consulting the doctor or nurse) and lingering afterwards "having a yarn." These observations led me to interpret the clinics as *de facto* community centres, analogous to the village market in other countries. The difference, however, was that the place (the clinics) and the time (clinic day) prompted a considerable amount of conversation that explicitly centred on health concerns (adapted from Kearns 1991b, 525).

around the kitchen to put away items that had just been washed, and this in turn often led to being shown around the house—a more "natural" way of being given a tour than if she had asked. For Tara, sometimes, these informal (non-interview schedule) interactions took more time than the formal questions, but they offered much in terms of building rapport and allowing informal observation. In other words, observation is least conspicuous when one is interacting most naturally with the research subjects. The lesson is that, as researchers, we are our own most crucial creative tool in facilitating observational opportunities in the field.

Recording Data

A clipboard or audio recorder are the standard means of recording information in qualitative approaches involving face-to-face communication such as interviews. Sometimes in observational research a recorder is useful. For example, when

BOX 15.6

Talking and Listening: The Embodied Observer

My embodiment as researcher was central to my construction of knowledge of Hokianga health services. Within the waiting area I had to position myself so as to neither be threatening (and thus inhibit conversation by gazing at others) nor overly welcoming of conversational engagement (hence my "hiding" behind a newspaper). Such choreography was bound to break down, and did. On one occasion two "locals" offered to help with the crossword puzzle I was half-heartedly completing, and on another, a *kuia* (female elder) entered, kissed, and welcomed all present to "her" clinic, including myself. My corporeality within the observed arena of social interaction thus rendered binary constructs of researcher/researched and subject/object thoroughly porous. In this *kuia*'s clinic, my conceptions of being an "autonomous self" were dissected and (re)embodied within their rightful web of socio-cultural relations (Kearns 1997, 5).

observing participants' choices and movements during photo elicitation interviews, Tara Coleman (2013) used a handheld recorder because of the "busyness" of the field encounter with seniors on Waiheke Island:

> There was much going on—taking photographs, trying to observe gestures, gauging emotions, trying to remember comments made by the participant. I found that rather than creating distance between me (the researcher) and the older participants, the recorder and camera equipment actually helped to make the photo elicitation interviews a mutual and shared experience. This was because participants ended up helping me to carry around the recorder and camera equipment as they could see it was a lot for me to manage. They seemed to take on a "co-researcher" role as a result. This really helped them to set up the scenes that we photographed. And they seemed to open up more because of it. Whereas, initially, photographing scenes with me had seemed quite odd to them, they ended up feeling empowered by it and gave a lot to it.

However, because these tools can in some instances disrupt the flow of conversation or interaction, there is often a greater reliance on recollection and a need to work on detailed note-taking after a period of field encounters. At the end of any day or session of observation, one is likely to feel tired and not inclined to take out a pen or go to a computer to record reflections. However, developing a discipline for such "homework" is a key part of field observation: notes are invaluable sources of data and prompts to further reflection. Field notes become a personal text for the researcher to refer to and analyze. They represent the process of transforming observed interaction into written *communication*. Jean Jackson (1990, 7) describes field notes creatively as "ideas that are marinating." Preliminary field notes may be taken on any materials at hand (even the margins of a newspaper; Kearns 1991b); however, back at "home base," jottings should be typed up. Keeping backup files and/or printouts of your field data is a crucial precaution (see Appendix 1 for an example of field notes).

Analysis

Analyzing the results of any period of observation will vary according to the purposes for which it was undertaken. Observations that have involved counting, or carefully recording each instance of some phenomenon, lend themselves to tables that enumerate occurrences and express data as frequencies or percentages. Observation that has complemented a more explicitly structured research design commonly leads to the presentation of quotations or descriptions that assist in our interpretation of findings derived from other sources. For example, an assessment of the impact of a "walking school bus" scheme on children's and parents' commuting patterns involved not only a calculation of car trips saved but also verbatim narratives from children gathered while the lead author walked with them to school (Kearns, Collins, and Neuwelt 2003).

When observation is embedded in an attempt to reach contextual understanding, a considerable volume of text will typically be accumulated, and strategies for the storage, classification, and analysis of information will need to be carefully considered (for a discussion of some of these issues, see Chapter 18). A number of useful texts are also available—for example, Bryman and Burgess (1994). This scale of observational engagement also suggests the merits of using qualitative software, such as NVivo, now employed by geographers (see Chapter 18).

Ethical Obligations

What are our obligations to those whom we have observed? Clearly, if observation has been fleeting and devoid of personal contact (for example, watching street-level pedestrian behaviour), any attempt at gaining permission or return of research results to the individuals observed would be impractical and unnecessary. Indeed the "observed" in this example are only nominally "participants" in the research. However, observations such as "reading" the presence of graffiti in the built environment (see Lindsey and Kearns 1994) have the potential to be a little more challenging. Observing the "what" and "where" of graffiti is unproblematic, since "tagging" is part of the publicly observed landscape. But because the act of inscribing graffiti is invariably illegal, what are our obligations if we witness "taggers" at work: report their breach of the law or preserve their anonymity? Arguably, our role as citizens might take precedence, and the former stance would hold sway.

When observation entails involvement in a place-based community or social group, there is surely an ethical imperative to maintain contact after the formal research period. This imperative may be formalized by the requirements of university ethics committees, but the stronger influence should surely come from the researcher, especially if pre-existing relationships have been developed (or reactivated) through the research process. As Tara Coleman (2013) discovered in her work with seniors on Waiheke Island, practical as well as ethical obligations can make the process of "exit" from social relationships developed through the research problematic. (See also Chapter 2.)

Such obligations may be taken for granted when pre-existing ties are involved but require careful consideration when social situations are entered into, or generated, for the sake of research. There is widespread agreement that cross-cultural fieldwork is particularly problematic (see Chapters 3 and 4 for a full discussion). This is the case for two reasons, both of which may be linked to the "field" metaphor: first, the researcher is potentially venturing onto another's turf, and second, because fieldwork involves a researcher working in a field of knowledge, there is the risk of overlooking local understandings and priorities and of (possibly unintended) one-way traffic of knowledge from the field (periphery) to the academy (centre). Jody Lawrence's (2007) work with the Auckland Somali community had very porous boundaries that saw her attending weddings and sewing groups and completing grant applications on the community's behalf. Why did she do this? The answer has as much to do with relationship as it does with her research, even though it was the latter activity that initially took her into that community. Box 15.7 draws on these ideas to complete the series of examples from my early Hokianga research. In such situations, the development of "culturally safe" research practice (Dyck and Kearns 1995) is important. Such practice recognizes the ways in which collective histories of power relations may affect individual research encounters. It also stresses the need for appropriate translation of materials into everyday language and the return of knowledge to the communities that provided or generated it (Kearns 1997). (See also Chapters 3, 4, and 17.)

Reflecting on the Method

As observers, our goal should be to achieve seeing that is, as the saying goes, believing. However, this chapter has argued that believable observation is the outcome of more than simply seeing; rather, it requires cognizance of the full sensory

Completing the Circle: Cultures, Theory, and Practice

While on an overseas sabbatical leave, I had an opportunity to recount the story of place and health in Hokianga and to connect its plot lines to the coordinates of other struggles for identity and turf. Drawing on Feinsilver's (1993) Cuban experience, for instance, I could identify Hokianga community action as a "narrative of struggle" in which there is a greater symbolism to health politics than just a defence of local services. However, I was left searching for a rationale for the place of the researcher in the narrative of struggle.

To impose theory (upon observation) without reference to the community would be as foolish as the unfettered importation of exotic wildlife or viruses into New Zealand. Clearly, the use of theory must be regulated. My own form of self-regulation has been to return draft papers to people involved with Hokianga's health trust for comment. In one such draft, I had interpreted their struggle as a "postmodern politics of resistance" (Kearns 1997). The returned manuscript was annotated with the comment from a health trust worker that my words sounded like "undigested theory." On reflection, this was a fair assessment. I had unthinkingly used theory imported from recent visits to conferences in Los Angeles and Chicago. It was time to return, to digest theory, and to reconnect with the source of the story. Such returns are made easier through adoptive *whanau* (extended family) relationships in which research is, at times, indistinguishable from the *aroha* (love, affection) and *korero* (purposeful talk) among friends. Hokianga has taught me about being a bicultural geographer, a role that requires respecting the rituals we can never fully enter while reforming the rituals of our own research (Kearns 1997, 6).

experience of being in place. If the questionnaire is the tool for survey researchers and the audio recorder for key informant interviewers, researchers themselves are the tool for participant observation. If we are to observe, we must take the time to do it properly. Indeed, as Laurier and Philo (2006, 194) remark, "If being academic researchers allows us certain privileges, one of them is to take the time to attend carefully and patiently to spatial phenomena." To reflect, we need prompts to keep the field experience fresh in our minds, so keeping a research log can be an important aid to recollection. A log book and its field notes can become data in themselves: reflections on the researcher's experience, as well as participants' meanings and perceptions.

The question remaining is: how do we know whether the fruits of participant observation are valid? Evans (1988) reminds us of the very important point that any method is, to a degree, valid when the knowledge that it constructs is considered by stakeholders to be an adequate interpretation of the social phenomena that it seeks to understand and explain (see Chapter 6).

Are there disadvantages to reliance on observation? One danger, perhaps, is privileging face-to-face interactions over less localized relations that remain beyond the view of the researcher in the field (Gupta and Ferguson 1997). To believe only what we see would be to make the serious mistake of denying the existence of structures such as social class or communicative processes such as virtual relationships that occur "off-stage."

Whether we seek to count, to gather complementary information, or to understand the context of place more deeply, the key to taking observation seriously is being attentive to detail as well as acknowledging our positions as researchers. These recognitions imply that we are aware of both our place within the social relations we are attempting to study and the reasons that we have the research agendas we do (White and Jackson 1995).

Key Terms

complete observer
complete participation
controlled observation
field notes
field relations
masculine gaze
observation
observer-as-participant
panopticon
participant-as-observer
participant observation
photovoice
primary observation
secondary data
secondary observation
soundscape
surveillance
uncontrolled observation

Review Questions

1. What are some of the ethical considerations that arise from observation?
2. In what ways might access to a specific social setting (for example, a sports club, an industrial workplace) be achieved for research purposes?
3. In what ways does observation involve more than seeing?
4. Suggest some ways in which our presence might influence interactions in the research setting.

5. Why, and in what circumstances, might one opt for participant observation as a research method?
6. In what types of circumstances might observational research be usefully aided by still or video photography?

Review Exercises

1. Choose a public or semi-public space in which "strangers" gather and in which you already need to spend a period of time over the coming days. This could be anywhere from a doctor's waiting room to a bus stop or railway car, but the key is that people are gathered and waiting to get somewhere or for something to happen. While there, develop your observational skills. Be attentive to who is there (ages, apparent intentions) and how people are interacting. What is the space like? Is it designed to align with the purpose for which it is being used? What signs and symbols are observable? Are there sounds and/or smells associated with the place? To what extent were you a participant in the activity? On return to your desk, jot down all you observed over the course of ten minutes. Think about how this evidence could contribute to an account of the social and cultural meanings of your chosen place.
2. After the journey between home and campus by your usual means of transport, take ten minutes and jot down a list of key observations. Then the next day choose an alternative means of transport (walking, driving, bus, train?) at least part of the distance between the same points. Compare any differences in what you were able to see, hear, and smell. Consider the ways means of mobility and associated speed and degree of "containment" from the outside world affects your ability to read the urban landscape. What might be the advantages of interviewing someone while walking so that conversation and observation can be combined?

Useful Resources

A helpful guide for students using participation in dissertation research is provided by Ian Cook's (2005) chapter in *Methods in Human Geography* (by R. Flowerdew and D. Martin, eds, Pearson). "Ethnography and participant observation" is ably discussed in a chapter by Annette Watson and Karen Till in the 2010 *Sage Handbook of Qualitative Geography* (edited by D. De Lyser, S. Herbert, S. Aitken, M. Crang, and L. McDowell, pp. 121–37).

16 New Media

Jamie Winders

Chapter Overview

New media—the on-demand information and interactivity associated with the Internet and the technologies and devices used to access that information—increasingly shape and mediate many people's experiences with and in the world. How, though, do new media change qualitative research in human geography? This chapter examines new media and their impacts on the qualitative study of human geographies. In doing so, it explores two questions: first, what new research topics do new media raise for geographers?; and, second, how can geographers use new media themselves as research tools? Themes in the chapter include the methods used to study new media, new media's impact on cultural and social identities and practices, its role in social movements and migration, and new media as research tools. As this discussion shows, thinking about the relationship between new media and qualitative geographic research means accounting for how new media shape social lives and become the ways that we experience *and* encounter the world.

Introduction

> If digital devices mediate and are in considerable measure the stuff of social, cultural, economic and governmental lives in contemporary northern societies, then what does this mean for our methods for knowing those lives? (Ruppert et al. 2013, 24)

It goes without saying that **new media**—the on-demand information associated with the Internet and the technologies and devices used to access that information—have changed the way that many people see and understand the world (Kitchin et al. 2013). As early as 2003, 88 per cent of the British population between the ages of 15 and 24 owned a mobile phone (Green and Singleton 2007, 506). In the US, young people spend significant portions of their days online, and the majority of households (54 per cent) have smartphones (World Internet Project 2013). In Cyprus, 62 per cent of the population used the Internet in 2012; in Mexico, 80 per cent of households had at least one user (World Internet Project 2013). Nearly 40 per cent of the Russian population goes online every day (World Internet Project 2013). Less than one-quarter of South African households have computers, but nearly 85 per cent have mobile phones. In 2012, Indonesia was Facebook's second-largest market (Oman-Reagan 2012). In all these ways, in all these places, and for a growing number of people, the social world is increasingly "saturated" with digital devices and the on-demand information and interactivity they enable (Ruppert et al. 2013, 23).

Of course, as has been well documented (e.g., Wilding 2006; Crampton et al. 2013; Graham 2013a), access to that on-demand information is geographically uneven across countries, between rural and urban locales, and from neighbourhood to neighbourhood. It is also socially differentiated along lines of race, income, gender, and age. Globally, more men than women use the Internet, and Internet usage positively correlates with education and income yet negatively correlates with age (World Internet Project 2013). These social and spatial differences in access are important starting points in thinking about the role of new media in qualitative geographic research. As nearly all scholars of new media and the virtual realm stress, there is no doubt that the Internet and the digital devices used to access it have created "a moment of radical shift from the past": what is less clear is "the nature of this shift" (Wilding 2006, 126).

This chapter examines that shift by looking at new media's impacts on the qualitative study of human geographies. As other chapters attest, the Internet age, and especially the increasing centrality of new media, have changed the ways that human geographers think about qualitative research. Chapter 2 addressed questions raised by the digital age concerning power dynamics, subjectivity, and ethics in qualitative research. Chapter 11 reflected on how the increasing digitization of archives and the growing use of digital cameras to capture archival material have reconfigured both the where and how of historical geography (see also DeLyser et al. 2004; Lorimer 2010). Together, they highlight the growing ubiquity of new media in the study of human geographies and the ways that new media shape and mediate people's experiences with and in the world.

Despite increasing awareness of how new media reconfigure human geographies, geographers have only recently begun to reflect on how the digital era might affect the ways we conduct qualitative research (see Madge 2007). Many questions remain about the relationships among new media, **social media**, and the Internet and how interchangeable studies of these topics are. Within research on new media, there is debate over whether to focus on new media as technologies, digital devices, or content, as well as on whether the technologies or the content they provide holds emancipatory or democratic possibilities (see Kitchin et al. 2013).

Although geographers acknowledge the complex interactions between the material and virtual worlds (Graham 2013a), there is no consensus about how human geographers should proceed vis-à-vis the digital realm or new media themselves. No one can deny that new media have changed the way that many people interact with one another and with places around the world. Key questions remain, however, about how geographers can incorporate new media into the study of the spatialities of human life. *What new research topics do new media raise for geographers? How might geographers use new media as research tools?*[1]

This chapter is organized around these two questions. Between these two approaches to new media—as a topic to study and as a way to conduct research—the latter has yet to be explored by geographers. Nonetheless, the expanding role of new media in the everyday lives of many people means that it is time to consider how they might be used as part of qualitative research. As Evelyn Ruppert and her co-authors (2013, 24) note, "digital devices and the data they generate are both the *material* of social lives and form part of many of the apparatuses for *knowing* those lives." Thus, thinking about the relationship between new media and qualitative geographic research means accounting for how new media shape social lives *and* become the ways that we experience and encounter the world. As this chapter will stress, new media force us to rethink key assumptions about conducting qualitative research at the same time that they reinforce key tenets in the qualitative study of the "offline" world.

The following sections lay out different ways to think about the relationship between new media and qualitative research in human geography. The first explores definitions of new media and how they fit vis-à-vis existing ideas in human geography, especially the link between social practices and material culture. The chapter then considers what human geographers can do with new media, reviewing innovative studies in geography and other fields. From there, it reflects on what new media as technologies and devices can themselves do, particularly vis-à-vis social movements and migration. The chapter's last sections examine questions of identity and community wrapped up in new media's dual meaning, before stepping back to identify new themes and possibilities in studying new media and using new media to conduct qualitative geographic research itself.

What Are New Media?

Before looking at new media as a topic and tool of qualitative research in human geography, it is worth discussing what new media are. We can think of new media as reflecting the increasingly enmeshed nature of virtual spaces, social lives, and technological devices, as recognition of "the hybrid and augmented ways in which the internet is embedded into our daily lives" (Graham 2013a, 180). In practice, new media include *both* the on-demand access to information and interactivity provided by a range of digital devices—computers, smartphones, tablets, and so on—and the devices themselves. Thus, the Internet, as well as interactive video games and phone apps, can be considered new media, which are understood here as anything that enables digital interactivity—the devices, the software or applications, and the online sites they access.[2] Of course, what constitutes new media today will not be new in the near future.[3] Nevertheless, in much of the world, digital devices are "increasingly the very stuff of social life…, reworking, mediating, mobilizing, materializing and intensifying social and other relations" (Ruppert et al. 2013, 24). While the kinds of digital devices we use to access and mediate our online lives and practices will change, the *idea* of new media as the way we do so will not. For this reason, new media, and the interactivity they enable through digital devices, should be of interest to human geographers as both a subject of and a tool for qualitative research.

The line between new and traditional media is not always clear. When you watch a television show on your tablet or smartphone, are you using new media (your smartphone and the on-demand information it provides) or more traditional media (television shows, film, etc.) or both? When you post a photograph on Instagram, are you using the traditional medium of photography or the new media of image manipulation and digitally shared images? How might either of these practices (streaming videos and taking photographs) become ways to conduct qualitative research?

Answering these questions means thinking about both the technology in use (phone app, operating system, Internet) and the objects used to access that technology (phone, tablet, or desktop computer). This dual meaning raises perplexing questions for researchers. Should scholars study Facebook or Skype for the connections and interactions they facilitate among users or for the ways that people access Facebook or Skype—whether they use smartphones or laptops, where and when they use these technologies, and so on? Should scholars focus more on the "what" of new media or the "how"? In many ways, geographers are ideally suited to this simultaneous focus on practice and material objects. From its beginnings, human geography has been interested in *both* social practice *and* material culture, in what people do (and why) and in what objects they use to do it. Geography blends a focus on discourse and materiality, on practice and things, in its study of social and spatial dynamics (Schein 1997). Thus, geographers should be well positioned to think critically about how new media reconfigure understandings and experiences of space. The question is, *how*?

Thinking Spatially about and with New Media

Geography's study of new media has a short history. In 1997, Derek Alderman and Daniel Good raised some of the discipline's first questions about "electronic mass media" in a study of "the Virtual South—the production of the idea of a distinct American South through the cultural discourse and new electronic folklore of cyberspace" (Alderman and Good 1997, 21).[4] What, they asked, does the Virtual American South look like? Where on the Web can it be found? By examining websites that promoted US southern cities and ideas about southern identity, they tried to identify "the possible differences and similarities between the cyber-representation of place and more traditional mediums of geographic portrayal." Through an analysis of how websites represented ideas about the American South, Alderman and Good endeavoured to describe "the cultural geography of the Web" (1997, 21) and, thus, laid the groundwork for what has become an exciting, if still developing, line of scholarship on the relationships between the virtual and the material and the devices and technologies used to mediate these relationships.

Since this early work, geographers have continued to interrogate online practices and digital devices. Dydia DeLyser and her co-authors (2004), for example, have shared their experiences of buying historical materials on eBay and turning eBay itself into a research tool and archive. For human geographers more wedded to field-based research, the digital era poses a number of thorny questions. When people are "on" their smartphones, where exactly are they? What are those spaces/places like, and how can geographers study them? How do we think geographically about social media like Facebook and Instagram that represent places, create a sense of community, and function as gathering sites for groups of people but are not "real" spaces in any obvious sense, aside from corporate headquarters and server farms? Can geographers study new media from their laptops, tablets, or phones? Can they study human geographies *with* these devices? Can geographers study users of new media without observing those users in person? What would this kind of research look like?

Across the social sciences, scholars have asked these and other questions of new media and the wider digital domain. Does distance matter in the age of the Internet (Mok et al. 2010; Wilding 2006)? (It does.) How do the Internet and new media change face-to-face interactions (Mok et al. 2010; Green and Singleton 2009)? (Less than you think.) Why are some parents fearful of what their children discover online (Valentine and Holloway 2001), and how do other parents use technologies like Skype to manage relationships with their kids (Longhurst 2013)? How quickly are new technologies of communication and interaction adopted and rendered passé (Wilding 2006)? How important are new media devices to the individuals who own them (Singleton and Green 2009)? How does the emergence of what some call the "fifth estate" linked to new social media shape understandings of the multiple publics and audiences accessed and produced through interactive technologies like **blogging** (Kitchin et al. 2013)?

Attempts to address such questions have involved a range of methods. In many cases, scholars have used in-depth interviews to map out with whom people communicate on a regular basis and through what technologies and devices they do so (e.g., email/computers, mobile phones, face to face) (Mok et al. 2010).

Others have used interviews to examine how online practices are interpreted by parents, teenagers, immigrants, and other groups (Valentine and Holloway 2001; Arnado 2010). Still others have found it "necessary to investigate what happens *in practice*" (Valentine and Holloway 2001, 80, emphasis added) when people interact with new media. This attention to practice can focus on how people spend time online (e.g., Dean and Laidler 2014), as well as how they interact and respond to new media like video games. Geographer James Ash (2010), for example, ethnographically studied the testing of video games at a design company, conducting interviews with game designers and keeping a research diary while he tested games for two years. Interested in the ways that affect could be manipulated for economic ends (in this case, more profit for game designers), Ash focused on how game designers tried to elicit particular responses and feelings from users and how users, especially game testers, became part of the process of manufacturing affective responses that made people play the game longer and more frequently.

Attention to practice vis-à-vis new media can also highlight how **information and communication technologies** (ICT) that link us to virtual spaces reshape the ways people use and design *material* spaces. Bjorn Nansen and his co-authors explored this question in a study of "the relation between people, their media stuff, and their homes" (2011, 697). What, they asked, happens to the idea of a media room in the house when people can watch television programs and movies on their phones, tablets, or laptops? How do the layout and use of domestic spaces shift as family members rely on new technologies for work, study, and play? To answer these questions, Nansen and his co-authors (2011) followed the ICT use of four families in Melbourne, Australia, from 2004 to 2007. Each family was given cameras to photograph different rooms in their house, scrapbooks to annotate and discuss those photographs, colour-coded stickers to tag different forms of ICT, and diaries and maps to trace how ICT moved in and out of their house over time. Through this detailed look at the ways that families incorporated new technologies into their homes, Nansen et al. (2011, 712) highlighted "the reciprocal relationship between ICT *stuff* and the structure they lived in." As they discovered, ICT does not just change how, where, or with whom we watch television or do homework. It also reshapes the very ways we use the spaces of our homes, leading to complete redesigns in some cases.

In a related vein of work, several scholars have examined how new media, especially mobile phones, shape social practices and identities. Eileen Green and Carrie Singleton (2007), for instance, have written about the "mobile selves" that young Pakistani-British men and women in northern England associate with and experience through their mobile phones. In 2004 and 2005, they conducted focus groups with Pakistani-British men and women between 15 and 25 years old to examine the gendered dynamics of phone use and "whether young women and men experience mobiles differently in everyday life" (2007, 508). As has been documented in other studies of mobile phones and youth, Green and Singleton (2007) found that participants often viewed their phones "as indispensable components of everyday life" (511), as extensions of themselves in some cases.[5] Of interest to geographers, they also discovered that young Pakistani-British men and women used mobile phones to create "virtual and private 'spaces of their own'" (508), but in different ways. Men tended to report shorter phone calls, to use texts to make arrangements with male friends, and to describe women's use of mobile phones as "gossip" that required longer time commitments. Women, by contrast, reported longer conversations, often at night, and the use of texts to forward chain messages, rather than arrange meet-ups. Despite these differences, both groups communicated in multiple languages through their mobile phones, mixing local and global influences to create a specific youth culture experienced and expressed through new media.

Of course, mobile phones are not just objects that allow contact with friends and family or even create new cultural practices of communication. They are also objects that confer status and earn social capital (Green and Singleton 2007). Phones can be personalized through cases, wallpaper, ring tones, and charms and, thus, can be studied for the messages they convey about understandings of self and others. What is more, mobile phones can be traded, sold, given away, lost, or stolen, becoming part of informal, formal, and even illicit exchanges bound up with and themselves shaping local cultures of commerce. These multiple ways that people perceive, use, and exchange mobile phones highlight the complexity of studying new media. As Ruppert et al. (2013, 31–2) suggest:

> We need to attend to the lives and specificities of devices and data themselves: where and how they happen, who and what they are attached to and the relations they forge, how they get assembled, where they travel, their multiple arrangements and mobilizations, and, of course, their instabilities, durabilities and how they sometimes get disaggregated too.

Julia Pfaff (2010) tackled many of these questions in her study of mobile phones, or of one particular mobile phone, in Tanzania. Through close attention to the social life of this phone, she interrogated broader ways of life for Swahili people and the role of trade within Swahili culture. Engaging with the phone as an object, Pfaff investigated the cultural meaning of trade by following one phone from person to person across Tanzania. This approach to the study of new media was innovative in many ways. Coming across the same phone twice in Tanzania, Pfaff traced its ownership back in time, identifying who had owned it, why and how it had changed hands, how each owner altered its use and functioning, and why it came to rest in a vendor's display case. Through this "mobile-phone biography," she brought a "follow-the-thing" approach to the study of new media, trying to understand "the actual movement of an object and, with it, the relationships between people as well as the connections between places" (2010, 345).

Although many of the studies discussed to this point mobilized common research approaches to study new media (interviews, focus groups, study of material culture, etc.), Ruppert and her co-authors suggest that the digital era also calls key aspects of social-science research into question. Whereas traditional research methods, especially qualitative ones, are "deeply implicated in the formation of human subjects" (2013, 33), the digital era may complicate this focus on a fixed human subject. Methods like surveys or interviews presuppose identifiable research subjects who participate in them (although see Rose 1997). When these methods shift into an online context, the idea of a knowable, clear subject "behind" the screen becomes much hazier. In his study of the Occupy Wall Street (OWS) movement in Indonesia, for example, Michael Oman-Reagan (2012) was never certain who his research subjects were. Indonesia's OWS movement took place online because activists faced harsh consequences if they protested in the streets. When Oman-Reagan (2012, 41) began to interact with Indonesian activists through their online profiles, he wondered "if I was meeting the shadows, the puppets, or the puppet masters, and how I would know the difference." For Omar-Reagan, qualitative research conducted through new media raised difficult questions about participants' identities, especially how to assess and understand the research subject accessed and encountered virtually.

Such questions create a quandary for researchers accustomed to seeing the identities of research participants as complex but at least "true" and apparent in some sense. How, for example, do you obtain informed consent from participants in research conducted online (Madge 2007)? How do you "know" who is generating the content you examine in virtual correspondence? Because of such questions, Ruppert et al. (2013, 34) urge us to consider the ways that digital devices "*observe and follow activities and 'doings'*—often, but not always or exclusively, those of people." In other words, "instead of tracking a subject that is reflexive and self-eliciting, they track the *doing subject*" (Ruppert et al. 2013, 35). This idea of a "doing subject" that moves beyond humanist notions of the subject resonates with growing interests in more-than-human geographies in ways that merit greater attention but are beyond the scope of this chapter. At a minimum, however, it forces us to think carefully about with whom we are interacting when we move research into the context of new media.

What New Media *Do*

The research described above pushes human geographers to consider what new media mean in and to people's lives. New media, however, also push us to consider the connections and events they enable (i.e., how new media change what people do). Jeffrey Juris (2012), for instance, has looked at how social media like Twitter, and especially the ability to access social media through smartphones, played key roles in the OWS movement. Through Twitter's shared temporality, OWS protesters in more than a thousand places around the world were able to exchange images and updates in October 2011 as part of a global protest produced through hand-held phones. The "global" reach of OWS, however, was an overstatement, since protesters in places like Egypt often lacked Internet connections and since even in sites that were fully wired, face-to-face contact remained key to OWS's success. Thus, although new media are reconfiguring how social movements function, it is important to remember that "places, bodies, face-to-face networks, social histories, and the messiness of offline politics continue to matter" (Juris 2012, 260) and to shape the contours of new media.

As this example shows, geographers interested in new media can examine the kinds of practices they facilitate—in this case, new patterns of protest and new ways to link networks of global justice activists. In this regard, new media seem similar to listservs and the Internet, which played key roles in social movements like the Zapatistas in Mexico in the mid-1990s (Froehling 1999). In both cases, new media created visibility for social movements beyond the local context and quickly mobilized advocates and resources in a range of places. "Microbroadcasting" updates and images through Twitter and Facebook (Juris 2012, 266), OWS took advantage of the scale-jumping and rapid connections that new media allowed protestors. At the same time, though, rapid bursts of information in 140 characters and images that went viral could not accommodate longer discussions and debates of OWS and its future that older technologies like listservs provided. This fact is an important reminder that while new media create new possibilities for social movements, they also create new challenges.

Rob Kitchin and his co-authors (2013) present yet another way that new media—in their case, blogging—change how people interact with a wider world. As they showed, their collective blog about Ireland's financial crisis not only enabled them, as academics, to reach a wider and different audience than did their standard academic publications but also produced a different kind of knowledge itself. Blogging as an instance of new media changed how they understood and interrogated the Irish financial crisis. It brought them closer to their readership, through things like comment sections on blogs, but also loosened control over how their ideas were picked up, reworked, and reproduced elsewhere both online and in traditional media. In this way, blogging changed how these geographers worked with and engaged a wider public, altering their relationship with readers, with their own arguments, and ultimately, with the kind of writing they generated.

New media also enable new kinds of connections among people living apart. Particularly in research on long-distance migration, scholars have begun to study how families use new media to stay in touch and, in the process, how new media change family dynamics (Wilding 2006). Geographer Robyn Longhurst (2013), for example, has examined whether the real-time communication offered through Skype is qualitatively different from written, text, and phone communication and, thus, whether it is changing mother-child relationships. Working with a group of mothers in Hamilton, New Zealand, she found that Skype reoriented how mothers and their adult children experienced distance and how mothers assessed their children's well-being. In some cases, Skype came to be seen as "part of the family," in the words of one mother, because it had "the capacity to bring [her] son and other children into focus" (Longhurst 2013, 670). In other cases, mothers felt restricted in using Skype, which forced them to sit still and speak to their children in front of their computers, rather than to talk on the phone and move around. This and other studies of family and new media clearly show that new media both enable families to feel closer and, thus, mediate a sense of separation *and* create new tensions around uneven adoptions of technologies and new expectations of constant availability through them.

As should be clear by this point, new media have the potential to reshape how things from social movements to families unfold, take place, and interact. In each case discussed, new media affected, but did not determine, the ways social, cultural, and political practices were "done" and created new opportunities and new obstacles for users. One task for human geographers interested in new media, thus, is to interrogate what new media allow users to do and how that doing transforms social, cultural, and political geographies and practices. Another task is sorting out who we are and become, as well as whom we are with, when we use new media.

Who Are We on New Media? Whom Are We With?

Several scholars have explored the question of identity online—more specifically, how users of new media, from instant messaging to social media, understand and present a sense of self through these technologies. Koen Leurs and Sandra Ponzanesi (2011) studied instant-messaging (IM) practices of Moroccan-Dutch teenage girls in the Netherlands. To examine "how gender, diaspora, youth culture and technologies intersect and influence each other" (56), they surveyed over 1500 teenagers in 2009 and 2010 on their use of digital media. They also interviewed Moroccan-Dutch girls about what they did and how they presented themselves online and analyzed the IM transcripts of six participants who agreed to save their IM conversations for two months. Through this work, Leurs and Ponzanesi found that Moroccan-Dutch teenage girls used IM as a key "communicative space of their own" and "a relatively safe playground" (2011, 56) where they tried out different relationships and identities (see also Dean and Laidler 2014).

Echoing findings from Green and Singleton's research on mobile-phone use, Moroccan-Dutch girls in Leurs and Ponzanesi's study used IM to create and manage public and private identities and relationships.[6] In the "*backstage* of IM" (2011, 58), young women engaged in private conversations, where they felt freer to talk about health, body issues, and discrimination with their friends and away from the watchful eyes of parents. "*Onstage*" in IM, through the selection of display names, photos, and the people identified on their "buddy lists," young women presented a different, more public sense of self. Also echoing Green and Singleton, Leurs and Ponzanesi noted that IM created a way for Moroccan-Dutch teenagers to feel part of a Muslim global youth culture and to find a sense of belonging amidst sometimes hostile reception in the Netherlands. In this way, IM became a way of being in the world for these teenagers and "a space where they can negotiate several issues at the crossroads of national, ethnic, racial, age and linguistic specificities" (2011, 56). Through IM and the digital devices that enabled it, Moroccan-Dutch teenage girls could experience being Muslim as "belonging to a particular Dutch as well as global youth-subculture grouping" (2011, 63), all from their bedrooms and in ways that partially challenged the gendered restrictions they faced in everyday life as young Muslim women.

Shifting questions of identity from the individual to a larger group, some scholars have studied how social media, such as Twitter, create a sense of community among users. Although research shows that most people who interact online also interact in person, newer platforms like Twitter may challenge this claim. Since Twitter works differently than other social media like Facebook (Twitter users do not have to follow those who follow them), Anatoliy Gruzd and his co-authors (2011) wondered whether it created a "new form of community . . . , in which spatial proximity seems to play a minimal role" (1297)? What can a study of Twitter show us about "how people integrate information and communication technologies (ICTs) to form new social connections or maintain existing ones" (Gruzd et al. 2011, 1313)? To study the kind of community that Twitter forms, Gruzd and his co-authors examined one of their own Twitter accounts. Counting followers and retweets of the author's posts and trying to identify "high centers" (2011, 1313) of popular individuals around whom Twitter users congregate, Gruzd et al. (2011) examined the virtual connections of Twitter as instances of "real" community. In doing so, they posed a question dear to the hearts of geographers: what kind of places do new media create?

Putting It All Together

Mark Graham (2013a, 177) argues that "Geographers should take the lead in employing alternate, nuanced and spatially grounded ways of envisioning the myriad ways in which the internet mediates social, economic and political experiences." The question this chapter has tried to address is: how should geographers do so in the context of new media? What do geographic concepts like space, place, and scale mean in new media? If "Our 'place' is wherever our computer and phone are" (Mok et al. 2010), what does that location mean to and for geographic research?

Some of these questions are beyond the scope of this chapter, but we can point to emerging themes from this discussion of new media. Perhaps the clearest theme is that the virtual and material realm complement, rather than supplement, each other in practice and, thus, should do so in our research. Online and offline aspects of social relations co-exist in complicated, ever-changing relationships that geographers must figure out how to investigate. We have also seen the ways that "social and technical competencies codevelop" (Valentine and Holloway 2001, 80). As we learn to work with new media like Instagram, we learn new ways to interact with people. These co-developing social relations and technologies, of course, are deeply gendered (Green and Singleton 2009; Dean and Laidler 2014), as well as raced, classed, and sexed in ways that geographers are just beginning to explore. Given this recursive relationship between the virtual and the material *and* between the social and the technological, our research must examine both the meanings attributed to the devices of new media and the materiality of those devices themselves, both what new media enable and how they materially do so. New media facilitate and inhibit various relationships. Our study of them must attend to how new media rework social dynamics and relations, as people themselves learn to use new media.

A second theme is that the *practices* of new media, from friending to tweeting to up-voting someone's Reddit post, create "a distinct writing style with its own 'Internet-speak' norms" (Leurs and Ponzanesi 2011, 58) and, thus, create new cultural and social practices. Whether the Twitterspeak of 140 characters, the use of codes like "mh" (Mom's here) in IM sessions among Muslim youth (Leurs and Ponzanesi 2011), or the "pinning" of aspirational images in Pinterest, new media shape how users communicate and, in the process, create new means of communication. We saw this most clearly in Kitchin et al.'s (2013) discussion of how blogging changed their work as academics, but the question of how new media generate new social, cultural, and political practices must be part of any study of it.

Third, the research discussed here has shown that it is important to learn as much as possible about whatever digital device, online practice, or virtual interactivity is under study. In their discussion of how historical geographers might use eBay, for example, DeLyser and her co-authors (2004) devoted much time to learning about how eBay works, how many users it has, how long it has been in existence, and how different groups of users interact with it (see also Gruzd et al. 2011). In this regard, studying new media is no different from studying the "offline" world. Knowing as much as possible about one's topic is crucial to any study, and this knowing comes from both reading about one's research and spending time with it oneself.

Finally, as in qualitative research conducted offline, research on and with new media brings its own set of ethical issues. In a virtual context, many of the same ethical mandates apply, as Clare Madge (2007) has noted: informed consent, confidentiality, privacy, debriefing, and "netiquette" (656). Figuring out how to meet those mandates, however, is more complicated, and this point in particular will continue to merit attention from geographers and other scholars. As more geographers turn to new media as something to study and to conduct research *with*, more questions about research ethics will surface and demand consideration.

What, though, about using new media to conduct qualitative geographic research itself? This question has yet to be explored, but to conclude this chapter, let me discuss some tentative ideas. New media like mobile phones can facilitate qualitative fieldwork in all sorts of ways. Texting, for example, can be a way to take fieldnotes without disrupting social norms.[7] Today, using a phone is socially acceptable nearly everywhere (except the classroom!), so mobile phones can be used to unobtrusively take "jottings" of field observations that can later be the basis of extended fieldnotes (Emerson et al. 1995). Many of us are always with our phones, so mobile phones themselves become convenient research tools. With programs like Evernote, fieldnotes made on phones, tablets, or other wireless devices can be automatically added to field diaries stored virtually and backed up regularly, thus reducing the chance of lost fieldnotes that plague researchers dependent on hard copies (or even hard drives).[8]

Second, phones with cameras or video capabilities can be used by researchers or research subjects to document and describe events and spaces that are central to a study—landscapes, social movements, monuments, neighbourhoods, and so on. With new media and especially with the ease of geo-tagging photographs (Jones and Evans 2012), the task of asking research participants to document their daily lives or key events through images becomes nearly instantaneous and, through the ease of sending photographs, can be done at great distances. In this way, research sites can be visually documented in multiple ways, and by multiple subjects, almost instantly. Montages can be created by either the researcher or the study participants with little effort, producing objects of discussion for individual or group interviews (in person or online), focus groups, or even online fora. Crowdsourcing the meaning of a place through user-generated photographs or videos is just one of many possibilities here.

Third, ICT like Skype can be used to conduct interviews or focus groups at a distance, "internationalizing research without adding costs to the funding body" (Madge 2007, 656). Of course, interviewing online is not the same as interviewing in person, but new media like Skype or other video-conferencing technologies open new possibilities for qualitative researchers, particularly those who, for various reasons, are unable to be in the field for extended periods of time or at great distances from home.[9] Skype can convene focus groups across places, bringing individuals with shared experiences but different locations together to discuss their experiences.[10] Skype and similar technologies can even facilitate participatory research by enabling researchers to stay in closer contact and dialogue with research participants. Virtual "returns" to the field to present preliminary findings or conduct follow-up interviews can become much easier, at least in places and among groups with access to these technologies.

Fourth, the popularity of sites like Twitter, Facebook, and Pinterest makes them avenues for informal surveys or group dialogues on various topics related to a given study. Obviously, there are substantial selection biases that limit how representative any one person's Facebook friends or Twitter followers will be. Additionally, as Madge (2007) discusses, there are serious ethical considerations for online research of this kind, ranging from whether online conversations are public or private to how informed consent online is obtained and confidentiality ensured. Even with these issues, though, the social networks and communities of interest that new media enable create exciting possibilities for rethinking the "where"—and, thus, the how—of qualitative geographic research.

Finally, the GPS-enabled devices that many of us carry with us hold much potential for examining daily experiences and lives, as well as innovative uses of volunteered geographic information in our research. Software programs like Google Earth can be used to conduct preliminary fieldwork before in-person visits to research sites, allowing researchers to become familiar with places before leaving home.[11] As Taylor Shelton and his co-authors (2014) showed in a study of tweeting patterns associated with Hurricane Sandy in 2012, many digital devices create "data shadows," or "imperfect representations of the world derived from the digital mediation of everyday life" (167). These geographic data (what some call the geoweb) can be useful for both qualitative and quantitative study in human geography. Geo-tagged tweets, photos, Facebook postings, and so on can be the basis of geographic analyses of, among other things, how places are represented, how those representations are reproduced and contested by different groups, and how places are linked or separated through social relations among residents. In the process, these spatial aspects of new media—these geographic relations produced through new media—can help us as scholars visualize and interrogate different social geographies that might not be evident through other methods, such as landscape analysis, interviews, or discourse analysis of non-virtual texts and images.

These possibilities are only the tip of the iceberg that is new media. The platforms, technologies, and devices discussed here—Skype, Facebook, Twitter, Instagram, mobile phones, email, and so on—capture and materially shape different kinds of relationships and geographies. They also shape "the pathways that guide how we use information" (Graham 2013b, 78) and, thus, are important not only as topics of study for human geographers but also as tools to conduct those studies. How much new media reshape how human geographers conduct qualitative research depends on how willing we are to think creatively about the practice of qualitative research and to think critically about the questions of ethics, power, subjectivity, and practice that new media raise for us.

Key Terms

blogging
information and communication technologies (ICT)
new media
social media

Review Questions

1. What new possibilities and new challenges do new media raise for qualitative research in human geography? How do new media change, and not change, how human geographers think about key elements of research (fieldwork, data collection, data analysis, etc.)?
2. How might a qualitative study in geography incorporate all of the different components of new media (digital devices, software, on-demand information, interactivity, etc.)? What might such a study look like?
3. How do new media change the way we think about research subjects (their identities and subjectivities) in qualitative research in human geography?

Review Exercises

1. Select a group of new media users that you are interested in studying (youth, a particular immigrant group, men, college students, etc.) and an aspect of new media (a digital device, an online practice, etc.). Create five questions to use in interviews or focus groups that examine how the aspect of new media you selected shapes social and spatial dynamics and relations for your group and how it is itself made meaningful by members of that group.
2. Pick one of the other methods discussed in this book (participatory action research, interviewing, questionnaires, etc.). List three ways that you could use new media to conduct this kind of qualitative research. What would new media add to this method? What might be challenging about using new media in this capacity?

Useful Resources

Green, E. and C. Singleton. 2009. "Mobile connections: An exploration of the place of mobile phones in friendship relations." *Sociological Review* 57 (1): 125–44.

Longhurst, R. 2013. "Using Skype to mother: Bodies, emotions, visuality, and screens." *Environment and Planning D: Society and Space* 31 (4): 664-79.

Nansen, B., M. Arnold, M. Gibbs, and H. Davis. 2011. "Dwelling with media stuff: Latencies and logics of materiality in four Australian homes." *Environment and Planning D: Society and Space* 29 (4): 693–715.

Pfaff, J. 2010. "A mobile phone: Mobility, materiality and everyday Swahili trading practices." *cultural geographies* 17 (3): 341–57.

Ruppert, E., J. Law, and M. Savage. 2013. "Reassembling social science methods: The challenge of digital devices." *Theory, Culture and Society* 30 (4): 22–46.

Shelton, T., A. Poorthius, M. Graham, and M. Zook. 2014. "Mapping the data shadows of Hurricane Sandy: Uncovering the sociospatial dimensions of 'Big Data.'" *Geoforum* 52: 167–79.

Notes

1. For a discussion of the implications of digital devices and data for social science methods in general, see Ruppert et al. 2013.
2. For a topic like new media that constantly changes, Wikipedia is a good place to start for up-to-date definitions and understandings of the term. See http://en.wikipedia.org/wiki/New_media. Accessed 23 May 2014.
3. It is important to note the ongoing debate over whether the digital age is an epochal shift or a more gradual change that supplements, rather than substitutes for, a material (i.e., non-virtual) world (Ruppert et al. 2013).
4. For a critique of the metaphor of cyberspace, see Graham 2013a.
5. Green and Singleton (2007, 514) also found a few resisters to the idea of "perpetual contact" and constant availability. Those people who chose to be unavailable or who did not always answer their phones, however, faced censure from peers more committed to the social expectations "that you will be always available and on hand" (515).
6. This finding echoes a wider characteristic of the Internet: that "there is no clear agreement about what is public and what is private" in its content or experience (Madge 2007, 661).
7. I thank Ricardo Millhouse for sharing this practice.
8. I thank Jesse Quinn for bringing this use to my attention.
9. Chris Gibson and Leah Gibbs (2013, 89) also stress that new media, especially things like blogging, can "overcome geographical marginality" and increase international visibility for scholars who might otherwise face the friction of distance in places like the Antipodes.
10. Beverly Mullings is currently exploring this strategy to both conduct qualitative research and create opportunities for connection and dialogue between groups separated by distance but linked by global systems of oppression.
11. I thank Jessie Speers for bringing this use to my attention.

Empowering Approaches: Participatory Action Research

Sara Kindon

Chapter Overview[1]

Participatory action research (PAR) is becoming more common in human geography. It involves academic researchers in research, education, and socio-political action with members of community groups as co-researchers and decision-makers in their own right (Thomas-Slayter 1995). As such, it is quite different from many other approaches to research and demands different types of attitudes and behaviours from a researcher. In this chapter, I discuss the overall process of PAR and its cycles of action and reflection. I also discuss different types of relationships and various strategies and techniques that can be used to enable the involvement of research participants in all stages of the process. These strategies and techniques aim at establishing a more democratic and ideally, empowering research process: a process that respects and builds co-researchers' capacity and generates appropriate knowledge for community change. If done well, PAR has many benefits for human geographers, particularly for those committed to challenging unequal power relationships and increasing **social justice**. PAR can also be challenging to carry out: the emphasis on power, relationships, and change is a potent mix. Finally, I consider how we can present PAR-generated research information to a range of audiences in effective and ethical ways.

What Is Participatory Action Research?

Participatory action research seeks to:

> engage people in a learning process that provides knowledge about the social injustices negatively influencing their life circumstances. The knowledge about social injustice includes understanding methods for change and thus organizing skills necessary to remedy the injustice. (Cammarota and Fine 2008, 4–5.)

PAR has been evolving since at least the 1970s and, as the above quote states, it is fundamentally different from many other approaches to social science research because its goal is not just to describe or analyze social reality but to help change it (Pratt 2000). This change occurs through the active involvement of research participants in the focus and direction of the research itself (Kindon, Pain, and Kesby 2007a; Kindon 2009). To clarify, as an academic researcher thinking about embarking on PAR, you would not usually determine a research agenda independently. Rather, you would work with a social group to define the issues facing them (see also Chapters 3 and 4). Together, you would then generate and analyze information, which you hope would lead to action and, ultimately, positive change for those involved (Cahill et al., 2008; Cameron and Gibson 2005; Pain and Askins 2011). In short, a PAR researcher does not conduct research *on* a group but works *with* them to achieve change that *they* desire.

PAR has its origins in **action research** (Lewin 1946) in the United States, which sought to inform change by testing theory through practical interventions and action, and **participatory research** (Hall 2005), which emerged out of Africa, India, and Latin America, where educators and others involved in community development devised a new **epistemology** grounded in people's struggles and local knowledges. When combined, participatory action research sought to "develop new alternative institutions and procedures for research that could be **emancipatory** and foster radical social change" (Kindon, Pain, and Kesby 2007b, 10).

PAR has many similarities with **community-based research** (CBR), which has evolved in public health as medical and health-care practitioners seek ways to better involve community members in research to improve health outcomes. Many universities, particularly in North America, have adopted CBR to reconnect academic interests with the communities in which they are located. Often CBR informs service-learning courses involving undergraduate students in research to benefit their local communities (see Cope 2009).

Today, diverse PAR generally involves a number of stages through which academic and social group members work together to define, address, and reconsider the issues facing them (Kindon, Pain, and Kesby 2007a; Parkes and Panelli 2001). Such issues commonly include the lack of access to information or resources, the threat of removal of services or subsidies, or the need to respond to and mitigate unanticipated events. The emphasis on an iterative cycle of **action–reflection** is one of the key distinguishing features of PAR. It can also enable multiple perspectives of different **stakeholders** to be taken into account throughout the research, which can lead to more informed decision making and more equitable and potentially sustainable outcomes.

PAR attends to power relations (Kesby, Kindon, and Pain 2007) and as such can be challenging, particularly in the context of an undergraduate research project. It often involves the researcher in a facilitative rather than "extractive" role and demands that he or she pay considerable attention to ethics and issues of representation. That said, PAR can be very rewarding, and even if it is not possible to involve research participants deeply in every step of a research project, it may be possible to make your research more participatory by adopting some of the ideas discussed in this chapter.

Conducting "Good" Participatory Action Research

PAR is an approach that ideally grows out of the needs of a specific context, research question, or problem and the relationships between researcher and research participants. It is more about the value orientation of the work and its approach (epistemology) than about the specific techniques used, although participatory techniques are certainly important (see Kesby, Kindon, and Pain 2005 for discussion of deep versus other forms of **participation**). It is also an approach that values the process as much as the product so that the "success" of a PAR project rests not only on the quality of information generated but also on the extent to which skills, knowledge, and participants' capacities are developed (Chatterton, Fuller, and Routledge 2007; Cornwall and Jewkes 1995; Kesby, Kindon, and Pain 2005; Maguire 1987).

According to some practitioners (see Chambers 1994), the most important aspects of participatory work are the attitudes and behaviours of "outside" researchers (usually academics or practitioners). These attitudes and behaviours affect the relationships formed with research participants and the outcomes achieved. Whether you, as a researcher, respect people's knowledge or perpetuate unequal power relations and extract information largely for your own benefit depends to an extent on your attitudes and behaviours and the nature of the research relationships you establish. To illustrate this point further, Box 17.1 shows common connections between the attitudes of the researcher (illustrated here by things a researcher may say to a researched group), the kind of research relationships they form, the resultant mode of participation possible, and the relationship between the researched group and the research itself.

Researchers using PAR generally strive to adopt and practise the attitudes and behaviours that result in people's **co-learning** and **collective action**. They also generally follow an iterative process of action–reflection (see Box 17.2), although the specifics of what actually happens, how, and when vary depending on the particular context and circumstances of those involved.

The cycles of action–reflection outlined in Box 17.2 ideally involve researchers and collaborators in each stage. However, this may not be possible within the confines of an undergraduate research project (see Pain et al. 2013 for a helpful discussion here). Do not be put off, since this "ideal" process can and should be adapted collectively to suit the particular needs and constraints of the situation.

BOX 17.1 The Importance of Attitudes to Relationships within PAR

Attitude of researcher and example of attitude reflected in what researcher might say to researched group (RG)	Relationship between researcher and researched group (RG)	Mode of participation	Relationship between research and researched group (RG)
Elitist "Trust and leave it to me. I know best."	Researcher designs and carries out research; RG representatives chosen but largely uninvolved; no real power-sharing.	Co-option	ON
Patronizing "Work with me. I know how to help." (i.e., I know best)	Researcher decides on agenda and directs the research; tasks are assigned to RG representatives with incentives; no real power-sharing.	Compliance	ON/FOR
Well-meaning "Tell me what you think, then I'll analyze the information and give you recommendations." (i.e., I know best.)	Researcher seeks RG opinions but then analyzes and decides on best course of action independently; limited power-sharing	Consultation	FOR/WITH
Respectful "What is important to you in the research? How about we do it together? Here's my suggestion about how we might go about this."	Researcher and RG determine priorities, but responsibility rests with researcher to direct the process; some power-sharing.	Cooperation	WITH
Facilitative "What does this mean for you? How might we do the research together? How can I support you to change your situation?"	Researcher and RG share knowledge, create new understandings, and work together to form action plans; power-sharing.	Co-learning	WITH/BY
Hands-off "Let me know if and how you need me."	RG sets their own agenda and carries it out with or without researcher; some power-sharing.	Collective action	BY

Source: Adapted from Parkes and Panelli (2001).

BOX 17.2

Key Stages in a Typical PAR Process

Phase	Activities
Getting started	• Assess information sources. • Scope problems and issues. • Initiate contact with researched group (RG) and other stakeholders. • Seek common understanding about perceived problems and issues. • Establish a mutually agreeable and realistic time frame. • Establish a memorandum of understanding (MoU) if appropriate.
Reflection	On problem formulation, power relations, knowledge construction process.
Building partnerships	• Build relationships and negotiate ethics, roles, and representation with RG and other stakeholders. • Establish team of co-researchers from members of RG. • Gain access to relevant data and information using appropriate techniques (see Box 17.3). • Develop shared understanding about problems and issues. • Design shared plans for research and action.
Reflection	Reformulation, reassessment of problems, issues, information requirements.
Working together	• Implement specific collaborative research projects, • Establish ways of involving others and disseminating information (see Box 17.3).
Reflection	Evaluation, feedback, re-participation, re-planning for future iterations.
Looking ahead	Options for further cycles of participation, research, and action with or without researcher involvement.

Source: Adapted from Parkes and Panelli (2001, 98).

What is most important in PAR is that the design and process are negotiated with the researched group and carried out in ways appropriate to the context, the time available, and the people involved, including yourself (Manzo and Brightbill 2007). Adopting even some of the strategies above will enable "partial" participation to occur. This will bring benefits to your project and some of your participants (see Chapters 3 and 4), particularly if you combine them with some of the participatory techniques discussed below (also see Kesby, Kindon, and Pain 2005).

The strategies and techniques used within a PAR process can involve and adapt some of the other methods discussed in other chapters of this book if they occur in the context of reciprocal relationships. Common methods and techniques are interviewing, and visual methods (including **participatory diagramming** or **participatory mapping** in which participants create diagrams, pictures, and maps to explore issues and relationships [see Box 17.3]; also see Alexander et al. 2007 for a discussion of participatory diagramming and Sanderson et al. 2007 for a discussion of participatory cartographies, including mapping). These methods and techniques emphasize shared learning (researcher and researched group), shared knowledge, and flexible yet structured collaborative analysis (*PLA notes* 2003). They embody the process of **transformative reflexivity** in which both researcher and researched group reflect on their (mis)understandings and negotiate the meanings of information generated together (see Crang 2003b, 497).

Some Strategies and Techniques Used within PAR

Establish a support base and platform for your research

- Find and critically review secondary data. Secondary data can help to establish the direction of the research and identify where gaps or contradictions in understanding exist.
- Involve those who are experts about specific issues and processes. Local experts always exist and can help facilitate the participation of others and inclusion of their knowledges.
- Negotiate and establish a memorandum of understanding (MoU) for the research team.[2] MoUs and the process that leads to them clarify research expectations, agreed-upon norms of behaviour, modes of interaction, and "ownership" of information generated and are important for sustaining participatory partnerships and realistic expectations.

Get involved with people and their lives

- Observe directly (see for yourself). Visiting people and places is essential. Taking a walk through an area with members of the researched group enables you to observe firsthand and question things directly. It might also be helpful to use a video camera as you go (see Chapter 13).

- Do-it-yourself. By living like the people you are working with, you can learn something—although never all—of their realities, needs, and priorities. You can also learn how they communicate with each other and get things done.
- Work with groups. Groups can be casual or encountered "randomly"; **focus groups** that are representative or structured for diversity; or community, neighbourhood, or specific social groups. Group interviews are usually a powerful and efficient way of generating and analyzing information (see also Chapter 10).

Use interviewing and storytelling approaches

- Collect case studies and stories. Focusing on specific events or cases, such as a household profile or history or how a group coped with a crisis, is a helpful way of teasing out issues. A variety of cases can reveal and illustrate common themes and important differences.
- Use open-ended questions and key probes. Asking questions that start with "what," "when," "where," "who," or "how" can generate specific information without leading respondents to particular answers. Probing (i.e., "what happens when . . . ?" and "why is that?") can identify key issues, local rationales, and current activities and procedures (see Chapter 8).

Use types of participatory mapping and diagramming

- Maps. Mapping, drawing, and colouring using locally appropriate materials can represent resources, issues, and relationships in ways that enable more democratic participation than verbal discussions alone. Maps might focus on tangible resources such as land, forests, houses, or services. Diagrams can tease out relationships between people, institutions, and resources (see below). Do remember—as Howitt and Stevens note in Chapter 3—that participatory mapping and diagramming may sometimes be suspect if used in the wrong context.
- Timelines and trend/change analysis. Locally defined chronologies of events showing approximate dates; people's accounts of how customs, practices, and things have changed; ethno-biographies or local histories of particular crops, animals, or trees; and changes in land use, population, migration, fuel uses, education, health, credit, and so forth may enable analysis of cause-and-effect factors over time.
- Seasonal calendars. Focusing on seasonal variation of particular factors (for example, rain, crop yields, workload, travel) can enable

insights into matters such as climatic variation, labour patterns, migration, diet, and local decision-making processes.

- Daily time-use analysis. Indicating the relative amount of time, degree of drudgery, and level of status associated with various activities may reveal local power relations and identify the best times for research activities.
- Institutional or Venn diagramming. Drawing out the relationships between individuals and institutions using overlapping circles signifying the importance or closeness of the relationships enhances understanding of power relations, local and surrounding contexts, and the identification of where there are blocks to or possibilities for change.
- Well-being grouping (or wealth ranking). Grouping or ranking households according to local criteria, including those considered poorest and worst off, can be a helpful lead into discussions about the livelihoods of the poor and how they cope depending on the particular cultural context.
- Matrix scoring and ranking. Drawing matrices of resources, such as different types of trees, soils, or methods of health provision, then using seeds or beads to score or rank how they compare according to different criteria (such as productivity, fertility, or accessibility) can reveal local preferences (what scores highly) and the aspects that inform decision-making strategies.

Use audiovisual technologies and the Internet to engage people and disseminate information

- Use cellular phones as a helpful tool for involving people, particularly if this is their usual mode of communication and/or they are highly mobile. Phones can be used to text or call co-researchers or participants with invitations to meetings, to invite responses to questions, to inform process and analysis, and to share results or outcomes.
- Use mobile phones or digital video cameras to engage people in mobile interviews and analysis of local issues. Short clips can then be shared with others to stimulate discussion and further analysis.
- Establish Internet chat sites, wiki, or blogs to create spaces in which co-researchers and participants can share ideas and responses to emerging analysis. As with all group spaces, these need to be carefully managed and moderated.
- Design and promote websites as a productive means of engaging participation and establishing a core identity for a group, as well as

disseminating the process, findings, and outcomes of participatory action research.
- Use sites like YouTube to disseminate information widely, but take care because once posted on the site, images and videos can be used and appropriated by others for their own means, with little regard for the original ethical agreements or MoUs negotiated between collaborating parties.

Engage people in joint analysis, reflection, and future planning

- Shared presentations and analysis. Involving people in the presentation and analysis of the maps, diagrams, and information generated throughout the research shares power and allows information to be checked, corrected, and discussed.
- Contrast comparisons. Asking group A to analyze the findings of group B and vice versa can be a useful strategy for raising awareness and establishing dialogue, particularly between different groups. This technique has been used for gender awareness, asking men, for instance, to analyze how women spend their time.

Further information on these and other strategies and techniques, with examples of their use, can be found in Pretty et al. (1995).

While the above list of techniques may seem exhaustive, it is not prescriptive. Integrating one or several of them, where appropriate, will enhance your research. However, as Richie Howitt and Stan Stevens suggest in Chapter 3, a few participatory techniques will not, in and of themselves, make your project PAR. For this to occur, the open negotiation of the research design and methodology with the people with whom you are working is critical, as is an emphasis on supporting people's capacity to do their own research and analysis.

For more ideas about what this might mean in practice, the work of Caitlin Cahill and the Fed Up Honeys is instructive. Over several weeks, a US academic researcher (Cahill) met with six young women (also known as the Fed Up Honeys) to carry out a PAR project, "Makes Me Mad: Stereotypes of Young Urban Women of Color, in New York City's Lower East Side." The process involved a period of time deciding first on the research focus and then on the approach they would adopt. They decided that each of them would engage in reflective journal writing and analysis of their own thoughts, plus the reading and discussion of each others' writing, to explore how they had internalized racist and sexist stereotypes and applied the same representations to others. It was an intense and at times frustrating process for participants because of the project's close attention to detail and the challenge it presented to their personal beliefs. The project required dedicated and reflexive facilitation by Cahill and commitment from each participant to enable them to work through their differences. It was worthwhile, however, for the depth of friendships it produced, the degree of increased awareness it provoked, and the level of political action that the participants engaged in through their development of a website and their implementation of a sticker campaign (see Cahill 2004 and the Fed Up Honeys website for more details: http://www.fed-up-honeys.org/mainpage.htm). Box 17.4 summarizes some ideas about how to approach doing PAR. The points integrate and reinforce ideas about the attitudes, behaviours, relationships, research design, process, and techniques discussed in this chapter.

The Value and Rewards of Participatory Action Research

Through attention to attitudes and behaviours, as well as the use of appropriate strategies and techniques to support people's participation in collaborative research and action, it is possible to examine and challenge forms of oppression and inequality. PAR is used most frequently by geographers with an activist agenda to work for social change because it offers a tangible way of putting the aims and principles of **critical geography** into practice (Kesby 2000). Often, this means specifically addressing issues of racism, ableism, sexism, heterosexism, and imperialism (Ruddick 2004, 239) and how they are manifested through people's unequal access to and control over resources or their positions within inequitable social relationships (See Box 17.5).

Because of this activist orientation, PAR can build capacity and alliances within a community. For example, in Canada, geographer Geraldine Pratt worked for more than 14 years with the Kalayaan Centre in Vancouver using a PAR approach. Together, they conducted four main projects exploring different dimensions of the lived experiences of Filipina migrants. The first project involved collecting women's stories of being live-in care-givers to Canadian nationals and mobilized them to change their working conditions. The second used storytelling to explore how the same women's lives had changed over time as a means of understanding long-term migrant integration. The third project adopted a focus group approach with first- and second-generation young people to record their stories of racism and to begin to foster a greater sense of collective identity and belonging. The fourth project involved women and young people to collect the stories of transnational families in an effort to develop and share collective community stories, to challenge individual feelings of isolation and despair (Pratt et al. 2007).

In my own work with *Te Iwi o Ngaati Hauiti* in the central North Island of Aotearoa/New Zealand, several *iwi* (tribe) members established themselves as a

Some Ways to Promote Participation in Geographic Research

- Involve/be involved with the group with whom you are working as equal decision makers to define the research questions, goals, and methods and as co-researchers and analysts of information generated.
- Show awareness that you are an outsider to the group you are researching, even if you are working together as co-researchers.
- Be clear about the potential impacts people's involvement may have and what will happen to the information generated (ideally through an MoU).
- Take care not to promise too much or inflate people's expectations of what might happen as a result of the research.
- Develop facilitation skills, which can stimulate initiative and sensitively challenge the status quo without imposing your own agenda.
- Work at fostering participatory processes and research techniques that will release creative ideas and enthusiasm but not take too much time for those involved (see Box 17.3).
- Seek out the perspectives and participation of the most vulnerable and marginal people.
- Find ways of limiting the dominance of interest groups and powerful people (including yourself, where appropriate).
- Acknowledge that process is as important as product (and sometimes more important), and factor in enough time to involve people appropriately at various stages of the research, including time for reflection.
- Support the group with whom you are working to share the benefits of their involvement with others and to take initiative to address their concerns.
- Involve the group with whom you are working in the writing and dissemination of relevant information; at the very least, acknowledge their contributions to any sole-authored work.
- Practise honesty, integrity, compassion, and respect at all times.
- Keep a sense of humour!

Source: Adapted from Botes and van Rensburg (2000, 53–4); Kesby, Kindon, and Pain (2005); and the author's own experiences.

Selected Geographers and Some of Their PAR Projects

Geographer	PAR work with . . .
Caitlin Cahill	Young women of colour—the Fed Up Honeys, USA
Jenny Cameron and Katherine Gibson	Economically "depressed" communities, Australia
Sarah Elwood	Urban community groups using participatory GIS, USA
J.K. Gibson-Graham	Women in mining communities, Australia
Robin Kearns	Primary schools and parents, Aotearoa/New Zealand
Mike Kesby	Young people with HIV/AIDS, Zimbabwe
Sara Kindon	Indigenous women and men in Indonesia and Aotearoa/New Zealand
Rob Kitchin	People with disabilities, Ireland
Fran Klodawsky	Homeless people and women in local government, Canada
Audrey Kobayashi	Immigrant and ethnic minority groups, Canada
Jan Monk	Women's and non-governmental organizations, USA and Mexico
Carolyn Moser and Cathy McIlwaine	Communities coping with violence, Colombia and Guatemala
Alison Mountz	Transnational communities, USA and El Salvador
Karen Nairn	Young people, Dunedin, Aotearoa/New Zealand
Rachel Pain	Locally born and asylum-seeking young people, UK
Linda Peake	Women's handicraft cooperative, Guyana
Geraldine Pratt	Philippine Migrant Workers Collective, Canada
Maureen Reed	Women in logging communities, Canada
Kevin St Martin and Madeleine Hall-Arbor	New England fishing communities, USA
Janet Townsend	Female rainforest settlers, Colombia and Mexico

Sources: Kindon, Pain, and Kesby (2007a); Pain (2003); Ruddick (2004).

community video research team to explore the relationships between place, cultural identity, and social cohesion (Hume-Cook et al. 2007; Kindon 2012). They undertook training in video production and community research with an academic colleague and me and then carried out video interviews with other members of the *iwi*. Sometimes I was a co-interviewer and my colleague a co-videographer; at other times they worked independently and later shared their tapes and analysis with us (Kindon 2003; Kindon and Latham 2002). We have also been involved in the collaborative editing of a documentary of a key event in the *iwi*'s efforts to foster social cohesion and cultural identity—a five-day rafting journey down their most significant river.

PAR enables rich and varied information embedded within specific "communities" to be shared, analyzed, and evaluated collectively (Cooke 2001). This information may be more accurate and relevant for other uses than had a researcher worked alone. In PAR work with rural communities in Bali, Indonesia, in the early 1990s (Kindon 1995; 1998), men associated with one village used information generated through our regular participatory research meetings and focus groups (see Chapter 10) to develop an action plan for their community. The plan addressed the need for better roads to open up access to markets. In the four years that followed, these men established a savings/credit fund and liaised with government agencies to raise enough money to seal a remote road and enable more efficient transport of produce to market. Elsewhere in the village, government planners acted on PAR information generated by women's groups about the need for more accessible health-care facilities and established a local health post. On my return to the village in 1998, these two key development needs had been met by the collective actions of these men and women and the support received from government agencies.

In summary, people's participation in their own research may challenge prevailing biases and preconceptions about their knowledge on the part of others in positions of power (Sanderson and Kindon 2004), such as government officials and policy makers. In addition, PAR can bring about desired change more successfully than "normal" social science research methods (Brockington and Sullivan 2003; Kesby 2000) and often results in improvements in living or working conditions for those involved (Kesby, Kindon, and Pain 2005; Kindon, Pain, and Kesby 2007a; Pain 2003, 2004; Parkes and Panelli 2001).

Challenges and Strategies

There are challenges associated with collaborative research endeavours as well (Monk, Manning, and Denman 2003), especially when participatory techniques are involved (Pain and Francis 2003). While participation is becoming increasingly popular, not all researchers are doing it well (Kindon 2010; Parnwell 2003).

Most PAR takes place in a group setting, which is both a strength and a weakness. Particular techniques (like the participatory mapping and diagramming techniques listed in Box 17.3) require group participation, often in public spaces. This spatial aspect of PAR shapes the construction of knowledge (Brockington and Sullivan 2003; Cooke 2001; Kesby 2007; Kindon 2012; Mosse 1994), and it usually tends to generate knowledge that reflects dominant power relations in wider society (Kothari 2001). It is therefore important to pay attention not only to *who* participates but *where* they participate and *how*. Keeping a field diary with this information can be helpful when you come to analyze the products of group work. If funds permit, involving a colleague or friend to keep notes on the process or make video recordings can provide a detailed and more dispassionate record of where and how people participated (yourself included) for later analysis (Kindon 2003; 2012).

In terms of the participatory techniques themselves, because they often appear to be quick and easy to use (Leurs 1997), it can be tempting to use them repeatedly in the same ways rather than adapt and modify them to the particular contexts involved (Chambers 1994). Certainly, many current participatory processes associated with development projects involve a sequence of participatory techniques such as community mapping, social well-being ranking, agricultural land-use transects, and institutional diagramming, regardless of the particular context or issues being assessed. This formulaic and researcher-led use of participatory techniques can result in what some academics have called the "tyranny of participation" (Cooke and Kothari 2001) in which any inequalities, particularly between the researcher and researched group, are reinforced (Parkes and Panelli 2001; Wadsworth 1998).

A way around these difficulties is to consider who defines participation or who initiates what activities at each stage of the PAR process. Being open to sharing facilitation and innovating techniques in response to specific contexts can play a vital role in helping to monitor who is framing participation and how (Hickey and Mohan 2004; Williams et al. 2003). Discussing these aspects early on is a particularly useful way of clarifying expectations, establishing greater collaboration, and specifying roles and responsibilities (Kindon and Latham 2002; Manzo and Brightbill 2007).

Participatory techniques can generate information quickly, but they are not a substitute for more in-depth social research methods (Kesby 2000) such as those discussed elsewhere in this book. Understanding the contexts within which information is generated is critical to our ability to rigorously analyze it. For this reason, you might wish to consider first undertaking PAR within a community or location already familiar to you. Within an undergraduate dissertation, this could provide you with some of the necessary contextual information, freeing you to focus more energy on the process.

Sometimes our desire to avoid exploitation or extractive research relationships can mean that we become so involved with our co-researchers that we are unable to work effectively for change. Establishing outside support networks (see Bingley 2002) can help to prevent this situation and sustain our endeavours.

A related point is that long-term relationships, even friendships, with participants and co-researchers commonly develop through PAR, and while some studies may become lifetime projects, we typically have to leave the group with whom we have been working. Investing time into a sensitive and appropriate leaving strategy at the beginning of the research can help to avoid raising expectations and assist in navigating the changing status of relationships (Kindon and Cupples 2014). Formal meetings, celebrations, feedback sessions, visits, and the exchange of gifts may all be appropriate mechanisms to assist with closure.

In other cases, it may be academically and professionally important to maintain a sustained engagement long after the official research project is over—particularly to ensure the spread and impacts of any empowering change (Kesby 2005) and/or to practise what Paul Chatterton, Duncan Fuller, and Paul

An Example of Participatory Publishing

The following text is from US geographer Sarah Elwood, who worked for many years with urban-based community groups and municipal councils in Chicago to support research, education, and community capacity-building within processes of neighbourhood redevelopment. It is a collaboratively authored chapter and represents one way in which multiple research partners' voices can be respected and shared.

> In this chapter, we provide a collectively authored discussion of our use of GIS as a negotiating tool, and the lessons we have learned about sustaining community-university PGIS (**participatory geographic information systems**) projects. Our author group includes a university-based researcher (Sarah), the executive directors of the Near Northwest Neighbourhood Network (NNNN) and the West Humboldt Park Family and Community Development Council (the "Development Council") (Bill and Eliud), a university-based research assistant (Nandhini), and past and present staff members of NNNN and the Development Council (Kate, Reid, Niuris, Lily and Ruben). . . .

Routledge (2007) have called "solidarity action research." Your engagement may be in the capacity of support person, community board member, fund writer, publicist, or campaigner. There may be ethical challenges if your status changes from co-researcher to "friend" or "resisting other" (Chatterton, Fuller, and Routledge 2007), and being clear about what you can commit to in any of these relationships is vital. Overall, a key way to manage this and other challenges associated with PAR is to be realistic with yourself, your co-researchers, and other stakeholders about what is possible within the time and resources available to you.

Presenting Results

In Chapter 3, Richie Howitt and Stan Stevens observed that representing people with whom we work is no easy undertaking, even if we involve them in the process. In their work with urban communities in Latin America, Cathy McIlwaine and Caroline Moser (2003) propose that a balance is needed with respect to the

From the perspective of the community organisation staff, PGIS partnerships are likely to be sustainable and effective if they produce immediately applicable results, and if university partners do not dictate results. One staff participant writes:

> There has to be a clear understanding that the action and the research produced are usable for the community, not just something to publish.

Both directors argue that leaders of organizations have a special role to play in sustaining GIS resources and skills. One writes:

> Part of the job of executive director is recognizing the importance of the project, but also finding a way to keep it for the organisation, and not let it go.

In sum, for community organizations a great deal of the power and impact of PGIS stems from the diversity of ways it can serve as a process for community change.

Source: Elwood et al. (2007, 170–8).

presentation of information. Information should influence policy makers (to effect change) and should be meaningful to those involved in the initial research. In practice, this requires culturally appropriate ways of sharing knowledge and may, for example, require the production of several reports or presentations for different audiences by different members of the research teams using different media (see Cameron and Gibson 2005).

Project or policy reports are powerful advocacy tools to advance action plans developed during the research. Clear, simple presentations work best, providing policy makers with a sense of the process and how it generated reliable, meaningful "data" upon which practicable policy can be developed. Presenting the results of PAR at public meetings, conferences, or other gatherings with co-researchers can be appropriate and often enjoyable. If their direct participation is not possible, then discussing what they would like you to emphasize in a presentation can go some way towards addressing the power imbalance and lend you some authority to speak on their behalf. Taking the findings "to the streets" in accessible media (for example, newsletters, magazine articles, plays, posters, websites, art [see Chapter 20]) should ideally be part of the iterative process of PAR and can have some of the greatest impact at the local level (see Cahill and Torre 2007; and mrs c kinpaisby-hill 2011 for more discussion).

For you as a student, the results of your work also need to meet the requirements of the academy. It can be challenging to present the "results" of an iterative and participatory research process within the context of a typical thesis or dissertation but certainly not impossible (Klocker 2012).[3] Engaging in PAR provides you with the opportunity to negotiate explicitly how you will use information and how you will represent others' experiences and/or views (Kindon and Latham 2002). Although sometimes time consuming, such steps can temper your powerful position as the sole author in what has been, until now, a collaborative process. Sharing your choices and discussing how you intend to construct your argument continues the participatory process and goes some way towards ensuring that your final product respects the people and diversity of issues involved. An MoU item about this at the beginning of the research can prevent misunderstandings when you later want to quote people or include maps or diagrams produced through the research process.

Within your dissertation, thesis, or research project itself, including direct quotations from a range of people, which tease out common or disparate threads, can illustrate the multiple perspectives in circulation. Citing a disagreement or exchange between people can highlight where there are tensions or differences of perspective. However, take care to contextualize and analyze them adequately or they could be overwhelming to the reader. In addition, discussing aspects of methodology—so important to the participatory process—can honour people's involvement, acknowledge that process is as important as product, and enrich the analysis of the "results" produced.

It may be appropriate and courteous to include co-researchers as co-authors on any papers that may emerge from PAR (see for example, Box 17.6; Hume-Cook et al. 2007; Peake 2000; Pratt et al. 1999; Townsend et al. 1995). We may work behind the scenes to enable our co-researchers to publish or disseminate their understandings independently of us. Or we can establish a collective name under which we write and publish. These strategies can demonstrate to others a collaborative and collective approach to research and writing (for example, see mrs kinpaisby 2008 and mrs c kinpaisby-hill 2008; 2011; 2013). However, with increasing pressures on academics to demonstrate their scholarship through publications and citation indexes, we need to weigh up the benefits and potential limitations of not having our names identified as an author. Overall, having multiple research products written by different combinations of people is often an appropriate strategy. It can enrich the knowledge produced and be critically important if ongoing action is to be sustained, while meeting institutional requirements and agendas.

Reflecting on Participatory Action Research

A key question of PAR is often "to whom is the research relevant?" (Pain 2003, 651). For us as researchers, if we accept that we have an opportunity and an obligation to co-construct responsible geographies (McClean et al. 1997; Williams et al. 2003), then PAR offers us an exciting means of undertaking relevant, change-oriented research. While academe does not usually reward such **activism**, the central role of space in many people's oppression (Ruddick 2004) means that human geographers are uniquely positioned, and morally beholden, to adopt ways of researching that build collaborative communities of inquiry (Reason 1998, cited in Hiebert and Swan 1999, 239) and challenge oppression.

Fortunately, certain parts of human geography, such as social geography, have a rich tradition of activism (Kindon 2009; Panelli 2004). In addition, PAR is becoming more common within geographic research, providing a growing body of work and experience from which to draw. PAR is not without its challenges, particularly within the confines of student research projects, but it is possible to adopt many of the principles discussed in this chapter to enable a rigorous research approach that also results in tangible benefits for those involved. Perhaps the greatest challenge of all is for academics, including undergraduate researchers, to "cross boundaries of privilege and confront their personal stake in an issue, and the ways they are positioned differently from members of the [groups] they work with" (Ruddick 2004, 239). I hope that this chapter has given you some ideas with which to begin this journey within your own work and some methodological resources for respectfully and ethically facilitating others' participation throughout the process.

Key Terms

action–reflection
action research
activism
co-learning
collaborative research
collective action
community-based research
critical geography
emancipatory
epistemology
focus groups
memorandum of understanding (MoU)
participation
participatory action research (PAR)
participatory diagramming
participatory geographic information systems (PGIS)
participatory mapping
participatory research
social justice
stakeholder
transformative reflexivity

Review Questions

1. Why is participation important in qualitative research?
2. Find an example of a participatory approach to geographic research in a recent book or journal. How is rigour established and maintained?
3. Given the importance of facilitation in PAR, make a note of the skills and attributes needed to be an effective PAR researcher. How might you develop and/or strengthen these skills and attributes in yourself?
4. What are some of the major challenges associated with doing PAR in geography? Make a list, and then devise strategies to manage these challenges productively.

Review Exercises

1. Constructing a Draft Memorandum of Understanding (MoU)

MoUs and the process that leads to them clarify research expectations, agreed-upon norms of behaviour, modes of interaction, and "ownership" of information generated at the outset of the research process. This exercise asks you to think about a hypothetical PAR project that will involve student academic researchers carrying our research with and for residents in a low-income, ethnically diverse city housing estate. The aim of the research has been determined by the local city council who wish to identify actions that they can take to improve health outcomes for the residents.

- Identify the key groups or stakeholders who have an interest in the project and its outcomes.

- Allocate stakeholder roles within your class or group so that there is more than one person representing each group's perspectives and expectations.
- Take 10 minutes to brainstorm within your group what you want to get out of the process of this PAR project (think about how you want the research to be carried out as well as what outcomes you would like, including attention to aspects of communication, ownership, and dissemination).
- Come back together and negotiate the MoU that will guide your project relationships. Write up your agreement.
- Now go back into your group and reflect on how the process was for you.
- Share with the wider class or group your insights into the politics and power relations involved in negotiating an MoU that places the community residents centrally. Are there insights that you will carry with you into your own work? Are there lessons you can learn from to mitigate challenges or conflicts?
- Reflect more widely on what the implications of this process might be for institutional processes in your city/university. How feasible would it be to facilitate a similar process as part of your own work? What challenges might you face?

2. Enhancing the participatory orientation of your current research project

Working either individually or in a small group, take 20 minutes to work through your current research proposal, paying attention to the implications of your research design for the following questions:

- Who is involved and in what capacity? Does your design enable access and input into decision making that will shape and focus the research?
- What kinds of methods will be used and who gets to propose or determine them? How might these enable different kinds of people to participate, or hinder their involvement?
- How long might each phase take and will you be able to sustain people's engagement through and across different phases?
- How will people be involved in the analysis of information generated? How will you accommodate or acknowledge differing interpretations?
- Once generated, how will information will be used, presented, and protected?
- If you make more aspects of your project participatory, are there resource or time costs that you need to consider?

After 20 minutes, share your insights and/or methodological adjustments with a classmate or other group, and discuss their ethical and operational implications.

Useful Resources

Cahill, C. 2004. "Defying gravity: Raising consciousness through collective research." *Children's Geographies* 2 (2): 273–86.

Cameron, J., and K. Gibson. 2005. "Participatory action research in a post-structuralist vein." *Geoforum* 36 (3): 315–31.

Fed Up Honeys. 2008. "Welcome." http://www.fed-up-honeys.org/mainpage.htm.

Kesby, M., S. Kindon, and R. Pain. 2005. "'Participatory' diagramming and approaches." In R. Flowerdew and D. Martin, eds, *Methods in Human Geography*, 2nd edn, 144–66. London: Pearson.

Kindon, S. 2010. "Participation." In S. Smith, R. Pain, S. Marston, and J.P. Jones III, eds, *The Handbook of Social Geography*. London: Sage.

Klocker, N. 2012. "Doing participatory action research and doing a PhD: Words of encouragement for prospective students." *Journal of Geography in Higher Education*, 36 (1), 149–63.

Manzo, L., and N. Brightbill. 2007. "Towards a participatory ethics." In S. Kindon, R. Pain, and M. Kesby, eds, *Participatory Action Research Approaches and Methods: Connecting People, Participation and Place*, 33–40. London: Routledge.

Mountz A., I. Miyares, R. Wright, and A. Bailey 2003. "Methodologically becoming: Power, knowledge and team research." *Gender, Place and Culture*, 10 (1), 29–46.

mrs c kinpaisby-hill 2011. "Participatory praxis and social justice: Towards more fully social geographies." In V. Del Casino, M. Thomas, P. Cloke, and R. Panelli, eds, *A Companion to Social Geography*, 214–234, London: Blackwells.

mrs c kinpaisby-hill. 2013. "Participatory approaches to authorship in the academy." In A. Blunt, ed., *Publishing and Getting Read: A Guide for New Researchers in Geography*, Section 4.2, p. 24. London: Wiley-Blackwell.

Pain, R., M. Finn, R. Bouveng, and G. Ngobe. 2013. "Productive tensions: Engaging geography students in participatory action research with communities." *Journal of Geography in Higher Education*, 37 (1), 28–43.

Notes

1. Readers of this chapter are strongly encouraged to also read Chapters 2, 3 and 4, which provide useful, complementary discussions.
2. The research team may consist of researchers only, researched people only, or both researched people and researchers.
3. If you are using PAR in a thesis, for example, and because universities typically expect a thesis or dissertation to be the "original work" of the student alone, you should discuss matters of authorship with your supervisor when preparing your project.

PART

"Interpreting and Communicating" Qualitative Research

18 Organizing and Analyzing Qualitative Data

Meghan Cope

Chapter Overview

This chapter discusses some ways qualitative data can be organized and analyzed systematically and rigorously to produce new knowledge. We begin with some comments on how to make sense of your data, and explore practices of "memoing," concept mapping, and coding. Much of the subsequent discussion revolves around coding as a process of distilling data and identifying themes. The chapter reviews different types of codes and their uses, as well as several ways to get started with coding in a qualitative project. Specifically, a distinction is drawn between *descriptive* codes, which are category labels, and *analytic* codes, which are thematic, theoretical, or in some way emerge from the analysis. The building of a "codebook" is also discussed, stressing the importance of looking critically at the codes themselves, identifying ways in which they relate, minimizing overlap between codes, and strengthening the analytical potential of the coding structure. Finally, several related issues are covered, such as coding with others, the use of **computer-aided qualitative data analysis software (CAQDAS)**, and integrating coding and mapping.

Introduction: Making Sense of Data

As is evidenced by this volume and the notable expansion, over the past decade, of publications on qualitative methods generally in the field of geography, scholars and students are increasingly engaged not only in *doing* qualitative research but also in thinking and writing critically about methodologies, including the ways that we evaluate, organize, and "make sense" of our data (see also Clifford, French, and Valentine, 2010; DeLyser et al. 2010; Gomez and Jones, 2010). As the "qualitative turn" in geography and related fields has matured, the techniques for gathering and producing data have expanded in scope and depth; this, in combination with critical reflection on issues of methods, context, and researcher positionality, has produced ever more thoughtful, reflexive, and creative research projects and, ultimately, a more robust production of knowledge. It is an exciting time to be doing qualitative geographic research, and, appropriately, the standards are high, requiring practitioners to be well-informed and attentive to principles of rigorous scholarship.

Towards that goal of rigorous scholarship, the practices of organizing and analyzing one's data deserve special attention. The job of the researcher is, in effect, one of synthesis and translation; we are tasked with observing and engaging with the world, making sense of the resulting data, and representing (re-*presenting*) the facts, stories, ideas, and events that were shared with us in a coherent manner. The focus for this chapter will be on the middle piece of this—making sense of data—while recognizing that it cannot be wholly separated from the other parts of the process.

Increasingly, qualitative data are taking more diverse forms, well beyond the classical text-based transcribed interviews, oral histories, archives, or diaries (though these are all still very important sources), into visual data (photos, video, maps, sketches); sources resulting from participatory engagement with community partners (collaborative maps, exhibits, activism); and varied creative techniques such as walk-along interviews, artwork, theatrical productions, photovoice, self-directed video, vernacular mapping, material culture analysis, and other methods that foster less hierarchical interaction with participants. In the context of such an expansion of research techniques, particularly those based less and less on text, it is natural that we should revisit some ideas about "making sense of data." I propose, however, that many of the same principles that have appeared in previous editions of this volume remain valuable: organizing and reducing data to manageable chunks, identifying themes, and paying attention to rigorous interpretation. Towards that end, I review some common techniques for the purposes of description, classification, and connection (Dey 1993) of qualitative data; these techniques are memoing, concept mapping, and coding.

Making Meaning I: Memos

A **memo** (short for "memorandum") is usually a short note to oneself or research collaborators jotted to capture a quick insight, to serve as a reminder of a future task, or to draw connections between multiple referenced items. In **ethnography**, researchers typically rely on heavy use of field notes to record the events, dialogues, and observations made while engaging with a community, yet ethnographers often use memos as a sort of intermediate-level mechanism to remind themselves of something, to reflect on patterns or connections, to contextualize events, and to forge new links between emerging themes. For example, Annette Watson and Karen Till (2010) provide a particularly rich example of the use of memos as a first stab at interpretation and reflection on their experiences and the material in their field notes:

> Both of us understand our writing as a safe space to explore, think through, and represent knowledges, desires, and fears. In memos, we question our experiences and assumptions, pay attention to processes, respond to our embodied and emotional presences, consider the material and visual cultures that constitute what is being studied, scrutinize various relationships with research fields and partners, and elaborate upon our insights. We also make connections to other studies of previous work in memos, and raise critical questions that inform future theoretical readings. (Watson and Till 2010, 128)

The crucial elements of memos, as demonstrated by Watson and Till, can be summarized as follows. First, memos are quick and informal. They are jottings that by their very nature allow the researcher space and freedom to explore possible connections and viewpoints, to put an idea down in its earliest undeveloped form, and to remind oneself to pay attention to this or that in the future. Second (and despite its informal nature), memoing constitutes a valuable *interpretive* practice towards making meaning of the data. Along with margin comments, sticky notes, sketches, and other annotations researchers commonly make on and about their data, memos are useful for sorting out ideas, identifying patterns and similarities, recording "Aha!" moments, and generally beginning the process of organizing and analyzing. Finally, memos are *reflexive*. That is, they foster critical review and contemplation, including self-reflection; memos represent a chance to step back and consider alternative interpretations, to critique one's own role in the research process, and identify—however tentatively—linkages between events and discussions that might not have been seen while in the midst of research mode.

The benefits of memoing are not limited to participant-observation research practices, of course. Memos can be useful for any research project across the spectrum of qualitative, quantitative, and mixed-methods scholarship because they serve the important functions of reminding us of ideas to return to, as well as connecting, processing, and critiquing the process and early findings of the research endeavour. Reflecting the value of memoing, most CAQDAS programs include memo functions, including some that allow one to make memos about memos!

Making Meaning II: Concept Mapping

Despite the word "mapping" in the name, there is little formal discussion of concept mapping in the qualitative geography literature, yet it is an important organizational and analytic strategy that most researchers probably do without really thinking too much about, perhaps because there is a certain intuitiveness to the process. At its most basic level, **concept mapping** involves visualizing data and their relationships. This could be performed through the process of categorizing, that is, sorting data into piles (literal or digital) that have some internal cohesion. For example, in studying teen mobility my research partner and I used open-ended survey questions to query parents' perceptions of their teens having driver's licences; we immediately saw that "safety" and "convenience" were major themes and, in one analytic exercise, began to sort survey responses along those lines, with subordinate factors contributing to the main concepts. By sorting and re-grouping responses on a large whiteboard, using statements from parents who mentioned the convenience of having teens drive, we were able to quickly identify some clusters: "working parent(s)" signified that both parents were employed full time (or that a single parent was employed full time); "rural/remote" indicated that these families tended to live in more rural areas and at farther distances away from school, home, and workplaces; and "younger children" indicated the households in which teens had younger siblings who also needed chauffeuring around to activities and events. Discerning these kinds of connections through the relatively quick process of concept mapping—really a visual brainstorming exercise—allowed us to form some initial insights that were then followed up with more systematic examinations of the correlation between "convenience" and the household situations of those parents and teen drivers. While sorting and piling are common components of concept mapping, other methods are also frequently employed, such as word clouds, flow charts, relationship trees, and non-hierarchical graphical plottings that begin to drill down into key associations within data.

Although the example mentioned here was intended as a relatively quick, exploratory, and coarse-level analysis, concept mapping has evolved into a tool that is much more formalized, and the results can in themselves be part of the final representation of the research; the latter has gained traction in such fields as planning, organizational research, and evaluation studies, but could lend itself to diverse disciplinary fields. Kane and Trochim (2006) acknowledge both the informal notion of concept mapping (as just discussed) and the much more systematized methodology in the very first sentences of their book on the topic:

> *Concept mapping* is a generic term that describes any process for representing ideas in pictures or maps. [Here] however, we use the term only to refer to one specific form of concept mapping, an integrated approach whose steps include brainstorming, statement analysis and synthesis, unstructured sorting of statements, multidimensional scaling and cluster analysis, and the generation of numerous interpretable maps and data displays. (Kane and Trochim 2006, 1. Emphasis in original)

In this method, which combines qualitative and quantitative analysis, researchers begin their analysis of, say, open-ended survey questions by isolating unique "statements" (single-topic fragments of longer responses), then sorting those statements into groups based on their substantive relationships. This "sort/pile" part of the method could be done by hand using slips of paper and actual piles or using CAQDAS functions digitally (for example, using the "network" functions in *Atlas.ti*, or the "code map" function in *HyperRESEARCH*). This in and of itself may be very helpful to researchers and, indeed, my research partner and I stopped at this point, having identified some valuable conceptual–empirical linkages that warranted further exploration. Kane and Trochim, however, go on to perform analyses of the clusters that result from the sort/pile method, including multidimensional scaling (which assesses clusters depending on the strength of their relationship or similarity) and cluster analysis (to assess connectivity, conceptual distance, and so forth). For those who are interested in these latter, more statistical, functions, the work of Kane and Trochim (2006) and their various research partners is recommended.

Ultimately, concept mapping can be as informal or formal as is suitable for the goals of making sense of data. For many geographers and other "visual" thinkers, it can be a helpful way of seeing data and relationships in different dimensions, it works well as a group analysis exercise with multiple researchers and research assistants, and, most importantly, concept mapping leads to new insights and connections being made.

Making Meaning III: Coding

Coding social data (text, images, talk, interactions) is sometimes derided as tedious, but if you think of it as a kind of detective work, it can be intriguing, exciting, and very valuable to the research process. The purposes of coding are partly **data reduction** (to help the researcher get a handle on large amounts of data by distilling key **themes**), partly organization (to act as a "finding aid" for researchers sorting through data), and partly a substantive process of data exploration, analysis, and theory-building. Further, different researchers use coding for different reasons depending on their goals and epistemologies; sometimes coding is used in an exploratory, inductive way such as in **grounded theory** in which the purpose is to generate theories from empirical data, while other times coding is used to support a theory or hypothesis in a more deductive manner. Several approaches are discussed here, with pointers on how to organize and begin the coding aspect of a research project.

A short caveat is necessary here: it is important to recognize that coding is not always the best or only way to rigorously understand qualitative data. Even in his coding manual, Saldaña points out, "No one, including myself, can claim final authority on coding's utility or the 'best' way to analyze qualitative data... *there are times when coding the data is absolutely necessary, and times when it is most inappropriate for the study at hand*." (2012, 2. Emphasis in original). Narrative analysis and discourse analysis are two methods that are more appropriate for many projects and are covered in Chapter 13 of this volume as well as other texts (see Dixon 2010; Doel 2010).

Types of Codes and Coding

One common type of coding is **content analysis**, which is essentially a *quantitative* technique and by no means represents the full extent of coding for qualitative research. Content analysis can be done by "hand" or by computer (see discussion of CAQDAS programs below), but either way it is a system of identifying terms, phrases, or actions that appear in a text document, audio recording, or video and then counting how many times they appear and in what context. For example, a researcher might be interested in how many times the word "democracy" is used in newspaper articles from a particular country or she/he might be interested in how places are portrayed in a television program. Frequently in content analysis, sampling is used in similar ways to quantitative analysis of populations; perhaps only front-page newspaper stories are included in the analysis, or a television program is sampled for five minutes out of each hour. Similarly, researchers using content analysis typically subject their coded findings to standard statistical analysis to determine frequencies, correlations, variations, and so on. There are many good guidebooks and instructions for conducting content analysis, including some available on the Internet and broader methods texts. While content analysis has its place, the practices of coding qualitative data go well beyond merely *quantifying* them; we consider those practices next.

There are many ways to approach coding (in geography, see Crang 2005a for additional strategies), but to simplify, I discuss two main types of codes—descriptive and analytic. **Descriptive codes** reflect themes or patterns that are obvious on the surface or are stated directly by research subjects. Descriptive codes can be thought of as category labels because they often answer "who, what, where, when, and how" types of question. Examples of descriptive codes that might interest geographic researchers include demographic categories (male, female, young, elderly), site categories (home, school, work, public space), or even scale identifiers (local, regional, national, global).

One special type of descriptive code is called ***in vivo* codes**; these are descriptive codes that come directly from the statements of subjects or are common phrases found in the texts being examined (Strauss and Corbin 1990). For example, Jacquie Housel did interviews with elderly women about their daily routines and spaces in Buffalo, NY, and they repeatedly mentioned concern with crime in their neighbourhoods (see Housel 2009). In this instance, "crime" became an *in vivo* descriptive code because the term is used by and describes something important to the subjects. *In vivo* codes are a good way to get started in coding, particularly in projects that are designed to be inductive (moving from data to theory) or exploratory. Other descriptive codes are generated by the researchers' interactions with the data as they sort through the sources, but they tend to—as their name implies—be fairly superficial.

Qualitative researchers also develop **analytic codes** to code text (and other forms of data) that reflect a *theme* the researcher is interested in or that has already become important in the project. Analytic codes typically dig deeper into the processes and into the context of phrases or actions. For example, in Housel's work it became apparent that the elderly white women just mentioned were experiencing a change of status in the neighbourhood as its demographics shifted to younger Black families: while in the past the women had been seen as the matriarchs of the local community, they increasingly experienced what they perceived as being ignored, disrespected, or even threatened by "newcomers" marked by race and cultural differences. Based on this shift, "erosion of white privilege" served as an analytic code (Housel 2009, 134). This code might then be applied to the rest of the data to identify other instances of fear and loss, perceptions of particular social actors (e.g., young men, Black families moving into the area), and the women's other negatively perceived experiences in the disinvested and deteriorating public spaces of the neighbourhood.

Often, descriptive codes bring about analytic codes by revealing some important theme or pattern in the data or by allowing a connection to be made (for example, crime, fear of youth in public, shifting meanings of "race"), while other times analytic codes are in place from the beginning of the coding process because they are embedded in the research questions. For instance, if Housel was interested from the start in how elderly women navigate urban spaces, their personal mobility and perception of their neighbourhoods would be themes reflected in the analytic codes from the project's very beginning. The recursive strength of coding lies in its being open to new and unexpected connections, which can sometimes generate the most important insights.

The Purposes of Coding

Three main purposes for coding qualitative material can be readily identified: data reduction, organization and the creation of searching aids, and analysis. As the prolific French theorist Henri Lefebvre noted, "Reduction is a scientific procedure designed to deal with the complexity and chaos of brute observations" (Lefebvre 1991, 105). Qualitative research often produces masses of data in forms that are difficult to interpret or digest all at once, whether the data are in the form of interview transcripts, hours of video, or pages of observation notes. Therefore, some form of reduction, or **abstracting**, is desirable to facilitate familiarity, understanding, and analysis. Coding helps to reduce data by putting them into smaller "packages." These packages could be arranged by topic, such as "instances in which environmental degradation was mentioned," or by characteristics of the participants such as "interviews with women working part-time," or by some other feature of the research context or subjects such as "observations in public spaces." By reducing the "chaos of brute observations," data reduction helps us get a handle on what we have and allows us to start paying special attention to the contents of our data.

The second purpose of coding is to create an organizational structure and finding aid that will help us make the most of qualitative data. Similar to data reduction, the organizational process mitigates the overwhelming aspects of minutiae and allows analysis to proceed by arranging the data along lines of similarity or relationship. An important background step is constructing and maintaining a complete record of sources, dates (of participant observation, interviews, or focus groups, for example), subject contact information, and other relevant information. While this database is not part of the coding process *per se*, it is a valuable part of organizing qualitative material for coding and analysis and also allows the researcher to find specific data more easily. For example, interview transcripts might be coded not only for their content but also by their circumstances—was the interview conducted in the participant's home? were others present? did the subject seem nervous?—which can help organize information. With better computer-assisted qualitative data analysis software (CAQDAS) available now (see below), organizing and searching within electronic documents is greatly simplified. Additionally, coding itself is also an important aspect of organizing and searching because it is essentially a process of categorizing and qualifying data. "The organizing part will entail some system for categorizing the various chunks [words, phrases, paragraphs], so the researcher can quickly find, pull out, and cluster the segments relating to a particular research question, hypothesis, construct, or theme" (Miles and Huberman 1994, 57). While the development of the **coding structure** is by no means a simple process, it is one that—if done well—enables the data to be organized in such a way that patterns, commonalities, relationships, correspondences, and even disjunctures are identified and brought out for scrutiny. Good organization also means the process will be more rigorous, an important consideration for defending one's work as reliable, transparent, and trustworthy (Baxter and Eyles 1997), even if it is not "generalizable" in the statistical sense.

The final, and principal, purpose of coding is analysis. While strategies for analytical coding will be examined in greater detail in the next section, at this point it is sufficient to note that the *process* of coding is an integral part of analysis. Rather than imagining that analysis of the data is something that begins after the coding is finished, we should recognize that coding *is* analysis (and is probably never truly "finished"!). Coding is in many ways a recursive juggling act of starting with **initial codes** that come from the research questions, background literature, and categories inherent in the project and progressing through codes that are more interpretive as patterns, relationships, and differences arise.

Coding, as with memoing, also opens the opportunity for **reflexivity**, that critical self-evaluation of the research process (see Chapter 2). By recursively reviewing data and the connections between codes, researchers can also come to see elements of their own research practice, subjects' representations, and broader strategies of knowledge construction that had not previously been apparent. Even if it is difficult to be self-critical in the midst of fieldwork or data collection, the process of coding is inherently more contemplative and analytical and thus offers a moment ripe for reflection.

Getting Started with Coding

The preceding discussion of types of codes addressed two main approaches to coding, which may be seen as descriptive and analytic codes, although other terms are also used in the literature (for example, initial codes and **interpretive codes**). The key distinction is that one type of code is fairly obvious and superficial and is often what the researcher begins with, such as simple category labels. The other type of code is interpretive and analytic, and has more connections to the theoretical framework of the study; it tends to come later in the coding process after some initial patterns have been identified. When coding is done manually, researchers develop a **codebook**—a list of codes that are categorized and organized repeatedly. Although current qualitative software packages typically do not use the term *codebook*, it is a useful concept that has relevance whether the codebook is actually a tangible item in manual coding or merely an abstraction in electronic coding.

To start a codebook, it is easiest to begin with the most obvious qualities, conditions, actions, and categories seen in your data and use them as descriptive codes. These elements will emerge quite rapidly from background literature, your own proposal, other research-planning documents, and the themes that stick out for you from gathering qualitative data (for example, memorable statements in interviews, notable actions seen while doing participant observation, key words that jump out in first readings of historical documents). For example,[1] in my work on how urban children conceptualized city spaces in Buffalo, New York, one of my original interests that was heavily present in my grant proposal was how children in the 8- to 12-year-old age group define "neighbourhood" and "community" (see Box 18.1). These terms are obvious starting points for my codebook.

However, codes can also be too general and become cumbersome. Because much of my children's urban geography research was centred around issues of neighbourhood and community, I found that I needed to break each of them into more specific codes, such as codes for the particular neighbourhoods the children

BOX 18.1 29 November 2003. After-School Program Observation Notes, Children's Urban Geographies Research

Text: Field Notes from the Quilt Project	Descriptive and Category Codes	Analytic Codes and Themes	Notes
As I was setting up, Jakob*, Mariana, and Ari came over and then Izzy and Salomé (a new girl I hadn't met before). We set up at a round table in between the bench and the "café," near the pool table.	Jacob Mariana Ari Izzy Salomé Relations: Izzy and Salomé are friends		Early release day from school—kids seemed wild and bored.
The noise level was very high and I had a hard time hearing the children at my table. Next to us, three or			Tape recording would not have worked here!
four younger boys (Stefan et al.) were playing a war board game and making	Stefan	Relationship between gender and violent play?	
lots of terrible noises (at one point I asked them to be quieter).	Research setting conditions: loud		
After I explained what I had planned (to use the materials to show your house or apartment building and family), I asked the children what a "neighbourhood" is.	"Neighbourhood"		Gave very loose instructions to allow children freedom within the project's scope.
There were varying answers immediately, mostly around the idea of "a bunch of houses next to each other." Izzy said, "It's when you have one house and then another one and you all	Play		

get together to play." Mariana said, "I don't live in a neighbourhood, I'm part of the West Side Community"(!)	"Community"	Difference between "neighbourhood" and "community"	Mariana seems proud of her West Side identification.
I couldn't hear very well, so I got out my notebook and went around the group to write down answers. Ari's answer was very long and complicated with something to do with your "home friend."	Tactic: attention		Ari (age 5) seems to crave attention.
I'd like to revisit the question of what is a neighbourhood in video interviews.	Technology		Future work–video interviews??
Then I got out my digital camera and took pictures of the group (all five gave full permission for this).	Nate		The children seem to love technology and the gizmos we bring in to get a lot of attention.
Nate came up and wanted to "see" the camera, which I didn't want to let him do because he is so volatile and unpredictable.	Instructions: Nate's bullying influences my perception of hm	Children's identification with a community or neighbourhood	
Reluctantly, I let him take a picture of me with the children working on the quilt and retrieved the camera from him immediately. [Ironically, the photo Nate took is one of my favourites of this project!]		Our relationships with specific children	Review and code photos.

*All children's names have been changed

referred to, the use of both "neighbourhood" and "community" to mean "local" (such as in reports from the city newspaper), and the way that school curriculum materials defined "community." This is a frequent characteristic of coding: an initial category becomes overly broad and must be refined and partitioned into multiple codes.

Bear in mind that the opposite also occurs—some codes die a natural death through lack of use. For instance, in my project I had naively expected the children, who were for the most part in low-income families, to talk about a lack of money or not being able to afford something they wanted. However, after four years in the project, I found little evidence of children discussing their own poverty (though that absence is itself an interesting research question). So, while other materials in the project necessitated that I keep "low income" as a code, it was rarely used in analyzing direct quotes from children. As Miles and Huberman said, "some codes do not work; others decay. No field material fits them, or the way they slice up the phenomenon is not the way the phenomenon appears empirically. This issue calls for doing away with the code or changing its level" (1994, 61).

Thus, the first step is to make a list of what you think are the most important themes up front, with the understanding that some of them will be split into finer specifics while others will remain largely unused. But how do you know what is important? Anselm Strauss, one of the founders of grounded theory, had a helpful system for beginning this awesome task (best represented in Strauss and Corbin 1990). He suggested paying attention to four types of themes:

- conditions
- interactions among actors
- strategies and tactics
- consequences

"Conditions" might include geographical context (both social and physical), the circumstances of individual participants, or specific life situations that are mentioned or observed (for example, losing a job, becoming a parent, a child changing schools). By thinking along the lines of "conditions" and coding only for them, the coding process is easily started, and you may learn a lot about your data in a short time.

The same is true for limiting your scope to "interactions among actors"—if you focus on relationships, encounters, conflicts, accords, and other types of interactions, a series of powerful codes will emerge that will be helpful throughout the research. For example, in her research on adolescent girls in the southern United States, Mary Thomas (2004) found that young (14-year-old) African-American girls' interactions with peers were strongly implicated in the type and level of their sexual activity. Thus, Thomas might have coded her interview transcripts regarding peer factors by *whether*, *how*, and *where* girls engaged in sexual activity, as well as *whom* they were influenced by or interacted with.

"Strategies and tactics"[2] is a little more complicated than Strauss's first two types of themes because it requires a deeper understanding of the things (events, actions, statements) you observe and how they relate to broader phenomena, and it suggests a certain level of purposeful intent among the research subjects that may demand additional inquiry on your part. For example, feminist geographers are often interested in women's survival or "livelihood" strategies in different areas of the world (see, for example, Oberhauser and Yeboah 2011; Jones and Murphy 2011). Noting that women in certain economic contexts tend to use particular types of financial survival tactics (for example, growing food products for sale in a local market) can begin to illuminate broader economic, social, and political processes that shape women's options and actions, which is a valuable insight for geographic research. Other types of strategies or tactics might involve education and career decisions, political activism, housing choices, family negotiations, or even subversion.

Coding for strategies and tactics can be straightforward (and descriptive) in instances when respondents say something like "I moved in with my mother so that she could care for my baby while I finished job training" or "I got involved with a local group of residents to raise awareness of environmental contamination in our neighbourhood because I was concerned about property values." Note the words "so" and "because" in these statements, which are good tip-offs that a strategy or tactic is embedded in the text.

Other times, coding for strategies and tactics may be more subtle—and more analytical—as when respondents do not explicitly state their reasons for certain actions but a connection emerges through observation, review of interview text, or other data. For instance, many geographers (for example, Blumen 2002; Lee, Kim, and Wainwright 2010; Lees 2012; Secor 2004) have paid attention to ways that people engage in *resistance* against diverse forms of oppression, which may be seen as strategies for empowerment, rights, or merely survival. For example, Orna Blumen (2002, 133) took "dissatisfaction articulated in subtle terms" by ultra-orthodox Jewish women as small but significant indicators of the women's resistance to their families' economic circumstances and, more broadly, to the status and roles of women in that community. For the women in Blumen's study, then, referring to fatigue, hoping their husbands would soon find paying work, and "minor, personal, nonconformist remarks suggestive of ambivalence" (2002, 140) could all be coded as tactics of resistance, in part because Blumen—through careful qualitative work—had sufficiently analyzed the broader context of the women's lives and goals to recognize them as such.

Similar to the above, "consequences" is a slightly more complicated code. On the surface, there are descriptive indicators for consequences, including terms such as "then," "because," "as a result of," and "due to" that may be used in subjects' statements and can be good clues to consequences and as a first run could certainly be used in this way. Again, however, there are also more analytically sophisticated ways of discovering and coding consequences that are dependent on the unique empirical settings and events of each study. Some consequences will be matters of time passing and actions taking place that result in a particular outcome—the passage of a law, a change in rules or practices, and so on. However, other consequences are more subtle and personal, or they are not the result of changes over time and therefore may be trickier to identify and code as such. For example, when Anna Secor (2004) heard from young Kurdish women living in Istanbul that they felt uncomfortable in some areas of the city, she might have coded her focus group transcripts for the consequences of feeling out of place due to the women's identity as an oppressed minority in Turkey. Coding for "consequences" of this kind requires sensitivity to both the subjects and their community context but is potentially a rich source of analysis and insight if done with care.

As an example of what a sample of coded material looks like with both descriptive and analytic codes, Box 18.1 demonstrates a small selection of field notes from my project on children's urban geographies along with codes, themes, and notes. Even this fairly short piece of text reveals several relationships (friends, cousins, bullies), tactics (ways of getting attention), and conditions (disorganization, noise level) that stimulated further examination in other project analysis. Additionally, several analytic themes or questions are seen emerging here: the possible relationship between gender and violent play, some children's pride in perceived community membership (despite living in a disinvested physical environment), and the importance of play in defining what a neighbourhood is among the children. Subsequent to the quilt project represented here and in combination with other Children's Urban Geographies Project data, I generated a theory of how children define and ascribe meaning to the idea of "neighbourhood" (Cope 2008); theory-building, after all, is an important goal of most qualitative research.

Developing the Coding Structure

Using the four types of themes reviewed here will take you a long way towards constructing a codebook, and you may find other types of themes that are helpful to you, such as "meanings," "processes," or "definitions." Using the combination of descriptive and analytic codes, you may well have more than 100 codes at some point, which is unwieldy at best and counterproductive at worst. Lists of codes that have not been categorized, grouped, and connected will be hard to remember, have too much overlap and/or leave uneven gaps, and will not enable productive analysis. Therefore, the next step is to develop a coding structure whereby codes themselves are grouped together according to their similarities, substantive relationships, and conceptual links. This process requires some amount of work but is well worth the effort, both for ease of coding your material and for discerning significant results from your findings.

Developing the coding structure can proceed in various ways, and there are many resources available that demonstrate different approaches (see Denzin and Lincoln 2011; Miles and Huberman 1994; Silverman 2001), but the main purpose is to organize the codes—and therefore the data and the analysis process. Some codes will automatically cluster; for example, codes relating to the *setting* of interviews (such as home, office, public space, clinic), *characteristics* of subjects (race, age, gender), or other *categories* (for example, occupations, leisure activities, life events). Other codes seem to fit together because of their *common issues*; for example, you might have a group of codes related to people's goals or intentions or a group of codes related to people's experiences of oppression. Finally, codes based on the *substantive content* of text or actions—and most likely related to the analytic themes you are developing—will create another cluster of codes, for example: perceptions, meanings, places, identities, memories, difference, representations, and associations.

Once the codebook is relatively comfortable (I hesitate to say "complete") and the coding structure is devised, you will want to go through much of your data again to capture connections that may have been missed the first (or second or third) time around. Remember that coding is an iterative process that feeds back on itself—only you can decide when it is time to move on. As Miles and Huberman (1994) point out, it is sometimes simpler when time or money pressures put some finality on projects that otherwise could always benefit from "one more case study" or endless additional tweaking of the coding structure.

Coding with Others

Depending on the size and resources of the research project, there may be a case for using multiple coders for the data, which can add considerable complexity to the process. There is an inherent tension in using multiple coders on a project: is the goal to make everyone code as consistently as possible, or is the goal to allow each coder to interpret data in her or his way within the bounds of the coding structure in order to capture many diverse meanings? The answer will depend on the project and the epistemological leanings of the lead researcher(s), but in fact both of these goals are important. In the first instance, reliability of the data is undoubtedly enhanced when several coders independently code a piece of data the same way—a common interpretation of data means that there is agreement on its meaning. For the sake of time and data reduction, having multiple coders can certainly be helpful, but only if they are truly consistent in their coding, which is rare but could be accomplished by achieving conformity on the meanings of codes and providing thorough definitions for each code. On the other hand, text and video—as social data sources—are inherently subject to multiple interpretations and understandings, all of which may be correct or "true." While there may be some interpretations that are farfetched or extreme, in general we as social researchers will be interested in capturing diverse understandings, and having multiple coders can be a great benefit for the project to make deeper and broader connections from the data.

Computer-Aided Qualitative Data Analysis Software (CAQDAS) and Qualitative GIS

Anyone who has coded research material "by hand" can understand the attraction of computer assistance in this endeavour—the idea that one's codes could be organized and employed using software holds many potential benefits in terms of time-savings, consistency, and data security, particularly when compared to stacks of notecards. However, there are also some cautions that should be kept in mind before embarking on the investment of time and resources to use CAQDAS. This brief overview is not intended as instructional or comprehensive, but rather, identifies some key considerations and points the reader towards further resources.

There are numerous CAQDAS packages on the market today, ranging in price, sophistication, functions, and utility. However, the basic premise of all of them is the same: to bring coding and analysis into a computerized operation. In fact, some of the most basic functions of coding can be completed in word processors or using spreadsheets, and for smaller or fairly basic analyses, these might be preferable. For example, a first-run content analysis is easily accomplished using the "find" function in word processors, and pieces of text can be highlighted and commented upon using review functions such as "track changes" in MS-Word. Lists of codes and memos, as well as notes about when to use which codes, are easily stored and managed in spreadsheet programs or "table" options in word processors. For more complex projects with more codes, more data, and/or multiple researchers investing in learning how to use a CAQDAS (and, sometimes, investing in purchasing a licence, which many universities already have available) can be well worth it.

I highly recommend reading the review and self-critique Bettina van Hoven (2010) wrote about her experiences learning and using CAQDAS in her dissertation project. She reveals several "advantages" of using CAQDAS, including managing large quantities of data, convenient coding and retrieving, and quick identification of deviant cases (advantages) as well as "concerns" that include obsessions with volume of data, mechanistic data analysis and a "taken-for-granted" mode of data handling, alienation from one's data, and an over-emphasis on grounded theory (van Hoven 2010, 462). No matter how much packages change, these advantages and concerns are likely to be timeless.

Some examples of this type of software being used by geographers include NVivo (a cousin of QSR N4), Atlas.ti, MaxQDA, and HyperRESEARCH. All of these have demo versions and online manuals in PDF on their respective websites, and most have lower-cost student versions; therefore one can easily "test-drive" a package before committing fully to it. Because packages are constantly being updated and changing in their functionality, an assessment of the relative strengths and merits of these would be quickly outdated, and therefore they will not be reviewed here *per se*; I recommend searching the web and online forums for reviews of the latest versions of CAQDAS to see what might suit a given project, the capacities of the researchers, and the available time and budget. However, there are some features that are worth mentioning here that are relevant to consider, regardless of what new offerings emerge.

First, CAQDAS has three primary functions: *text retrieval* (especially helpful for the "constant comparison" approach to analysis), *coding*, and *theory-building*. It is, of course, essential to realize that the program does not *do* the coding for you, except in the most rudimentary way based on a set of instructions you provide; the researcher still must engage in thoughtful and thorough building of codes and applying them to text segments, photo elements, or video footage clips. However, the CAQDAS is an excellent organizational tool for storage, retrieval, and interpretation of data, and all packages have at least some "memoing" capacity. Targeted searches or browsing by coded material can reveal connections that would have been difficult to see with paper index cards, and the **theory-building functions** (such as hierarchies or concept maps that show relations between data, or even between codes) are useful for analysis and the generation of original interpretations and conceptual understanding.

The second feature to consider is the degree to which a CAQDAS can be geographically linked (if this is something that is relevant for your research). These capabilities are pushing forward the practice of **qualitative GIS** (Cope and Elwood, 2009) by increasingly developing the integrated analysis of qualitative and geographically referenced data (Gilbert, Jackson, and di Gregorio, 2013; Jung, 2014; Verd Pericas and Porcel, 2012). For example, Fielding and Cisneros-Puebla (2009) review the capacities of both Atlas.ti and MaxQDA for integrating geographical maps (that is, not just concept maps!) into analysis. Atlas.ti, for instance, allows the importing of KML files, and Google maps can be opened from within the program; MaxQDA allows hyperlinks to Google Earth images. In what Fielding and Cisneros-Puebla identify as a "technological convergence" these developments should be promising for many geography scholars and students. Further, with the rapid growth of online mapping (both vernacular and professional), volunteered geographic information, crowdsourcing, and spatially referenced social media, the possibilities for qualitative research of, with, and across "big data" are growing as fast as the Internet itself (see, for example, Crampton, Graham, Poorthuis, Shelton,

Stephens, Wilson, and Zook 2013). These raise the important question of how the massive saturation of spatially referenced digital data will affect qualitative research in geography and beyond.

New work is emerging that pushes the boundaries of coding and leads to creative analyses and representations of qualitative geographic data. With the rise of participatory research in geography (Kesby, Kindon, and Pain 2005; see also Chapter 17 in this volume), qualitative GIS (Cope and Elwood 2009; Kwan and Knigge 2006, Preston and Wilson, 2014), participatory mapping, mixed methods, and other integrative practices, we need to stay attuned to how coding can keep up with new research processes and technologies. One example of this is Jin-Kyu Jung's (2009) experiments with using codes as a bridge between qualitative analysis using CAQDAS and spatial data analysis using GIS. In his work, the code literally serves as a software-level link between databases, allowing analysis programs to "speak" to each other in a platform he calls **CAQ-GIS** (**computer-aided qualitative geographical information systems**). At a conceptual level in Jung's work, the code also serves as an analytical connection between social contextual data and spatially referenced data, allowing researchers to develop new understandings of social–spatial relations. Further, these innovations have been increasingly taken up outside of geography, for example in urban sociology and in public health research, where the potential for integrating spatially referenced numerical data with qualitative data is especially relevant. Along these lines, Fielding and Cisneros-Puebla (2009, 349) argue that "there is an emergent convergence of methodologies and analytical purposes between qualitative geography and qualitative social science" that represents an exciting new realm of inquiry. With the many possibilities for mixed-methods and qualitative research facilitated by better and faster digital technologies, the necessity for researchers to be well-trained, rigorous, reflexive, and ethical has never been greater.

Being in the World, Coding the World

By way of conclusion, let me point out that interpreting and analyzing the social world is not a mysterious process that must be learned from scratch but rather is one that we are all already actively practising in our everyday lives. The recognition that we are all constantly "coding" and making sense of the world around us may be a helpful realization for getting started in a research project and can also assist us in critiquing our own practices of data reduction, organization, and analysis. As Silverman (1991, 293) points out, there are many ways of "seeing" and interpreting the world, and—as social beings—we never really shut those lenses off, so why not embrace diverse perspectives and turn our gaze to the process of interpretation?

> How we code or transcribe our data is a crucial matter for qualitative researchers. Often, however, such researchers simply replicate the positivist model routinely used in quantitative research. According to this model, coders of data are usually trained in procedures with the aim of ensuring a uniform approach. . . . However, ethnomethodology reminds us that "coding" is not the [sole] preserve of research scientists. In some sense, researchers, like all of us, "code" what they hear and see in the world around them [all the time]. . . . The ethnomethodological response is to make this everyday "coding" (or "interpretive practice") the object of inquiry. [Silverman 1991, 293]

Being in the world requires us to categorize, sort, prioritize, and interpret social data in all of our interactions. Analyzing qualitative data is merely a formalization of this process in order to apply it to research and to provide some structure as a way of conveying our interpretations to others.

Key Terms

abstracting
analytic code
CAQDAS (computer-aided qualitative data analysis software)
CAQ-GIS (computer-aided qualitative geographical information systems)
codebook
coding
coding structure
concept mapping
content analysis
data reduction
descriptive code
ethnography
grounded theory
initial codes
interpretive codes
in vivo code
memo
qualitative GIS
reflexivity
theme
theory-building functions

Review Questions

1. What is the difference between descriptive and analytic codes, how do they relate to one another, and what are their respective uses in coding qualitative data?
2. Why does the author state that coding *is* analysis?
3. What are some potential benefits and potential problems with using a CAQDAS in a project?

4. In what ways do we "code" events, processes, and other phenomena in everyday life? How might thinking about these ways help us to become better qualitative researchers?

Review Exercises

1. *Memoing:* Have the class write a short, first-person account of some type of experience ("my journey from home to the classroom today" works well!) and ask each person to go back over their own account making memos about why certain events, conditions, interactions, or thoughts were important.
2. *Concept mapping:* Collect all the first-person accounts from Exercise 1 and print them single-sided, then lay them out on a large table or tack them to a wall. Have students review each account and identify key statements, copying each key statement onto a note card (remember to mark which original document it came from). Then use the "sort and pile" method to identify common statements, build themes, and identify connections between concepts. The notecards can then be arranged in clouds, networks, hierarchies, or other visual patterns to further analyze relationships.
3. *Practise coding:* Take a newspaper story or a short video clip (5–10 minutes) and work singly or in small groups to identify some basic *in vivo* (present in the original text/video) and descriptive codes that would suit an initial coding of that material. Each person writes her/his codes on large note cards or sticky notes. Collect all the codes from class members, tack them to the wall, and collectively construct a coding structure by eliminating duplicates, grouping similar codes together, and identifying outliers or gaps.

Useful Resources

For an excellent step-by-step guide to coding, see *The Coding Manual for Qualitative Researchers* by Johnny Saldaña (2012). Additionally, there are several examples of coding and "making sense" of data by geographers, including collections by Kitchin and Tate (2000); Clifford, French, and Valentine (2010); Flowerdew and Martin (2005); Limb and Dwyer (2001); and Moss (2002b).

Notes

1. While it is always difficult to convey examples of coding without recounting the entire scope of the research, it is hoped that these examples from a real research project are sufficiently illustrative to demonstrate different coding approaches.

2. Despite the similar pairing of these two words, I am not referring here to Michel de Certeau's (1984) notion of "strategy" (a technique of spatial organization employed by "the powerful") and "tactic" (an everyday means of "making do," typically used by those with few options), although there are certainly potential connections. Rather, I am using the terms in their most literal sense as they are employed in Strauss and Corbin (1990) to convey ideas about how people conceptualize what they want and what they do to try to arrive at those goals.

Writing Qualitative Geographies, Constructing Meaningful Geographical Knowledges

Juliana Mansvelt and Lawrence D. Berg

Chapter Overview

In this chapter, we examine the process of "writing-up the results" of qualitative research in human geography. Our aim, however, is to contest simplistic understandings of the relationship between research, writing, and the production of knowledge that arises from describing the process in this manner. The very phrase *writing-up* implies that we are somehow able to unproblematically reproduce the simple truth(s) of our research in our writing. In this chapter we challenge a view of writing that suggests it is a mirror that innocently reflects the reality of research "findings." Instead, we draw upon post-structuralist approaches to argue that writing is not merely a mechanical process that reflects the "reality" of qualitative research findings but rather shapes in part how and what we know about our research. Writing is thus not so much a process of writing-*up* as one of writing-*in*, a perspective that has significant implications for how research is conceptualized.[1]

Styles of Presentation

Part of our argument in this chapter revolves around the idea that a number of powerful **dichotomies**—such as subject/object, researcher/researched, data/conclusions, and research/writing—currently exist and that these dichotomies structure our understanding of research. These dichotomies have developed historically as part of a **neo-positivist** framework that, we suggest, limits our understanding of the writing process. We thus feel it is important to present readers with a very brief history of the development of positivist thought in human geography before moving on to discuss an alternative approach to writing human geography.

Positivist and Neo-Positivist Approaches: Universal Objectivity

The philosophical approach to scientific knowledge known as positivism was founded by Auguste Comte (1798–1857), a French philosopher and sociologist (see Gregory 1978; Kolakowski 1972). Comte argued that scientific knowledge of the world arises from observation only. The notion of a singular truth, a central truth that provides explanations for how the world is structured and understood was associated with this. Consequently, positivists of Comte's tradition had significant difficulties with issues and questions arising from the relationship between truth and phenomena such as religion, ethics, and morals.

The logical positivists, whose work developed as a critique of strict positivism during the 1920s, differed from the Comtean positivists in their conception of scientific knowledge. While Comte allowed only statements arising from empirically verifiable knowledge (that is, information available to the senses) as the basis of factual knowledge (or truth), the logical positivists accepted the validity of more than just empirical observation as the basis for truth. Logical positivists acknowledged the existence of analytical and synthetic statements. With analytical statements, truth is contained in their internal logic (for example, if A = B and B = C, then A = C), and synthetic statements, the truth of which is proven through recourse to empirical observation (for example, if we observe high levels of homelessness in areas with low levels of public housing, our empirical observations could lead us to conclude that the incidence of homelessness is a consequence of particular state housing policies). As you may appreciate from the previous example, numerical correlation is not the same as causation. Assessing the validity of analytical statements derived from empirical observations is thus problematic, and particularly so in social science.

Karl Popper (1959) recognized this and developed an alternative form of neo-positivist thought he termed "critical rationalism." Unlike the logical positivists, who focused upon verification as the basis of knowledge, Popper argued that **falsification** should be the basis for making decisions about the validity of factual statements. His argument was based on the notion that we can never know for sure whether a particular hypothesis is true or not (such as whether a lack of state housing does result in more homelessness), but we do have the ability to ascertain whether such statements are false (for example, by finding examples of a low incidence of homelessness in areas with minimal public housing). Under critical rationalism, then, we never prove something, but instead we can either disprove something (or prove it to be false through the process of falsification) or accept it to be contingently "true" (until it has been proved false). According to Popper, scientific knowledge is much more contingent than that conceptualized by the logical positivists, since present "truths" are always open to future falsification.

Although few geographers have ever fully taken up the ideas of any single school of positivist thought, positivism of one sort or another has played a foundational role in the way that many geographers have come to understand the world. This became most explicit during the so-called **quantitative revolution** of the 1960s (Gregory 1978; Guelke 1978), which saw geography develop as what we might broadly term a neo-positivist "science." Geographers maintained a strict distinction between facts and values (Gregory 1978), giving emphasis to observational statements over theoretical ones (Berg 1994) and often universalizing their findings across all contexts (Barnes 1989). They also made a strict distinction between objective and subjective knowledge.

Objective knowledge is seen as "scientific," rigorous, and detached and consequently valid. It is constituted in opposition to **subjective** knowledge, which is personal, value-based, non-scientific, and non-academic (and therefore unacceptable as a basis for establishing "the" truth). Objective knowledge is founded on interrelated and highly gendered notions of rationality, disembodied reason, and universality (Berg 1994; 1997; Bondi 1997). Trevor Barnes and Derek Gregory (1997, 15) suggest that "scientific geographers" thus imagined themselves:

> as a person—significantly, almost always a man—who had been elevated above the rest of the population, and who occupied a position from which he could survey the world with a detachment and clarity that was denied to those closer to the ground (whose vision was supposed to be necessarily limited by their involvement in the mundane tasks of ordinary life).

Knowledge from the vantage point of the objective researcher looks the same from any perspective—it is monolithic, universal, and totalizing. This concept of un-located and disembodied rational knowledge draws on powerful metaphors of mobility (the researcher can move to any and all perspectives) and transcendence (the researcher is not part of the social relations she or he is examining but instead can rise above them to see everything) for its rhetorical power to convince readers of its claims to the truth (see Barnes and Gregory 1997; Haraway 1991).

In adopting a broadly neo-positivist model for their work, geographers also developed a specific approach to writing their research. They developed commonly used approaches or forms of writing (or what we call **tropes**) in their academic studies that attempted to erase the authorial self from their written work. Similarly, they tried to create, through their writing, distance between themselves as researcher/author and their research objects. These tropes are most evident in the practice of writing in the third person—a practice that is still prevalent in much academic writing today. Many geography undergraduate students, for example, are still required to use the formal **third-person narrative** form in their essay assignments. The notion of universal knowledge underpins writing in the third person, and it is also intended to erase the imprint of the author on the text and to ensure the distancing of the author from the research. In other words, third-person narratives are needed to maintain the appearance of impartiality and objectivity, two cornerstones of the neo-positivist model. The third-person narrative constructs an **objective modality** (Fairclough 1992). This effectively removes the author from his or her writing while at the same time implying the author's full agreement with the statements being made. In so doing, it does the work of transforming interpretative statements into factual statements (see Box 19.1).

Writing in the objective mode is often accompanied by **nominalization**, a process that further removes the writer from the text (see Box 19.1). Nominalization involves the transformation of adjectives and verbs into nouns. It occurs, for example, when, in the process of researching, researchers and research subjects are rhetorically transformed into things such as "the research" or when actions and processes are given particular kinds of argumentative power as nouns, such as can be found in statements such as "the analysis suggests" (Fowler 1991). Nominalization deletes a great deal of helpful and important information from sentences. For example, it removes information about the participants—normally the agent or researcher (the person "doing" something), and the affected participant (or the person having something "done to" them). It also changes the **modality**—that is, the implicit indication of the writer's degree of agreement with the statements being made in a text. Nominalization creates **mystification** because it permits concealment, hiding the participants, details of time and place, and the stance of the participants involved. It also results in **reification**, whereby complex, uncertain, and often contradictory processes, such as are involved in human geographic research, assume the much more certain status of "things." Thus, the complex processes of research are smoothed over into a one-dimensional and simple thing: "the research" (see, for example, Fowler 1991, 80). Ironically, while positivist epistemology is founded on the idea that any statement can be contested through recourse to empirical evidence and logic, the language used to communicate research findings within this framework implies an unquestionable accuracy that is a very poor approximation of the messiness of social life.

Interestingly, writers drawing on neo-positivist approaches have shown ambivalence with regard to their own writing practices. On the one hand, they have explicitly acted as if language has no impact on meaning, yet ironically they have implicitly acknowledged—through their insistence on writing in the third person—the significant role that language plays in constructing knowledge and meaning.

Post-Positivist Approaches: Situated Knowledges

Although the positivist-oriented science model dominated during the quantitative theoretical period of the 1960s and 1970s, its **hegemony** (or conceptual dominance) began to be contested in geography during the late 1980s and early 1990s by post-positivist approaches such as humanism, Marxism, political economy, and post-structuralism (for example, Barnes 1993; Berg 1993; Dixon and Jones 1996; England 1994; Massey 1993). Perhaps the strongest critiques from within geography arose in response to the works of feminist **post-structuralist**

Removing the Writer—Third Person and Nominalization

Writing in third person and a process called nominalization help to create a more formal written text. This has the effect of distancing the author from the narrative being constructed and creating a sense of neutrality and objectivity.

As you read the sentences below, consider how the use of third person, by removing terms that describe oneself (such as "I") and substituting descriptors of others (for example, "the investigator") creates a sense of distance from the research and the research subjects, and makes the statement appear more factual.

> "The investigator examined five types of place-based knowledge."
>
> "The researcher demonstrated that understandings of home are complex."
>
> "One tries not to influence the responses gained from interview questions."

Nominalization is the process of turning verbs into nouns. It involves taking actions and events and changing them into objects, concepts, or things—making sentences seem more formal and abstract. Like writing in third person, nominalization contributes to a sense of detachment from the text (and the research), removing both personal reflections and the context of the subject being discussed.

For example, the verb "developed" is nominalized to the noun "development":

> "I *developed* a framework for analyzing interviews" (**first-person narrative**) might become "the *development* of a framework for analyzing interviews."

Two more examples of text follow. In the informal text, few verbs have been nominalized. The second paragraph is an example of more formal research writing in which verbs and verb phrases have been nominalized. As you read these two texts, think how the agency of the researcher appears to be removed in the second, concealing the researcher's emotions and experiences. Note, for example, how the first sentence in the formal text suggests that there is much to be learned from the research (now assumed to have the status of a thing) rather than from the researcher-led processes that constituted it, as implied by the more informal excerpt.

Informal text:

Designing and conducting the research was a learning process for me. Because I did not explain the purpose of my study clearly, the people I approached for interviews often refused to participate. My structuring of the interview questions was also too complex, and my participants struggled to answer them. Consequently I had difficulty in analyzing the information I had obtained from my interviews and my insights were limited. I learnt a lot through the process but concluded that my research project was ill-conceived!

Formal text:

Much was learned from the design and conduct of the research. A higher-than-expected level of refusal to participate was a consequence of a lack of clarity in the explanation of the research. The interview questions were too complex, and negatively influenced participant responses. The research analysis produced limited insights. The research was an ill-conceived learning exercise.

The formality achieved through the use of third person in the second excerpt has the effect of producing the author as an all-seeing and all-knowing (seemingly objective) surveyor of the research rather than an active participant, complicit in the production of both text and research. Such distancing is an illusion, negating the ways in which power and subjectivity are constituted through the research process.

writers in the wider social sciences (Anzaldúa 1987; Frankenberg and Mani 1993; Mohanty 1991) who were keen to confront the universalism, mastery, and disembodiment inherent in positivist notions of objectivity, criticizing masculinist and Eurocentric concepts of universal knowledge. This has led to a rich vein of work in geography that examines the relationship between geographers as situated knowledge workers and their relationship to the objects of knowledge that they "produce" in their research (e.g., Berg 2004; 2012; Clough and Blumberg 2012; de Leeuw, Cameron, and Greenwood, 2012; Pain 2014; Radice 2013). All of these authors would agree that simplistic notions of "objectivity" play a significant role in marginalizing those who do not fit into dominant conceptions of social life.

Donna Haraway's (1991) evocative metaphor of **situated knowledges** provides perhaps the most useful approach to contest universalist forms of knowledge. She argues that within dominant ideologies of scientific knowledge, objectivity must be seen as a "God Trick" of seeing everything from nowhere. She proposes a different concept of "objectivity," one that attempts to situate knowledge by making knowers accountable to their *positions* (see also Chapter 2). All knowledge is the product of specific embodied knowers, located in particular places and spaces: "there is no independent position from which one can freely and fully observe the world in all its complex particulars" (Barnes and Gregory 1997, 20). Research that draws upon situated knowledges is thus based on a notion of objectivity much different from that posed by positivists, and this conception of objectivity also requires a different form of writing practice.

Work by Isabel Dyck (1997) provides an excellent example of the importance of "position" and the kind of "truths" situated knowledge might produce (also see England 1994). Dyck reflects on the importance of her own position as a white middle-class Canadian academic and the impact it had on two different research projects she undertook to examine the time–space strategies adopted by Indo-Canadian immigrant women in Vancouver. She found that immigrant women were more willing to speak with her about certain aspects of their lives than about others because she was not seen as a threat to their own social networks and relationships within the Indo-Canadian community in Vancouver. At the same time, as an outsider she was occasionally excluded from aspects of Indo-Canadian women's lives that were defined as culturally sensitive. Her research thus points to the specificity of position and the importance of recognizing the politics of position in research processes.

Post-structuralists and feminists also contest approaches to inquiry that conceptualize writing and language as simple reflections of "reality." They argue that "language lies at the heart of all knowledge" (Dear 1988, 266). It should be made clear, however, that such arguments do not assume that language and ideas are the same as "real" phenomena, objects, and material things. Instead, arguments about the centrality of language express the fact that all processes, objects, and things are understood by humans through the medium of language. Thus, while we might experience the very material process of hitting our "funny bone" on a table in ways that do not necessarily entail language (for example, as a very visceral sensation we know as "pain"), we come to understand the process and the objects involved through language (with categories such as table, funny bone, pain, and so on). Accordingly, language must be seen as not merely reflective but instead as *constitutive* of social life (for example, Barnes and Duncan 1992; Bondi 1997; Dear 1988).

The post-structuralist critiques of both the **mimetic** concept of language and of disembodied concepts of universal objectivity have significant implications for writing practices. If language is constitutive of knowledge and meaning, then it would seem to matter *how* we write our knowledges of the world. Likewise, if we are to *locate* our knowledge, then we must locate ourselves as researchers and writers within our own writing. Accordingly, post-structuralist writers reject the ostensibly "objective" modality of writing their work in the third person. Instead, they opt for locating their knowledge-defining objectivity as something to be found not through distance, impartiality, and universality but through contextuality, partiality, and **positionality**. However, as Gillian Rose (1997) has argued, given the difficulty in completely understanding the "self," it may be virtually impossible for authors to fully situate themselves in their research. Notwithstanding such difficulties, it is possible for authors to go some way towards locating themselves within their work. Certainly, the first step is to reject the third-person narrative, replacing it with a first-person narration of our essays.[2] At the same time, it is not enough to merely adopt the first-person narrative form. Instead, it is important to both reflect upon and analyze how one's position in relation to the processes, people, and phenomena we are researching actually affects both those phenomena and our understanding of them. Again, Isabel Dyck's (1997) study of "two research projects" just discussed provides an excellent example of how we might undertake such analyses. Thus, rather than writing ourselves *out* of our research, we write ourselves back *in*. This is, perhaps, one of the most important distinctions to be made between what we will refer to as the writing-*up* (distanced, universal, and impartial) and the **writing-*in*** (located, partial, and situated knowledge) models. Another significant difference arises from the ways that post-positivists conceptualize the relationship between observation and theory.

Balancing Description and Interpretation—Observation and Theory

The Role of "Theory" and the Constitution of "Truth"

We argue in this section that there currently exist a number of powerful dichotomies—observation/theory, subject/object, researcher/researched, data/conclusions, and research/writing—that structure our understanding of research. It is important to remember that these dichotomies are not recent developments in

Western thought. Instead, they arose within the long history of dualistic thinking in Western philosophy (Berg 1994; Bordo 1986; Derrida 1981; Foucault 1977b; Jay 1981; Le Doeff 1987; Lloyd 1984; Nietzsche 1969). These dichotomies became racialized and gendered through a long historical process of developing a singular Eurocentric and masculine concept of rational thought. For example, through a process that Susan Bordo (1986) terms the "Cartesian masculinisation of thought," Descartes's mind–body distinction came to define appropriate forms of knowledge. The mind was conceptualized as rational, and it came to be a property of European men. The body was seen as irrational (have you ever heard the phrase "mind over matter"?) and was associated with everything that was not European or masculine: women, racial minorities, and sexual dissidents, for example. In other words, the mind was unmarked, but the body was a mark of difference. Further, Descartes's dualistic philosophy of knowledge formed the basis for the dominant present-day conception of objectivity as impartial, distanced, and disembodied knowledge. Accordingly, Cartesian dualistic thinking forms the foundation of positivist thought (Karl Popper, of whom we spoke earlier, for example, was a well-known adherent of the mind–body dualism).

As we have already discussed, positivist-inspired geographers make a rigid distinction between objective and subjective knowledge and between theory and observation. As with the observation/theory binary, the so-called "objective" is valued at the expense of the subjective. Such hierarchically valued dichotomies form parts of a whole series of other binary concepts—including (but not limited to) mind/body, masculine/feminine, rationality/irrationality, and research/writing—that are interlinked through complex processes of signification (Derrida 1981; Jay 1981; Le Doeff 1987; Lloyd 1984). In the observation/theory binary, observation is equated with objectivity, mind, masculinity, rationality, and research. Theory is constituted as lacking, and it is associated with all those other negatively valued concepts: the subjective, the body, the feminine, and the irrational.

Conceptualizing one side of the binary as a *lack* of the other leads to a devaluation of the subordinate term. Thus, in the case of the positivists who conceive of theory as a lack of empirical observation, the ways in which theory constitutes our understanding of empirical "reality" (in addition to explaining it) are underestimated. Indeed, positivist constructions of factual knowledge as phenomena that are available to the senses (empirically observable) tend to efface the very theoretical nature of positivist thinking itself. As we have already suggested, positivism has both a history and a geography associated with Europe. It is an epistemology—a theory of knowledge—that has developed relatively recently and has come to dominate contemporary intellectual life in the West. Nonetheless, it is not the only theory of knowledge; rather, it is one of many competing theories of knowledge. However, because it is dominant, or hegemonic, it rarely has to account for its own epistemological frameworks. With this in mind, we argue for "recognition that we cannot insert ourselves into the world free of theory, and neither can such theory ever be unaffected by our experiences in the world" (Berg 1994, 256). Observations are thus *always already* theoretical, just as theory is always touched by our empirical experience. Recognition of this relationship has important consequences for the way we write-*in* our work (and *work*-in our writing).

Writing and Researching as Mutually Constitutive Practices

Metaphors that allude to research as "exploration" or "discovery" are hard to avoid, so taken for granted are their meanings (how often have you heard lecturers speak of their research in terms of "exploring," "examining," "discovering," and "uncovering"?). This is particularly the case with the so-called "writing-up" of the research. The term *writing-up* powerfully articulates the written aspect of the research process in a way that engenders it as somehow less significant and/or less problematic than other aspects of the research process. Writing-up is usually seen as a phase that occurs at the end of a research program; indeed, many textbooks about conducting research (this one included) include the section on "writing-up" at the end of the book (for example, Bryman, 2012; Flowerdew and Martin 2005; Kitchin and Tate 2000). Writing is also often seen as a neutral activity, although ironically, scientific and positivist modes of writing and thinking are subject to rhetoric, creativity, and intuition (Bailey, White, and Pain 1999a).

The "writing-up phase" may be discussed in such a way that it appears to be merely a matter of presenting the results and conclusions in an appropriate format at the *end* of a research program. We argue here that in writing research, the researcher is not so much presenting her or his findings as *re-presenting* the research through a particular medium. Rather than reflecting the outcome of a particular research endeavour, we believe the act of writing is a means by which the research is constituted—or given form—and that this process occurs throughout the research process. For example, we can see that any attempt to write research involves a process of selecting categories and language to describe complex phenomena and relationships. But we are getting at much more than that here. Research and writing are iterative processes, and writing helps to shape the research as much as it reflects it. In addition, the way we conceptualize the author (as distant and impartial or as involved and partial, for example) has significant implications for the ways that the very processes of research itself can be understood. Perhaps just as important, writing involves very clear decisions to include some narratives and to exclude others. It thus brings some ideas into existence while (implicitly) denying the existence of others (Ely 2007). For us, then, writing is not so much a matter of writing-*up* as of writing-*in*, a perspective that has considerable implications for how qualitative research is conceptualized and undertaken.

Writing is not devoid of the political, personal, and moral issues that are a feature of undertaking research, nor is writing devoid of our embodied emotions as we sense and feel the narratives we construct. As a creative process, writing can be enjoyable for authors and readers (Bradford 2003), opening up new possibilities for representation, narrative, and engagement with those who may participate in and reflect on the research. Further, the separation between "fieldwork" and writing is artificial (Denzin 2008). Whatever the qualitative research technique utilized, some means of recording the researcher's interpretations, impressions, and analysis must be used, and although such accounts may be recorded as digital audio or video files, the words with which they are constructed are an integral part of the research, not simply a result, recollection, or recording of it. The research cannot be separated from the labels, terms, or categories used to describe it and interpret it, because it is through them that the research is made meaningful.

For example, after I (Juliana) had conducted several qualitative interviews with local authority economic development officers for my PhD research, I realized that using the word "traditional" as a label for a certain form of local economic initiative was problematic. This was because definitions of "traditional economic initiatives" were contested by officers and because the term appeared to position local authorities who undertook this type of initiative as "old-fashioned" or "not progressive." I learned a valuable lesson about the power embodied in words I had simply drawn from the literature on local economic development and about the kind of assumptions embedded in my own research agenda. This understanding enabled me to construct my questions (and consequently my entire research project) in a different way.

I (Lawrence) had a similar experience of reorienting research agendas in a participatory action research project I undertook with a number of colleagues (Berg et al. 2007). This project involved the fusion of Indigenous methodologies, participatory action research, and white studies (Evans et al. 2009) to understand the exclusion of urban Aboriginal people from the so-called "universal health-care system" in Canada. However, before we even started our formal research, we had to come to grips with the hegemonic ideology among white Canadians that Aboriginal people do not live in cities in Canada but instead are to be found out in the countryside, on rural reserves, disengaged from modernity. Such categories are especially problematic given that, as of the 2006 census, more than 51 per cent of all Aboriginal people now live in urban areas. What it means to be "Aboriginal" in Canada is thus contested (Berg et al. 2007) and so was our own understanding of our research project as a "participatory" partnership with urban Aboriginal people (Berg et al. 2007; Evans et al. 2009). It should be clear that any attempts to write about such work involve significant politics in the construction of our categories and in the decisions about what to say (and not to say) about any specific events. Accordingly, the "writing-up" of the research is only the initial phase of an iterative process of negotiating the production of knowledge between the researchers and the participants. I am not suggesting here, however, that negotiated knowledge is intrinsically better than other forms of knowledge. Instead, what I want to point out is that the process of making explicit the act of negotiation helps to make the research accountable in ways that are appropriate given the specificity of positions and power relations. All research is caught up with power relations, and to deny this is to deny an important aspect of knowledge production (see Chapters 2, 3, 4, and 17). Taking this process seriously has enabled me to rethink the role of writing in my intellectual endeavours.

Thoughts, observations, emotions, and interpretations that occur during the research become important components of any research endeavour, not because they record events or ideas but because they are **signifiers** of them (in this sense, they act to "define" complex constellations of ideas and thoughts about the research in more simplified categories of knowledge). Whether interpretations are noted by way of a personal diary, log, video, or audio recording, they can provide insight into the researcher's own speaking position and how it is articulated, challenged, and modified through the research journey.

Writing-*in* is not a matter of "telling"—it is about knowing the world in a certain way. The process of writing constructs what we know about our research, but it also speaks powerfully about who we are and where we speak from. As we suggested in a previous section, the detached third-person writing style so common in academic journals and reports implies that the researcher is omnipotent—that he or she has a perspective that is all-seeing and all-knowing. However, what may appear to be the truth spoken from "everywhere" is actually a partial perspective spoken from some*where* and by some*one*. Knowledge does not, according to a post-structuralist perspective, exist independently of the people who created it—knowledges are partial and geographically and temporally located. As the researcher writes and inscribes meaning in the qualitative text, she or he is actually constructing a particular and partial story. Richardson and St Pierre (2008) suggest that writing creates a particular view not only of what we are talking about (and what we do not say) but also of ourselves. Power is connected with speaking position through text, so qualitative researchers should consider not only their standpoint in constructing a research account but also the implications of their interpretations for those who may have been involved in the research.

Because the practice of writing is not neutral, the voices of qualitative researchers do not need to hide behind the detached "scientific" modes of writing. Such modes of writing position the researcher as a disembodied observer of the truth rather than as a (re)presenter and creator of a particular and partial truth. The researcher is an instrument of the research, and accordingly we suggest that researchers should acknowledge their position in ways that demonstrate the connection between the processes of research and writing. **Reflexivity** is the term often used for writing self into the text. Kim England (1994, 82) defines this as "self-critical sympathetic introspection and the self conscious analytical scrutiny of self as researcher." A reflexive approach can make researchers more aware of their necessary connection to the research and their effects on it and of asymmetrical (i.e., where a researcher has more social power and influence than their participants) or exploitative relationships, but it cannot remove them (England 1994, 86; also see Rose 1997 and Chapters 2, 3, and 4 of this volume). One way in which reflexivity can be encouraged in the writing-*in* of qualitative research is by the use of personal pronouns (for example, "I," "we," "my," "our").

However, it is important that the use of personal pronouns does not become merely an emotive tool. Alison Jones (1992) suggests that in academic spheres, the use of "I" may result in the insertion of "emotion" as a replacement for "reason," thereby creating a work of "fiction" hiding power relations as much as it might make them explicit. Employing reflexivity through use of first person should instead make explicit the politics associated with the personal voice and draw attention to assumptions embedded in research texts. Reflexivity is also concerned with constructing research texts in a way that gives consideration to the voices of those who may have participated in the research. Reflexivity is about writing critically, in a way that reflects the researcher's understanding of his or her position in time and place, particular standpoint, and the consequent partiality of the researcher's perspective. Writing situated accounts may involve acknowledging the role of embodied emotions in research (Bondi, Davidson, and Smith 2005; Widdowfield 2000), thinking about how one's positionality is "mutually constituted through the relational context of the research process" (Valentine 2003, 377) and considering the ways in which our texts might constitute performances, enacting and valorizing certain metaphors and meanings and silencing others in the spaces where they might circulate (Denzin 2003). This understanding of the **dialogic nature of research and writing** (in the sense of a "dialogue" between various aspects of the research process) enables qualitative researchers to acknowledge in a meaningful way how their assumptions, values, and identities constitute the geographies they create. It also provides an opportunity to play and experiment with writing as a way of knowing and representing.

Just as there are many ways of knowing, there are also numerous ways in which qualitative researchers may construct their research narratives as "re-presentations"—constructions that evoke what we as researchers have lived and learned (Ely 2007). Richardson and St Pierre (2008) believe that to write "mechanically" shuts down the creativity and sensibilities of the researcher. They encourage researchers to explore text and genre in the (re)presentation of qualitative research through a variety of media, including oral and visual (see, for example, Latham and McCormack 2007 and Sanders 2007 on how students can use digital photography in presenting their research) and to experiment with diverse forms of the written word (prose, poetry, play, autobiography). We support Richardson and St Pierre's (2008) metaphorical construction of writing as "**staging a text**" and encourage geographical researchers to consider how they are (re)presenting the research "actors," creating the plot, action, and dialogue of a research "tale"; how they are constructing the stage, the setting of the "research" play; and to whom (i.e., the audience) the "production" is aimed.

Of course, most undergraduate geography students will be required to write their research within a given format—the essay or report (for some advice on the conventions associated with these forms of writing, see Bradford 2003; Hay 2012; Hay, Bochner, Blacket, and Dungey 2012; Kitchin and Tate 2000; and Chapter 20 of this book). Nevertheless, it may be possible to persuade your lecturer to let you produce another form of geographic representation: a play, a video, a poem, a short story, a poster-board, to name a few options. For example, Lawrence's third-year students in his "Culture, Space, and Politics" course create two- to three-minute videos involving critical analyses of "landscapes" as part of their major assignment for the class and then upload these videos to YouTube for public viewing, and Juliana's "Consumption and Place" postgraduate students are asked to create an auto-ethnography (Butz, 2010) of their learning in the course during the year presented through any medium they choose. Despite writing constraints and "staging" conventions imposed by self and audience (such as for an academic publication or a thesis), we argue that there is no single correct way to "stage" a text (see Box 19.2). By exploring the varied ways in which the text can be staged and how in such staging different stories may be emphasized and other voices may come to the fore, the researcher has the potential to create dynamic and interesting research pieces that engage and challenge both writer and reader.

Issues of Validity and Authenticity

The interpretative nature of qualitative research has given rise to a considerable amount of debate concerning how the **validity** and authenticity of qualitative research accounts might be assessed (see Bailey, White, and Pain 1999a; Baxter and Eyles 1997). The reflexive writing-*in* of research experiences and assumptions is not a licence for sloppy research or monographs based solely on personal opinion. Validity, integrity, and honesty in writing-*in* are no less important in qualitative research than they are in quantitative research. Works by writers such as Anna Tarrant (2010) and Wilton, DeVerteuil, and Evans (2014) provide telling examples of critical qualitative research and carefully considered analysis of socio-spatial relations.

The truth and validity of knowledge arising from geographical and social science research has been the subject of discussion for almost four decades (see for example Harrison and Livingston 1980). Post-structuralist thinking has challenged

Writing-In—Alternative Representations

The unconventionality of Marcus Doel and David Clarke's (1999) staging of an "academic" article is conveyed by its unusual title, "Dark Panopticon. Or, Attack of the Killer Tomatoes." In this paper, which appeared in *Environment and Planning D: Society and Space*, the authors write as if in a dream world. Their piece is simultaneously fragmentary and powerful. The form of writing is cleverly indicative of post-structuralist perspectives through which various arguments and thoughts are constructed. In it, narratives of dream, text from other academics, and authorial thoughts are presented as a complex reflexive tapestry. The narratives contained within the article are disjointed and struggled over, and the readers are invited on hallucinatory journeys through real and imagined spaces. These are spaces that reflect the ways in which depersonalization, de-individuation, and fatigue characterize both ways of thinking about the world and commodity relations. As you read the following excerpt, consider how theoretical stance and textual staging are integrated through this "alternative" approach:

> Flash-darks touch the flesh of the world, giving rise to all manner of ontic unease. To be touched by the Dark Panopticon is exemplified for us in the display of greengrocery that greets countless millions as they flow into the world's markets, supermarkets, and hypermarkets. And what we feel here—is it only us?—is an unprecedented and almost unbearable brutality against the flesh of the world: not simply the flesh of humans or animals, but above all else the flesh of vegetables and soft fruit. In particular we are horrified by violence against tomatoes. Yet as we shall tease out in due course, tomatoes, like all objects in the consumer society and data in the surveillance society, take their revenge. The tomatoes strike back . . . (Doel and Clarke 1999, 429)

the assumption of a singular truth and the privileging of certain claims to knowledge. Associated with this is what has been termed the "crisis of representation" (Marcus and Fisher 1986). That is, doubts have arisen over researchers' authority to speak for others in the conduct and communication of research (Alvermann, O'Brien, and Dillon 1996). In recent years, an increased sensitivity to power and control on the part of some qualitative researchers has encouraged a rethinking of research design and implementation (Evans et al. 2009). It has also meant a growing concern over how researchers appropriate and assume participants' voices in the writing of research accounts (Opie 1992). Moreover, it has become important to acknowledge that one's research writing might have an active role in constructing and reconstructing research relations in unintended ways and that research accounts are in themselves a kind of creative invention (Freeman 2007).

Baxter and Eyles (1997) have argued that geographers need to be more explicit about how rigour has been achieved throughout the research process. The process of writing-*in* qualitative research requires, therefore, that writers explicitly state the criteria by which a reader may assess the "trustworthiness" of a given piece of research (see Chapters 6 and 7 for a discussion). Addressing this issue is difficult, because it involves qualitative writers grappling with the tensions between the complexity and richness of information that emerges from qualitative research and the need to produce some sort of "standardized" evaluation criteria (Baxter and Eyles 1997).

Post-structuralist thinking casts doubt on foundational arguments that seek to anchor a text's authority in terms such as reliability, validity, and generalizability. The notion of multiple truths, and a growing awareness of issues of representation, do present difficulties for assessing the validity of qualitative research. Much of the debate surrounding the validity of qualitative writing rests upon how terms such as *rigour*, *validity*, *reliability*, and *truthfulness* are defined. A related issue is whether these definitions, which have often been used as "objective" measures of the quality of quantitative research, are applicable to qualitative endeavours (see Chapters 6 and 7). A debate between two geographers in 1991 and 1992 issues of *Professional Geographer* highlights some of the issues surrounding reading and evaluating texts and the validity of qualitative approaches. The exchange between Erica Schoenberger (1991; 1992), and Linda McDowell (1992b) is significant because it was written at a time when in-depth interviewing was not used extensively in industrial and economic geography. The debate is interesting not only because it centred upon contested definitions of validity but also because it raised issues of meaning and interpretation, audience, and representation through the language that constituted these articles. It demonstrates powerfully the multiple reading of texts and the care needed in constructing qualitative interpretation. Schoenberger sought to argue for the legitimacy of qualitative forms of interviewing but in doing so suggested how quantitative definitions of reliability (defined by Schoenberger as the stability of methods and findings) and validity (accuracy and truthfulness of findings) might be applied to qualitative research. McDowell disagreed with Schoenberger's concept of validity as interpretation that is verifiable and that corresponds to some external truth. She suggested that Schoenberger's definition was based on a continued adherence to a positivist-inspired quantitative research model, as demonstrated by the language through which her arguments were constructed. Schoenberger (1992) addressed McDowell's criticisms by referring to issues of intended audience (positivist geographers), power, and interpretation, suggesting that McDowell had misinterpreted her claims.

If concepts such as validity are contested, how then is it possible to construct truthful research texts? Clifford Geertz (1973) has argued that good qualitative research comprises **thick description**. Such descriptions take the reader to the centre of an experience, event, or action, providing an in-depth study of the context and the reasons, intentions, understandings, and motivations that surround that experience or occurrence. While it may not be possible to assess the authenticity of such partial descriptions, the interpretations upon which they are constructed can be articulated. Mike Davis (1990, 253–7), for example, provides a particularly compelling description of the makeshift prisons and holding centres for the urban underclass who make up the more than 25,000 prisoners in "the carceral city" found in a three-mile radius of Los Angeles city hall (see Box 19.3).

Writing Fortress LA

The demand for law enforcement *lebensraum* in the central city, however, will inevitably bring the police agencies into conflict with more than mere community groups. Already the plan to add two highrise towers, with 200–400 new beds, to County Jail on Bauchet Street downtown has raised the ire of planners and developers hoping to make nearby Union Station the center of a giant complex of skyscraper hotels and offices. If the jail expansion goes ahead, tourists and developers could end up ogling one another from opposed highrises. One solution to the conflict between carceral and commercial redevelopment is to use architectural camouflage to finesse jail space into the skyscape. If buildings and homes are becoming more prison- or fortress-like in exterior appearance, then prisons ironically are becoming architecturally naturalized as aesthetic objects. Moreover, with the post-liberal shift of government expenditure from welfare to repression, carceral structures have become the new frontier of public architecture. As an office glut in most parts of the country reduces commissions for corporate highrises, celebrity architects are rushing to design jails, prisons, and police stations.

An extraordinary example, the flagship of an emerging genre, is Welton Becket Associates' new Metropolitan Detention Center in Downtown Los Angeles, on the edge of the Civic Center and the Hollywood

His descriptions, meticulously researched and referenced, are as much novel-like evocations of city life as they are academic descriptors. In the end, we are trying to understand the *meaning* of particular aspects of social and spatial life, and meaning is always a matter of interpretation.

Communicating qualitative research is thus about choices—for example, about how we show the workings of our research, how we present and convey others' voices, how we locate our own subject position, and what effects our writing may have for those who may read it and who may want to engage with it (Denzin 2008; Holliday 2007). Although such choices are not always conscious and not necessarily made in circumstances of our own choosing, researchers can attempt to take responsibility for their perspective and position in research (sometimes referred to as "transparency") through reflexively acknowledging and making explicit those choices that have influenced the creation, conduct, interpretation, and writing-*in* of the research. Such choices are likely to be guided by principles

Freeway. Although this ten-story Federal Bureau of Prisons facility is one of the most visible new structures in the city, few of the hundreds of thousands of commuters who pass it by every day have any inkling of its function as a holding and transfer center for what has been officially described as the "managerial elite of narco-terrorism." Here, 70% of federal incarcerations are related to the War on Drugs. This postmodern Bastille—the largest prison built in a major US urban center in generations—looks instead like a futuristic hotel or office block, with artistic charms (like the high-tech trellises on its bridge-balconies) comparable to any of Downtown's recent architecture. But its upscale ambience is more than mere facade. The interior of the prison is designed to implement a sophisticated program of psychological manipulation and control—barless windows, a pastel color plan, prison staff in preppy blazers, well-tended patio shrubbery, a hotel-type reception area, nine recreation areas with Nautilus workout equipment, and so on. In contrast to the human inferno of the desperately overcrowded County Jail a few blocks away, the Becket structure superficially appears less a detention than a convention center for federal felons—a "distinguished" addition to Downtown's continuum of security and design. But the psychic cost of so much attention to prison aesthetics is insidious. As one inmate whispered to me in the course of a tour, "Can you imagine the mindfuck of being locked up in a Holiday Inn?" (Mike Davis 1990, 256–7).

of ethics and truthfulness (see Chapters 2 and 3). Transparency may make researchers, and the audiences for whom they write, more aware of the constraints on interpretation, of the limitations imposed by the "textual staging," and of the implications of the former for research participants. For example, the choices surrounding the use of research participants' quotes in a written text comprise far more than a simple matter of how, where, how many, and in what form participants' voices are to be included. The inclusion of quotations raises issues of representation, authority, appropriation, power, and participation. Kay Anderson (1999, 83–4), for example, provides a particularly compelling description of the juxtaposition of identities and spaces in Redfern, an inner-city suburb of Sydney closely identified with spaces of "Aboriginality" (see Box 19.4). Anderson weaves together quotes from research participants, evidence gathered from archival research, and rich theoretical writing to argue for understanding Redfern as a hybrid space of porous and fluid identities and spaces. Her work manages quite subtly to acknowledge issues of power in research and representation while simultaneously providing us with a vivid description of life on the Block.

Jamie Baxter and John Eyles (1997) suggest that the criteria of credibility, transferability, dependability, and confirmability are useful general principles

BOX 19.4 Rethinking Redfern—Writing Qualitative Research

On the Block [a small area of Redfern, an inner-city suburb of Sydney] itself, in 1994 the dominant language group was Banjalang, but a wide range of other place-based dialect groups were present, including Eora (Sydney region), Wiradjuri (Nowra), Kamlaroi (Dubbo and Moree), and many other groups from throughout New South Wales and Queensland. In addition, approximately two-thirds of the total number of rent-payers on the block were women, a significant minority of whom (among those interviewed) were married or partnered to Tongans, Fijians, Torres Strait Islanders, and members of other ethnicities. In most cases the men did not live with the women, who supported their children and funded the periodic visits of husbands or partners, friends, and relatives. Sociability has always been both fractious and friendly. Some tenants saw the Block as "home"; others perceived their place of birth as home; most considered they had multiple homes, including Redfern. In the words of a tenant known on the Block as a community elder and who has since refused to leave it: "Redfern is an Aboriginal meeting place. People come from all over the country to get for guiding an evaluation of the rigour (trustworthiness) of a piece of qualitative research. They see these categories as broadly equivalent to the concepts of validity, generalizability, reliability, and objectivity that have been used to evaluate the quality of quantitative research endeavours. We believe it is important to keep in mind the constructedness of these concepts, to avoid using them as universal assessment criteria, and to avoid engaging in comparative analysis of qualitative studies. Notwithstanding this, Baxter and Eyles's principles have much to commend them in that they may assist in evaluating the internal consistency and trustworthiness of a piece of research. It is important to note, however, that we avoid using the term "rigour" in our approach to constructing pieces of qualitative research, since we believe the term is too closely tied to quantitative approaches that are predicated on systematic exposure of pre-existing truth. Instead, we prefer to use the term "trustworthiness," which speaks much more directly to qualitative geographical research as a reflexive practice that constitutes and understands meanings in place: one that actively recognizes that knowledge is constructed, open-ended, and fluid. (see Chapter 6, for example, for a different interpretation of rigour). Writing qualitatively should encourage researchers to explore and make explicit their own research agendas and assumptions and

news of friends and family." Now a member of the housing coalition that has been formed to fight the AHC [Aboriginal Housing Company], the same woman recently stated that Redfern has long been a place where children taken from their relatives encountered relatives or information that would lead to those relatives (*Melbourne Age*, 1 February 1997). Another woman had this to say in an interview for a recent documentary: "People who come from interstate or wherever make to Sydney. Their first aim is the Block because they gotta know who's who and where's where and where to go from here" (ABC TV 1997).

The "traffic of relating" on the Block opens up fresh ways of thinking about home and community in a context of more widely invoked images of "flight," "flow," "crossings," "travel," and transnational exchange in contemporary cultural geography. Models of mixing that work with the idea that cultures are porous and fluid are particularly apt in relation to this case study. There is, as I have suggested, no "pure" culture at Redfern, no crisp boundaries of inside and outside, even for so stigmatised an area. This is not only a methodological issue. It is also an epistemological problem in that the boundaries of researchable communities are not secure and areas never exist as discrete entities (Anderson 1999, 83–4).

to elaborate on how they believe their research text constitutes the "truth" about a particular subject (see Box 19.5). While transparency is not in itself ultimately achievable, if conscious reflexive writing produces qualitative texts that are open to scrutiny by research participants and audience, and if they present challenges to taken-for-granted ways of seeing and knowing and provoke and promote

Auto-Ethnography: A Method for Reflexive Writing

Ellis (2004, xix) defines auto-ethnography as

> research, writing, story, and method that connect the autobiographical and personal to the cultural, social, and political. Auto-ethnographic forms feature concrete action, emotion, embodiment, self-consciousness, and introspection portrayed in dialogue, scenes, characterization, and plot. Thus auto-ethnography claims the conventions of literary writing.

Ian Cook has encouraged his students to use auto-ethnography as a means of writing reflexively and critically, bringing the insights from their reading, visual, and aural material research and their felt experiences, emotions, and knowledges into their assessment for a course on geographies of material culture (Cook et al. 2007). Similarly, David Butz and Katharine Besio (2004) have used auto-ethnography to better understand the implications (for research and knowledge production) of their own positionality as white, Western researchers working in the Karakoram. Auto-ethnographies provide one way by which the emotions, experiences, contradictions, and inconsistencies in our research journeys (and those whose lives and experiences intersect with them) can be re-presented. The use of literary rather than objective modalities and the active inclusion of self means that one's writing,

> far from being a dispassionate process of producing what was, is instead a product of the present, and the interests, needs, and wishes that attend it. This present, however—along with the self whose present it is—is itself transformed in and through the process at hand. [Freeman 2007, 137–8]

questions about "place" in the world, then perhaps this goes some way towards establishing the "validity," credibility, and trustworthiness of a qualitative research text. Communicating qualitative research is as much about how we know as it is about what we know, as examples of **auto-ethnographic** writing demonstrate (see Box 19.5).

Writing thus becomes a method of knowing. Examine this piece of auto-ethnographic writing by Sarah, one of Ian Cook's students:

> It's evening. To go out-on-the town I need an image refinement. Understatement! My cybernetic mask thickens with the make-up and lip-gloss, hair spray. . . . A reach for the wardrobe and a selection of T-shirts. Removing my glasses I read the small print—a third "gaze": no glasses—close-up-focus but not *understanding.* Black tops from Hong Kong and EEC. Or should I go to Greece? Or with the red tops—more classy—to Thailand, England (no?!), Indonesia? I must emphasize never having looked at the origin of my clothes before. Think I will wear the silver one which has no label. Safer to not know where it comes from. Would "made in Syria" really make a difference? I still wouldn't know how, where, when it was made and by whom.[7] And it goes with my chain—connecting my outfit as well as me with the world.[8]

[7] Again, the concept of "commodity fetishism" is important. As a cybernetic being I am embedded in global networks and yet I am unaware of those that I connect.

[8] Through the clothes that I wear I am connected not only to the people who made them but to the machines themselves that manufactured my outfit. As a cyborg, are those machines a part of me? The amalgam of man [sic] and machine, to which Haraway's (1991) "cybernetic organism" alludes, where does it end? What is the extent of my hybridity?

What effects might such writing have? Sarah's writing, like Doel and Clarke's earlier "staging" (see Box 19.2), might be seen as opaque, diffuse, and uncertain. While auto-ethnographic pieces might seem disconcerting, even confusing, relative to conventional objective modalities of writing research, such narratives can provoke an awareness of the situated problematics and politics of knowing, representing, and transforming knowledge. Ian Cook believes this approach can have radical effect—an effect that arises out of a less straightforward or didactic connection between what is known and how it is interpreted. What do you think?

Conclusion

We have discussed the importance of language in the social construction of knowledge, how power is articulated through dichotomies, and how meaning is inscribed in language. We believe models of writing that construct the writer as a disembodied narrator are inappropriate for communicating qualitative research. Our focus has been on written rather than visual texts, since writing remains the predominant means of communicating qualitative research. Breaking down dichotomies—through an interpretative understanding of writing and researching as mutually constitutive processes—and an understanding of which principles might guide valid qualitative research are critical to writing "good" qualitative research. It is also crucial to understand how power and meaning are inscribed in the words that we use (and those we choose not to use) to constitute the research process, to recognize our subjectivities, standpoint, and locatedness (shifting and partial though they might be), and to acknowledge the voices of those with whom we undertake research. We believe that doing this enables qualitative researchers to have confidence in the "validity" and truthfulness of their interpretations. Consequently, this chapter has not been a "how to" guide but a means of raising important issues that are inherent in the writing-*in* process.

Key Terms

auto-ethnography
dialogic nature of research and writing
dichotomy
falsification
first-person narrative
hegemony
mimetic
modality
mystification
neo-positivism
nominalization
objective
objective modality
positionality
positivism
post-structuralism
quantitative revolution
reflexivity
reification
signified/signifier
situated knowledge
staging a text
subjective
thick description
third-person narrative
trope
validity
writing-*in*

Review Questions

1. What implications do post-structuralist perspectives have for writing qualitative research?

2. In what ways is the term *writing-up* misleading?
3. Why should writing be seen as an integral part of the entire research process?
4. How can a researcher endeavour to produce "trustworthy" research?
5. Why can writing be seen as a method of knowing?

Review Exercises

1. In this chapter we have stressed the importance of using words with care. Accordingly, we have drawn upon specific terms/concepts that underpin the conceptualization and writing of geographical research. The table below has been incorrectly formulated. In groups, see how quickly you can match the correct term, explanation and example (eg 1, H, I). You may find it more fun to copy and cut up a copy of the table, but don't destroy this book!

	Term/concept		Explanation		Example
1	Positivism	A	A practice of writing that involves ongoing and active reflection throughout a research project, which acknowledges how the act/style of writing shapes the researcher's conceptualization, conduct, and analysis of the research.	a	*"The research was conducted in accordance with standardized ethics procedures." Tells the reader little about where/what and how ethical procedures were employed.*
2	Falsification	B	A conceptualization of knowledge that explicitly recognizes that it is written, spoken, felt, sensed, and experienced differently by people who are situated and positioned in particular spaces and places.	b	*"I wote notes as part of conducting my participant observation" becomes "the conduct of participant observation involved written field notes."*
3	Trope	C	A variant of positivism which argues that truth statements (e.g. hypotheses and "facts") can be disproved, but not verified through empirical observation.	c	*Writing reflects on research practice and is the end product of "meaningful" research. "I am on to writing now; it should be easy from here as I have done all the research."*

	Term/concept		Explanation		Example
4	Objective modality	D	The process of transforming verbs and adjectives into nouns.	d	*For example, first-or third-person narratives, or figurative conventions found in writing such narratives of "love as war," "knowledge as building," or "theory as soft."*
5	Third-person narrative	E	A practice generally seen to occur at the end of a research project, where the researcher writes an account of the research project and the significance of the research findings. Here writing is simply a means of reflecting rather than constructing meaning.	e	*"Globally, the health of all older people declines as one ages." The example above represents a truth statement that is easier to falsify than prove, by finding evidence of older people whose health has not deteriorated with age.*
6	First-person narrative	F	A form of writing that distances the author from the statements being made which implies objectivity and a universal "perspective."	f	*As an author of the text I recognize the meaning and interpretation of the research is constructed through the writing process. Writing consequently has an important role in producing rather than simply reflecting research outcomes.*
7	Mystification	G	A common trope in scientific writing that suggests the author is not a participant in the production of the text.	g	*"My positioning as a younger woman interviewing older men influenced the conversation in interviews. As we talked, I sensed there were places and feelings that were uncomfortable for us—barriers to conversation that I felt I could not transgress."*

	Term/concept		Explanation		Example
8	Reification	H	A series of approaches used in the scientific production of knowledge that centre on empirical observation as the basis for establishing "the truth."	*h*	*An example of this trope is: "The researchers explained in detail the basis of informed consent to potential survey volunteers."*
9	Nominalization	I	The use of language to conceal or hide the participants, their stance, and details of time and place.	*i*	*Comte's tradition, logical positivism, falsification.*
10	Writing-in	J	A trope in which the self is explicitly acknowledged and embedded in the construction of the text.	*j*	*Where researchers, participants, stakeholders, practices, methods, and equipment become the "The research" or "The report."*
11	Writing-up	K	Where the complexity of the interactions between the social and physical worlds and the embodied human beings involved in the research are reduced to the status of things.	*k*	*An example of this trope is: "I explained in detail the basis of informed consent to my potential survey volunteers."*
12	Embodied and situated knowledge	L	Commonly used forms, conventions, and styles of writing.	*l*	*"The research was well conceptualized, and produced results that support the existence of poverty in large cities." Illustrates a "God's eye" view of the research and implies universality of the research findings.*

2. Writing modalities—Using first-person narrative, write a paragraph describing your experiences of eating dinner last night. Swap your account with a friend and re-write the paragraph using third-person narrative. Together discuss your paragraphs: which version seems more "truthful," and why?

Useful Resources

Baxter, J., and J. Eyles. 1997. "Evaluating qualitative research in social geography: Establishing 'rigour' in interview analysis." *Transactions of the Institute of British Geographers* 22 (4): 505–25.

Bondi, L. 1997. "In whose words? On gender identities, knowledge and writing practices." *Transactions of the Institute of British Geographers* 22: 245–58.

Chandler, D. n.d. "Semiotics for beginners." http://visual-memory.co.uk/daniel/Documents/S4B/semiotic.html. This site is a comprehensive introduction to some post-structuralist understandings of the power of language.

Cook, I. 2009. "Geographies of food—afters." http://food-afters.blogspot.com. This is Ian Cook's blog for co-authoring an article for the journal *Progress in Human Geography*. Ian is an associate professor of geography at the University of Exeter and has been active in various forms of participatory qualitative research. The blog gives a wonderful insight into many of the issues discussed in this chapter.

Cook, I., et al. 2007. "'It's more than just what it is': Defetishising commodities, expanding fields, mobilising change . . .". *Geoforum* 38: 1113–26.

Ellis, C., T.E. Adams, A.P. Bochner. "Autoethnography: An Overview." Forum Qualitative Sozialforschung / *Forum: Qualitative Social Research,* [S.l.] 12 (1). Available at: http://www.qualitative-research.net/index.php/fqs/article/view/1589/3095. Date accessed: 22 Oct. 2013. Presents a useful summary of the origins and practice of auto-ethnography.

Jones, A. 1992. "Writing feminist educational research: Am 'I' in the text?" In S. Middleton and A. Jones, eds, *Women and Education in Aotearoa.* Wellington: Bridget Williams Books.

McNeill, D. 1998. "Writing the new Barcelona." In T. Hall and P. Hubbard, eds, *The Entrepreneurial City: Geographies of Politics, Regime and Representation.* London: John Wiley.

Richardson, L.. Professor Emeritus of Sociology, The Ohio State University. http://sociology.osu.edu/people/richardson.9 A key advocate for creative writing in research, Laurel Richardson's CV provides a substantial list of her publications.

Richardson, L., and E.A. St Pierre. 2008. "Writing: A method of inquiry." In N.K. Denzin and Y.S. Lincoln, eds, *Collecting and Interpreting Qualitative Materials Research*, 3rd edn. Los Angeles, London, New Delhi, and Singapore: Sage.

Rose, G. 1997. "Situating knowledges: Positionality, reflexivities and other tactics." *Progress in Human Geography* 21: 305–20.

Notes

1. Readers seeking specific "how to" advice on stylistic conventions associated with the presentation of research are advised to consult Chapter 20 of this volume, Hay (2012), Hay, Bochner, and Dungey (2012), Kneale (1999), or Stanton (1996) in conjunction with the conceptual material of this chapter.
2. It is appropriate to note, however, that it can be difficult institutionally to present some forms of research this way. For instance, social or environmental impact statements prepared by government departments or consulting firms will frequently not list authors, in which case the "we" would be poorly defined. Moreover, neither group would want individuals to be sued for their opinions.

Exercise 1 Answers: 1H*i*, 2C*e*, 3L*d*, 4F*l*, 5G*h*, 6J*k*, 7I*a*, 8K*j*, 9D*b*, 10A*f*, 11E*c*, 12B*g*

20 Communicating Qualitative Research to Wider Audiences

Eric Pawson and Dydia DeLyser

Chapter Overview

This chapter aims to help beginning qualitative researchers communicate their research effectively to wider audiences. To do so, it explores different modes of communication, including written work, public presentations, websites, and social media, and considers audiences, including supervisors and examiners, research participants and respondents, and the public at large. It asks who the audience is, as well as how to reach the audience, and how to convince the audience. It considers options for organizing and reporting the kinds of information and insights that come from qualitative research. To do so, it describes ways to focus creativity, along with techniques that can help in the imaginative expression of ideas. Finally, it addresses the power and credibility of qualitative research, and suggests how to harness the results of such research in or for public discussion and debate.

Introduction

How to **communicate** the findings of qualitative research to an audience? This is a central and often troubling issue for any researcher. Yet with the exception of a short chapter on "writing-up," it is a topic about which most books say very little. Even the phrase "writing-up" is itself misleading, implying that writing is something that occurs straightforwardly at the very end of the research process. As Chapter 19 of this book has argued, writing and research are mutually constitutive practices, with writing being continuously woven throughout the research process, and research-driving ideas occurring throughout the writing process as well. Even so, writing—a conventional and often fruitful means for communicating research—is today far from our only avenue for sharing our research with broader audiences. Public presentations, group talks, web pages, and social media are just some of the other means that can be employed to present and represent the richness of qualitative work.

Many of the most vexing contemporary issues lend themselves particularly well to qualitative methods, for the rich and focused insights of qualitative research can give voice to difference and convey meaningful insights into events and experiences (Braun and Clarke 2013). For example, in 2010, what was to become a more than three-year sequence of thousands of earthquakes began to strike the city of Christchurch in New Zealand. Several particularly violent events destroyed buildings and caused ground liquefaction (where water-saturated soil fails to support the weight of structures built upon it, acting like a liquid, rather than a solid) across much of the urban area. Quantitative studies revealed the profound numerical extent of the devastation in numbers of buildings, numbers of homeless, acres of land rendered useless. The real devastation, though, was not just to buildings but to *homes*, and not only to a city, but to *people's lives and livelihoods*. While many different kinds of research are needed to understand and recover from complex tragedies like these, in this case qualitative researchers—many of them students, or groups of students based at universities in the city or far from it—have mustered meaning-filled qualitative research to make meaningful contributions to disaster recovery efforts (Wilson 2013).

This chapter uses a range of examples, and suggests a number of techniques, to show how qualitative research, which can speak so powerfully to the lives of individuals and communities, can be mobilized to communicate effectively and to reach a broader public. It draws on experiences of environmental disasters to underline its main points. It also identifies and discusses issues like how to persuade audiences of the power and credibility of qualitative research, and how to use qualitative research to contribute to or advance public debate. To do so, it is necessary to first consider the nature of the audiences that we as researchers wish to reach.

Understanding Audiences

To reach an audience effectively we must first know something about who the audience is or might be. Just as a personal conversation or email is shaped by knowledge of the person we interact with, so must the audience for a piece of research—which is likely to have involved far more preparation and effort than most conversations or emails—be considered. But "who is the audience?" is not a straightforward question, and the answer is different for every project, and even for different expressions of each project. The initial answer may be surprising: we first communicate our research with ourselves. We write first for and to ourselves, because the very act of writing reveals what it is that we think and know and what we do not (Richardson and St Pierre 2005). In other words, writing is an iterative process (Kitchin and Tate 2000), with reading, writing, and research being interlinked phases and each activity clarifying the direction that needs to be taken in the others. Reading makes clearer what to write and vice versa. Writing helps to elucidate what we think, and being clearer on what we think enables more effective communication of those thoughts to others. Understood in this way, it becomes evident that the writing process itself is *formative*; writing is an essential part of clarifying thinking (Becker 2007) because "Writing *is* thinking" (Wolcott 2001, 22).

Since academic life takes place within social networks and involves social responsibilities, in order to participate fully in the academic world, many qualitative researchers contribute in different ways both to academic life and to the communities that have provided the research opportunities (Becker 2007), and here **audience** becomes a critical issue. Students as scholars have a wide public: being accountable within the academy (to meet course or degree requirements or to peers), as well as having obligations beyond the university (for example to those who gave access to networks or responded to requests to participate). Depending on which audience or which part of the audience we wish to communicate with, the ground rules vary. Universities, departments, or instructors publicize their expectations for class presentations, term papers, dissertations, or theses, and such rules are best adhered to. Here practice and feedback from peers will go a long way: to ensure that a talk does not run over time, or that chapter drafts are clear and polished.

More challenging is the issue of obligations to wider audiences, particularly those who have helped in the research. Qualitative inquiry in general seeks to reduce power differences and encourages sharing of meaning-making between researchers and participants. In this it differs from the traditional conception of quantitative research, where the researcher, as the ultimate source of authority, does not encourage participants' active contribution to the research process (Karnieli-Miller et al. 2009). **Sharing the field** also serves "as a means of giving back something to those who have helped us" (Cupples and Kindon 2003, 224). One way in which this can be done is by returning interview transcripts, or those parts of a written account that use their words, to respondents. Such strategies need careful consideration however: although often motivated by ideals of participant co-ownership of the research and participant empowerment, they can as likely result in "surprise and embarrassment" (Forbat and Henderson 2005, 1118). Because it is people's realities, feelings, and lives that that are represented in qualitative research, this must be done sensitively.

This sensitivity begins before the research even starts, for some populations feel heavily (even too heavily) researched by people from outside. This can easily arise in small towns that are popular for university field trips, or with particular suburbs that attract frequent attention. In post-disaster areas, well-meaning researchers acting without coordination run the risk of asking similar questions time and again. The result can be "research fatigue." In 2005 Hurricane Katrina flooded much of New Orleans; many who stayed behind or returned to their city faced not only the devastation of the storm but also an inundation by well-intentioned researchers (Robinson 2007). Such "over-researching" has also long been felt in Indigenous communities. Inuit residents of Arctic Canada, for example, have been known to compare researchers from the south with the annual visit of the snow geese: they arrive en masse, carry on, then depart leaving behind a mess (Stewart 2009). Linda Tuhiwai Smith characterizes this sort of behaviour as a reproduction of colonizing attitudes. Don't be a snow goose: "sharing is the responsibility of research" (Smith 1999, 161).

In Christchurch during and after the earthquakes, student researchers faced many similar issues. One way forward has been to engage research energies in an upper-level service-learning course. In this course, groups of senior geography undergraduates begin by negotiating the topics for research *with* community representatives, and then take responsibility for ensuring that the findings are communicated between the parties in multiple ways (O'Steen et al. 2011). One of these ways has been through an off-campus conference where each group presents a fifteen-minute illustrated talk. Such talks are an effective way of involving an audience and giving them the opportunity to participate through listening, questions, and discussion. Box 20.1 outlines the key things that will engage an audience, as well as those that will have the opposite effect.

Organizing and Representing Qualitative Findings

Whether we choose to present our qualitative research through verbal, visual, or written means, the basic academic fundamentals remain critical. These include cogent thinking, good preparation, and transparent **structure**. Developing a transparent structure, one that is clear to both presenters and listeners/readers, is worth both extra effort and the extra time that effort demands. It cannot be achieved if research questions, aims, and goals are not lucid and convincing to the researcher or research team first—so that means a well-planned outline must signpost or map the contents of the presentation from the outset. There are various ways of doing this, but many find going analog with a piece of paper or whiteboard the most effective. The whiteboard is great for brainstorming among members of a group and for working out a collective and agreed plan. A well-planned outline that is clear to us as authors will more likely to be clear to the audience as well.

This heavy emphasis on structure may seem unconvincing, or even annoying. Writers facing projects like term papers or theses might feel that any externally imposed form or structure hinders creativity, forcing the work into "boring" moulds or "stifling" formats. But the opposite proves true. Highly restrictive forms like the *haiku* or the sonnet reveal that such structures are stimulants to creativity—after all, no writer today would deny the creativity in Shakespeare's work. Contemporary examples of equally restrictive structures bear this out as well. Twitter, whose 140-character restriction puts a premium on identifying the point of the message, changed the world in the Arab Spring of 2011 (Simon 2011). Meanwhile, pecha-kucha presentations, in their purest form, restrict story telling

For Successful Public Presentations

Consider:

- *Preparation.* If you know your material, you will be less nervous about presenting it, and that will give your audience more confidence in what you are saying.
- *Content.* Your audience has come to hear what you have to say and only secondarily how you say it.
- *Visuals.* Use them to illustrate your points, ensuring that they add value to the talk.
- *Practice.* Even in an empty room, practice builds confidence and helps ensure the presentation is the right length.
- *Interaction.* Engage the audience by including them. It is often easiest to look at two or three people in particular while representing; then leave time for questions.

Avoid:

- *Rambling.* Both you and your audience will understand what is being said if you have a clear structure and keep to it.
- *Poor time-keeping.* Make sure you can see a clock or some means of telling the time, and keep to the allotted length.
- *Mumbling.* Ask the audience if they can hear you. Voice clarity for the audience comes from speaking clearly, loudly enough, and not too quickly.

to a sequence of 20 PowerPoint slides, each used for 20 seconds, for a total of 6 minutes and 40 seconds (20 × 20 = 6:40). Here the objective is to keep presentations brief and focused, at the same time as stimulating discussion.

More productive then is an understanding that structure harnesses creativity. Consider the now-well-known TED (Technology, Entertainment, Design [www.ted.com/talks] talks, growing in popularity since the 1980s. Their maximum length of eighteen minutes again puts a premium on clarity and purpose—though they may seem spontaneous, invariably they have been carefully planned. Likewise, it is not possible to make a film without a tightly written script. In standard screenplay format, one page of script becomes one minute in the final film. Because most cinema films run for around 90 to 110 minutes in length, the script will run about that many pages. Just as in a novel, it is necessary for the screenwriter to hook the audience in the first ten or so of those pages, showing them why they will want to watch it to the end. This means, simply, that nearly every film we've ever seen—from action and horror to drama and romantic comedy—has followed such a rigid structure.

Successful academic writers and speakers also use structure in carefully considered ways. Box 20.2 highlights some of the key techniques that enable the presentation of the richness of qualitative research and the expression of originality within clear structures. In specific ways, the pointers in the box apply to different options for organizing and reporting qualitative findings, such as written reports, talks, web pages, or conference papers. Clarity of communication comes first through attention to the basics of structure.

Today PowerPoint is often the staple of an academic presentation to a class, at a conference, or in a public meeting. But the standard sequence of slides does not of itself provide structure and meaning. Rather, using PowerPoint makes the focus, clarity, and storyline of a presentation even more important. The alternative is the kind of "death by PowerPoint" too often seen in lecture halls: overloaded slides heaving with bullet points that no one remembers by the halfway mark, let alone at the end. Excellent resources like *Beyond Bullet Points* (Atkinson 2011) and *presentationzen* (Reynolds 2011) can take straightforward simplicity to the

Using Structure to Communicate

- *Begin with clear research questions and aims.* These focus attention for both authors and speakers by sharpening their creative abilities, and focus the audience on the key elements of what is to be said.
- *Develop an outline.* Essential to clear organization, outlining ensures that the main points are made, and in the most appropriate order. The outline can change and grow as preparation proceeds, providing it does not become detached from the research questions or aims.
- *Use headings and subheadings.* When ideas need to shift, call attention to this with a new heading on the slide or a subheading in the text. Headings and subheadings are also invaluable tools for organizing and clarifying the argument in early drafts.
- *Develop clear ways to order the argument into sections or slides.* In written work, this is the purpose of the paragraph: locate the thesis statement (the "point" of the paragraph) at the beginning or end, and use the middle to build that idea.

level of art form. But "simplicity" in this context does not mean oversimplification, "rather, it comes from an intelligent desire for clarity that gets to the essence of an issue" (Reynolds 2011, 115).

Simplicity in this sense is not easy to achieve. It requires the kind of hard work in advance that mobilizes planning, focus, and structure. To illustrate the true challenge of this kind of simplicity Reynolds asks, "What is my absolutely central point?" "If the audience will remember only one thing (and you'll be lucky if they do), what do you want it to be?" (2011, 63). Identifying that central point comes readily from an open process of discussion with oneself, collaborators, or members of the community, allowing input and focusing ideas. Again the sketch-pad or the whiteboard can be an excellent tool for this. The more effort put into preparation, the better the end result, and the more likely that we will carry the audience with us (Box 20.3).

An alternative means of communicating research results is via a website, a tool effective in a range of circumstances and with varying audiences. Setting up web pages while doing research is another aspect of the formative process of writing. Blogs (web logs) and websites that encourage public comment are useful as a means of enabling research respondents to give feedback on the initial results, assisting in their verification (James and Rashed 2006). They have proved valuable in northern Canada (e.g., northern Manitoba and Nunavut) where community-based research, consultation, and feedback are now expected (Stewart and Draper 2009). They provide an opportunity to let you help respondents feel that they are valued for their roles and are part of the research process rather than merely being "the researched." In other words, such outreach can demystify the research process and contribute to the maintenance of meaningful research relationships in particular places.

Some choose to describe their research not through individually designed websites but through a social networking utility like Facebook. Others elect to make information about their research topics available through publicly editable online wiki pages. Still others update followers of their progress via Twitter. Carefully and attractively done, such sites and postings may draw an audience in—they can give you contacts working in similar fields or potential respondents; they may also help you to make your research available to those outside your department and to give something back to those you worked with in the field. A web presence is also a way of giving something back to your department: it may attract others to think about studying there in the future.

Writing for the web requires design as careful as any other means of communication. Balance words with images, but be aware of the politics of visual representation: it is only possible to use images for which you have permission to do so, and research subjects should not be misrepresented by picturing them insensitively—i.e., portraying them in ways devoid of context or placing them in

For Success with PowerPoint

Use it creatively:

- Minimize the number of slides and bullets: simplicity is the ally of the audience.
- What is the central point? If the audience is to remember just one thing, what do you want it to be?
- Use the slides to reinforce your words, not to repeat or anticipate them. The slides should convey your message in fewer words than you speak.
- Choose images carefully and consider the representational issues.
- Avoid dizzying dissolves, spins, or other amped-up transitions; avoid anything that distracts from your message.

Death by PowerPoint:

Think back to all the bad presentations that you've witnessed:

- with far too many slides and bullet points, often that are at odds with or overload the message
- with the speaker repeating the slides word for word and adding nothing to them
- with complex background schemes or silly colour choices that distract the audience
- with illustrations that detract from the point of the presentation
- that confuse quantity of information with clarity of message

To present well, we must do things differently, avoiding these pitfalls.

inappropriate juxtapositions (Rose 2008). Whether the chosen medium is words, sounds, or images, questions of representation have to be considered. Chapter 19 argued that writing research re-presents findings, thereby underlining the mediated character of communication. One aspect of this is the use of photographs, the staple of many forms of geographical communication, be they field reports, theses, web pages, social media, or PowerPoint presentations. It is vital to think carefully and critically about what photographs mean, the purpose for which they were taken, and how a specific selection of photographs reflects or silences particular points of view (Hall 2009).

The Power of Qualitative Research

Some wonder how qualitative research can be used effectively to address important issues when it cannot be quickly summarized in tables or statistics and must be presented and engaged in full (Richardson and St. Pierre 2005). The power of qualitative research lies precisely in this rich interaction with human stories that matter and in the fact that it cannot be reduced to a row of numbers but must retain the lives of its participants. So although some qualitative researchers worry that their research needs to be "defended" against those of a quantitative persuasion, and some books use this wording (Marshall and Rossmann 1999), the value of a qualitative approach should rather be actively promoted.

Box 20.4 outlines the valuable attributes of qualitative research and shows how, if good academic procedures are employed to ensure its **credibility**, it will readily convince your audience. With its focus on experience, and giving voice to those who otherwise might not be heard, it yields invaluable insights into lived social worlds. For example, in Christchurch, it was widely assumed that New Zealand's official Earthquake Commission (EQC), which provides natural-disaster insurance for residential-property owners, would take care of people in the wake of disaster. But the sheer extent of property damage overwhelmed the EQC, which had to increase its staff nationally from 26 before the earthquakes to around 1600 to handle the 467,000 claims that it received (Marsh 2014). Many of these claims exceeded the maximum coverage that EQC is permitted to provide, resulting in lots of householders becoming mired in lengthy assessments and disputes with more than one insurer, at the same time as having to deal with the physical and mental stress of broken houses and streets.

None of this anguish is conveyed by the widely available statistics of the extent of damage. The three biggest Christchurch shakes, on September 4, 2010, February 22, 2011, and June 13, 2011 were respectively the fifth, the third, and the tenth costliest "insured earthquakes" in the world to date (Marsh 2014). But only qualitative research can reveal the human realities of anxiety, frustration, and loss of security that lie behind such numbers. This is one aspect of what Mason means when she writes that qualitative research "should be formulated around an intellectual puzzle—that is, something which the researcher wishes to explain" (Mason 2002, 7). In this case, the puzzle is why, despite apparently generous insurance schemes, the long-term recovery from disasters is often a slow process (Gordon 2004). Four years after the first earthquake many people still did not know when their homes would be repaired.

This means that qualitative research can also be used effectively *in conjunction with* quantitative research, and the two should not necessarily be construed as in opposition to each other (see Chapter 1 for a discussion). Hence the most skillful politicians, who like to underline their achievements with carefully selected

The Power and Credibility of Qualitative Research

What's valuable about qualitative research?

- It provides insight into and renders meaning from human experiences.
- It focuses on how complex social worlds are produced, experienced, and interpreted.
- It is flexible and sensitive to the social context in which the data are created.
- It is based on methods of analysis and explanation that seek understandings of complexity, detail, and context in our dynamic social worlds.

Promoting its value to your audience

- Qualitative research allows respondents to speak for themselves.
- In moving beyond mere description, qualitative research can provide explanations to intellectual puzzles.
- When data has been gathered according to principles of qualitative research and analyzed in ways consistent with a body of theory or concepts (as argued in the previous chapters), it is clear that work was conducted systematically and rigorously.
- Triangulation, that is, drawing data from different sources to corroborate, elaborate, or further illuminate the argument you seek to forward, is convincing.
- When these points are met, the findings will have wider resonance, and may be applicable to other circumstances and other places (quantitative researchers call this "generalizability") (see Chapter 7 for a discussion).
- When your research subjects are able to recognize themselves and their experiences in your work, they will receive your work as credible.

statistics, also draw simultaneously on people's stories. In the sort of post-disaster situations that have been a backdrop to this chapter, it is also often the case that challenging individual experiences, convincingly told by academic or media reporters, can lead to political action. This chapter concludes with a renowned example.

Conclusion

Each year the sitting American President gives a public presentation known as the "State of the Union Address," which reaches an audience of many millions. Here a specific form has evolved over decades that begins with statistics highlighting political accomplishments, citing, for example, numbers of new jobs created or figures about an administration's response to a disaster. Each State of the Union speech is, in this way, jammed with statistics. But though such lists can be impressive, they can also be dull, and they mask the real lives of the people involved. So presidents then turn to specific, individual examples by inviting Americans who have lived the experiences the statistics describe to sit in the House of Congress's First Lady's Box and have their stories told as part of the Address.

In his 2013 State of the Union speech, President Barack Obama spoke just months after Hurricane ("Superstorm") Sandy had devastated much of the US eastern seaboard and upper midwest, particularly parts of the states of New York and New Jersey, causing an estimated US$65 billion in damage (Joseph 2013). Obama alluded to global climate change and the widespread predictions of worsening storms as a possible cause. But to make the example more real, and to urge Americans to reflect on ways to contribute in such circumstances, he also drew from the experience of just one woman, a registered nurse in a New York City neonatal unit. Menchu Sanchez, a Filippina immigrant living in Seacaucus, New Jersey, knew her *home* must be flooding, but when her *hospital* began flooding she kept her focus on the twenty premature babies in her care. As night fell the hospital lost power, leaving the babies with just hours to reach another hospital before back-up batteries in their ventilators and other medical devices would fail. Administrators searched for options, but it was nurse Sanchez who devised a human plan to safely evacuate these most fragile babies from the ninth floor: with non-medical personnel lighting the dark stairwells by cell phone and flashlight, and with medical staff in trail holding the bottles and machines serving as each infant's lifeline, Menchu, along with other nurses and doctors carefully *carried* the babies in their arms down to waiting ambulances. With Menchu making multiple trips herself, all twenty babies reached safety and survived. Menchu, Obama said, was one whose example we should all follow (Gretchen 2013).

This pattern, of giving meaning to statistics with the experiences of real people, has become so powerful that even invitations to the First Lady's Box are mobilized to draw into focus a positive and human side of often otherwise dark national and international issues. In ways such as these, qualitative data can reveal the complex and changing truths of human lives and social worlds, conveying them through real struggles and real lives that can find currency for those far outside the original story's location. Thus the power of qualitative research often lies in the ways it can lend the insight of human experience to challenging social issues and vexing ordeals—even, or perhaps especially, when the sample size is just *one*.

Key Terms

audience
communication
credibility
sharing the field
simplicity
structure

Review Questions

1. Why does the audience matter?
2. How is writing also thinking?
3. In what ways can the field be shared?
4. Why does structure help to harness creativity?
5. How would you promote the value of qualitative research to an audience?

Review Exercises

1. Using a current class assignment, list the five or six key points that could form the core of a PowerPoint presentation about it.
2. In a small group, work together to brainstorm the strengths of both qualitative and quantitative approaches to a particular issue.

Useful Resources

A wide and constantly changing array of academic and commercial websites is available to help with writing and presentation skills. An Internet search will yield many. Some sites that we found helpful are listed below, together with useful published resources.

Atkinson, C. 2011. *Beyond Bullet Points. Using Microsoft PowerPoint to Create Presentations that Inform, Motivate and Inspire.* 3rd edn. Redmond, WA: Microsoft Press.

Becker, H.S. 2007. *Writing for Social Scientists: How to Start and Finish Your Thesis, Book, or Article.* 2nd edn. Chicago: University of Chicago Press.

Braun, V. and V. Clarke. 2013. *Successful Qualitative Research. A Practical Guide for Beginners.* Los Angeles: Sage.

DeLyser, D. 2010. "Writing qualitative geography." In D. DeLyser et al., eds, *Handbook of Qualitative Geography*, 341–358. London: Sage.

Giesbrecht, W., and W. Denton. 2014. "Academic writing guide." http://www.library.yorku.ca/ccm/rg/preview/academic-writing-guide.en. York University's helpful guide to writing for a variety of academic purposes includes advice on subjects such as preparing literature reviews, research reports, and compare-and-contrast essays.

Hay, I. 2012. *Communicating in Geography and the Environmental Sciences.* 4th edn. Toronto: Oxford University Press.

Joseph, M. 2013. *Fluid New York. Cosmopolitan Urbanism and the Green Imagination.* Durham, NC: Duke University Press.

Presentation magazine, 2014. http://www.presentationmagazine.com/. Offers guidance and resources for effective speeches and presentations, from voice-improvement techniques to thousands of PowerPoint templates.

Proctor, M. 2014. "Advice on academic writing." http://www.writing.utoronto.ca/advice. A helpful and comprehensive review of academic writing, geared to students.

Reynolds, G. 2011. *presentationzen: Simple Ideas on Presentation, Design and Delivery*, 2nd edn. Berkeley, CA: New Riders.

Smith, L.T. 1999. *Decolonizing Methodologies: Research and Indigenous Peoples.* London: Zed Books.

Stewart, E.J., and D. Draper. 2009. "Reporting back research findings: A case study of community-based tourism research in northern Canada." *Journal of Ecotourism* 8 (2): 128–43.

Weebly. 2014. "Weebly—Web creation made easy." http://www.weebly.com.

Writing Center at UNC Chapel Hill. 2014. https://writingcenter.unc.edu/handouts/. Offers comprehensive writing advice on topics ranging from audience to argument, from commas to conclusions.

Appendix

Field Notes Example: *Interviewing Sam*

These notes comprise extracts from a PhD student's record of fieldwork. Tara Coleman (2013) interviewed older residents of Waiheke Island, 15 kilometres off the coast of Auckland, in order to understand the challenges and opportunities of island life. She was especially interested in residents' relationships with their dwellings, including the views out the sea. Note the way these reflections have been drafted at three points in time and the way they reflect on the researcher's feelings, and make note of her observations of artifacts that may be of significance but did not feature in the interview itself. They serve not just as a record of an encounter but also as text that can stimulate further ideas and potential in the broader research project. In this case these notes in part were the springboard to develop a paper after her PhD was complete—on the significance of blue spaces (sea, sky) in the lives of older people on Waiheke Island (Coleman and Kearns, 2015).

Before Interview with Sam

On my way to do Sam's second interview. He has texted to say how much he's looking forward to catching up today. Again I've got to be careful re attachment . . . Will use my next appointment with Frank as a full stop on my time with Sam and will tell. Its a beautiful day on the ferry. I feel very glad to be arriving at Waiheke but worried about stepping into Sam's world without being without being very sensitive to the fact he is lonely. It will be different perhaps seeing Frank who is always more open and realistic about discussing the research as a process perhaps with some friendship thrown in.

Reflecting on the Interview

Breathing laboured. Lots of pills on the table. House very messy. Lots of guitars and amps around. Sat in chair by window with music (guitar, CDs) all around him. Makes bread, showed me bread-maker, flour all around kitchen. Was very dressed up for interview and had bought quite expensive biscuits. Letters all around the house, some pinned are on the wall. He looked at view and letters as he talked. Talked a lot about achievements (showed me boat logs) and what works in house before interview officially started. House smelled damp.

After Interview with Sam

Sam talked a lot about his lounge room, view from the window there and having a place to play his guitar. He repeatedly emphasised these things through the interview. He also talked about wanting the research to help other seniors and not just those on Waiheke Island. . . . Sam talked a lot about his house being in excellent condition but it appears quite run down. There is a lot of deterioration especially around his wooden windows and some are taped up for insulation. The steps are steep and look unstable in places especially the railing. There is a lot of mould. But it was very hard to engage Sam in talking about maintenance issues. He was very keen to discuss what is rather than what is not working at his house. He seemed to work hard in the interview to present his success in maintaining himself and daily life. I will need to find different ways to discuss maintenance issues next time. Maybe I could use photography method to assist this? I seem to be spending a lot of time building rapport but this also raises the issue of attachment.

Source: Tara Coleman, Field notes towards (2013) PhD at the University of Auckland, NZ.

Glossary

abstracting Reducing the complexity of the "real world" by generating summary statements on the basis of common processes, experiences, or characteristics in the data. (See also *data reduction.*)

accidental sampling See *convenience sampling.*

action–reflection Periods of action followed by times when participants reflect on what they have done and what can be learned. The learning informs the next phase of action, creating an iterative cycle of action and reflection. This process enables change to occur throughout the research process. (See also *reflexivity.*)

action research A term coined by sociologist Kurt Lewin in the 1940s to talk about the idea of an iterative (or repeated) cycle of action and reflection in research, oriented towards solving a problem or improving a situation.

activism Political and practical action usually intended to bring about social, economic, or other change. (See also *applied people's geography* and *critical geography.*)

Actor Network Theory (ANT) An approach to social theory and research in which objects (i.e., nonhumans) are treated as part of social networks.

aide-mémoire A list of topics to be discussed in an interview. May contain some clearly worded questions or key concepts intended to guide the interviewer. Alternative term for *interview guide.*

analytical generalization A strategy for creating in-depth, rich, and credible concepts/theory. Rather than achieving generalization through large probability samples, analytical generalization is focused on the qualitative notion of transferability, specifically: (a) the careful selection of informative cases and (b) the creation of theory that is neither too abstract nor too case-specific. Readers of research narratives must be able to see how the concept might apply to other phenomena or in other contexts.

analytical log Critical reflection on substantive issues arising in an interview. Links are made between emergent themes and the established literature or theory. (See also *personal log.*)

analytic code A code that is developed through analysis and is theoretically informed; a code based on themes that emerge from relevant literature and/or the data. (See also *interpretive codes*; compare with *descriptive code.*)

anecdote A story, often personalized to the author or presenter and directly related to the point of the paper or presentation, that captures the attention of an audience and persuades them of the importance, relevance, and/or interest of what they are reading or hearing.

applied peoples' geography Term coined by David Harvey to refer to geographical research that is consciously "part of that complex of conflictual social processes which give birth to new geographical landscapes" (Harvey 1984, 7). (See also *activism* and *critical geography.*)

archival research Research based on documentary sources (for example, public archives, photographs, newspapers).

archives Narrowly defined as the non-current records of government agencies but also includes company and private papers. Typically managed by a specialist in a government agency dedicated to the records' long-term use and preservation. A distinction can be made between archives as surviving records and archives as the institution dedicated to their preservation.

archivist Professional curator of non-current records who has expertise in the accession, arrangement, and preservation of such records (in contrast to the current files of a central or local government agency that are controlled by records managers).

asymmetrical power relation A research situation characterized by an imbalance in power or influence between researcher and participant. Sometimes used to refer specifically to relationships in which informants are in positions of influence relative to the researcher. (See also *potentially exploitative power relation*, *reciprocal power relation*, and *studying up.*)

asynchronous groups Online discussions in which participants contribute at different times. Used in focus group research with groups lasting from several days to several months and with up to 40 participants. (See also *online focus group* and *synchronous groups.*)

asynchronous interviewing When the answers in an interview do not occur immediately after the questions—that is, there is some extended delay between the putting of a question and the receipt of answer. It is most common in interviews undertaken using email in which there is a delay between the sending of a question and the making of a reply, since the researcher and informant are unlikely to be undertaking the interview at the same time. (Compare with *synchronous interviewing.*)

audience The people with whom you wish to communicate the results of your research. Identifying who constitutes the audience is fundamental to the success of communicating with them. In discourse analysis, the term "audience" refers to the social process through which different collective social categories are forged, say, "American" or "tourist." Audiences are conceived to emerge through how texts are produced, circulated, intersected, and interpreted. Hence, discourse analysis sees an audience "taking shape" rather than being a pre-given category. For example, a presidential inauguration speech may help to create an audience fashioned by shared values of the collective "we" of a nation. Alternatively, infomercials selling beauty products may rely on cultural norms of attractiveness and femininity to generate an audience (market) interested in purchasing their products.

auto-ethnography A qualitative method involving the explicit writing of an embodied and situated self into the research. It often entails writing in literary fashion, involving autobiographical narratives in which the author/researcher actively reflects on her or his choices, emotions, and knowledges as a vital part of the construction of the research.

bias Systematic error or distortion in a data set that might emerge as a result of researcher prejudices or methodological characteristics (for example, case selection, non-response, question wording, interviewer attitude).

blogging Short for "web logging." Blogging is the act of creating discrete posts or entries published online. Blogs can have individual or multiple authors and typically include space for reader comments and interactions.

border pedagogy Henry Giroux's concept of teaching that challenges, crosses, and reconfigures taken-for-granted boundaries and creates borderlands where students—and intellectuals—create new identities, histories, learning, and intellectual spaces.

bulletin board An electronic medium devoted to sending and receiving messages for a particular interest group (for example, the about.com Geography Bulletin Board at http://geography.about.com/mpboards.htm?once=true&).

CAQDAS See *computer-assisted qualitative data analysis software.*

CAQ-GIS (computer-aided qualitative geographical information systems) The integration of computer-assisted qualitative data analysis software (CAQDAS) with geographic information systems (GIS) using techniques that bridge the two software systems and move towards the goal of creating complementary analysis of qualitative and spatially referenced data.

case Example of a more general process or structure that can be theorized. (See also *case study.*)

case study Intensive study of an individual, group, or place over a period of time. Research is typically done in situ.

CATI See *computer-assisted telephone interviewing.*

chain sampling See *snowball sampling.*

Chicago School of Sociology Both a body of work and a group of University of Chicago researchers from the 1920s and 1930s involved in pioneering work in urban ethnography. The Chicago School helped to establish the in-depth case study as a legitimate and powerful means for conducting relevant social science. Notable researchers from this era are William Thomas (1863–1947), Robert Park (1864–1944), Ernest Burgess (1886–1966), and Louis Wirth (1897–1952).

closed questions Questions for which respondents are offered a limited series of alternative answers from which to select. Respondents may be asked, for example, to select one or more categories, to rank items in order of importance, or to select a point on a scale measuring the intensity of an opinion. (Compare with *open questions* and *combination questions*.)

CMC **interviews** See *computer-mediated communications (CMC) interviewing*.

co-constitution of knowledge The idea, informed by post-structural theory, that knowledge does not exist "out there" ready to be discovered by objective researchers using discerning research methods but instead is built intersubjectively through interaction between the researcher and the research subjects. Qualitative research encounters are thus seen as social relationships (however fleeting) in which the researcher and the researched are closely involved in the process of construction of knowledge. (See also *co-learning*.)

codebook An organizational tool for keeping track of the codes in a project, including their meanings and applications, as well as notes regarding the coding process.

coding The processes of assigning qualitative or quantitative "values" to chunks of data or categorizing data into groups based on commonality or along thematic lines for the purposes of describing, analyzing, and organizing data.

coding structure The organization of codes into meaningful clusters, hierarchies, or categories.

co-learning A philosophy of teaching and learning that positions researchers and participants as equals in a process of mutual inquiry and education. Co-learning challenges traditional assumptions that the researcher has more knowledge and is dominant in the research relationship. It requires considerable reflexivity on the part of the researcher and a genuine desire to facilitate a process in which researcher and participants collectively create a learning community. (See also *co-constitution of knowledge*.)

collaborative research Research designed, conducted, interpreted, and disseminated by a team of local and non-local members, with local members directing the process or sharing equally in decision-making (see also *participatory action research* and *participatory research*).

collective action A process in which a group of individuals take action together to effect social or political change. This action may be the outcome of research and analysis in which individuals have come to understand the ways in which they are each implicated or affected by wider inequalities. The desire to take collective action reflects an interpersonal commitment and what some authors have called a "we intention."

colonial research Imposed, often exploitative research in both imperial and non-imperial contexts that maintains distance from, and domination of, the marginalized "others" that it seeks to study and that denies the validity of their knowledge, ways of knowing, experience, and concerns. (Compare with *decolonizing research*, *inclusionary research*, and *post-colonial research*.)

combination questions Questions made up of both *closed* and *open* components. Their closed component offers respondents a series of alternative answers to choose between, while their open component allows respondents to suggest an additional answer not listed in the closed component or to elaborate on the reason why a particular option was selected in the closed component.

common questions Asked of each participant in oral history interviews. They build up varying views and information about certain themes. (Compare with *orientation questions*, *specific questions*, and *follow-up questions*.)

communication The process of sharing research ideas, evidence, and analysis effectively with an audience.

community-based research CBR focuses on research priorities identified by community members rather than outside researchers. It aims to involve all partners in a collaborative approach to inquiry and social change through the use of multiple knowledge sources and research methods.

community supervision Procedures through which Indigenous peoples and other communities authorize research and oversee researchers and research programs.

comparative analysis A form of analysis used in case study research that compares similarities and differences across multiple instances of a phenomenon to enhance theoretical/conceptual depth. It is also known as *comparative case study* or *parallel case study*.

comparative case study See *comparative analysis*.

complete observation A situation in which observation is overwhelmingly one-way and the researcher's presence is masked such that she or he is shielded from participation.

complete participation A situation in which the researcher's immersion in a social context is such that she or he is first and foremost a participant. As a result of this level of immersion, the researcher may need to adopt critical distance to achieve an observational stance. That critical distance might be gained by reflection out-of-hours in the field or through short-term exits from the field.

computer-assisted cartography Any hardware or software that is used to facilitate map-making. *Geographic information systems* (GIS) belong under this heading.

computer-assisted qualitative data analysis software (CAQDAS) Both a general acronym and the specific acronym for the CAQDAS network based in Surrey, United Kingdom.

computer-assisted telephone interviewing (CATI) Questionnaire/interview conducted by telephone with questions read directly from a computer file and responses recorded directly onto a computer file.

computer-mediated communications (CMC) interviewing Interviews that use the medium of the Internet. They can be asynchronous, taking the form of either email exchanges or postings to a web-based platform. Less commonly, they can be synchronous through the use of chat rooms or instant messaging boards. They do not have the direct access that a face-to-face interview has or the direct voice contact of a telephone interview. (See also *asynchronous interviewing* and *synchronous interviewing*.)

concept building Refers to the process of entering and *coding* data in a systematic way that relates to the research question being asked. The ability of a software package to support the systematic organization of concepts leads to that software being categorized as theory-building software. (See also *concept mapping*.)

concept mapping As for *concept building* but refers to the specific ability to visually represent data in some form. For example, QSR NVIVO™ software uses *hierarchical tree* structures, and Atlas.ti uses network diagrams. Inspiration and Decision Explorer are purpose-built conceptual mapping programs for qualitative research.

conceptual framework Intellectual structure underlying a research project that emerges from an integration of previous literature, theories, and other relevant information. The conceptual framework provides the basis for framing, situating, and operationalizing research questions.

conditions of use form Sometimes known as an informed consent form, a form outlining what will happen to the material research participants share with you—what their rights are, who will own copyright, where recordings will be stored and for how long, what they will be used for, and so on. (See also *memorandum of understanding*.)

confederate Someone thought by other research study participants to be another participant but who is in fact part of the research team.

confirmability Extent to which results are shaped by respondents and not by researcher's biases.

constructionist approach An approach for challenging assumptions of coherence and truth within positivist knowledge (sometimes referred to as either rationalist, objectivist, or Cartesian knowledge (see *positivism*). Draws attention to social practices in the production of all knowledge, including scientific knowledge.

content analysis See *latent content analysis* and *manifest content analysis*.

controlled observation Purposeful watching of worldly phenomena that is strictly limited by prior decisions in terms of scope, style, and timing. (Compare with *uncontrolled observation*.)

convenience sampling Involves selecting cases or participants on the basis of expedience. While the approach may appear to save time, money, and effort, it is unlikely to yield useful information. Not recommended as a *purposive sampling* strategy.

copperplate script A writing style used in the nineteenth century (and earlier) produced using a sharp metal-nibbed pen. It is characterized by an elegant looping script in which the lettering is thicker on the heavy downward stroke and thinner on the upward loops. Typewriters typically replaced copperplate handwritten records in most government records from the 1880s and 1890s. (See also *modern hand* and *secretary hand*.)

co-researcher In cross-cultural research, an individual of different, "other" social/cultural identity and positionality with whom one carries out research collaboratively with equal participation in decision-making. (See also *Other*, *positionality*, and *collaborative research*.)

corroboration A strategy for guarding against threats to the credibility of a theory or concept. Corroboration often involves the process of checking that a concept/theory makes sense to the participants in a case study (i.e., participant checking). Longitudinal studies provide a useful context for corroboration whereby concepts developed through an intensive case study are checked for enduring relevance in later time periods.

creative interviewing Defined by Douglas (1985) as informant focused and driven. This approach to interviewing has been positioned as the opposite of professional interviewing in which the researcher uses the pre-interview to establish status and respect. Douglas's creative interviewing style involves the humbling of the researcher and the ceding of power and status to the informant. (Compare with *professional interviewing*.)

credibility The plausibility of an interpretation or account of experience.

criterion sampling Choosing all cases that satisfy some predetermined standard.

critical consciousness The process by which a group becomes aware of its cultural oppression and colonized mentality, and by doing so discovers that it has a popular culture, identity, and societal role.

critical geography Various ideas and practices that are committed to challenging unequal power relationships, developing and applying critical theories to geographical problems, and working for political change and social justice. (See also *activism* and *applied people's geography*.)

critical inner dialogue Constant attention to what an informant is saying, including in situ analysis of the themes being raised and a continual assessment of whether the researcher fully understands what is being said.

critical reflexivity See *reflexivity*.

cross-case comparison A strategy for comparative analysis that compares different case studies. Such comparisons help to develop richly detailed conceptual explanations of phenomena.

cross-sectional case study A case study conducted at one point in time. Contrast with *longitudinal case study*.

culturally safe Having knowledge of the history, beliefs, and practices of minority groups and maintaining awareness of these factors.

cultural protocols Local-, community-, or group-defined codes of appropriate behaviour, interaction, and communication to which outsider researchers are expected to adhere.

data reduction Using categorization and qualification to lump data together into larger packages, thereby reducing the complexity and number of data points but increasing the level of understanding of trends, processes, or other insights. (See also *abstracting*.)

debriefing Procedure by which information about a research project (some of which may have been withheld or misrepresented) is made known to participants once the research is complete.

decode To analyze in order to understand the hidden meanings.

decolonizing research Research whose goals, methodology, and use of research findings contest imperialism and other oppression of peoples, groups, and classes by challenging the cross-cultural discourses, *asymmetrical power relationships*, and institutions on which they are based. (See also *applied people's geography*, *inclusionary research*, and *post-colonial research*.)

deconstruction A method for challenging assumptions of coherence and truth within a *text* by revealing inconsistencies, contradictions, and inadequacies (for example, when matters that are problematical have been naturalized).

deduction Reasoning from principles to facts. (Compare with *induction*).

deep colonizing Colonialism perpetuated through relationships and practices that are embedded in supposedly decolonizing institutions and practices.

dependability Minimization of variability in interpretations of information gathered through research. Focuses attention on the researcher-as-instrument and the extent to which interpretations are made consistently.

descriptive code A *code* describing some aspect of the social data, typically aspects that are fairly obvious. (See also *manifest message* and *initial code*. Compare with *analytic code*.)

deviant case sampling Selection of extraordinary cases (for example, outstanding successes, notable failures) to illuminate an issue or process of interest.

dialogic nature of research and writing Research and writing are dialogic in that they are relational. In this sense, research is always informed and constructed by writing, just as writing is always already informed and constructed by research. The dialogic nature of research and writing extends beyond their relationality to each other and also alludes to their relationality to the wider set of social and spatial relations within which academic researcher–writers work.

diary interviews Diary interviews and diary photographs are interrelated techniques encouraging respondents to take photos of their daily routines and record their experiences, encounters, feelings, and other thoughts, the specific and the general, in written or montage form. These texts constitute the basis for one or more in-depth interviews. Commenting on this approach, Latham (2003; 2002) asserts that the diaries and interviews themselves become "a kind of performance" enabling deeper insights into the lived reality of specific spatial engagements.

diary photographs See *diary interviews*.

dichotomy A division or binary classification in which one part of the dichotomy exists in opposition to the other (for example, light/dark, rich/poor). In most dichotomous thinking, one part of the binary is also more positively valued than the other.

disclosure When a researcher reveals information about herself or himself or the research project or when research participants reveal information about themselves.

disconfirming case Example that contradicts or calls into question researchers' interpretations and portrayals of an issue or process.

discontinuous writing A form of longitudinal journal kept by a research respondent but which does not require entries to be made on a strict daily basis. Rather, the respondent records certain events and their reactions to and/or feelings about them as they occur or in reflection. The events/feelings to be recorded are determined by the researcher, typically in negotiation with the respondent. Such diaries are also referred to as *solicited diaries*, denoting the relationship between researcher and respondent. Meth (2003) provides a thorough discussion of the uses and limitations, especially matters of ethics, associated with discontinuous writing. See also Bell (1998) on solicited diaries.

discourse There is no fixed meaning for "discourse." The term has accrued a number of meanings that are in circulation in both academic and popular cultures. Even among cultural theorists, whose ideas human geographers draw upon, "discourse" is used in differing ways (for example, Mikhail Bakhtin's double-voiced discourse). In this book, "discourse" is generally understood as it was used by Michel Foucault. To make matters more complex, Foucault employed the term in at least three different ways: (1) as written/visual texts or statements that have meaning and effect, (2) an individual system or group of texts or statements that have meaning, and (3) a regulated practice of rules and structures that govern particular texts or statements. Particular attention in this book is given to his third definition because it is this rule-governed quality of discourse that is of primary importance to geographers. This definition of discourse evokes how it shapes social practices, influencing our actions, attitudes, and perceptions. (See also *discourse analysis*.)

discourse analysis Method of investigating rules and structures that govern and maintain the production of particular written, oral, or visual *texts*. (See also *discourse*.)

discursive structures or formations A key concept of Foucauldian *discourse analysis*. The rules and structures governing the production of *discourse* that affect the way individuals think, act, and express themselves—for example, through travel, comportment, clothes, make-up.

document In archival research, an individual archived item such as a memorandum or letter, handwritten or typed, that constitutes a single item or part of a larger file.

emancipatory A term used to describe a process by which someone is freed from political or other restrictions.

embodied knowledge Refers to how bodies are experienced. The ways people make sense of their experiences, and themselves, cannot be separated from competing and contradictory discourses through which bodies are given meaning. For example, in the context of the uneven gendered social relationships of a public bar, there are normative assumptions about what women should do and wear to become "attractive" to men. Hence, because of the possibilities of feeling self-confident, attractive, and feminine, some women may make a deliberate decision to make their bodies more visible by wearing a low-cut top and push-up bra, for example. Others may make deliberate decisions to make their bodies less visible to avoid being marked as sexual objects, experiencing inappropriate sexist behaviour, and being disrespected.

emoticon Used commonly in Internet communications as well as texting or SMS (short messaging service). The word is an amalgam of the concepts of emotion and icon. Emoticons began as text symbols that approximate crude graphics, such as a colon and a right-side parenthesis, to approximate a smiley face, if looked at from side-on :). Most word processing systems now convert those two characters into the smiley graphic ☺. These crude graphics can efficiently convey emotions that would otherwise take some time to convey in written word. Popular abbreviations and acronyms have similarly become popular in Internet and SMS communication (e.g., "lol" for laugh out loud).

episteme In the writings of Michel Foucault, episteme refers to the whole sets of discursive structures/formations within which a culture thinks. An episteme refers to the social processes by which certain statements about the world are considered as knowledge and others are dismissed. Episteme therefore requires critically addressing the range of methodologies that a culture employs at a particular time as "common sense" to allow certain statements to become knowledge about particular people, events, and places. (See also *discursive structures or formations*.)

epistemology Ways of knowing the world and justifying belief. (See also *ontology*.)

empowerment The process of increasing the social, political, spiritual, economic, and/or psychological potential of individuals and communities.

ethnographic research See *ethnography*.

ethnography A research method dependent on direct field observation in which the researcher is involved closely with a social group or neighbourhood. Also an account of events that occur within the life of a group, paying special attention to social structures, behaviour, and the meaning(s) of them for the group.

extensive research Research typically involving large-scale questionnaires or other standardized or semi-standardized methods to identify regularities, patterns, and distinguishing features of a population and to yield descriptive generalization. (Compare with *intensive research*.)

external validity See *generalizability*.

extreme case sampling See *deviant case sampling*.

facilitator The person who encourages or moderates the discussion in a *focus group*. In *participatory action research*, a person who helps others to learn by guiding an appropriate process rather than imposing their own agenda, using techniques that enable people's self-reflection and analysis.

falsification Derived from the writings of Karl Popper (1902–94), the concept of falsification suggests that it is possible to demarcate scientific theory from non-scientific theory on the basis that scientific theories could be falsified or proven untrue by empirical observation and testing. Popper noted that it was not possible to prove a scientific theory through recourse to empirical evidence, since scientific knowledge is always contingent (that is, since scientific observations cannot include all aspects of a phenomenon, we can never be sure that a scientific theory covers the total population of a phenomenon). Accordingly, Popper developed his concept of falsification, which allowed for the contingency of scientific knowledge and allowed scientists to state that a particular theory had yet to be proven false and thus that theory could still be considered valid. Popper's famous example is that it only takes one black swan to falsify a theory that "all swans are white." Although this term is used largely by quantitative researchers, in the context of qualitative studies falsification may help to develop more robust concepts and/or open up new areas of inquiry by exploring negative cases.

field notes An accumulated written record of the fieldwork experience. May comprise observations and personal reflections. (Compare with *research diary.*)

field relations Refers to the ways in which we relate to those individuals and groups we encounter while undertaking fieldwork. One challenge in maintaining integrity in field relations can be achieving a balance between ethical conduct and rigorous procedures in the collection of qualitative data.

fieldwork diary See *field notes.*

files A set of archived papers—typically held together by a paperclip—created by an official agency and relating to a common topic or theme. Usually organized in reverse chronological order with the newer material overlaying older documents.

finding aids *Archives* equivalent to a library catalogue, typically taking the form of a list of accession of *files* by name as originally organized by creating agencies. Some archives now have computer-based systems that allow material to be located by use of keywords.

first-person narrative In the context of a research narrative (e.g., research report, journal publication, public talk), the presentation of the researcher(s) as "I" or "we." In first-person narrative, researchers or "narrators" are able to insert their stance, beliefs, emotions, and assumptions explicitly into the text. In so doing, they become more accountable, and their knowledge can be better "situated." (Compare with *third-person narrative.*)

focus group A research method involving a small group of between 6 and 10 people discussing a topic or issue defined by a researcher, with the researcher facilitating the discussion.

follow-up questions Sometimes known as *prompts*, these are questions that permit the interviewer to ask the participant to elaborate on certain elements of an earlier response.

funnel structure Interview question ordering such that the topics covered move from general issues to specific or personal matters. (Compare with *pyramid structure.*)

generalizability The degree to which research results can be extrapolated to a wider population group than that studied. (See also *transferability.*)

genre In discourse analysis, a subcategory of source forms. For example, oral sources can be subdivided into genres including oral histories, semi-structured interviews, focus groups, conversations, and life narratives. Remaining alert to the genre of the source form is crucial because particular accounts of the world are articulated through differences among the intended audience, social relationships, and technologies of production.

geographic information system (GIS) A generic title for a number of integrated computer tools for the processing and analyzing of geographical data, including specialized software for input (digitizing) and output (printing or plotting) of mappable data. GIS is not the name of a specific software package.

geovisualization A combined form of the terms "geographic" and "visualization." Geovisualization refers to the analysis of geospatial data through the use of interactive visualization. By depicting geospatial data visually, its interpretation may be enhanced and extended.

GIS See *geographic information system.*

go-along interviews As the name suggests, the go-along interview involves the researcher accompanying the respondent within the "field" and engaging in a direct discussion of spatial engagement. This technique combines aspects of well-established techniques such as *interviewing*, *oral history*, and *participant observation*.

grounded theory A systematic inductive (data-led) approach to building theory from empirical work in a recursive and reflexive fashion. That is, using a method of identifying themes or trends from the data, then checking through the data (or collecting more), then refining the themes using repeated checks with the data to build theory that is thoroughly "grounded" in the real world. Initiated by Glaser and Strauss (1967) and reinterpreted and refined greatly since then by them and other authors.

hegemony A social condition in which people from all sorts of social backgrounds and classes come to interpret their own interests and consciousness in terms of the *discourse* of the dominant or ruling group. The hegemony of the dominant group is thus based, in part at least, on the (unwitting) consent of the subordinate groups. Such consent is created and reconstituted through the web of social relations, institutions, and public ideas in a society.

hermeneutic circle The circle (or more broadly, process) of interpretation of qualitative information, which accounts for the point that no such interpretation is free from the values, experiences, attitudes, and ideas of the observer or researcher. Implicit in this realization is a need for the researcher to be clear about his or her position and to ensure that interpretation is participatory and iterative—that is, involves participants and is done in one or more collaborative rounds.

hermeneutics The study of the interpretation of meaning in *texts*, whether there is assumed to be a single dominant meaning or a multiplicity of meanings.

idiographic An approach to knowledge that highlights the particular, subjective, and contingent aspects of the social world. Cases are understood holistically. Credible and authentic descriptions are emphasized instead of statistical *generalizability*. Contrast with *nomothetic*.

inclusionary research *Decolonizing research* projects that empower marginalized and oppressed peoples, groups, and classes with training and tools that they can use to transform their situations and conditions. (See also *post-colonial research*.)

indexicality Indexicality of photographs means that photographs have different meanings in different occasions. For example, this could be temporal (someone engaging an image of the American Civil War in 1864 would derive a different meaning than a viewer of the same image in 2016); it could be social: viewers of the same image will derive different meanings based on their social identity such as gender, sexual identity, ethnicity, class, national identity and so on.

Indigeneity The term describes the evolving pan-Indigenous movement and corresponding identity among peoples who, despite often considerable cultural divergence, share similarities founded in an ancestral birthright in the land, a common core of collective interests, and the shared experience of dispossession precipitated by ongoing colonialism.

Indigenous methods Methods of research conceived and articulated from Indigenous worldviews. (See also *Indigenous methodologies*.)

Indigenous methodologies In many Indigenous settings, the term "research" is deeply enmeshed in colonizing processes and experiences. In developing approaches to contemporary research, many Indigenous groups are exploring ways of conceptualizing, executing, disseminating, and evaluating research against their own explicit criteria. These methodologies affirm Indigenous peoples' ways of knowing, research purposes, and protocols as well as critiquing and adapting "Western" research paradigms. Adoption of Indigenous methodologies shifts the balance of power, responsibility for ethical oversight, and judgment about the value, meaning, and utility of research away from traditional research institutions such as universities and funding agencies and towards contemporary structures of Indigenous governance, decision-making, and accountability—including new structures and processes within some universities and agencies.

induction Process of generalization involving the application of specific information to a general situation or to future events. (Compare with *deduction*.)

informant Person interviewed by a researcher. Some refer to those who are interviewed as "subjects" or "respondents." Others argue that someone who is interviewed, as opposed to simply observed or surveyed, is more appropriately referred to as an informant. That is because an interview informant is likely to have a more active and informed role in the research encounter.

information and communication technologies (ICT) Sometimes known as information technology (IT). ICT refers to the convergence of audio-visual and telephone technologies and networks with computer networks. It is often a shorthand for the digital and analog devices in our daily lives.

informed consent Informant/subject agreement to participate in a study having been fully apprised of the conditions associated with that study (for example, time involved, methods of investigation, likely inconveniences, and possible consequences).

informed consent form See *conditions of use form.*

initial codes Codes that are pre-determined in some way, usually because they are a prominent theme in the research questions or are inherent in the topic of the research. (See also *descriptive code.*)

insider A research position in which the researcher is socially accepted as being "inside" or a part of the social groups or places involved in the study. (Compare with *outsider.*)

intensive research Research typically of individual agents or small groups, involving semi-standardized or unstructured methods (e.g., interviews, oral histories, ethnography). Focuses on causal processes and mechanisms underpinning events and particular cases. (Compare with *extensive research.*)

interpretive codes Codes that emerge from analysis and interpretation of the data along themes and toward theory generation. (See also *analytic codes.*)

interpretive community Involves established disciplines with relatively defined and stable areas of interest, theory, and research methods and techniques. Influences researchers' choice of topics and approaches to and conduct of study.

intersubjectivity Meanings and interpretations of the world created, confirmed, or disconfirmed as a result of interactions (language and action) with other people within specific contexts. (See also *subjectivity* and *objectivity.*)

intertextuality The necessary interdependence of a *text* with those that have preceded it. Any text is built upon and made meaningful by its associations with others.

interview A means of data collection involving an oral exchange of information between the researcher and one or more other people.

interview guide A list of topics to be covered in an interview. May contain some clearly worded questions or key concepts intended to guide the interviewer. (Compare with *interview schedule.*)

interview schedule Ordered list of questions that the researcher intends to ask informants. Questions are worded similarly and are asked in the same order for each informant. In its most rigid form, an interview schedule is a questionnaire delivered in face-to-face format. (Compare with *interview guide.*)

interviewer effects When a person is interviewed, they are not doing their normal everyday activities. An interview is a formal data-gathering process. This formality, and the unusual discursive style of an interview, can have an influence on what informants say and how they say it. This is one example of an interviewer effect. There could also be effects that flow from the demeanour, dress, accent, and physiology of an interviewer. More broadly, in ethnographic work, they are referred to as researcher effects.

***in vivo* codes** Codes that emerge from the body of the work being examined; phrases and terms used by respondents in the course of ethnographic research or already appearing in examined texts that suggest a theme worthy of analysis. (See also *coding.*)

landscape Landscape is used broadly to mean a built, cultural, or physical environment (and even the human body), which can be "read" and interpreted.

latent content analysis Assessment of implicit themes within a text. Latent content may include ideologies, beliefs, or stereotypes. (Compare with *manifest content analysis.*)

latent message The underlying or implied meanings of data; compare with *manifest message.*

legitimacy Approval and respect accorded researchers, research projects, and research methodologies and methods that are considered appropriate and are valued and welcomed by the people(s) and communities with whom researchers work.

life history An interview in which data on the experiences and events of a person's life are collected. The aim is to gain insights into how a person's life may have been affected by institutions, social structures, relations, rites of passage, or other significant events. (Compare with *oral history.*)

listening posts Small columns, poles, or other parts of a designed display containing audio devices or speakers—such as one might find in a museum—that allow playback of recordings (e.g., from interviews or oral histories) and other sounds.

literature review Comprehensive critical summary and interpretation of resources (for example, publications, reports) and their relationship to a specific area of research.

local research authorization Review of researchers, research projects, and research methodologies and methods by Indigenous peoples, communities, or local groups whose permission is sought or required prior to research with them or in their territories.

longitudinal case study A case study that involves a revisit whereby the researcher returns to the case after an intervening time period during which no appreciable research is done.

lurking Observing—typically anonymously and perhaps with the purpose of gathering data—and not contributing to chat rooms, listserves, or other online platforms. Tends to be regarded as anti-social behaviour.

manifest content analysis Assessment of the surface or visible content of text. Visible content may include specific words, phrases, or the physical space dedicated to a theme (for example, column centimetres in a newspaper or time in a video). (Compare with *latent content analysis.*)

manifest message The plainly visible content of the data; compare with *latent message.*

margin coding A simple system of categorizing material in transcripts. Typically involves marking the transcript margin with a colour, number, letter, or symbol code to represent key themes or categories.

masculine gaze This term speaks to the ways in which a viewer looks upon the people either present or represented (e.g., via photography, painting). Feminist theory has added to understanding by speaking of the masculine gaze to express an asymmetric (or unequal) power relationship between viewer and the person or population viewed.

maximum variation sampling Form of *sampling* based on high diversity aiming to uncover systematic variations and common patterns within those variations.

member checking See *participant checking.*

memoing In qualitative software systems, a process whereby the researcher may write memos or reflections on the research process as she or he works and then incorporate these memos as data for further investigation.

memorandum of understanding (MoU) A document specifying the aims, process, roles, responsibilities, and rights of parties involved in a research project. (See also *conditions of use form.*)

metaphor An expression applied to something to which it is not literally applicable in order to highlight an essential characteristic.

method The means by which data are collected and analyzed (e.g., in-depth interviewing).

methodology The philosophical and theoretical basis for conducting research that is much broader and sometimes more politically charged than method alone (e.g., feminist methodology).

mimetic Miming or imitating.

mixed methods A combination of techniques

for tackling a research problem; the term is often used specifically to mean a combination of quantitative and qualitative methods.

modality See *objective modality* and *subjective modality*.

moderator See *facilitator*.

modern hand A way of referring to the array of more simplified ways of forming linked letters constituting writing, as opposed to printing, from the later nineteenth century onwards. Initially produced by metal-nibbed pens dipped in ink and later with fountain pens. From an archival researcher's perspective, this is often a sort of "dark age" when handwriting standards slip and file material can be very difficult to read and may be badly smudged. It is, however, important not to ignore handwritten notes and comments in favour of typewritten records because the former can often provide vital clues as to the concerns underpinning decisions. (See also *copperplate script* and *secretary hand*.)

multiple interviews A series of interviews recorded with the same interviewee in separate sessions over a period of time.

multiple ontologies The appreciation that different societies (or groups and individuals in society) might have different views of the world, or have diverse ways of being in the world. (See also *ontology*).

mystification In a research narrative (e.g., research report, journal publication, public talk), the concealment of particular details in ways that give other details legitimacy and/or coherence. For example, scientific third-person writing and grammatical structure may sometimes involve obscuring details of thought, feeling, emotion, time, place, embodiment, and stance in ways that mystify the social agents and participants being described. Third-person writing is particularly useful for mystifying the social location, partiality, and embeddedness of authors (researchers). Haraway refers to this as the "God trick" of being everywhere and nowhere simultaneously.

neo-positivism Variants of positivism as they have evolved from the original axiom of "logical positivism" developed by the Vienna Circle in the 1920s and 1930s. Neo-positivists have responded to various critiques (especially Popper's critique that logical positivism's reliance on "verifiability" was too strong a criterion for science and instead we should rely on falsifiability as a primary criterion for scientific knowledge) and incorporated these critiques into their work. Neo-positivism was most strongly represented in geography during the quantitative revolution of the 1960s and 1970s.

new media The on-demand information and interactivity associated with the Internet and the technologies and devices used to access that information. New media include anything that enables digital interactivity—the devices, the software or applications, and the online sites they access.

nominalization The transformation of verbs and adjectives into nouns. Nominalization reduces information available to readers, and it mystifies social processes by hiding actions and the identity of actors.

nomothetic An approach to knowledge that emphasizes generalizability for understanding the social world. Social phenomena are reduced to variables for the purposes of generating statistically generalizable findings. The credibility, authenticity, and holistic understanding of each sub-unit studied is of lesser importance. (Contrast with *idiographic*.)

NUD*IST™ A software system for qualitative data analysis developed by Richards and Richards at Qualitative Research Solutions in Australia. The acronym stands for Nonnumerical, Unstructured Data: Indexing, Searching, Theorising. Superseded by *NVivo*.

NVivo A software package to help organize and analyze qualitative data. This is a specific form of computer-assisted qualitative data analysis software (CAQDAS). Allows importing and coding of textual data, text editing, coded data retrieval and review, word and coding pattern searches, and data import/export to quantitative analysis software. (See also *computer-assisted qualitative data analysis software* and *NUD*IST™*.)

objective/objectivity Unaffected by feelings, opinions, or personal characteristics. Often contrasted with *subjectivity*. (See also *intersubjectivity*.)

objective modality A form of writing that implicitly hides the writer's presence in the text (for example, third-person narrative form) but that clearly signals agreement with the statement being made. (Compare with *subjective modality.*)

observation Most literally, purposefully watching worldly phenomena. Increasingly broadened beyond seeing to include apprehending the environment through all our senses (for example, sound, smell) for research purposes.

observer-as-participant A research situation in which the researcher is primarily able to observe but in so doing is also participating in a social situation. (Compare with *participant-as-observer.*)

online focus group Focus groups that are conducted online using real-time technology such as chat rooms or asynchronous technology such as bulletin boards. (See also *synchronous groups* and *asynchronous groups.*)

ontology Beliefs about the world. Understanding about the kinds of things that exist in the universe and the relations between them. (See also *epistemology* and *multiple ontologies.*)

open questions Questions in which respondents are able to formulate their own answers, unrestricted by having to choose between pre-determined categories. (Compare with *closed questions* and *combination questions*).

opportunistic sampling Impromptu decision to involve cases or participants in a study on the basis of leads uncovered during fieldwork.

oral history A prepared *interview* conducted in question-and-answer format with a person who has firsthand knowledge of a subject of interest. (Compare with *life history.*)

oral methods Verbal techniques, such as *interviews* or *focus groups*, as opposed to written methods for seeking information.

Orientalism As used by Edward Said (1978), a key post-colonial term (see *post-colonialism*) referring to Western (mis)representations and construction of an imagined Orient in discourses that serve to produce and legitimize imperialism.

orientation questions Used in *oral history* interviews to establish the participant's background. (Compare with *common questions, specific questions,* and *follow-up questions*).

Other The non-Self, groups and peoples perceived as fundamentally different from one's self and against which a person might compare themselves and establish their own social position, meaning, and identity. Also taken to mean that which is oppositional to the mainstream—marginal or outside the dominant ideology. Initially developed by Simone de Beauvoir in her 1949 book *The Second Sex* to characterize patriarchal representations and subjugation of women, the term was extended by Franz Fanon (1967) and Edward Said (1978) to the cross-cultural representations and relationships that underlie colonialism.

outsider A research position in which the researcher is rendered "outside" a social circle or feels "out of place" on account of differences such as visible appearance, unfamiliarity, or inability to speak the language or vernacular used. (Compare with *insider.*)

over-disclosure Can occur in an interview or focus group when research participants reveal personal, sensitive, or confidential information that goes beyond the scope of the research or that they may regret having mentioned after the interview or focus group. (Contrast with *under-disclosure.*)

panopticon A circular prison with cells surrounding a central guards' station. In the panopticon, inmates may be observed at any time, but they are not aware of the observation.

paradigm Set of values, beliefs, and practices shared by a community (e.g., members of an academic discipline) that provides a way of understanding the world.

paralinguistic clues Tacit signs perceptible in face-to-face interviewing. The tone of speech used by an informant is an important indicator of their emotional disposition when answering a question. It can also indicate an informant's level of comfort and the degree of rapport between the researcher and the informant. Other non-spoken clues include eye contact, fidgeting, furtive glances, and aggressive or defensive postures.

parallel case study The study of multiple cases at the same time for the purposes of *comparative analysis.*

participant Person taking part in a research project. Usually the *informant* rather than a member of the research team.

participant-as-observer A research situation in which the researcher is primarily a participant in a social situation or gathering place but in so doing can maintain sufficient critical distance to observe social dynamics and interactions. (Compare with *observer-as-participant.*)

participant checking Informant's review of the transcript of her or his contribution to an *interview* or *focus group* for accuracy and meaning. May also involve the informant reviewing the overall research output (for example, thesis, report). Also serves as a means of continuing the involvement of informants in the research process.

participant community One's research participants; the community may be known as such to one another—members of a formal or informal grouping—or may be seen as a community by the researcher.

participant observation A fieldwork method in which the researcher studies a social group while being a part of that group.

participation A process in which people play active roles in decision making and other activities affecting their lives.

participatory action research (PAR) An umbrella term covering a range of participatory approaches to action-oriented research involving researchers and participants working together to examine a situation and change it for the better. (See also *action research*, *participatory research*, *collaborative research*, and *decolonizing research.*)

participatory diagramming A technique whereby a group of people, with support from a *facilitator*, collectively produces a visual representation (for example, drawing, diagram, chart, mindmap, sketch) for subsequent analysis using locally appropriate materials (for example, stones, leaves, chalk, ground, pens, paper, whiteboards), criteria, and symbols. Diagrams usually convey relationships between key stakeholders, institutions, or resources, sometimes over different time periods. (See also *participatory mapping.*)

participatory geographic information systems (PGIS) A process that adapts GIS software to incorporate local expertise and knowledge, usually within a participatory action research framework to enable mapping and data analysis by non-professionals. PGIS emerged in response to criticisms that the high financial, time, and training requirements of GIS can discourage grassroots groups from using it to inform their own research and development. (See also *participatory mapping.*)

participatory mapping A technique whereby a group of people, with support from a *facilitator*, collectively produces a "map" for subsequent analysis using locally appropriate materials, criteria, and symbols. Maps usually focus on material aspects of life such as a watershed, a village, a body, or the distribution of particular resources within a particular area. (See also *participatory diagramming* and *participatory geographic information systems.*)

participatory research A community-based approach to research involving local people and their knowledges as a foundation for social change. Participatory research was developed by educators like Paulo Freire in Brazil and others in the global South to support consciousness-raising and political action. It now informs participatory action research and processes of participatory development within the global South.

"pencil only" rule A typical archival convention to ensure the protection of file materials, it requires that researchers make notes in pencil only. The rule is rendered somewhat redundant by the increasing use of personal computers.

performativity Employed in human geography to denote the manner in and through which ideas and concepts are performed or acted out. In doing so, roles and expectations are given physical expression. The acts we actually perform, are expected to perform, and/or struggle against performing are not natural but are complex engagements with power.

personal log Recorded reflections on the practice of an interview. Includes discussions of the appropriateness of the order and phrasing of questions and of the informant selection. Also contains assessments of matters such as research design and ethical issues. (See also *analytical log.*)

PGIS See *participatory geographic information systems.*

photo-elicitation A technique related to diary photographs and *diary interviews*. However, according to Harper (2002), the principal objective of this technique is to elicit alternative ways of seeing and understanding the same image among respondents. The outcome of this is to invoke a deep discussion of values and meaning.

photovoice A methodology that was developed within health education and has now been adopted and adapted within the social sciences, including geography. Participants are asked to represent their perspective by taking photographs, which subsequently offer a window into how they conceptualize their circumstances. Photography "gives voice" to otherwise marginalized groups and has been associated with social action.

pilot study Abbreviated version of a research project in which the researcher practises or tests procedures to be used in a subsequent full-scale project.

pluriversal world The idea that the world is made up of manifold, heterogeneous, dynamic ways of being and knowing.

polycentric epistemologies Different groups may have diverse ways of knowing, asking different types of questions about the world and transmitting them in varied ways. In other words, there are many different ways of conceptualizing knowledge. (See also *epistemology*).

population The larger group from which a *sample* has been selected for inclusion in a study. In quantitative research, based on *probability* (random) *sampling*, it is assumed that the sample has been selected such that the mathematical probability of sample characteristics being reproduced in this broader population can be calculated. In qualitative research where *purposive sampling* is used, no such assumption is made.

positionality A researcher's social, locational, and ideological placement relative to the research project or to other participants in it. May be influenced by biographical characteristics, such as class, race, and gender, as well as various formative experiences.

positionality statement A record of how a researcher is situated in the production of knowledge over time. This requires careful reflection on the researcher's points of connection and disconnection with a project. Positionality statements are not stand-alone testimonials. Instead, they document the co-constitution of researcher and project. (See also *co-constitution of knowledge*.)

positivism An approach to scientific knowledge based around foundational statements about what constitutes truth and legitimate ways of knowing. There are a number of variants of positivist thought, but central to all is the construction of a singular universal and value-free knowledge based on empirical observation and the scientific method.

postcolonialism An approach to knowledge that seeks to represent voices of the *Other*, especially colonized peoples and women, and to recognize knowledge that has been ignored through processes of colonization and patriarchy.

postcolonial research Research that rejects imperialism and the goals, attitudes, representations, and methods of imposed, "colonial" research and instead seeks to conduct research that is welcomed and that fosters egalitarian relationships and openness, values local knowledge and ways of knowing, and contributes to self-determination and locally defined welfare. (See also *decolonizing research* and *inclusionary research*. Contrast with *colonial research*.)

post-modernism A movement in the humanities and social sciences that includes *postcolonialism* and embraces the pluralism of multiple perspectives, knowledges, and voices rather than the grand theories of modernism. Individual interpretation is considered partial because it is to some degree socially contingent and constituted.

post-structuralism A school of thought that endeavours to link language, subjectivity, social organization, and power.

potentially exploitative power relation A research situation in which the researcher is in a position of power relative to the research participant. (See also *asymmetrical power relation*, *reciprocal power relation*, and *studying up*.)

power In Foucauldian discourse analysis, power is central to thinking about *discourse* as something that has an impact. It is through power that the elements of discourse have effects on what people do and think and how they express themselves. Yet power is not conceptualized in terms of acting upon people in an oppressive way. Rather, the individual is seen as an effect of power. That is, power makes things possible as well as restricting possible actions and attitudes. These possibilities are instances of power/knowledge relationships. (See also *power/knowledge*.)

power/knowledge A key concept of Foucauldian discourse analysis. Foucault argues that the relationship between power and knowledge is essential to thinking about the effects of *discourse*. He argues that statements that are accepted as knowledge are themselves the outcome of power struggles. For example, what has constituted geographical knowledge in universities has been a constant struggle over different versions of what constituted space/place.

praxis The use of research findings by researchers to make constructive social and environmental change.

preliminary meeting In *oral history*, a first meeting between interviewer and *respondent* designed to establish rapport, clarify ethical responsibilities and rights, explore the parameters of the pending interview(s), and work through other matters of mutual interest or concern; an important step for both parties after which additional preparations may be made for the interview.

pre-testing See *pilot study*.

primary observation Research in which the investigator is a participant in and interpreter of human activity involving her or his own experience. (Compare with *secondary observation*.)

primary question Interview question used to initiate discussion of a new topic or theme. (Compare with *probe*.)

primary sources From a historical geography perspective, primary sources in the narrower sense are public and private records created in an earlier period of time that is of interest to the researcher. They include letters, diaries, journals, original census returns, minutes of organizations, and files of government departments. In addition, original maps, survey plans, and photographs would be included. In some circumstances, period newspapers and published official documents might also be regarded as primary sources. (Compare with *secondary literature*.)

privileged discourse While there are always competing and sometimes contradictory cultural discourses to make sense of the world, privileged discourse is one that takes priority over others in shaping social, cultural, and political meanings. For example, there exists a range of conflicting and competing discourses about Hawai'i. Yet in Western society, that which is privileged portrays Hawai'i as an earthly paradise.

probability sampling Sampling technique intended to ensure a random and statistically representative *sample* that will allow confident generalization to the larger *population* from which the sample was drawn. (Compare with *purposive sampling*.)

probe A gesture or follow-up question used in an interview to explore further a theme or topic already being discussed. (See also *prompt*. Compare with *primary question*.)

professional interviewing A style of interviewing relationship in which the researcher seeks to establish and maintain the respect of the informant. The interview has a formal feel to it, and the relationship between the researcher and informant is intended to be professional. This style of interviewing relationship has been considered important when researchers are interviewing powerful people and the researcher believes that in order for him or her to be taken seriously, it is important to seek and maintain the respect of the informant (Schoenberger 1992) (Compare with *creative interviewing*.)

prompt A follow-up question in an *interview* designed to deepen a response (for example, "Why do you say that?," "What do you mean?"). (See also *probe*. Compare with *primary question*.)

provenance Organizational principle for *archives* that stresses the importance of the original internal arrangement of a collection of *files* and the order of information in files as devised by their creating agencies as a means of understanding past events.

psychoanalysis As a critical visual methodology

in geography, psychoanalysis is concerned mostly with subjectivity, sexuality, and the unconscious.

purposeful sampling See *purposive sampling.*

purposive sampling *Sampling* procedure intended to obtain a particular group for study on the basis of the specific characteristics they possess. (Compare with *probability sampling.*) Aims to uncover information-rich phenomena/participants that can shed light on issues of central importance to the study.

pyramid structure Order of interview questions in which easy-to-answer questions are posed at the beginning of the interview while deeper or more philosophical questions/issues are raised at the end. (Compare with *funnelling.*)

qualitative GIS An approach to mixed methods research involving the iterative, integrated analysis of qualitative and quantitative data through the capabilities of geographic information systems, such as through spatial analysis and mapping in concert with qualitative coding.

quantitative methods Statistical and mathematical modelling approaches used to understand social and physical relationships.

quantitative revolution A period in the mid-twentieth century (particularly from the late 1950s) during which transformations in information and computer technologies, developments in mathematical modelling, and sophisticated quantitative techniques influenced the form and nature of research being conducted. The quantitative revolution led to an increased use of statistical techniques for collating and analyzing large amounts of data, with these techniques being linked to empirical testing of models, theory, and hypotheses.

random sampling See *probability sampling.*

rapport A productive interpersonal climate between informant and researcher. A relationship that allows the informant to feel comfortable or confident enough to offer comprehensive answers to questions.

reciprocal power relation A research situation in which researcher and *informant* are in comparable social positions and experience relatively equal costs and benefits of participating in the research. (See also *asymmetrical power relation, potentially exploitative power relation*, and *studying up.*)

records A generic term for files, maps, plans, and other documents held in an *archive.*

recruitment The process of finding people willing to participate in a research project. Recruitment strategies can range from asking people "on the street" (perhaps to fill in a questionnaire) to inviting key individuals to participate (in a focus group, for example).

reflexivity Self-critical introspection and a self-conscious scrutiny of oneself as a researcher. (See also *action–reflection.*)

reification When the complexity of social life is reduced to concrete and simplified "things" in the construction of texts. For example, one might talk about "the research" or "the participant," thereby obscuring the way in which these "nouns" come to be constituted and expressed through a variety of social and spatial relations.

reliability Extent to which a method of data collection yields consistent and reproducible results when used in similar circumstances by different researchers or at different times. (See also *validity.*)

replicability Ability to be repeated or tested to see how general the particular findings of a study are in the wider *population.*

representation The way in which something (the world, human behaviour, a city, the landscape) is depicted, recognizing that it cannot be an exact depiction. An important insight from post-structuralist thinkers is that representations not only describe the social world but also help to shape or constitute it.

research diary A place for recording observations in the process of being reflexive. Contains thoughts and ideas about the research process, its social context, and the researcher's role in it. The contents of a research diary are different from those of the field notes, which more typically contain qualitative data, such as records of observations, conversations, and sketch maps.

research design The framework, encompassing

question, theory, method, and procedures that are used to conduct research.

research ethics Refers to the moral conduct of researchers and their responsibilities and obligations to those involved in the research.

respondent See *informant.*

rigour Accuracy, exactitude, and trustworthiness.

sample Phenomena or participants selected from a larger set of phenomena or a larger *population* for inclusion in a study.

sampling Means of selecting phenomena or participants for inclusion in a study. A key difference between qualitative and quantitative inquiry is in the logic underpinning their use of *purposive* and *probability* (random) *sampling* respectively.

sampling frame A list or register (for example, electoral roll, phone directory) from which respondents for an extensive study, such as a questionnaire, are drawn.

satisficing behaviour Conduct in which the decision-maker or agent acts in ways that yield satisfactory outcomes rather than optimal or "maximizing" outcomes.

saturation The point in the data-gathering process when no new information or insights are being generated. This is one method used by researchers to determine when to stop gathering data.

secondary data Information collected by people/agencies and stored for purposes other than for the research project for which they are being used (for example, census data being used in an analysis of socio-economic status and water consumption).

secondary literature From a historical geography perspective, an existing set of published material in the form of books, essays, and articles produced after the events that they discuss and based on primary sources and/or oral testimony and personal observations. In terms of a research exercise, it also refers to the existing academic writing on a specific topic. This literature may be obviously divided into groups that offer contrasting interpretations of the same events. (Compare with *primary sources.*)

secondary observation Research in which the data are the observations of others (for example, photographs).

secondary question Interview prompts that encourage the informant to follow up or expand on an issue already discussed. (See also *follow-up questions.*)

secretary hand A style of writing used by professional clerks that became increasingly widespread in England in the sixteenth and seventeenth centuries. Twenty-first-century researchers require some special skills to be able to decipher the calligraphy and the now archaic English prose. (See also *copperplate script* and *modern hand.*)

semiology See *semiotics.*

semiotics The system or language of signs (sometimes referred to as semiology). (See also *signifier* and *signified.*)

semi-structured interview Interview with some predetermined order but which nonetheless has flexibility with regard to the position/timing of questions. Some questions, particularly sensitive or complex ones, may have a standard wording for each *informant.* (Compare with *structured interview* and *unstructured interview.*)

series lists A list of individual files from a particular organization held in an archive. Typically, they replicate the referencing system used by the creating organization; they are usually numerical or alphanumeric and can include sub-series as well as large files that are broken into several individually recorded parts.

sharing the field Ways of sharing our work with participants and respondents and the decisions that have to be made concerning reciprocity and confidentiality (i.e., what do we give back to them in return for their involvement, and how do we protect confidences?).

signified The meaning derived from a *signifier* (or from a set of signifiers, such as a text).

signifier Images such as written marks or features of the landscape with which meaning is associated.

silence In discourse analysis, silence refers to how frameworks of understanding always conceal as much as they reveal about the world. Concealment inevitably occurs when all frameworks of understanding are conceived of as the outcome of a highly social process. For example, life narratives often frame events surrounding discriminatory legislation around personal accounts, often silencing the role of political organizations. In contrast, a spokesperson for a political organization may choose to frame the same events through ideas about the collective nation. Such official accounts may erase differences in lived experiences based on various intersections of gender, sexuality, class, race, and so on.

simplicity Using clarity intelligently in order to get to the essence of an issue.

situated knowledge A metaphor that evokes recognition of the positionality (or contextual nature) of knowledges. The inscription and creation of knowledge is always partial and "located" somewhere.

situated learning Learning that occurs in the same context as that in which it is applied.

snowball sampling A sampling technique that involves finding participants for a research project by asking existing informants to recommend others who might be interested. From one or two participants, the number of people involved in the project "snowballs." Also known as *chain sampling*.

social justice A situation in which there is an equitable and respectful negotiation of social and spatial difference and a fair distribution of resources.

social structure See *structure*.

sound document The outcome and output from an *oral history* interview.

soundscape A compound word joining "sound" and "landscape" to create a term that captures the way a sound or combination of sounds can contribute to the sensory environment. Sounds can be "natural" (e.g., water) or contrived (e.g., music), and their apprehension can be understood as a non-visual form of observation.

specific questions Relate to interviewees' individual experiences and are developed through follow-up work. (Compare with *orientation questions*, *common questions*, and *follow-up questions*.)

staging a text A theatrical metaphor for *writing-in* research that encourages geographical researchers to consider how the construction of a research text is actually a form of cultural production. The author of the text is a creator, director, and performer in the particular narrative that he or she is constructing.

stakeholder Any individual or group that has an interest in a project because of the way it may benefit, harm, or exclude them.

standardized questions A uniform set of questions that are repeated for all *interviews* or *focus groups* in a research project, in contrast to spontaneous questions that develop out of the conversational flow of an interview or focus group.

strategy of conviction Refers to the processes by which particular social realities or ways of making order of the world become accepted within a particular spatial and historical context as common sense or as truth. The term is derived from the work of Michel Foucault. In particular, the term *strategy of conviction* is derived from his argument that knowledge and power are inseparable. According to Foucault, the process by which particular social realities are produced, circulated, and maintained requires thoughtful consideration of the intersection between authorship, technologies, and the type of text.

structure A functioning system (for example, social, economic, or political) within which individuals are located, within which all events are enacted, and that is reproduced and transformed by those events.

structured interview Interview that follows a strict order of topics. Usually the order is set out in an *interview schedule*. The wording of questions for each interview may also be predetermined. (Compare with *unstructured interview* and *semi-structured interview*.)

studying up An asymmetrical power relationship in which the research participant is in a position of power relative to the researcher. (See also *asymmetrical power relation*, *potentially exploitative power relation*, and *reciprocal power relation*.)

subaltern Oppressed, exploited, marginalized minority peoples and groups. The term derives from the work of Antonio Gramsci and from the "subaltern studies" project undertaken by Indian historians since the early 1980s that endeavoured to write history from "below" but is often used in a wider sense.

subjective/subjectivity Refers to the insertion of the personal resources, opinions, and characteristics of a person into a research project. Often contrasted with *objectivity*. (See also *intersubjectivity*.)

subjective modality A form of writing that explicitly acknowledges the writer's presence in the text (e.g., *first-person narrative* form) and clearly signals their agreement or disagreement with the statement being made. (Compare with *objective modality*.)

surveillance Involves monitoring the behaviour of people or objects so as to observe (and sometimes record) deviation from conformity. In French, "surveillance" means "watching over," and in one sense this covers all types of all observation. Contemporary technology such as closed-circuit television epitomizes surveillance as social control.

synchronous groups Real-time online discussions used in focus group research. Groups last for several hours and have 6 to 10 participants. (See also *asynchronous groups* and *online focus group*.)

synchronous interviewing Interview in which the informant's answers immediately follow the questions. Face-to-face and telephone interviews are common examples. Some forms of Internet-based interviewing can also be synchronous, such as conversations that occur within chat rooms and other online platforms. (Compare with *asynchronous interviewing*.)

synergistic effect A key feature of focus groups that occurs when participants are prompted and provoked by the things others say. Generally results in a lively discussion as participants respond to each other and add new thoughts and ideas. Best summed up by the phrase "the whole is greater than the sum of its parts."

text Traditionally synonymous with the written page but now used more broadly to refer to a range of source forms such as oral texts (including *semi-structured interviews* and *oral histories*) and images (including painting, photographs, and maps) as well as written and printed texts (including newspapers, letters, and brochures).

textual analysis Reading and constant reinterpretation of texts as a set of *signs* or signifying practices.

theme In *coding*, an important process, commonality, characteristic, or theory that emerges from the data and can be used to analyze and abstract the data.

theoretical (analytical) generalization In contrast to statistical generalization that relies on sampling large numbers from a population, theoretical generalization refers to creating concepts that may plausibly apply to other cases as yet unstudied. A key requirement is that those concepts must not be too case-specific. (See also *transferability*.)

theory-building software Computer programs that deal with relationships between data categories to develop higher-order classifications and categories and to formulate and test propositions or assertions. Examples include AQUAD, ATLAS.ti, HyperRESEARCH, and N6.

theory generation The inductive process of identifying concepts to explain a phenomenon. One of the major advantages of qualitative research is that it facilitates the generation of new or revised theory. The theory may be as simple as a few loosely connected concepts or as complex as a large number of tightly integrated concepts (e.g., Marx's theory of capitalism).

theory testing A process whereby concepts or theories are compared to empirical data to test for congruence between the concepts and the empirical world. It is often conducted formally as quantitative statistical hypothesis testing but may also be done qualitatively using non-quantified textual data. (See also *deduction* and *induction*.)

thick description A term made common by anthropologist Clifford Geertz, which involves not just describing an event, occurrence, or practice but the detailed context in which it occurs. This allows researchers to interpret the situated nature of the "event," reflecting on the ways in which the subject or object of study is constructed symbolically in relation to broader cultural and social relations and discourses.

third-person narrative In the context of a research narrative (e.g., research report, journal publication, public talk), involves constructing the narrative without reference to the researcher's thoughts, opinions, or feelings. This consequently conveys a distanced and seemingly neutral, omniscient, and "objective" point of view. Third-person narratives have dominated in the presentation of "scientific" research. (Compare with *first-person narrative.*)

timed tape logs A summary of the contents of an interview that notes the topics discussed and key words and the time at which they occur within the recording. These logs allow users to locate desired audio or video segments easily.

transcript Written record of speech (for example, interview, focus group proceedings, film dialogue). May also include textual description of informant gestures and tone.

transferability Extent to which the results of a study might apply to contexts other than that of the research study. (See also *generalizability.*)

transformative reflexivity A process through which a researcher and researched group reflect on their (mis)understandings and negotiate the meanings of the information generated together. The shared process has the potential to transform each person's own understandings.

triangulation Use of multiple or mixed methods, researchers, and information sources to confirm or corroborate results.

trolling A form of anti-social behaviour within online environments. There are two aspects to trolling. First, there is the non-participatory collection of data from online communities (see also *lurking*). Second, and more commonly, trolling refers to the baiting of participants by posting controversial or purposefully incendiary statements that are intended to incite heated exchanges (e.g., racist or sexist comments). Internet trolls appear to delight in generating discord, and their comments distract user groups and online communities from their common topics of discussion. Researchers using the Internet need to monitor for the presence of trolls within their own discussions and must ensure that their own postings and prompts are not seen as trolling.

trope Figure of speech that allows writers or producers of other forms of *text* to say one thing but mean something else. Also used to describe recurring themes or motifs in creative works. May involve use of *metaphor* or metonymy.

trustworthiness A principle in ethics and description applying to those who are given trust by others and who do not violate that faith; qualitative data and insights gained from them need to be trustworthy in order to be reliable and generalizable.

typical case sampling Selection of samples that illustrate or highlight what is considered typical or normal.

uncontrolled observation Purposeful watching of worldly phenomena that is relatively unconstrained by restrictions of scope, style, and time. (Compare with *controlled observation.*)

under-disclosure Occurs when participants in an interview or in focus groups provide very little information. (Contrast with *over-disclosure.*)

unstructured interview Interview in which there is no predetermined order to the issues addressed. The researcher phrases and raises questions in a manner appropriate to the informants' previous comment. The direction and vernacular of the interview is informant-driven. (Compare with *structured interview* and *semi-structured interview.*)

validity The truthfulness or accuracy of data compared with acceptable criteria. (See also *reliability.*)

vernacular Occurring in the location where it originated. Vernacular language is the language of a place.

voice capture Voice capture software can be used in questionnaire survey work conducted by telephone or face-to-face mode and is particularly useful for recording responses to open-ended questions. Software packages are available to digitally record responses in the respondents' actual voice, capturing information about, for instance, the intensity, emotion, or intonation of the response. The responses can then be played back for electronic coding to produce numeric tables and charts for further analysis.

voice recognition computer software Computer software that converts spoken words into text. It has three principal uses: (1) to support dictaphone use; (2) for accessibility purposes (e.g., so that people who are vision-impaired or who cannot use keyboards can use Internet communications); and (3) for the "transcription" of research interviews into text. Examples include Dragon Naturally Speaking™, Windows Speech Recognition™, and MacSpeech Dictate™. (See also *transcript.*)

warm-up A set of pre-interview techniques intended to enhance rapport between interviewer and *informant.* May include small talk, sharing food, or relaxed discussion of the research.

"wicked problem" A term deployed by Rittel and Webber (1973, 160) to mean challenges that were not "themselves ethically deplorable" but that were or could be malignant, viciously circuitous, or aggressive in effect. Rittel and Webber argued that wicked problems are simultaneously unique and the symptom of other problems. They have no "stopping rule," and solutions to them have neither immediate nor ultimate tests of success, there being no capacity to learn by trial and error "because every attempt counts."

word processing A generic concept that includes the use of computer capacity to create, edit, and print documents.

writing-*in* The active and situated process of writing in which the author engages with the ways in which meanings are constructed through the creation of his or her text.

References

Abbott, D. 2006. "Disrupting the 'whiteness' of fieldwork in geography." *Singapore Journal of Tropical Geography* 27, 326–341.

ACME: An International E-Journal for Critical Geographies. 2003. 2 (1): special issue on Practices in Feminist Research).

Adelman, C., ed. 1981. *Uttering, Muttering: Collecting, Using and Reporting Talk for Social and Educational Research*. London: Grant McIntyre.

Agar, M., and J. MacDonald. 1995. "Focus groups and ethnography." *Human Organization* 54 (1): 78–86.

Agius, P., et al. 2004. "Comprehensive native title negotiations in South Australia." In M. Langton et al., eds, *Honour among Nations? Treaties and Agreements with Indigenous People*. Melbourne: Melbourne University Press.

Agius, P., R. Howitt, and S. Jarvis. 2003. Different visions, different ways: Lessons and challenges from the native title negotiations in South Australia. Paper presented to the Native Title Conference, Alice Springs, June.

Ahmed, S. 2004. *The Cultural Politics of Emotion*. Edinburgh: Edinburgh University Press.

Aitken, S. 1996. "Textual analysis: Armchair theories and couch-potato geography." In R. Flowerdew and D. Martin., eds., *Methods in Human Geography*. Harlow: Longman.

Aitken, S., and Craine, J. 2014. "A brief history of mediated, sensational and virtual geographies." In S. Mains, J. Cupples, and C. Lukinbeal, eds, *International Handbooks of Human Geography: Mediated Geographies/ Geographies of Media*. London: Routledge.

Alder, P.A., and P. Alder. 1994. "Observational techniques." In N.K. Denzin and Y.S. Lincoln, eds, *Handbook of Qualitative Research*. Thousand Oaks, CA: Sage.

Alderman, D., and D. Good. 1997. "Exploring the virtual South: The idea of a distinctive region on 'the Web.'" *Southeastern Geographer* 37 (1): 20–45.

Alderman, D.H., and E.A. Modlin. 2013. "Southern hospitality and the politics of African American belonging: an analysis of North Carolina tourism brochure photographs." *Journal of Cultural Geography* 30, 6–31.

Alexander, C., et al. 2007. "Participatory diagramming: A critical view from north east England." In S. Kindon, R. Pain, and M. Kesby, eds, *Participatory Action Research Approaches and Methods: Connecting People, Participation and Place*, 112–21. London: Routledge.

Alvermann, D.E., D.G. O'Brien, and D.R. Dillon. 1996. "On writing qualitative research." *Reading Research Quarterly* 31: 114–20.

American Indian Culture and Research Journal. 2008. 32 (3): special issue on Mainstreaming Indigenous Geography).

Anderson, K.J. 1993. "Place narratives and the origins of inner Sydney's Aboriginal settlement, 1972–73." *Journal of Historical Geography* 19 (3): 314–35.

———. 1999. "Reflections on Redfern." In E. Stratford, ed., *Australian Cultural Geographies*. Melbourne: Oxford University Press.

Andrews, G.J., et al. 2006. "Their finest hour: Older people, oral histories, and the historical geography of social life." *Social and Cultural Geography* 7 (2): 153–77.

Anfara, V.A., K.M. Brown, and T.L. Mangione. 2002. "Qualitative analysis on stage: Making the research process more public." *Educational Researcher* 31 (7): 28–38.

Ansell, N. 2014. "Challenging empowerment: 'AIDS-affected southern African children and the need for a multi-level relational approach.'" *Journal of Health Psychology* 19 (1): 22–33.

Ansell, N., and L. van Blerk. 2005. "Joining the conspiracy? Researching around ethics and emotions of AIDS in southern Africa." *Ethics, Place and Environment: A Journal of Philosophy and Geography*, 8 (1): 61–82.

Antipode. 1995. 27 (71–101): special issue on Discussion and Debate: Symposium on Feminist Participatory Research).

Anzaldúa, G. 1987. *Borderlands/La Frontera: The New Mestiza*. San Francisco: Spinsters/Aunt Lute Press.

Arnado, J. 2010. "Performances across time and space: Drama in the global households of Filipina transmigrant workers." *International Migration* 48 (6): 132–54.

Ash, J. 2010. "Architectures of affect: Anticipating and manipulating the event in processes of

videogame design and testing." *Environment and Planning D: Society and Space* 28: 653–71.
Askew, L., and P.M. McGuirk. 2004. "Watering the suburbs: Distinction, conformity and the suburban garden." *Australian Geographer* 35: 17–37.
Association of American Geographers. 2005. *Statement on Professional Ethics*. http://www.aag.org/Publications/Other%20Pubs/Statement%20on%20Professional%20Ethics.pdf.
Atkinson, C. 2011. *Beyond Bullet Points. Using Microsoft PowerPoint to Create Presentations that Inform, Motivate and Inspire*. 3rd edn. Redmond, WA: Microsoft Press.
Atkinson, P., and M. Hammersley. 1984. "Ethnography and participant observation." In N.K. Denzin and Y.S. Lincoln, eds, *Handbook of Qualitative Research*. Thousand Oaks, CA: Sage.
Attanapola, C., C. Brun, and R. Lund. 2013. "Working gender after crisis: Partnerships and disconnections in Sri Lanka after the Indian Ocean tsunami." *Gender, Place and Culture* 20 (1): 70–86.
Babbie, E. 1992. *The Practice of Social Research*. 6th edn. Belmont, CA: Wadsworth.
———. 1998. *The Practice of Social Research*. 8th edn. Belmont, CA: Wadsworth.
———. 2001. *The Practice of Social Research*. 9th edn. Belmont, CA: Wadsworth.
———. 2013. *The Practice of Social Research*. 13th edn. Belmont, CA: Wadsworth.
Bailey, C. 2001. "Geographers doing household research: Intrusive research and moral accountability." *Area* 33 (1): 107–10.
Bailey, C., C. White, and R. Pain. 1999a. "Evaluating qualitative research: Dealing with the tension between 'science' and 'creativity'." *Area* 31 (2): 169–83.
———. 1999b. "Response." *Area* 31 (2): 183–4.
Bailey, M.M., and R. Shabazz. 2013. "Gender and sexual geographies of Blackness: New Black cartographies of resistance and survival (part 2)." *Gender, Place & Culture: A Journal of Feminist Geography*, DOI: 10.1080/0966369X.2013.786303.
Baker, A.R.H. 1997. "The dead don't answer questionnaires: Researching and writing historical geography." *Journal of Geography in Higher Education* 21 (2): 231–43.
Bampton, R., and C.J. Cowton. 2002. "The e-interview." *Forum: Qualitative Social Research Sozialforschung* 3 (2): article 9.
Banaji, S., and D. Buckingham. 2010. "Young people, the internet, and civic participation: An overview of key findings from the CivicWeb Project." *International Journal of Learning and Media*, 2 (1): 15–24.
Barbour, R. 2007. *Doing Focus Groups*. Los Angeles: Sage.
Barnes, T.J. 1989. "Place, space and theories of economic value: Contextualism and essentialism." *Transactions of the Institute of British Geographers* NS 14, 299–316.
———. 1993. "Whatever happened to the philosophy of science?" *Environment and Planning A* 25: 301–4.
Barnes, T.J., and J. Duncan. 1992. "Introduction: Writing worlds." In T.J. Barnes and J. Duncan, eds, *Writing Worlds: Discourse, Text and Metaphor in the Representation of Landscape*. London: Routledge.
Barnes, T.J., and D. Gregory. 1997. "Worlding geography: Geography as situated knowledge." In T.J. Barnes and D. Gregory, eds, *Reading Human Geography: The Poetics and Politics of Inquiry*. London: Arnold.
Barr, S., G. Shaw, T. Coles, and J. Prillwitz. 2010. "'A holiday is a holiday': Practicing sustainability, home and away." *Journal of Transport Geography* 18 (3): 474–81.
Barrett, M. 1991. *The Politics of Truth: From Marx to Foucault*. Cambridge: Polity Press.
Baxter, J., and J. Eyles. 1997. "Evaluating qualitative research in social geography: Establishing 'rigour' in interview analysis." *Transactions of the Institute of British Geographers* 22 (4): 505–25.
———. 1999a. "The utility of in-depth interviews for studying the meaning of environmental risk." *Professional Geographer* 51 (2): 307–20.
———. 1999b. "Prescription for research practice? Grounded theory in qualitative evaluation." *Area* 31 (2): 179–81.
Baxter, J., and D. Lee. 2004. "Explaining the maintenance of low concern near a hazardous waste treatment facility." *Journal of Risk Research* 6 (1): 705–29.
Becker, H.S. 2007. *Writing for Social Scientists: How to Start and Finish Your Thesis, Book, or Article*. 2nd edn. Chicago: University of Chicago Press.
Beckett, C., and S. Clegg. 2007. "Qualitative data from a postal questionnaire: Questioning the presumption of the value of presence." *International Journal of Social Research Methodology* 10: 307–17.

Bedford, T., and J. Burgess. 2002. "The focus-group experience." In M. Limb and C. Dwyer, eds, *Qualitative Methodologies for Geographers: Issues and Debates*. London: Edward Arnold.

Bee, B. 2013. "Who reaps what is sown? A feminist inquiry into climate change adaptation in two Mexican *ejidos*." *ACME*, 12: 131–54.

Bell, D., J. Binnie, J. Cream, and G. Valentine. 1994. "All hyped up and no place to go." *Gender, Place and Culture* 1 (1): 31–48.

Bell, L. 1998. "Public and private meanings in diaries: Researching family and childcare." In J. Ribbens and R. Edwards, eds, *Feminist Dilemmas in Qualitative Research: Public Knowledge and Private Lives*, 72–86. London: Sage.

Bennett, K. 2002. "Interviews and focus groups." In P. Shurmer-Smith, ed., *Doing Cultural Geography*. London: Sage.

Berg, B.L. 1989. *Qualitative Research Methods for the Social Sciences*. Boston: Allyn and Bacon.

Berg, L.D. 1993. "Between modernism and postmodernism." *Progress in Human Geography* 17: 490–507.

———. 1994. "Masculinity, place, and a binary discourse of theory and empirical investigation in the human geography of Aotearoa/New Zealand." *Gender, Place and Culture* 1 (2): 245–60.

———. 1997. "Banal geographies." Paper presented to the Inaugural International Conference of Critical Geographers, Vancouver, 10–13 August.

———. 2004. "Scaling knowledge: Towards a critical geography of critical geography." *Geoforum* 35 (5): 553–8.

———. 2012. "Geography—(neo)liberalism—white supremacy." *Progress in Human Geography* 36 (4): 508–17.

Berg, L.D., M. Evans, D. Fuller, and the Okanagan Urban Aboriginal Health Research Collective. 2007. "Ethics, hegemonic whiteness, and the contested imagination of 'Aboriginal community' in social science research in Canada." *ACME: An International E-Journal for Critical Geographies* (special issue on participatory research ethics) 6 (3): 395–410.

Berger, T.R. 1991. *A Long and Terrible Shadow: White Values, Native Rights in the Americas*. Vancouver and Seattle: Douglas and McIntyre and University of Washington Press.

Bernard, H.R. 1988. *Research Methods in Cultural Anthropology*. Newbury Park: Sage.

Berreman, G.D. 1972. *Hindus of the Himalayas: Ethnography and Change*. 2nd edn. Berkeley: University of California Press.

Bertrand, J.T., J.E. Brown, and V.M. Ward. 1992. "Techniques for analyzing focus group data." *Evaluation Review* 16 (2): 198–209.

Billinge, M., D. Gregory, and R.L. Martin. 1984. "Reconstructions." In M. Billinge, D. Gregory, and R.L. Martin, eds, *Recollections of a Revolution: Geography as Spatial Science*. London: Macmillan.

Billo, E., and N. Hiemstra. 2013. "Mediating messiness: Expanding ideas of flexibility, reflexivity, and embodiment in fieldwork." *Gender, Place & Culture: A Journal of Feminist Geography* 20 (3): 313–28.

Bingley, A. 2002. "Research ethics in practice." In L. Bondi et al., eds, *Subjectivities, Knowledges and Feminist Geographies: The Subjects and Ethics of Research*. London: Rowman and Littlefield.

Binnie, J., and G. Valentine. 1999. "Geographies of sexuality—A review of progress." *Progress in Human Geography* 23: 176–87.

Bishop, P. 2002. "Gathering the land: The Alice Springs to Darwin rail corridor." *Environment and Planning D: Society and Space* 20: 295–317.

Bishop, R. 1994. "Initiating empowering research?" *New Zealand Journal of Educational Studies* 29 (1): 175–88.

———. 2005. "Freeing ourselves from neocolonial domination in research: A Kaupapa Māori approach to creating knowledge." In N. Denzin and Y. Lincoln, eds, *The Sage Handbook of Qualitative Research*, 3rd edn. 109–38. Thousand Oaks, CA: Sage.

Blaikie, P., and H. Brookfield. 1987. *Land Degradation and Society*. London: Methuen.

Blake, K. 2001. "In search of Navajo sacred geography." *The Geographical Review* 91 (4): 715–24.

Blanch, F. 2013. "Encountering the Other: One indigenous Australian woman's experience of racialisation on a Saturday night." *Gender, Place and Culture* 20: 253–60.

Blumen, O. 2002. "Criss-crossing boundaries: Ultraorthodox Jewish women go to work." *Gender, Place and Culture* 9 (2): 133–51.

Bogdan, R.C., and S.K. Biklen. 1992. *Qualitative Research for Education: An Introduction to Theory and Methods*. 2nd edn. Boston: Allyn and Bacon.

Bondi, L. 1997. "In whose words? On gender identities, knowledge and writing practices." *Transactions of the Institute of British Geographers* 22: 245–58.

———. 2003. "Empathy and identification: conceptual resources for feminist fieldwork." *ACME: an International E-journal for Critical Geographies* 2: 64–76.

Bondi, L., H. Avis, R. Bankey, A. Bingley, J. Davidson, R. Duffy, V.I. Einagel, A-M. Green, L. Johnston, S. Lilley, C. Listerborn, M. Marshy, S. McEwan, N. O'Connor, G. Rose, B. Vivat, and N. Wood. 2002. *Subjectivities, Knowledges and Feminist Geographies: The Subjects and Ethics of Social Research.* Lanham, MD: Rowman and Littlefield.

Bondi, L., J. Davidson, and M. Smith. 2005. "Introduction: Geography's 'emotional turn.'" In J. Davidson, L. Bondi, and M. Smith, eds, *Emotional Geographies*, 1–16. Aldershot: Ashgate.

Bordo, S. 1986. "The Cartesian masculinization of thought." *Signs* 11: 439–56.

Botes, L., and D. van Rensburg. 2000. "Community participation in development: Nine plagues and twelve commandments." *Community Development Journal* 35 (1): 53–4.

Botz-Bornstein, T. 2013. "From the stigmatized tattoo to the graffitied body: Femininity in the tattoo renaissance." *Gender, Place & Culture: A Journal of Feminist Geography* 44 (3): 236–52.

Bourdieu, P. 2003. "Participant observation." *Journal of the Royal Anthropological Institute* 9: 281–94.

Boyd, J. 2013. "'I go to dance, right?': Representation/sensation on the gendered dance floor." *Leisure Studies.* DOI: 10.1080/02614367.2013.798348.

Bradford, M. 2003. "Writing essays, reports and dissertations." In N.J. Clifford and G. Valentine, eds, *Key Methods in Geography*, 515–32. London: Sage.

Bradshaw, M. 2001. "Contracts and member checks in qualitative research in human geography: Reason for caution?" *Area* 33 (2): 202–11.

Brannen, J., ed. 1992a. *Mixing Methods: Qualitative and Quantitative Research.* Aldershot, US: Avebury.

———. 1992b. "Combining qualitative and quantitative approaches: An overview." In J. Brannen, ed., *Mixing Methods: Qualitative and Quantitative Research.* Aldershot, US: Avebury.

Braun, V., and V. Clarke. 2013. *Successful Qualitative Research. A Practical Guide for Beginners.* Los Angeles: Sage.

Brockington, D., and S. Sullivan. 2003. "Qualitative research." In R. Scheyvens and D. Storey, eds, *Development Fieldwork: A Practical Guide.* London: Sage.

Browne, K. 2003. Negotiations and fieldsworkings: Friendship and feminist research. *ACME: An International E-Journal for Critical Geographies*, 2 (2):.132–46.

Brunner, C. 2011. "Nice-looking obstacles: Parkour as urban practice of deterritorialization." *AI & Society* 26 (2): 143–52.

Bruno, G. 1987. "Ramble City: Postmodernism and 'Blade Runner.'" *October* 41: 61–74.

Bryant, R., and S. Bailey. 1997. *Third World Political Ecology.* New York: Routledge.

Bryman, A. 1984. "The debate about quantitative and qualitative research: A question of method or epistemology?" *The British Journal of Sociology* 35: 75–92.

———. 2006. "Integrating quantitative and qualitative research: How is it done?" *Qualitative Research* 6 (1): 97–113.

———. 2012. *Social Research Methods.* 4th edn. Oxford: Oxford University Press.

Bryman, A., and R.G. Burgess, eds. 1994. *Analyzing Qualitative Data.* London: Routledge.

Buang, A., and J. Momsen, eds. 2013. *Women and Empowerment in the Global South.* Bangi, Selangor, Malaysia: University of Malaysia Press.

Buchanan, E. (2011) "Internet research ethics: past, present, and future." In M. Consalvo and C. Ess, eds, *The Handbook of Internet Studies*, 83–107, Oxford: Wiley Blackwell.

Buckley, A., M. Gahegan, and K. Clarke. 2000. "Geographic visualization." In *Emerging Themes in GIScience Research.* University Consortium for Geographic Information Science White Papers.

Burawoy, M., et al. 1991. *Ethnography Unbound: Power and Resistance in the Modern Metropolis.* Berkeley: University of California Press.

Burgess, J. 1988. "Exploring environmental values through the medium of small groups: 2. Illustrations of a group at work." *Environment and Planning A* 20 (4): 457–76.

———. 1996. "Focusing on fear: The use of focus groups in a project for the Community Forest Unit, Countryside Commission." *Area* 28 (2): 130–5.

Burgess, J., and P. Wood. 1988. "Decoding Docklands: Place advertising and the decision-making strategies of the small firm." In J. Eyles and D.M. Smith, eds, *Qualitative Methods in Human Geography*. Cambridge: Polity Press.

Burgess, R.G. 1982a. "Elements of sampling in field research." In R.G. Burgess, ed., *Field Research: A Sourcebook and Field Manual*. London: George Allen and Unwin.

———. 1982b. "Multiple strategies in field research." In R.G. Burgess, ed., *Field Research: A Sourcebook and Field Manual*. London: George Allen and Unwin.

———. 1982c. "The unstructured interview as a conversation." In R.G. Burgess, ed., *Field Research: A Sourcebook and Field Manual*. London: George Allen and Unwin.

Burgwyn-Bailes, E., L. Baker-Ward, B.N. Gordon,, and P.A. Ornstein. 2001. "Children's memory for emergency medical treatment after one year: The impact of individual difference variables on recall and suggestibility." *Applied Cognitive Psychology*, 15: S25–S48.

Burman, E., and I. Parker. 1993. "Against discursive imperialism, empiricism and constructionism: Thirty-two problems with discourse analysis." In E. Burman and I. Parker, eds, *Discourse Analytic Research: Repertoires and Readings of Texts in Action*. London: Routledge.

Butler, J. 2007. "Torture and the ethics of photography." *Environment and Planning D: Society and Space* 25: 951–66.

Butler, R. 1997. "Stories and experiments in social inquiry." *Organisation Studies* 18 (6): 927–48.

Butz, D. 2007. "Sustained by the guidelines: Reflections on the limitations of standard informed consent procedures for the conduct of ethical research." *ACME: An International E-journal for Critical Geographies* 7 (2): 239–59.

———. 2010. "Autoethnography as sensibility." In D. DeLyser, S. Herbert, S. Aitken, M. Crang, and L. McDowell, eds, *The SAGE Handbook of Qualitative Geography*, 138–156, London: SAGE Publications. DOI: http://dx.doi.org/10.4135/9780857021090.n10

Butz, D., and K. Besio. 2004. "The value of autoethnography for field research in transcultural settings." *Professional Geographer* 56 (3): 350–60.

Byrne, J. 2012. "When green is White: The cultural politics of race, nature and social exclusion in a Los Angeles urban national park." *Geoforum* 43 (3): 595–611.

Cahill, C. 2004. "Defying gravity: Raising consciousness through collective research." *Children's Geographies* 2 (2): 273–86.

Cahill, C., and M. Torre. 2007. "Beyond the journal article: Representations, audience, and the presentation of participatory action research." In S. Kindon, R. Pain, and M. Kesby, eds, *Participatory Action Research Approaches and Methods: Connecting People, Participation and Place*, 196–205. London: Routledge.

Cahill, C., M. Bradley, et al. 2008. "'Represent': Reframing risk through participatory video research." In M. Downing, and L. Tenney, eds, *Video Vision: Changing the Culture of Social Science Research*. Cambridge: Scholars Publishing.

Callon, M. 1986. "Some elements of a sociology of translation: Domestication of the scallops and the fishermen of St Brieuc Bay." In J. Law, ed., *Power, Action and Belief: A New Sociology of Knowledge?* London: Routledge.

Cameron, E. 2012. "New geographies of story and storytelling." *Progress in Human Geography* 36: 573–92.

Cameron, J. 1992. "Modern-day tales of illegitimacy: Class, gender and ex-nuptial fertility." (Department of Geography, University of Sydney, MA minor thesis).

Cameron, J., and K. Gibson. 2005. "Participatory action research in a poststructuralist vein." *Geoforum* 36 (3): 315–31.

Cameron, J., and S. Hendriks. 2014. "Narratives of social enterprise: Insights from Australian social enterprise practitioners." In H. Douglas and S. Grant, eds, *Social Innovation, Social Entrepreneurship and Social Enterprise: Context and Theories*. Victoria, Australia: Tilde University Press.

Cammarota, J., and M. Fine. 2008. "Youth participatory action research: A pedagogy for transformational resistance." In J. Cammarota and M. Fine, eds, *Revolutionizing Education: Participatory Action Research in Motion*. New York: Routledge.

Campbell, D., and J. Stanley. 1966. *Experimental and Quasi-experimental Designs for Research*. Chicago: Rand McNally.

Canadian Institutes of Health Research, Natural Sciences and Engineering Research Council of Canada, and Social Sciences and Humanities Research Council of Canada, 2010, *Tri-Council Policy Statement: Ethical*

Conduct for Research Involving Humans, December 2010. Available at: http://www.pre.ethics.gc.ca/pdf/eng/tcps2/TCPS_2_FINAL_Web.pdf

Canadian Geographer. 1993. 37: *Special Issue: Feminism as Method.* 48–61.

Canadian Geographer. 2012. 56 (2: special issue on Community-Based Participatory Research Involving Indigenous Peoples in Canadian Geography: Progress?)

Carey, M.A. 1994. "The group effect in focus groups: Planning, implementing and interpreting focus group research." In J.M. Morse, ed., *Critical Issues in Qualitative Research Methods*. Thousand Oaks, CA: Sage.

Carr, J. 2010. "Legal geographies—skating around the edges of the law: Urban skateboarding and the role of law in determining young peoples' place in the city." *Urban Geography* 31 (7): 988–1003.

Carroll, J., and J. Connell. 2000. "'You gotta love this city': The Whitlams and inner Sydney." *Australian Geographer* 31 (2): 141–54.

Carroll, P., K. Witten, R. Kearns, and P. Donovan. 2015. "Kids in the city: Children's use and experiences of urban neighbourhoods in Auckland, New Zealand." *Journal of Urban Design* 20 (4): 417–36.

Castree, N. 2005. "The epistemology of particulars: Human geography, case studies and 'context.'" *Geoforum* 36 (5): 541–44.

Ceglowski, D. 1997. "That's a good story, but is it really research?" *Qualitative Inquiry* 3 (2): 188–99.

Chacko, E. 2004. "Positionality and praxis: fieldwork experiences in rural India." *Singapore Journal of Tropical Geography* 25: 51–63.

Chakrabarti, R. 2004. "Constrained Spaces of Pre-natal Care: South Asian Women in New York City," Department of Geography, University of Illinois at Urbana-Champaign, unpublished research proposal, 2004, © Ranjana Chakrabarti. Reprinted with permission.

Chakrabarty, D. 2000. *Provincializing Europe: Postcolonial Thought and Historical Difference*. Princeton: Princeton University Press.

Chambers, R. 1994. "The origins and practice of participatory rural appraisal." *World Development* 22 (7): 953–69.

Chapin, M., and B. Threlkeld. 2001. *Indigenous Landscapes: A Study of Ethnocartography.* Center for the Support of Native Lands, Arlington, VA.

Charles Booth Online Archive. booth.lse.ac.uk/

Charmaz, K. 2000. "Grounded theory: Objectivist and constructivist methods." In N.K. Denzin and Y.S. Lincoln. eds, *Handbook of Qualitative Research*, 2nd edn. Thousand Oaks, CA: Sage.

Chatterton, P., D. Fuller, and P. Routledge. 2007. "Relating action to activism: Theoretical and methodological reflections." In S. Kindon, R. Pain, and M. Kesby, eds, *Participatory Action Research Approaches and Methods: Connecting People, Participation and Place*, 216–22. London: Routledge.

Chattopadhyay, S. 2013. "Getting personal while narrating the 'field': A researcher's journey to the villages of the Narmada Valley." *Gender, Place and Culture* 20 (2): 137–59.

Chen, P.J., and S.M. Hinton. 1999. "Realtime interviewing using the world wide web." *Sociological Research Online* 4 (3).

Chen, S., G. Hall, and M. Johns. 2004. "Research paparazzi in cyberspace: The voices of the researchers." In M. Johns, S. Chen, and G. Hall, eds, *Online Social Research: Methods, Issues and Ethics*, 157–75. New York: Peter Lang.

Chih, Y.W. 2011. "'Protest is just a click away!' Responses to the 2003 Iraq War on a bulletin board system in China." *Environment and Planning D: Society and Space* 29 (1): 131–49.

Cho, H., and R. LaRose. 1999. "Privacy issues in Internet surveys." *Social Science Computer Review* 17 (4): 421–34.

Chow, B.D.V. 2010. "Parkour and the critique of ideology: Turn-vaulting the fortresses of the city." *Journal of Dance and Somatic Practices* 2 (2): 143–54.

Christensen, J. 2012. "Telling stories: Exploring research storytelling as a meaningful approach to knowledge mobilization with Indigenous research collaborators and diverse audiences in community-based participatory research." *The Canadian Geographer* 56: 231–42.

Clarke, G. 2001. "From ethnocide to ethnodevelopment? Ethnic minorities and indigenous peoples in Southeast Asia." *Third World Quarterly* 22 (3): 413–36.

Clifford, N., S. French, and G. Valentine, eds. 2010. *Key Methods in Geography*, 2nd edn, London: Sage.

Clifford, N., and G. Valentine, eds. 2003. *Key Methods in Geography*. London: Sage.

———. 2009. *Key Methods in Geography*. 2nd edn. London: Sage.

Cloke, P., et al. 2004. "Talking to people." In *Practising Human Geography*. London: Sage.

Clough, A.R., et al. 2004. "Emerging patterns of cannabis and other substance use in Aboriginal communities in Arnhem Land, Northern Territory: A study of two communities." *Drug and Alcohol Review* 23 (4) 381–90.

Clough, N., and R. Blumberg. 2012. "Toward anarchist and autonomous Marxist geographies." *ACME: An International E-Journal for Critical Geographies* 11 (3): 335–51.

Cobb, S.K., and S. Forbes. 2002. "Qualitative research: What does it have to offer to the gerontologist?" *The Journals of Gerontology: Series A* 57 (4): 197–202.

Coe, N.M., P.F. Kelly, and H.W.C. Yeung. 2007. *Economic Geography*. Malden, MA: Blackwell.

Coleman, T. 2007. Out of place? Young people, education and sexuality. Thesis (MA-Geography), University of Auckland.

———. 2013. "Ageing-in-place on Waiheke Island: experiencing 'place', 'being aged' and implications for wellbeing." Thesis (PhD-Geography), University of Auckland.

Collier, M.J., and M. Scott. 2010. "Focus group research in a mined landscape." *Land Use Policy* 27 (2): 304–12.

Connell, R.W. 1991. "Live fast and die young: The construction of masculinity among young working-class men on the margin of the labour market." *Australia and New Zealand Journal of Sociology* 27 (2): 141–71.

Cook, I. 1997. "Participant observation." In R. Flowerdew and D. Martin, eds, *Methods in Human Geography*. Harlow: Addison Wesley Longman.

———. 2000. "'Nothing can ever be the case of "us" and "them" again': Exploring the politics of difference through border pedagogy and student journal writing." *Journal of Geography in Higher Education* 24 (1): 13–27.

Cook, I., et al. 2007. "'It's more than just what it is': Defetishising commodities, expanding fields, mobilising change . . .." *Geoforum* 38: 1113–26.

Cooke, B. 2001. "The social psychological limits of participation?' In B. Cooke and U. Kothari, eds, *Participation: The New Tyranny?* London: Zed Books.

Cooke, B., and U. Kothari. 2001. "The case for participation as tyranny." In B. Cooke and U. Kothari, eds, *Participation: The New Tyranny?* London: Zed Books.

Coombes, B. 2012. "Collaboration: Inter-subjectivity or radical pedagogy?" *The Canadian Geographer / Le Géographe Canadien* 56: 290–1.

Coombes, B., J.T. Johnson, and R. Howitt. 2012. "Indigenous geographies I: Mere resource conflicts? The complexities in Indigenous land and environmental claims." *Progress in Human Geography* 36 (6): 810–21.

———. "Indigenous geographies II: The aspirational spaces in postcolonial politics—reconciliation, belonging and social provision." *Progress in Human Geography* 37 (5): 691–700.

———. (2014). "Indigenous geographies III: Methodological innovation and the unsettling of participatory research." *Progress in Human Geography* 38 (6): 845–54.

Coombs, H.C. 1978. *Kulinma: Listening to Aboriginal Australians*. Canberra: Australian National University Press.

Coombs, H.C., et al. 1989. *Land of Promises: Aborigines and Development in the East Kimberley*. Canberra: Centre for Environmental Studies, Australian National University and Aboriginal Studies Press.

Cooper, A. 1994. "Negotiating dilemmas of landscape, place and Christian commitment in a Suffolk parish." *Transactions of the Institute of British Geographers* 19: 202–12.

———. 1995. "Adolescent dilemmas of landscape, place and religious experience in a Suffolk parish." *Environment and Planning D: Society and Space* 13: 349–63.

Cope, M. 2008. "Patchwork neighborhood: Children's real and imagined geographies in Buffalo, NY." *Environment and Planning A* 40: 2845–63.

Cope, M. 2009. "Challenging adult perspectives on children's geographies through participatory research methods: Insights from a service-learning course." *Journal of Geography in Higher Education* 33 (1), 33–50.

Cope, M., and S. Elwood, eds. 2009. *Qualitative GIS: A Mixed-Methods Approach*. London: Sage.

Cornwall, A., and R. Jewkes. 1995. "What is participatory research?" *Social Science and Medicine* 41: 1667–76.

Cosgrove, D., and S. Daniels. 1988. *The Iconography of Landscape*. Cambridge: Cambridge University Press.

Cousins, J.A., J. Evans, and J.P. Sadler. 2009. "'I've paid to observe lions, not road maps!': An emotional journey with conservation volunteers in South Africa." *Geoforum* 40 (6): 1069–80.

Cousins, M., and A. Houssain. 1984. *Michel Foucault*. Basingstoke: Macmillan.

Cowen, D., and A. Siciliano. 2011. "Surplus masculinities and security." *Antipode* 43 (5): 1516–41.

Craine, J., and S. Aitken. 2005. "Visual methodologies: what you see is not always what you get." In R. Flowerdew and D. Martin, eds, *Methods in Human Geography*, 2nd edn., 250–69, Harlow: Longman.

Craine, J., G. Curti, and S. Aitken. 2013. "Cosmopolitan sex, monstrous violence and networks of blood." In G. Curti, J. Craine, and S. Aitken, eds, *The Fight To Stay Put: Social Lessons Through Media Imaginings of Urban Transformation and Change*, 249–65. Mainz: Franz Steiner Verlag.

Crampton, J., J. Bowen, D. Cockayne, B. Cook, E. Nost, L. Shade, L. Sharp, and M. Jacobsen. 2013. "Whose Geography? Which Publics?" *Dialogues in Human Geography* 3 (1): 73–6.

Crampton, J., M. Graham, A. Poorthuis, T. Shelton, M. Stephens, M. Wilson, and M. Zook. 2013. "Beyond the geotag: situating 'big data' and leveraging the potential of the geoweb." *Cartography and Geographic Information Science* 40 (2): 130–39.

Crang, M. 1997. "Picturing practices: Research through the tourist gaze." *Progress in Human Geography* 21: 359–73.

———. 2003a. "The hair in the gate: Visuality and geographical knowledge." *Antipode*, 35(2): 238–243.

———. 2003b. "Qualitative methods: Touchy, feely, look-see?" *Progress in Human Geography* 27 (4): 494–504.

———. 2005a. "Analysing qualitative materials." In R. Flowerdew and D. Martin, eds, *Methods in Human Geography: a Guide for Students Doing a Research Project*, 2nd edn, 218–33. Harlow: Pearson Prentice Hall.

———. 2005b, "Qualitative methods: There is nothing outside the text?" *Progress in Human Geography* 29: (2): 225–33.

Crang, P. 1994. "It's showtime: On the workplace geographies of display in a restaurant in southeast England." *Environment and Planning D: Society and Space* 12: 675–704.

Crystal, D. 2006. *Language and the Internet*. 2nd edn. Cambridge: Cambridge University Press.

Cupples, J., and J. Harrison. 2001. "Disruptive voices and the boundaries of respectability in Christchurch, New Zealand." *Gender, Place and Culture* 8 (2): 189–204.

Cupples, J., and S. Kindon. 2003. "Returning to university and writing the field." In R. Scheyvens and D. Storey, eds, *Development Fieldwork: A Practical Guide*, 217–31. London: Sage.

Curtis, F., and K. Mee. 2012. "Welcome to Woodsie: Inverbrackie alternative place of detention and performances of belonging in Woodside, South Australia, and Australia." *Australian Geographer*, 43 (4): 357–75.

Daniels, S., and D. Cosgrove. 1988. "Introduction: Iconography and landscape." In D. Cosgrove and S. Daniels, eds, *The Iconography of Landscape: Essays on the Symbols, Representation, Design and Use of Past Environments*. Cambridge: Cambridge University Press.

Davidson, J., L. Bondi, and M. Smith. 2005. *Emotional Geographies*. London: Ashgate.

Davies, G., and C. Dwyer. 2007. "Qualitative methods: Are you enchanted or are you alienated?" *Progress in Human Geography* 31 (2): 257–66.

Davis, C.M. 1954. "Field techniques." In P.E. James and C.F. Jones, eds, *American Geography: Inventory and Prospect*. Syracuse, NY: Syracuse University Press for the Association of American Geographers.

Davis, M. 1990. *City of Quartz: Excavating the Future in Los Angeles*. London: Verso.

Davison, G. 2003. "History and archives in the 21st century." *Australian History Association Bulletin* 55: 50–7.

Dean, M., and K.A.J. Laidler. 2014. "Leveling the playing field through Facebook: How females construct online playspaces." *Journal of Youth Studies* 17 (1): 113–29.

Dear, M. 1988. "The postmodern challenge: Reconstructing human geography." *Transactions of the Institute of British Geographers* NS 13: 262–74.

de Certeau, M. 1984. *The Practice of Everyday Life*. Berkeley: University of California Press.

de Leeuw, S., E. Cameron, and M.L. Greenwood. 2012. "Participatory and community-based research, Indigenous geographies, and the spaces of friendship: A critical engagement." *The Canadian Geographer*. 56(2): 180–94.

Deloria, V., Jr. 1988 [1969]. *Custer Died for Your Sins: An Indian Manifesto*. Norman: University of Oklahoma Press.

DeLyser, D., et al., eds. 2009. *The Sage Handbook of Qualitative Research in Human Geography.* London: Sage.

DeLyser, D., S. Herbert, S. Aitken, M. Crang, and L. McDowell, eds. 2010. *The Sage Handbook of Qualitative Geography.* London: Sage.

DeLyser, D., R. Sheehan, and A. Curtis. 2004. "eBay and Research in Historical Geography." *Journal of Historical Geography* 30: 764–82.

DeLyser, D., and P.F. Starrs. 2001. "Doing fieldwork: Editors' introduction." *The Geographical Review* 91 (1–2): iv.

DeLyser, D., and P.F. Starrs, eds. 2001. *Geographical Review (Special Issue: Doing Fieldwork).* 91 (2).

DeLyser, D., and D. Sui. 2013. "Crossing the qualitative-quantitative chasm III: enduring methods, open geography, participatory research and the forth paradigm." *Progress in Human Geography,* DOI: 10.1177/0309132513479291

Denzin, N.K. 1978. *The Research Act: A Theoretical Introduction to Sociological Methods.* New York: McGraw-Hill.

———. 2003. "Reading and writing performance." *Qualitative Research* 3 (2): 243–68.

———. 2008. "Emancipatory discourses and the ethics and politics of interpretation." In N.K. Denzin and Y.S. Lincoln, eds, *Collecting and Interpreting Qualitative Materials Research,* 3rd edn, 435–71. Los Angeles: Sage.

Denzin, N.K., and Y.S. Lincoln. 1994. "Introduction: Entering the field of qualitative research." In N.K. Denzin and Y.S. Lincoln, eds, *Handbook of Qualitative Research.* Thousand Oaks, CA: Sage.

———, eds. 2000. *Handbook of Qualitative Research.* 2nd edn. Thousand Oaks, CA: Sage.

———, eds. 2005. *Handbook of Qualitative Analysis.* 3rd edn. Thousand Oaks, CA: Sage.

———, eds. 2008. *Collecting and Interpreting Qualitative Materials.* 3rd edn. Los Angeles: Sage.

———, eds. 2011. *Handbook of Qualitative Analysis.* 4th edn. Thousand Oaks, CA: Sage.

Derrida, J. 1981. *Dissemination.* tr. B. Johnson. Chicago: University of Chicago Press.

Desbiens, C., and E. Rivard. 2014. "From passive to active dialogue? Aboriginal lands, development and métissage in Québec, Canada." *Cultural Geographies,* 21: 99–114.

de Vaus, D. 2014. *Surveys in Social Research.* 6th edn. Sydney: Allen and Unwin.Dewalt, K.M., and B.R. Dewalt. 2002. *Participant Observation: A Guide for Fieldworkers.* Lanham, MD: Rowman Altamira.

Dey, I. 1993. *Qualitative Data Analysis: A User-Friendly Guide for Social Scientists.* London: Routledge.

DiBiase, D., A. MacEachren, J. Krygier, and C. Reeves. 1992. "Animation and the role of map design in scientific visualization." *Cartography and Geographic Information Systems,* 19(4): 201–214.

Dillman, D. 2007. *Mail and Internet Surveys: The Total Design Method.* 2nd edn. Hoboken, NJ: John Wiley.

Dillman, D., and D. Bowker. 2001. "The web questionnaire challenge to survey methodologists." In U. Reips and M. Bosnjak, eds, *Dimensions of Internet Science,* 159–78. Lengerich, Germany: Pabst Science Publishers.

Dillman, D., G. Phelps, R. Tortora, K. Swift, J. Kohrell, J. Berck, and B. Messer. 2009. "Response rate and measurement differences in mixed-mode surveys using mail, telephone, interactive voice response (IVR) and the Internet." *Social Science Research,* 38: 1–18.

Dittmer, J. 2005. "Captain America's empire: reflections on identity, popular culture, and post-9/11 geopolitics." *Annals of the Association of American Geographers,* 95 (3): 626–43.

———. 2006. "Of Gog and Magog: The geopolitical visions of Jack Chick and Premillennial Dispensationalism." *ACME: An International E-journal for Critical Geographies,* 6(2): 91–98.

———. 2010. "Comic book visualities: A methodological manifesto on geography, montage and narration." *Transactions of the Institute of British Geographers* 35: 222–36.

Dixon, D., 2010. "Analysing Meaning." In B. Gomez and J.P. Jones III, eds, *Research Methods in Geography,* 392–407. Chichester, UK: Wiley-Blackwell.

Dixon, D.P., and J.P. Jones III. 1996. "For a supercalifragilisticexpialidocious scientific geography." *Annals of the Association of American Geographers* 86: 767–79.

Doel, M., 2010. "Analysing Cultural Texts." In D. DeLyser, S. Herbert, S. Aitken, M. Crang, and L. McDowell, eds. *The Sage Handbook of Qualitative Geography,* 485–96. London: Sage.

Doel, M.A., and D.B. Clarke. 1999. "Dark Panopticon. Or, attack of the killer tomatoes." *Environment and Planning D: Society and Space* 17 (4): 427–50.

Dominey-Howes, D., A. Gorman-Murray, and S. McKinnon. 2013. "Queering disasters: On the need to account for LGBTI experiences in natural disaster contexts." *Gender, Place & Culture: A Journal of Feminist Geography*, DOI: 10.1080/0966369X.2013.802673

Donovan, J. 1988. "When you're ill, you've gotta carry it." In J. Eyles and D.M. Smith, eds, *Qualitative Methods in Human Geography*. Cambridge: Polity Press.

Douglas, J.D. 1985. *Creative Interviewing*. Beverly Hills, CA: Sage.

Driver, F. 1988. "Historicity and human geography." *Progress in Human Geography* 12 (4) 479–83.

Dryzek, J.S. 2005. *The Politics of the Earth: Environmental Discourses*. 2nd edn. Oxford: Oxford University Press.

Duffy, M., and G. Waitt. 2011. "Sound diaries: a method for listening to place." *Aether* 8: 119–36.

Dumas, A., and S. Laforest. 2009. "Skateparks as a health-resource: Are they as dangerous as they look?' *Leisure Studies* 28 (1): 19–34.

Dumbrava, G., and A. Koronka. 2006. "Writing for business purposes: Elements of email etiquette." *Annals of the University of Petrosani, Economics* 6: 61–4.

Duncan, J.S. 1987. "Review of urban imagery: Urban semiotics." *Urban Geography* 8 (5): 473–83.

———. 1992. "Elite landscapes as cultural (re) production: The case of Shaughnessy Heights." In K. Anderson and F. Gale, eds, *Inventing Places: Studies in Cultural Geography*. Melbourne: Longman Cheshire.

———. 1999. "Complicity and resistance in the colonial archive: Some issues in the method and theory of historical geography." *Historical Geography* 27: 119–28.

Duncan, J., and N. Duncan. 1988. "(Re)reading the landscape." *Environment and Planning D: Society and Space* 6 (2) 117–26.

Dunn, K. 1993. "The Vietnamese concentration in Cabramatta: Site of avoidance and deprivation, or island of adjustment and participation?" *Australian Geographical Studies* 31 (2): 228–45.

———. 2008. "Comparative analyses of transnationalism: A geographic contribution to the field." *Australian Geographer* 39 (1): 1–7.

Dunn, K.M., P.M. McGuirk, and H.P.M. Winchester. 1995. "Place making: The social construction of Newcastle." *Australian Geographical Studies* 33 (2): 149–66.

Dunn, K.M., and M. Mahtani. 2001. "Media representations of ethnic minorities." *Progress in Planning* 55 (3): 163–72.

Dyck, I. 1997. "Dialogue with difference: A tale of two studies." In J.P. Jones III, H.L. Nast, and S.M. Roberts, eds, *Thresholds in Feminist Geography: Difference, Methodology, Representation*. Lanham, MD: Rowman and Littlefield.

———. 2005. "Feminist geography, the 'everyday', and local-global relations: Hidden spaces of place-making." *The Canadian Geographer* 49 (3): 233–43.

Dyck, I., and R. Kearns. 1995. "Transforming the relations of research: Towards culturally safe geographies of health and healing." *Health and Place* 1 (3): 137–47.

Eden, S., C. Bear, and G. Walker. 2008. "Mucky carrots and other proxies: Problematising the knowledge-fix for sustainable and ethical consumption." *Geoforum* 39: 1044–57.

Edwards, E. 2003. "Negotiating spaces." In J.M. Schartz and J.R. Ryan, eds, *Picturing Place Photography and the Geographical Imagination*. London and New York: I.B. Tauris.

Eichler, M. 1988. *Nonsexist Research Methods: A Practical Guide*. Boston: Allen and Unwin.

Ellis, C. 2004. *The Ethnographic I: A Methodological Novel about Autoethnography*. Walnut Creek, CA: AltaMira Press.

Elwood, S., et al. 2007. "Participatory GIS: The Humboldt/West Humboldt Park Community GIS Project, Chicago, USA." In S. Kindon, R. Pain, and M. Kesby, eds, *Participatory Action Research Approaches and Methods: Connecting People, Participation and Place*, 170–8. London: Routledge.

Elwood, S., and A. Leszczynski. 2012. "New spatial media, new knowledge politics." *Transactions of the Institute of British Geographers* 38 (4): 544–59.

Ely, M. 2007. "In-forming re-presentations." In D.J. Clandinin, ed., *Handbook of Narrative Inquiry, Mapping a Methodology*, 567–98. Thousand Oaks, CA, and London: Sage.

Emerson, R., R. Fretz, and L. Shaw. 1995. *Writing Ethnographic Fieldnotes*. Chicago: University of Chicago Press.

England, K.V.L. 1993. "Suburban Pink Collar Ghettos: The Spatial Entrapment of Women?" *Annals of the Association of American Geographers*, 83 (2): 225–42.

———. 1994. "Getting personal: Reflexivity, positionality, and feminist research." *Professional Geographer* 46: 80–9.

———. 2006. "Producing feminist geographies: Theory, methodologies and research strategies." In S. Aitken and G. Valentine, eds, *Approaches to Human Geography,* 286–97. London: Sage.

Ergler, C.R. 2013. "The power of place in play: A Bourdieusian analysis of seasonal outdoor play practices in Auckland children's geographies." (PhD thesis in Geography, The University of Auckland).

Escobar, A. 1995. *Encountering Development: The Making and Unmaking of the Third World.* Princeton, NJ: Princeton University Press.

Eshun, G., and C. Madge. 2012. "'Now let me share this with you': Exploring poetry for postcolonial geography research." *Antipode* 44 (4): 1395–428.

Evans, J., and P. Jones. 2011. "The walking interview: methodology, mobility and place." *Applied Geography* 31: 849–58.

Evans, M. 1988. "Participant observation: The researcher as research tool." In J. Eyles and D.M. Smith, eds, *Qualitative Methods in Human Geography.* Cambridge: Polity Press.

Evans, M., R. Hole, L.D. Berg, P. Hutchinson, and D. Sookraj. 2009. "Common insights, differing methodologies toward a fusion of Indigenous methodologies, participatory action research, and White studies in an urban Aboriginal research agenda." *Qualitative Inquiry* 15 (5): 893–910.

Evans, V., and J. Sternberg. 1999. "Young people, politics and television current affairs in Australia." *Journal of Australian Studies* December: 103–9.

Eyles, J. 1988. "Interpreting the geographical world: Qualitative approaches in geographical research." In J. Eyles and D.M. Smith, eds, *Qualitative Methods in Human Geography.* Cambridge: Polity Press.

Eyles, J., and D.M. Smith, eds. 1988. *Qualitative Methods in Human Geography.* Cambridge: Polity Press.

Fairclough, N. 1992. *Discourse and Social Change.* Cambridge: Polity Press.

———. 2003. *Analysing Discourse: Textual Analysis for Social Research.* New York and London: Routledge.

Fan, W., and Z. Yan. 2010. "Factors affecting response rate of the web survey: a systematic review." *Computers in Human Behaviour,* 26, 132–39.

Faria, C. 2013. "Staging a new South Sudan in the USA: Men, masculinities and nationalist performance at a disaporic beauty pageant." *Gender, Place and Culture* 20: 87–106.

Feinsilver, J.M. 1993. *Healing the Masses: Cuban Health Politics at Home and Abroad.* Berkeley: University of California Press.

Fekete, E., and B. Warf. 2013. "Information Technology and the 'Arab Spring.'" *The Arab World Geographer* 16 (2): 210–27.

Ferbrache, F. 2013. "Le Tour de France: a cultural geography of a mega-event. " *Geography* 98: 144–151.

Fielding, N. 1999. "The norm and the text: Denzin and Lincoln's handbooks of qualitative method." *British Journal of Sociology* 50 (3) 525–34.

Fielding, N., and C.A. Cisneros-Puebla. 2009. "CAQDAS-GIS Convergence: Toward a New Integrated Mixed Method Research Practice?" *Journal of Mixed Methods Research* 3 (4): 349–70.

Fine, M., E. Tuck, and S. Zeller-Berkman. 2007. "Do you believe in Geneva? Methods and ethics at the global local nexus." In C. McCarthy, A. Durham, L. Engel, A. Filmer, M. Giadina, and M. Malagreca, eds, *Globalizing Cultural Studies: Ethnographic Interventions in Theory, Method, and Policy,* 493–525. New York: Peter Lang Publications.

Finer, C. 2000. "Researching in a contemporary archive." *Social Policy and Administration* 34 (4): 434–47.

Fish, S. 1980. *Is There a Text in This Class? The Authority of Interpretive Communities.* London: Harvard University Press.

Fisher, J.F. 1990. *Sherpas: Reflections on Change in Himalayan Nepal.* Berkeley: University of California Press.

Fisher, K.T. 2014. "Postionality, subjectivity, and race in transnational and transcultural geographical research." *Gender, Place and Culture,* DOI:10.1080/0966369X.2013.879097

Flick, U. 1992. "Triangulation revisited: Strategy of validation or alternative?" *Journal for the Theory of Social Behaviour* 2 (2): 175–97.

——. 2009. *An Introduction to Qualitative Research*, 4th edn. Thousand Oaks, London, New Delhi: Sage Publications.

Flowerdew, R., and D. Martin, eds. 1997. *Methods in Human Geography: A Guide for Students Doing a Research Project*. Harlow: Pearson.

——. 2005. *Methods in Human Geography: A Guide for Students Doing a Research Project*. 2nd edn. Harlow: Pearson.

Flyvbjerg, B. 1998. *Rationality and Power: Democracy in Practice*. tr. S. Sampson. Chicago: University of Chicago Press.

——. 2006. "Five misunderstandings about case-study research." *Qualitative Inquiry* 12 (2): 219–45.

Forbat, L., and J. Henderson. 2005. "Theoretical and practical reflections on sharing transcripts with participants." *Qualitative Health Research* 15 (8), 1114–28.

Forrest, J., and K. Dunn. 2013. "Cultural diversity, rural racialisation and the experience of racism in rural Australia: the South Australian case," *Journal of Rural Studies* 30: 1–9.

Foucault, M. 1972. *The Archaeology of Knowledge*. tr. A. Sheridan Smith. New York: Pantheon.

——. 1975. *The Birth of the Clinic*. New York: Vintage Books.

——. 1977a. *Discipline and Punish: The Birth of the Prison*. New York: Pantheon.

——. 1977b. *Language, Counter-memory, Practice: Selected Essays and Interviews*. Ithaca: Cornell University Press.

——. 1980. *Power/Knowledge*. Brighton: Harvester.

Fowler, F. 2002. *Social Survey Methods*. Thousand Oaks, CA: Sage.

Fowler, R. 1991. *Language in the News: Discourse and Ideology in the Press*. London: Routledge.

Frankenberg, R., and L. Mani. 1993. "Crosscurrents, crosstalk: Race, 'postcoloniality' and the politics of location." *Cultural Studies* 7: 292–310.

Frankfort-Nachmaias, C., and D. Nachmaias. 1992. *Research Methods in the Social Sciences*. 4th edn. London: Edward Arnold.

Frantz, K., and R. Howitt. 2012. "Geography for and with Indigenous peoples: Indigenous geographies as challenge and invitation." *GeoJournal* 77(6): 727–31.

Freeman, M. 2007. "Auto-biographical understanding and narrative inquiry." In D.J. Clandinin, ed., *Handbook of Narrative Inquiry, Mapping a Methodology*, 120–45. Thousand Oaks, CA, and London: Sage.

Freire, P. 1972a. *Cultural Action for Freedom*. Harmondsworth: Penguin.

——. 1972b. *Pedagogy of the Oppressed*. Harmondsworth: Penguin.

——. 1976. *Education: The Practice of Freedom*. London: Writers and Readers Publishing Cooperative.

——. 2000. *Pedagogy of the Oppressed*. New York: Continuum.

Fridolfsson, C., and I. Elander. 2013. "Faith and place: constructing Muslim identity in a secular Lutheran society." *Cultural Geographies*, 20 (3): 319–37.

Froehling, O. 1999. "Internauts and Guerilleros: The Zapatista Rebellion in Chiapas, Mexico and Its Extension into Cyberspace." In M. Crang, P. Crang, and J. May, eds, *Virtual Geographies: Bodies, Space, and Relations*. New York: Routledge, 164–77.

Furer-Haimendorf, C. von. 1964. *The Sherpas of Nepal: Buddhist Highlanders*. London: J. Murray.

——. 1975. *Himalayan Traders: Life in Highland Nepal*. London: J. Murray.

——. 1984. *The Sherpas Transformed: Social Change in a Buddhist Society of Nepal*. New Delhi: Sterling Publishers.

Fyfe, N.R. 1992. "Observations on observation." *Journal of Geography in Higher Education* 16 (2): 127–33.

Gade, D.W. 2001. "The languages of foreign fieldwork." *The Geographical Review* 1-2: 370–9.

Gale, G.F. 1964. *A Study of Assimilation*. Adelaide: Libraries Board of South Australia.

Gardner, R., H. Neville, and J. Snell. 1983. "Vietnamese settlement in Springvale." Environmental Report no. 14. Melbourne: Monash University Graduate School of Environmental Science.

Garlick, S. 2002. "Revealing the unseen: Tourism, art and photography." *Cultural Studies* 16 (2): 289–305.

Garvin, T., and K. Wilson. 1999. "The use of storytelling for understanding women's desires to tan: Lessons from the field." *The Professional Geographer* 51: 297–306.

Geertz, C. 1973. *The Interpretation of Culture: Selected Essays*. New York: Basic Books.

——. 1980. "Blurred genres." *American Scholar* 49: 165–79.

——. 1984. "Anti-anti-relativism." *American Anthropologist* 86: 263–77.

Gender, Place and Culture. 2002. 9 (3: special issue on Feminist Geography and GIS).

Geografiska Annaler: Series B: Human Geography. 2006. 88 (3: special issue on Encountering Indigeneity: Re-Imagining and Decolonizing Geography).

Geographical Research. 2007. 45 (2: special issue on Creating Anti-Colonial Geographies: Embracing Indigenous Peoples' Knowledges and Rights).

Geographical Review. 2001. Special issue on "doing fieldwork." 91 (1–2).

George, A., and A. Bennett. 2005. *Case Studies and Theory Development in the Social Sciences*. Cambridge, MA: MIT Press.

George, K. 1999a. *A Place of Their Own: The Men and Women of War Service Land Settlement at Loxton after the Second World War*. Adelaide: Wakefield Press.

———. 1999b. *City Memory: A Guide and Index to the City of Adelaide Oral History Collection*. Adelaide: Corporation of the City of Adelaide.

Gerring, J. 2004. "What is a case study and what is it good for?" *American Political Science Review* 98 (2): 341–54.

———. 2007. *Case Study Research: Principles and Practices*. Cambridge: Cambridge University Press.

Gibson, C. 2003. "Digital divides in New South Wales: A research note on social–spatial inequality using 2001 census data on computer and Internet technology." *Australian Geographer* 34: 239–57.

Gibson, K., J. Cameron, and A. Veno. 1999. *Negotiating Restructuring and Sustainability: A Study of Communities Experiencing Rapid Social Change*. Australian Housing and Urban Research Institute (AHURI) Working Paper. Melbourne: AHURI, Melbourne. Also available: http://www.communityeconomies.org/papers/comecon/comeconp4.pdf.

Gibson-Graham, J.K. 1994. "'Stuffed if I know!': Reflections on post-modern feminist social research." *Gender, Place and Culture* 1 (2): 205–24.

———. 2006. *A Postcapitalist Politics*. Minneapolis: University of Minnesota Press.

Gilbert, L., K. Jackson, and S. di Gregorio. 2013. "Tools for analyzing qualitative data: The history and relevance of qualitative data analysis software." In J. M. Spector, M.D. Merrill, J. Elen, and M.J. Bishop, eds, *Handbook of Research on Educational Communications and Technology*. New York: Springer.

Gill, R. 1996. "Discourse analysis: Practical implementations." In J.T.E. Richardson, ed., *Handbook of Qualitative Methods for Psychology and the Social Sciences*, 141–56. Leicester: British Psychological Society.

Gillham, B. 2000. *Developing a Questionnaire*. London: Continuum.

Ginn, F. 2014. "Death, absence and afterlife in the garden." *Cultural Geographies*, http://cgj.sagepub.com/content/early/2013/04/16/1474474013483220.

Glaser, B.G., and A.L. Strauss. 1967. *The Discovery of Grounded Theory: Strategies for Qualitative Research*. Chicago: Aldine.

Godlewska, A. and N. Smith, eds. 1994. *Geography and Empire*. Oxford: Basil Blackwell.

Godlewska, A.M.C., L.M. Schaefli, and P.J.A. Chaput. 2013. "First Nations assimilation through neoliberal educational reform." *The Canadian Geographer / Le Géographe canadien* 57 (3): 271–79.

Goeman, M. 2013. *Mark My Words: Native Women Mapping Our Nation*. Minneapolis: University of Minnesota Press.

Golafshani, N. 2003. "Understanding reliability and validity in qualitative research." *The Qualitative Report* 8 (4): 597–607.

Gold, J.R. 1980. *An Introduction to Behavioural Geography*. Oxford: Oxford University Press.

Gold, R.L. 1958. "Roles in sociological field observation." *Social Forces* 36: 219–25.

Gomez, B., and J.P. Jones III, eds. 2010. *Research Methods in Geography*. Chichester, UK: Wiley-Blackwell.

Gordon, R. 2004. "Community process and the recovery environment following emergency." *Environmental Health* 4 (1): 9–24.

Gorman-Murray, A. 2007. "Contesting domestic ideas: Queering the Australian home." *Australian Geographer* 38 (3): 195–213.

———. 2012. "Queer politics at home: Gay men's management of the public/private boundary." *New Zealand Geographer* 68 (2): 111–20.

Gorman-Murray, A., C. Brennan-Horley, K. McKlean, G. Waitt, and C. Gibson. 2010. "Mapping same-sex couple family households in Australia." *Journal of Maps* 11, 186–196.

Goss, J.D. 1993. "Placing the market and marketing place: Tourist advertising of the Hawaiian Islands, 1972–1992." *Environment and Planning D: Society and Space* 11: 663–88.

Goss, J.D., and T.R. Leinbach. 1996. "Focus groups as alternative research practice:

Experience with transmigrants in Indonesia." *Area* 28 (2): 115–23.

Gough, K. 1968. "Anthropology and imperialism." *Monthly Review* 19 (11): 12–27.

Gould, P. 1988. "Expose yourself to geographic research." In J. Eyles, ed., *Research in Human Geography*. Oxford: Blackwell.

Graham, M. 2013a. "Geography/internet: Ethereal alternate dimensions of cyberspace or grounded augmented realities." *The Geographical Journal* 179 (2): 177–82.

———. 2013b. "Social Media and the Academy: New Publics or Public Geographies?' *Dialogues in Human Geography* 3 (1): 77–80.

Grbich, C. 2007. *Qualitative Research in Health: An Introduction*. London: Sage.

Green, E., and C. Singleton. 2007. "Mobile selves: Gender, ethnicity and mobile phones in the everyday lives of young Pakistani-British women and men." *Information, Communication and Society* 10 (4): 506–26.

———. 2009. "Mobile connections: an exploration of the place of mobile phones in friendship relations." *Sociological Review* 57(1): 125–44.

Green, J., M. Frauquiz, and C. Dixon. 1997. "The myth of the objective transcription: Transcribing as a situated act." *TESOL Quarterly* 31 (1): 172–6.

Green, K. 2011. "It hurts so it is real: Sensing the seduction of mixed martial arts." *Social and Cultural Geography* 12 (4): 377–96.

Gregory, D. 1978. *Ideology, Science and Human Geography*. London: Hutchinson.

———. 1994. "Paradigm." In R.J. Johnston, D. Gregory, and D.M. Smith, eds, *The Dictionary of Human Geography*. Oxford: Blackwell.

Gregory, D., R. Johnston, G. Pratt, M. Watts, and S. Watmore, eds. 2009. *The Dictionary of Human Geography*. Revised 5th edn. London: Wiley Blackwell.

Gretchen. 2013. "Nurse Menchu Sanchez honored by President Obama for heroic rescue of 20 premature babies," *Seacaucus Home* News 1 March 2013. http://secaucushomenews.com/nurse-menchu-sanchez-honored-by-president-obama-for-heroic-rescue-of-20-premature-babies/ [accessed March 22, 2014].

Grimwood, B.S.R., N.C. Doubleday, G.J. Llubicic, S.G. Donaldson, and S. Blangy. 2012. "Engaged acclimatization: Towards responsible community-based participatory research in Nunavut." *The Canadian Geographer / Le Géographe canadien* 56 (2): 211–230.

Groves, R.M. 1990. "Theories and methods of telephone surveys." *Annual Review of Sociology* 16: 221–40.

Gruzd, A., B. Wellman, and Y. Takhteyev. 2011. "Imagining Twitter as an imagined community." *American Behavioral Scientist* 55 (10): 1294–318.

Guelke, L. 1978. "Geography and logical positivism." In D. Herbert and R.J. Johnston, eds, *Geography and the Urban Environment*, v. 1. New York: Wiley.

Gupta, A., and J. Ferguson. 1997. *Anthropological Locations: Boundaries and Grounds of a Field Science*. Berkeley: University of California Press.

Guthrie, G. 2010. *Basic Research Methods: An Entry to Social Science Research*. London: Sage.

Hackel, S., and A. Reid. 2007. "Transforming an eighteenth-century archive into a twenty-first century database: The Early Californian Population Project." *History Compass* 5 (3): 1013–25.

Hall, B. 2005. "In from the cold? Reflections on participatory research from 1970–2005." *Convergence* 38 (1): 5–24.

Hall, S., ed. 1997. *Representation: Cultural Representations and Signifying Practices*. London: Sage.

Hall, T. 2009. "The camera never lies. Photographic research methods in human geography." *Journal of Geography in Higher Education* 33 (3): 453–62.

Hallowell, N., P. Lawton, and S. Gregory. 2005. *Reflections on Research: The Realities of Doing Research in the Social Sciences*. Berkshire: Open University Press.

Hammersley, M. 1992. "Deconstructing the qualitative–quantitative divide." In J. Brannen, ed., *Mixing Qualitative and Quantitative Research*. Aldershot, US: Avebury.

Hammersley, M., and P. Atkinson. 1983. *Ethnography: Principles in Practice*. London: Tavistock.

Haraway, D. 1991. "Situated knowledges: The science question in feminism and the privilege of partial perspective." In D. Haraway, ed., *Simians, Cyborgs and Women: The Reinvention of Nature*. London: Routledge.

Harding, S. 2011. "Other cultures' sciences." In S. Harding S, ed., *The Postcolonial Science and Technologies Study Reader*, 151–8. Durham and London: Duke University Press.

Hares, A., J. Dickinson, and K. Wilkes. 2010. "Climate change and the air travel decisions of UK tourists." *Journal of Transport Geography*, 18 (3): 466–73.

Harley, J.B. 1992. "Deconstructing the map." In T. Barnes and J. Duncan, eds, *Writing Worlds: Discourse, Text and Metaphor in the Representation of Landscape*. London: Routledge.

Harper, D. 2002. "Talking about pictures: A case for photo elicitation." *Visual Studies* 17: 13–26.

———. 2003. "Framing photographic ethnography: A case study." *Ethnography* 4: 241–66.

Harris, C. 2001. "Archival fieldwork." *Geographical Review* 91 (1–2): 328–34.

Harris, L., and J. Wasilewski. 2004. "Indigeneity, an alternative worldview: Four R's (relationship, responsibility, reciprocity, redistribution) vs. two P's (power and profit). Sharing the journey towards conscious evolution." *Systems Research and Behavioral Science* 21 (5): 489–503.

Harrison, R.T., and D.N. Livingstone. 1980. "Philosophy and problems in human geography: A pre-suppositional approach." *Area* 12: 25–30.

Hartig, K.V., and K.M. Dunn. 1998. "Roadside memorials: Interpreting new deathscapes in Newcastle, New South Wales." *Australian Geographical Studies* 36 (1): 5–20.

Harvey, D. 1984. "On the history and present condition of geography: An historical materialist manifesto." *Professional Geographer* 36: 1–11.

Harvey, K. 2006. "From bags and boxes to searchable digital collections at the Dalhousie University Archives." *Journal of Canadian Studies* 40 (2): 120–38.

Hassini, E. 2006. "Student–instructor communication: The role of email." *Computers and Education* 47: 29–40.

Hawkins, H. 2011. "Dialogues and doings: Sketching the relationships between geography and art." *Geography Compass* 5 (7): 464–478.

Hawley, P. 2003. *Being Bright Is Not Enough*. 2nd edn. Springfield, IL: Charles C. Thomas.

Hay, I. 1998. "Making moral imaginations: Research ethics, pedagogy, and professional human geography." *Ethics, Place and Environment* 1 (1) 55–75.

———. 2003. "From 'millennium' to 'profiles': Geography's oral histories across the Tasman." *Proceedings of the 22nd Conference of the New Zealand Geographical Society*, 6–11 July, University of Auckland, New Zealand, 5–6.

———. 2012. *Communicating in Geography and the Environmental Sciences*, 4th edn. Melbourne: Oxford University Press.

Hay, I., D. Bochner, G. Blacket, and C. Dungey. 2012. *Making the Grade: A Guide to Successful Communication and Study*, 4th edn. Melbourne: Oxford University Press.

Hay, I., A. Hughes, and M. Tutton. 2004. "Monuments, memory and marginalisation in Adelaide's Prince Henry Gardens." *Geografiska Annaler B* 86 (3): 200–15.

Hay, I., and S. Muller. 2012. "That tiny, stratospheric apex that owns most of the world'—Exploring geographies of the super-rich. *Geographical Research* 50 (1): 75–88.

He, C. 2013. "Performance and the politics of gender: Transgender performance in contemporary Chinese films." *Gender, Place & Culture: A Journal of Feminist Geography*, DOI:10.1080/0966369X.2013.810595.

Head, L., and P. Muir. 2006. "Suburban life and the boundaries of nature: Resilience and rupture in Australian backyards." *Transactions of the Institute of British Geographers* 31 (4): 505–24.

Heath, A.W. 1997. "The proposal in qualitative research." *The Qualitative Report* 3 (1). http://www.nova.edu/ssss/QR/QR3-1/heath.html.

Heathcote, R.L. 1975. *Australia*. London: Longman.

Herbert, S. 2012. "A taut rubber band: Theory and empirics in qualitative geographic research." In D. DeLyser, S. Herbert, S. Aitken, M. Crang, and L.McDowell, eds, *The Sage Handbook of Qualitative Geography*. Los Angeles: Sage.

Herlihy, P.H., and G. Knapp. 2003. "Maps of, by, and for the peoples of Latin America." *Human Organization* 62 (4): 303–14.

Hershkowitz, I., and A. Terner. 2007. "The effects of repeated interviewing on children's forensic statements of sexual abuse." *Applied Cognitive Psychology* 21: 1131–43.

Herman, R.D.K. 1999. "The Aloha State: Place names and the anti-conquest of Hawai'i." *Annals of the Association of American Geographers* 89 (1): 76–102.

Herman, T., and D. Mattingly. 1999. "Community, justice, and the ethics of research: Negotiating reciprocal research relations." In J.D. Proctor and D.M. Smith, eds, *Geography and Ethics: Journeys in a Moral Terrain*. London: Routledge.

Herod, A. 1993. "Gender issues in the use of interviewing as a research method." *Professional Geographer* 45 (3): 305–17.

Hesse-Biber, S., and A. Griffin 2013. "Internet-mediated technologies and mixed methods research: Problems and prospect." *Journal of Mixed Methods Research* 7 (1): 43–61.

Hesse-Biber, S., and P. Leavy. 2004. *Approaches to Qualitative Research: A Reader on Theory and Practice.* Oxford: Oxford University Press.

Hewson, C., et al. 2003. *Internet Research Methods.* London: Sage.

Hibbard, M., M.B. Lane, and K. Rasmussen. 2008. "The split personality of planning: Indigenous peoples and planning for land and resource management." *Journal of Planning Literature* 23 (2): 136–51.

Hickey, S., and G. Mohan, eds. 2004. *Participation: From Tyranny to Transformation.* London: Zed Books.

Hiebert, W., and D. Swan. 1999. "Positively fit: A case study in community development and the role of participatory research." *Community Development Journal* 34: 356–64.

Hoelscher, S. 1998. *Heritage on Stage: The Invention of Ethnic Place in America's Little Switzerland.* Madison, WI: University of Wisconsin Press.

Hoggart, K., L. Lees, and A. Davies. 2002. *Researching Human Geography.* London: Arnold.

Holbrook, B., and P. Jackson. 1996. "Shopping around: Focus group research in North London." *Area* 28 (2): 136–42.

Holland, P., E. Pawson, and T. Shatford. 1991. "Qualitative resources in geography." *New Zealand Journal of Geography* 92 (special issue).

Holliday, A. 2007. *Doing and Writing Qualitative Research.* 2nd edn. Thousand Oaks, CA, and London: Sage.

Holloway, S.L, M. Jayne, and G. Valentine. 2008. "'Sainsbury's is my local': English alcohol policy, domestic drinking practices and the meaning of home,' *Transactions of the Institute of British Geographers* 33: 532–47.

Holloway, S.L., G. Valentine, and M. Jayne. 2009. "Masculinities, femininities and the geographies of public and private drinking landscapes." *Geoforum* 40: 821–31.

Holt-Jensen, A. 1988. "Multiple meanings: Shopping and the cultural politics of identity." *Environment and Planning A* 27: 1913–30.

Hones, S. 2011. "Literary geography: the novel as a spatial event." In S. Daniels, D. Delyser, J.N. Entrikin, and D. Richardson, eds, *Envisioning Landscapes, Making Worlds. Geography and the Humanities*, 247–255. London and New York: Routledge.

Hook, D. 2001. "Discourse, knowledge, materiality, history: Foucault and discourse analysis." *Theory and Psychology* 11 (4): 521–47.

hooks, b. 1990. *Yearning: Race, Gender, and Cultural Politics.* Boston: South End Press.

Hooper-Greenhill, E. 2000. *Museums and the Interpretation of Visual Culture.* London: Routledge.

Hoppe, M.J., et al. 1995. "Using focus groups to discuss sensitive topics with children." *Evaluation Review* 19 (1): 102–14.

Horowitz, I.L. 1967. "The rise and fall of Project Camelot." In I. Horowitz, *The Rise and Fall of Project Camelot: Studies in the Relationship between Science and Practical Politics.* Cambridge, MA: MIT Press.

Horvath, R.J. 1971. "The Detroit Geographical Expedition and Institute experience." *Antipode* 3 (1): 73–85.

Housel, J. 2009. "Geographies of whiteness: the active construction of racialized privilege in Buffalo, New York." *Social and Cultural Geography* 10 (2): 131–51.

Howitt, R. 1992. "The political relevance of locality studies: A remote Antipodean viewpoint." *Area* 24 (1): 73–81.

———. 1997. "Getting the scale right: The geopolitics of regional agreements." *Northern Analyst* 2: 15–17.

———. 2001. *Rethinking Resource Management: Justice, Sustainability and Indigenous Peoples.* London: Routledge.

———. 2002a. "Worlds turned upside down: Inclusionary research in Australia." *Proceedings of the Association of American Geographers Conference*, Los Angeles, March. http://www.es.mq.edu.au/~rhowitt/AAG2002.panel.htm.

———. 2002b. "Decolonizing research: Ethical and methodological issues." *Proceedings of the Institute of Australian Geographers Conference*, Australian National University, Canberra.

———. 2002c. "Scale and the other: Levinas and geography." *Geoforum* 33: 299–313.

———. 2003. "Scale." In J. Agnew, K. Mitchell, and G. Toal, eds, *A Companion to Political Geography.* Oxford: Blackwell.

———. 2011. "Knowing/Doing." In V.J. Del

Casino Jr, M.E. Thomas, P. Cloke, and R. Panelli, *A Companion to Social Geography*. 131–45. Chichester, Wiley-Blackwell.

Howitt, R., G. Crough, and B. Pritchard. 1990. "Participation, power and social research in Central Australia." *Australian Aboriginal Studies* 1: 2–10.

Howitt, R., K. Doohan, S. Suchet-Pearson, S. Cross, R. Lawrence, G.J. Lunkapis, S. Muller, S. Prout, and S. Veland. 2013. "Intercultural capacity deficits: Contested geographies of coexistence in natural resource management." *Asia Pacific Viewpoint* 54 (2): 126–40.

Howitt, R., and S. Jackson. 1998. "Some things do change: Indigenous rights, geographers and geography in Australia." *Australian Geographer* 29 (2): 155–73.

Howitt, R., and S. Suchet-Pearson. 2003. "Ontological pluralism in contested cultural landscapes." In K. Anderson, M. Domosh, S. Pile, and N. Thrift, eds, *Handbook of Cultural Geography*. London: Sage.

Hudson, B.J. 2013. "The Naming of Waterfalls." *Geographical Research* 51 (1): 85–93.

Hulme, P. 2012. "Writing on the land: Cuba's literary geography." *Transactions of the Institute of British Geographers* 37 (3): 346–358.

Hume-Cook, G., et al. 2007. "Uniting people with place using participatory video in Aotearoa/New Zealand." In S. Kindon, R. Pain, and M. Kesby, eds, *Participatory Action Research Approaches and Methods: Connecting People, Participation and Place*, 160–9. London: Routledge.

Hunt, S. 2014. "Ontologies of Indigeneity: The politics of embodying a concept." *Cultural Geographies* 21: 27–32.

Hutcheson, G. 2014. "Methodological reflections on transference and countertransference in geographical research: Relocation experiences from post-disaster Christchurch, Aotoearoa New Zealand." *Area* 45 (4): 477–84.

Isaacman, A., P. Lalu, and T. Nygren. 2005. "Digitization, history, and the making of a postcolonial archive of southern Africa liberation struggles: The Aluka project." *Africa Today* 52 (2): 54–77.

Israel, M., and I. Hay. 2006. *Research Ethics for Social Scientists: Between Ethical Conduct and Regulatory Compliance*. London: Sage.

Jackson, J. 1990. "'I am a field note': Fieldnotes as a symbol of professional identity." In R. Sanjek, ed., *Fieldnotes: The Makings of Anthropology*. Ithaca, NY: Cornell University Press.

Jackson, P. 1983. "Principles and problems of participant observation." *Geografiska Annaler* 65B: 39–46.

———. 1993. "Changing ourselves: A geography of position." In R.J. Johnston, ed., *The Challenge for Geography*. Oxford: Basil Blackwell.

Jackson, P., and B. Holbrook. 1995. "Multiple meanings: Shopping and the cultural politics of identity." *Environment and Planning A* 27: 1913–30.

Jacobs, J.M. 1999. "The labour of cultural geography." In E. Stratford, ed., *Australian Cultural Geographies*. Melbourne: Oxford University Press.

James, A.M., and T. Rashed. 2006. "In their own words: Utilizing weblogs in quick response research," in Natural Hazards Center, *Learning from Catastrophe: Quick Response Research in the Wake of Hurricane Katrina*. Boulder: Institute of Behavioural Science, 71–96.

James, N., and H. Busher. 2006. "Credibility, authenticity and voice: Dilemmas in online interviewing." *Qualitative Research* 6 (3): 403–20.

Jay, N. 1981. "Gender and dichotomy." *Feminist Studies* 7: 38–56.

Jenkins, M. 2013. "Parkour classes are helping pensioners stay agile and active." *The Guardian*, Wednesday 28 August, http://www.theguardian.com/society/2013/aug/28/parkour-classes-pensioners-agile-active, accessed on 11 October 2013.

Jenkins, S., V. Jones, and D. Dixon. 2003. "Thinking/doing the 'F' word: On power in feminist methodologies." *ACME: an International e-journal for critical geographies* 2 (1): 57–63.

Jick, T.D. 1979. "Mixing qualitative and quantitative methods: Triangulation in action." *Administrative Science Quarterly* 24 (December): 602–11.

Johnson, A. 1996. "'It's good to talk': The focus group and the sociological imagination." *The Sociological Review* 44 (3): 517–38.

Johnson, B.E. 2011. "The speed and accuracy of voice recognition software-assisted transcription versus the listen-and-type method: A research note." *Qualitative Research* 11: 91–7.

Johnson, J.T. 2012. "Place-based learning and knowing: Critical pedagogies grounded in Indigeneity." *GeoJournal* 77: 829–36.

Johnson, J.T., et al. 2007. "Creating anti-colonial

geographies: Embracing Indigenous peoples' knowledges and rights." *Geographical Research* 45 (2): 117–20.

Johnson, J.T., R.P. Louis, and A. Pramono. 2005. "Facing future: Encouraging critical cartographic literacies in indigenous communities." *ACME: An International E-Journal for Critical Geographies* 4 (1): 80–98.

Johnson, J.T., and B. Murton. 2007. "Replacing native science: Indigenous voices in contemporary constructions of nature." *Geographical Research* 45 (2): 121–29.

Johnston, L. 2001. "(Other) bodies and tourism studies." *Annals of Tourism Research* 28 (1): 180–201.

———. 2005. "Transformative tans? Gendered and raced bodies on beaches." *New Zealand Geographer,* 61: 110–16.

Johnston, R.J. 1978. *Multivariate Statistical Analysis in Geography: A Primer on the General Linear Model.* New York: Longman.

———. 1983. *Philosophy and Human Geography: An Introduction to Contemporary Approaches.* London: Edward Arnold.

———. 2000. "On disciplinary history and textbooks: Or where has spatial analysis gone?" *Australian Geographical Studies* 38 (2): 125–37.

Johnston, R.J., et al., eds. 2000. *The Dictionary of Human Geography.* Oxford: Blackwell.

Jones, A. 1992. "Writing feminist educational research: Am 'I' in the text?" In S. Middleton and A. Jones, eds, *Women and Education in Aotearoa.* Wellington: Bridget Williams Books.

Jones, A., and J. T. Murphy. 2011. "Theorizing practice in economic geography: Foundations, challenges, and possibilities." *Progress in Human Geography* 35 (3): 366–92.

Jones, J.P. III, H.J. Nast, and S.M. Roberts, eds. 1997. *Thresholds in Feminist Geography: Difference, Methodology, Representation.* Lanham, MD: Rowman and Littlefield.

Jones, J.P., H.J. Nast, and S.M. Roberts, eds. 1997. *Thresholds in Feminist Geography: Difference, Methodology, Representation.* Oxford: Rowman and Littlefield.

Jones, P., and J. Evans. 2012. "The spatial transcript: Analysing mobilities through qualitative GIS." *Area* 44 (1): 92–9.

Jorgensen, J. 1971. "On ethics and anthropology." *Current Anthropology* 12 (3): 321–56.

Joseph, M. 2013. *Fluid New York: Cosmopolitan Urbanism and the Green Imagination.* Durham, NC: Duke University Press.

Jung, J.-K. 2009. "Computer-aided qualitative GIS: A software-level integration of qualitative research and GIS." In M. Cope and S. Elwood, eds, *Qualitative GIS: A Mixed-Methods Approach,* 115–35. London: Sage.

———. 2014. "Code clouds: Qualitative geovisualization of geotweets." *The Canadian Geographer* 59 (1): 52–68.

Juris, J. 2012. "Reflections on #Occupy Everywhere: Social media, public space, and emerging logics of aggregation." *American Ethnologist* 39 (2): 259–79.

Kane, M., and W. Trochim 2006. *Concept Mapping for Planning and Evaluation.* London: Sage.

Karnieli-Miller, O., R. Strier, and L. Pessach. 2009. "Power relations in qualitative research." *Qualitative Health Research* 19 (2): 279–89.

Katz, C. 1994. "Playing the field: Questions of fieldwork in geography." *Professional Geographer* 46 (1): 67–72.

Kearns, R. 1987. "In the shadow of illness: A social geography of the chronically mentally disabled in Hamilton, Ontario." (Department of Geography, McMaster University, PhD dissertation).

———. 1991a. "Talking and listening: Avenues to geographical understanding." *New Zealand Journal of Geography* 92: 2–3.

———. 1991b. "The place of health in the health of place: The case of the Hokianga special medical area." *Social Science and Medicine* 33: 519–30.

———. 1997. "Constructing (bi)cultural geographies: Research on, and with, people of the Hokianga District." *New Zealand Geographer* 52: 3–8.

———. 2014. "Sounds, surrounds and wellbeing on planet WOMAD." In G. Andrews. P. Kingsbury, and R. Kearns, eds, *Soundscapes of Wellbeing in Popular Music,* 107–22. Aldershot, Surrey: Ashgate Press.

Kearns, R., and I. Dyck. 2005. "Culturally safe research." In D. Wepa, ed., *Cultural Safety in Aotearoa New Zealand,* 79–88. Auckland: Pearson/Prentice Hall.

Kearns, R., and J. Fagan. 2014. "Sleeping with the past? Heritage, recreation and transition in New Zealand tramping huts." *New Zealand Geographer* 70 (2): 116–30. .

Kearns, R.A., D.C.A. Collins, and P.M. Neuwelt. 2003. "The walking school bus: Extending children's geographies." *Area* 35: 285–92.

Kearns, R.A., P.M. Neuwelt, B. Hitchman, and M. Lennan. 1997. "Social support and psychological distress before and after childbirth." *Health and Social Care in the Community* 5: 296–308.

Kearns, R.A., C.J. Smith, and M.W. Abbott. 1991. "Another day in paradise? Life on the margins in urban New Zealand." *Social Science and Medicine* 33: 369–79.

Keen, J., and T. Packwood. 1995. "Case study evaluation." *British Medical Journal* 311 (7002): 444–8.

Keighren, I.M. 2012, "Fieldwork in the archive." In R. Phillips and J. Johns, eds, *Fieldwork for Human Geography*, 138–40. London: Sage.

Kellehear, A. 1993. *The Unobtrusive Researcher: A Guide to Methods*. Sydney: Allen and Unwin.

Kenny, A. 2005. "Interaction in cyberspace: An online focus group." *Methodological Issues in Nursing Research* 49 (4): 414–22.

Kesby, M. 2000. "Participatory diagramming: Deploying qualitative methods through an action research epistemology." *Area* 32 (4): 423–535.

———. 2005. "Retheorizing empowerment-through-participation as a performance in space: Beyond tyranny to transformation." *Signs: Journal of Women in Culture and Society* 30: 2037–65.

———. 2007. "Spatialising participatory approaches: The contribution of geography to a mature debate." *Environment and Planning A* 39 (12): 2813–31.

Kesby, M., S. Kindon, and R. Pain. 2005. "'Participatory' diagramming and approaches." In R. Flowerdew and D. Martin, eds, *Methods in Human Geography*, 2nd edn, 144–66. London: Pearson.

———. 2007. "Participation as a form of power: Retheorising empowerment and spatialising participatory action research." In S. Kindon, R. Pain, and M. Kesby, eds, *Participatory Action Research Approaches and Methods: Connecting People, Participation and Place*, 19–25. London: Routledge.

Kidder, J.L., 2012. "Parkour, the affective appropriation of urban space, and the real/virtual dialectic." *City & Community*, 11 (3), 229–53.

Kindon, S. 1995. "Exploring empowerment methodologies with women and men in Bali." *New Zealand Geographer* 51 (1): 10–12.

———. 1998. "Of mothers and men: Challenging gender and community myths in Bali, Indonesia." In I. Guijt and M. Kaul Shah, eds, *The Myth of Community: Gender Issues in Participatory Development*. London: Intermediate Technology Publications.

———. 2003. "Participatory video in geographic research: A feminist practice of looking?" *Area* 35 (2): 142–53.

———. 2010. "Participation." In S. Smith, R. Pain, S. Marston, and J.P. Jones III, eds, *The Handbook of Social Geography*. London: Sage.

———. 2012. "'Thinking-through-Complicity' with Te Iwi o Ngaati Hauiti: Towards a Critical Use of Participatory Video for Research," unpublished PhD thesis, Hamilton: University of Waikato.

Kindon, S., and J, Cupples. 2014. "Anything to declare? The politics of leaving the field." In R. Scheyvens, ed, *Development Fieldwork: A Practical Guide*, 2nd edn, 217–35. London: Sage.

Kindon, S., and A. Latham. 2002. "From mitigation to negotiation: Ethics and the geographical imagination in Aotearoa/New Zealand." *New Zealand Geographer* 58 (1): 14–22.

Kindon, S., R. Pain, and M. Kesby, eds. 2007a. *Participatory Action Research Approaches and Methods: Connecting People, Participation and Place*. London: Routledge.

———. 2007b. "Participatory action research: Origins, approaches and methods." In S. Kindon, R. Pain, and M. Kesby, eds, *Participatory Action Research Approaches and Methods: Connecting People, Participation and Place*, 9–18. London: Routledge.

Kinsley, S. 2013. "Beyond the screen: Methods for investigating geographies of life 'online" *Geography Compass* 7 (8): 540–55.

Kirby, S., and I. Hay. 1997. "(Hetero)sexing space: Gay men and 'straight' space in Adelaide, South Australia." *Professional Geographer* 49 (3): 295–305.

Kirk, J., and M. Miller. 1986. *Reliability and Credibility in Qualitative Research*. Beverly Hills, CA: Sage.

Kitchin, R., and N. Tate. 2000. *Conducting Research in Human Geography: Theory, Methodology and Practice*. London: Longman.

Kitchin, R., D. Leinhan, C. O'Callaghan, and P. Lawton. 2013. "Public geographies through social media." *Dialogues in Human Geography* 3 (1): 56–72.

Kitzinger, J. 1994. "The methodology of focus

groups: The importance of interaction between research participants." *Sociology of Health and Illness* 16 (1): 103–21.

Klocker, N. 2012. "Doing participatory action research and doing a PhD: Words of encouragement for prospective students." *Journal of Geography in Higher Education* 36 (1): 149–63.

Kneale, J. 2001. "Working with groups." In M. Limb and C. Dwyer, eds, *Qualitative Methodologies for Geographers*, 136–50. London: Arnold.

Kneale, P. 1999. *Study Skills for Geography Students*. London: Arnold.

Kobayashi, A. 1994. "Coloring the field: Gender, 'race', and the politics of fieldwork." *Professional Geographer* 46 (1): 73–80.

Kolakowski, L. 1972. *Positivist Philosophy: From Hume to the Vienna Circle*. London: Penguin.

Kong, L. 1998. "Refocussing on qualitative methods: Problems and prospects for research in a specific Asian context." *Area* 30 (1): 79–82.

Kothari, U. 2001. "Power, knowledge and social control in participatory development." In B. Cooke and U. Kothari, eds, *Participation: The New Tyranny?* London: Zed Books.

Kovach, M. 2009. *Indigenous Methodologies: Characteristics, Conversations and Contexts*. Toronto: University of Toronto Press.

Krueger, R. 1994. *Focus Groups: A Practical Guide for Applied Research*. 2nd edn. Thousand Oaks, CA: Sage.

———. 1998. *Analyzing and Reporting Focus Group Results*. Focus Group Kit 6. Thousand Oaks, CA: Sage.

Kuhn, T.S. 1962. *The Structure of Scientific Revolutions*. Chicago: University of Chicago Press.

Kusenbach, M. 2002. "Up close and personal: Locating the self in qualitative research." *Qualitative Sociology* 25 (1): 149–52.

———.2003. "Street phenomenlogy: the go-along as ethnographic research tool." *Ethnography*, 4: 455–85.

Kwan, M.P. 2008. "From oral histories to visual narratives: re-presenting the post-September 11 experiences of the Muslim women in the USA." *Social & Cultural Geography* 9 (6): 653–669.

Kwan, M.P., and L. Knigge. 2006. "Doing qualitative research using GIS: An oxymoronic endeavor?" *Environment and Planning A* 38: 1999–2002.

Lane, M.B., et al. 2003. "Sacred land, mineral wealth, and biodiversity at Coronation Hill, Northern Australia: Indigenous knowledge and SIA." *Impact Assessment and Project Appraisal* 21 (2): 89–98.

Lane, R. 1997. "Oral histories and scientific knowledge in understanding environmental change: A case study in the Tumut region, NSW." *Australian Geographical Studies* 35 (2): 195–205.

Lahiri-Dutt, K. 2013. "Bodies in/out of place, Masculinities and motherhood of Kamins in Indian coal mines." *South Asian History and Culture* 4: 213–29.

Larocque, E. 2013. "Long way from home." *Socialist Studies / Études socialistes* 9: 22–6.

Latham, A. 2000. "Urban renewal, heritage planning and the remaking of an inner-city suburb: A case study of heritage planning in Auckland, New Zealand." *Planning Practice and Research* 15 (4): 285–98.

———. 2003. "Research, performance, and doing human geography: Some reflections on the diary-photography, diary-interview method." *Environment and Planning A* 35: 1993–2017.

Latham, A., and D.P. McCormack. 2007. "Digital photography and web-based assignments in an urban field course: Snapshots from Berlin." *Journal of Geography in Higher Education* 31 (2): 241–56.

Latimer, B. 1998. "Masculinity, place and sport: Rugby Union and the articulation of the 'new man' in Aotearoa/New Zealand." (Department of Geography, University of Auckland, MA thesis.).

Laurier, E., and C. Philo. 2006. "Cold shoulders and napkins handed: Gestures of responsibility." *Transactions of the Institute of British Geographers* NS 31: 193– 207.

Law, J. 2004. *After Method: Mess in Social Science Research*. London: Routledge.

Law, L. 2000. *Sex Work in Southeast Asia: The Place of Desire in a Time of AIDS*. London and New York: Routledge.

Law, L., C.J.W.L. Wee, and F. McMullan. 2011. "Screening Singapore: The cinematic landscape of Eric Khoo's *Be With Me*." *Geographical Research* 49 (4): 363–74.

Lawrence, J. 2007. "Placing the lived experience(s) of TB in a refugee community in Auckland, New Zealand." (University of Auckland, New Zealand, PhD thesis [Geography].)

Lawrence, R., and M. Adams. 2005. "First Nations

and the politics of indigeneity: Australian perspectives on indigenous peoples, resource management and global rights." *Australian Geographer* 36 (2): 265–72.

Lawson, V. 1995. "The politics of difference: Examining the quantitative/qualitative dualism in post-structuralist feminist research." *Professional Geographer* 47 (4): 449–57.

Le Doeff, M. 1987. "Women and philosophy." In T. Moi, ed., *French Feminist Thought: A Reader*. Oxford: Basil Blackwell.

Lee, R. 2000. *Unobtrusive Methods in Social Research*. Buckingham: Open University Press.

Lee, S-O., S-J. Kim, and J. Wainwright. 2010. "Mad cow militancy: Neoliberal hegemony and social resistance in South Korea." *Political Geography* 29 (7): 359–69.

Lees, L. 2012. "The geography of gentrification: Thinking through comparative urbanism." *Progress in Human Geography* 36 (2): 155–71.

Lefebvre, H. 1991. *The Production of Space*. Oxford: Blackwell.

Leurs, R. 1997. "Critical reflections on rapid and participatory rural appraisal." *Development in Practice* 7 (3): 290–3.

Leurs, K. and S. Ponzanesi. 2011. "Communicative spaces of their own: Migrant girls performing selves using instant messaging software." *Feminist Review* 99: 55–78.

Levinas, E. 1969. *Totality and Infinity: An Essay on Exteriority*. tr. A. Lingis. Pittsburgh: University of Pittsburgh Press.

Levy, P. 1997. *Collective Intelligence: Toward an Anthropology of Cyberspace*. New York: Plenum Trade.

———. 1998. *Becoming Virtual: Reality in the Digital Age*. New York: Plenum Trade.

Lewin, K. 1946. "Action research and minority problems." *Journal of Social Issues* 1–2: 34–6.

Lewis, P.F. 1979. "Axioms for reading the landscape: Some guides on the American scene." In D.W. Meinig, ed., *The Interpretation of Ordinary Landscapes: Geographical Essay*, 11–33. New York: Oxford University Press.

Ley, D. 1974. *The Black Inner City as Frontier Outpost*. Monograph no. 7. Washington: Association of American Geographers.

Liamputtong, P. 2007. *Researching the Vulnerable: A Guide to Sensitive Research Methods*. London: Sage.

Liebow, E. 1967. *Tally's Corner: A Study of Negro Streetcorner Men*. Boston: Little Brown.

Limb, M., and C. Dwyer, eds. 2001. *Qualitative Methodologies for Geographers: Issues and Debates*. New York: Oxford University Press.

Lincoln, Y., and E. Guba. 1981. *Effective Evaluation*. San Francisco: Jossey-Bass.

———. 1985. *Naturalistic Inquiry*. Beverly Hills, CA: Sage.

———. 2000. "Paradigmatic controversies, contradictions, and emerging confluences." In N.K. Denzin and Y.S. Lincoln, eds, *Handbook of Qualitative Research*, 2nd edn. Thousand Oaks, CA: Sage.

———. 2002. "Judging the quality of case study reports." In M. Huberman and M. Miles, eds, *The Qualitative Researcher's Companion*. London: Sage.

Lindsay, J.M. 1997. *Techniques in Human Geography*. London: Routledge.

Lindsey, D., and R.A. Kearns. 1994. "The writing's on the wall: Graffiti, territory and urban space." *New Zealand Geographer* 50: 7–13.

Livingstone, D.N. 2005. "Science, text and space: Thoughts on the geography of reading." *Transactions of the Institute of British Geographers* 30 (4): 391–401.

Lloyd, G. 1984. *The Man of Reason: "Male" and "Female" in Western Philosophy*. London: Methuen.

Lockwood, M., J. Davidson, A. Curtis, E. Stratford, and R. Griffith. 2010. "Governance principles for natural resource management." *Society and Natural Resources* 23 (10): 986–1001.

Longhurst, R. 1995. "The geography closest in—The body . . . the politics of pregnability." *Australian Geographical Studies* 33: 214–23.

———. 1996. "Refocusing groups: Pregnant women's geographical experiences of Hamilton, New Zealand/Aotearoa." *Area* 28 (2): 143–9.

———. 2000. "'Corporeographies' of pregnancy: 'Bikini babes'." *Environment and Planning D: Society and Space* 18: 453–72.

———. 2005. "(Ad)dressing pregnant bodies in New Zealand: Clothing, fashion, subjectivities and spatialities." *Gender, Place and Culture*, 12 (4): 433–46.

———. 2009 "YouTube: a new space for birth?" *Feminist Review* 93: 46–63

———. 2012. "Becoming smaller: Autobiographical spaces of weight loss." *Antipode* 44 (3): 871–88.

———. 2013. "Using Skype to mother: Bodies, emotions, visuality, and screens."

Environment and Planning D: Society and Space 31: 664–79.

Lorimer, H. 2010. "Caught in the nick of time: archives and fieldwork." In D. DeLyser, S. Herbert, S. Aitken, M. Crang, and L. McDowell, eds, *Handbook of Qualitaative Research*. 249–73. Los Angeles: SAGE.

Louis, R.P. 2007. "Can you hear us now? Voices from the margin: using indigenous methodologies in geographic research." *Geographical Research* 45: 130–9.

Lowenthal, D., and H. Prince. 1965. "English landscape tastes." *Geographical Review* 47 (4): 449–57.

Lukinbeal, C., and S. Zimmerman, eds. 2008. *The Geography of Cinema—A Cinematic World*. Franz Steiner Verlag GmbH.

Lumsden, J. 2005. "Guidelines for the design of online questionnaires." National Research Council of Canada. Published as NRC/ERC-1127. 9 June, NRC48231. http://iit-iti.nrc-cnrc.gc.ca/iit-publications-iti/docs/NRC-48231.pdf.

Lunt, P., and S. Livingstone. 1996. "Rethinking the focus group in media and communications research." *Journal of Communication* 46 (2): 79–98.

Lutz, C.A., and J.L. Collins. 1993. *Reading National Geographic*. Chicago: University of Chicago Press.

McClean, R., L.D. Berg, and M.M. Roche. 1997. "Responsible geographies: Co-creating knowledges in Aotearoa." *New Zealand Geographer* 53 (2): 9–15.

Maccoby, E., and N. Maccoby. 1954. "The interview: A tool of social science." In G. Lindzey, ed., *Handbook of Social Psychology*. Cambridge, MA: Addison-Wesley.

McCoyd, J.L.M., and T.S. Kerson. 2006. "Conducting intensive interviews using e-mail: A serendipitous comparative opportunity." *Qualitative Social Work* 5 (3): 389–406.

McDowell, L. 1992a. "Multiple voices: Speaking from outside and inside the project." *Antipode* 24: 56–72.

———. 1992b. "Valid games? A response to Erica Schoenberger." *Professional Geographer* 44 (2): 212–15.

———. 1998. "Illusions of power: Interviewing local elites." *Environment and Planning A* 30: 2121–32.

McDowell, L., and G. Court. 1994. "Performing work: Bodily representations in merchant banks." *Environment and Planning D: Society and Space* 12: 727–50.

McFarlane, T., and I. Hay. 2003. "The battle for Seattle: Protest and popular geopolitics in *The Australian* newspaper." *Political Geography* 22: 211–32.

McGregor, G. 1994. *EcCentric Visions: Re-constructing Australia*. Waterloo, ON: Wilfrid Laurier University Press.

McGuirk, P.M. 2002. "Producing the capacity to govern in global Sydney: A multiscaled account." *Journal of Urban Affairs* 25: 201–23.

McGuirk P.M., and R. Dowling. 2011. "Rethinking urban politics: new directions in governing social reproduction and everyday life in the city." *Urban Studies* 48: 2611–2628.

McIlwaine, C., and C. Moser. 2003. "Poverty, violence and livelihood security in urban Colombia and Guatemala." *Progress in Development Studies* 3 (2): 113–30.

McKay, D. 2002. "Negotiating positionings: Exchanging life stories in research interviews." In P. Moss, ed., *Feminist Geography in Practice*. Oxford: Blackwell.

McKendrick, J.H. 1996. *Multi-method Research in Population Geography: A Primer to Debate*. Manchester: University of Manchester, Population Geography Research Group.

Mackenzie, S. 1989. *Visible Histories: Women and Environments in a Post-war British City*. Montreal: McGill-Queen's University Press.

McLafferty, S. 2010. "Conducting questionnaire surveys." In N. Clifford and G. Valentine, eds, *Key Methods in Geography*, 77–88. London: Sage

McLean, J., and S. Maalsen. 2013. "Destroying the joint and dying of shame? A geography of revitalised feminism in social media and beyond." *Geographical Research* 51: 243–56.

McLennan, S., and G. Prinsen. 2014. "Something old, something new: research using archives, texts, and virtual data." In R. Scheyvens, ed., *Development Fieldwork: A Practical Guide*, 81–100. London: Sage

McLuhan, M. 1994 (1964). *Understanding Media*. Cambridge, MA: MIT Press.

McNay, L. 1994. *Foucault: A Critical Introduction*. Cambridge: Polity Press.

Madge, C. 2007. "Developing a geographer's agenda for online research ethics." *Progress in Human Geography* 31: 654–74.

Maguire, P. 1987. *Doing participatory research: A feminist approach*. Amherst: Centre for

International Education, University of Massachusetts.

Mann, C., and F. Stewart. 2002. "Internet interviewing." In J.F. Gubrium and J.A. Holstein, eds, *Handbook of Interview Research: Context and Method*, 603–27. Thousand Oaks, CA: Sage.

Manning, K. 1997. "Authenticity in constructivist inquiry: Methodological considerations without prescription." *Qualitative Inquiry* 3 (1): 93–104.

Manzo, L., and N. Brightbill. 2007. "Towards a participatory ethics." In S. Kindon, R. Pain, and M. Kesby, eds, *Participatory Action Research Approaches and Methods: Connecting People, Participation and Place*, 33–40. London: Routledge.

Marcus, G.E., and M.M.J. Fisher. 1986. *Anthropology as Cultural Critique: An Experimental Moment in the Human Sciences*. Chicago: University of Chicago Press.

Markwell, K. 1997. "Dimensions of a nature-based tour." *Annals of Tourism Research* 24 (1): 131–55.

———. 2002. "Mardi Gras tourism and the construction of Sydney as an international gay and lesbian city." *GLQ* 8 (1–2): 81–100.

Markwell, K., and G. Waitt. 2009. "Festivals, space and sexuality: gay pride in Australia." *Tourism Geographies* 11 (2): 143–68.

Marsh. 2014. *Comparing Claims from Catastrophic Earthquakes*, February 2014, Marsh & McLennan Companies, http://deutschland.marsh.com/Portals/32/Documents/Marsh%20Risk%20Management%20Research_Earthquakes.pdf [accessed March 22, 2014]

Marshall, C., and G.B. Rossman. 1999. *Designing Qualitative Research*, 3rd edn. Sage: Thousand Oaks.

Martin, D. 2007. "Bureaucratizing ethics: Institutional review boards and participatory research." *ACME: An International E-journal for Critical Geographies* 6 (3): 319–28.

Marwick, M. 2001. "Postcards from Malta: Image, consumption, context." *Annals of Tourism Research* 28, 417–438.

Mason, J. 2004. *Qualitative Researching*, 2nd edn. London: Sage.

Mason, K., and T. Zanish-Belcher. 2007. "Raising the archival consciousness: How women's archives challenge traditional approaches to collecting and use—What's in a name?" *Library Trends* 56 (2): 344–59.

Massey, D. 1993. "Power-geometry and a progressive sense of place." In J. Bird, B. Curtis, G. Robertson, and L. Tickner, eds, *Mapping the Futures*. London: Routledge.

———. 1994. *Space, Place and Gender*. Cambridge: Polity Press.

Matheson, J. 2007. "The voice transcription technique: Use of voice recognition software to transcribe digital interview data in qualitative research." *The Qualitative Report* 12(4): 547–60.

Matless, D., J. Oldfield, and A. Swain. 2007. "Encountering Soviet geography: Oral histories of British geographical studies of the USSR and eastern Europe, 1945–1991." *Social and Cultural Geography* 8 (3): 353–72.

Matthews, H., M. Limb, and M. Taylor. 1998. "The geography of children: Some ethical and methodological considerations for project and dissertation work." *Journal of Geography in Higher Education* 22: 311–24.

May, T. 2011. *Social Research: Issues, Methods and Process*. 4th edn. Maidenhead: Open University.

Maya People of Southern Belize, Toledo Maya Cultural Council, and Toledo Alcades Association. 1997. *Maya Atlas: The Struggle to Preserve Maya Land in Southern Belize*. Berkeley, CA: North Atlantic Books.

Mayhew, R. 2003. "Researching historical geography." In A. Rogers and H. Viles, eds, *The Student's Companion to Geography*, 2nd edn. Oxford: Blackwell.

Mays, N., and C. Pope. 1997. "Rigour and qualitative research." *British Medical Journal* 310 (6997): 109–13.

Mee, K.J. 1994. "Dressing up the suburbs: Representations of western Sydney." In K. Gibson and S. Watson, eds, *Metropolis Now: Planning and the Urban in Contemporary Australia*. Sydney: Pluto Press.

———. 2007. "'I ain't been to heaven yet? Living here, this is heaven to me': Public housing and the making of home in Inner Newcastle." *Housing, Theory and Society*, 24, 207–28.

Mee, K., and R. Dowling. 2003. "Reading *Idiot Box*: Film reviews intertwining the social and cultural." *Social and Cultural Geography* 4 (2): 185–215.

Meek, D. 2012. "YouTube and social movements: a phenomenological analysis of participation, events and cyberplace." *Antipode* 44 (4): 1429–48.

Meho, L.I. 2006. "E-mail interviewing in qualitative research: A methodological discussion." *Journal of the American Society* for Information Science and Technology 57 (10): 1284–95.

Meinig, D.W. 1979. *The Interpretation of Ordinary Landscapes: Geographical Essay*. New York: Oxford University Press.

Merriman, P. 2012. "Human geography without time-space." *Transactions of the Institute of British Geographers* 37 (1): 12–27.

Merton, R.K. 1987. "The focussed interview and focus groups: Continuities and discontinuities." *Public Opinion Quarterly* 51 (4): 550–66.

Meth, P. 2003. "Entries and omissions: Using solicited diaries in geographical research." *Area* 35 (2): 195–205.

Metz, C. 1974. *Film Language: A Semiotics of the Cinema*. Ann Arbor: University of Michigan Press.

———. 1986 (1977). *Imaginary Signifier: Psychoanalysis and the Cinema*. Bloomington, IN: University of Indiana Press.

Miles, M.B., and A.M. Huberman. 1994. *Qualitative Data Analysis: An Expanded Sourcebook*. 2nd edn. Thousand Oaks, CA: Sage.

Mills, S. 1997. *Discourse*. London and New York: Routledge.

Minichiello, V., et al. 1995. *In-depth Interviewing: Principles, Techniques, Analysis*. 2nd edn. Melbourne: Longman Cheshire.

Mitchell, H., R.A. Kearns, and D.C.A. Collins. 2007. "Nuances of neighbourhood: Children's perceptions of the space between home and school in Auckland, New Zealand." *Geoforum* 38: 614–27.

Mohammad, R. 1999. "Marginalisation, Islamism and the production of the 'Other's' 'Other'." *Gender, Place and Culture* 6 (3): 221–40.

Mohanty, C.T. 1991. "Cartographies of struggle: Third world women and the politics of feminism." In C. Mohanty, A. Russo, and L. Torres, eds, *Third World Women and the Politics of Feminism*. Bloomington: University of Indiana Press.

Mok, D., B. Wellman, and J. Carrasco. 2010. "Does distance matter in the age of the Internet?" *Urban Studies* 47 (13): 2747–83.

Monk, J., and S. Hanson. 1982. "On not excluding half of the human in human geography." *Professional Geographer* 34: 11–23.

Monk, J., P. Manning, and C. Denman. 2003. "Working together: Feminist perspectives on collaborative research and action." *ACME: An International E-Journal for Critical Geographies* 2 (1): 91–106.

Morgan, D.L. 1996. "Focus groups." *Annual Review of Sociology* 22: 129–52.

———. 1997. *Focus Groups as Qualitative Research*. 2nd edn. Thousand Oaks, CA: Sage.

Morin K. 2008. *Civic Discipline: Geography in America, 1860–1890*. Farnham: Ashgate.

Morrison, C-A. 2012. "Solicited diaries and the everyday Geographies of heterosexual love and home: reflections on methodological process and practice." *Area* 44 (1): 68–75.

Moss, P. 1995. "Reflections on the 'gap' as part of the politics of research design." *Antipode* 27 (1): 82–90.

———. 2002. *Feminist Geography in Practice*. Oxford: Blackwell.

———, ed. 2002a. "Taking on, thinking about and doing feminist research in geography." In P. Moss, ed., *Feminist Geography in Practice: Research and Methods*, 1–17. Oxford: Blackwell.

———, ed. 2002b. *Feminist Geography in Practice: Research and Methods*. Oxford: Blackwell.

———. 2005. "A bodily notion of research: power, difference and specificity in feminist methodology." In L. Nelson and J. Seager, eds., *A Companion to Feminist Geography*, 41–59. Oxford: Blackwell.

Moss, P., and A-H.K. Falconer, eds. 2008. *Feminisms in Geography. Rethinking Space, Place and Knowledges*. Plymouth: Rowman and Littlefield.

Moss, P., and A-H.K. Falconer. 2010. "Rhizomatic encounters and encountering possibilities." *Thirdspace: Journal of Feminist Theory and Culture*, 9 (1), http://www.thirdspace.ca/journal/article/viewArticle/385/295 (Accessed July 2013).

Mosse, D. 1994. "Authority, gender and knowledge: Theoretical reflections on the practice of participatory rural appraisal." *Development and Change* 25: 497–526.

Mostyn, B. 1985. "The content analysis of qualitative research data: A dynamic approach." In M. Brown, J. Brown, and D. Canter, eds, *The Research Interview: Uses and Approaches*. London: Academic Press.

Mould, O. 2009. "Parkour, the city, the event." *Environment and Planning D: Society and Space* 27 (4): 738–50.

mrs kinpaisby. 2008. "Taking stock of participatory geographies: Envisioning the communiversity." *Transactions of the Institute of British Geographers* 33: 292–9.

mrs c kinpaisby-hill. 2008. "Publishing from

participatory research." In A. Blunt, ed., *Publishing in Geography: A Guide for New Researchers*, 45–7. London: Wiley-Blackwell.

———. 2011. "Participatory praxis and social justice: Towards more fully social geographies." In V. Del Casino, M. Thomas, P. Cloke, and R. Panelli, eds, *A Companion to Social Geography*, 214–34. London: Blackwell.

———. 2013. "Participatory approaches to authorship in the academy." In A. Blunt, ed., *Publishing and Getting Read: A Guide for New Researchers in Geography*, Section 4.2, p. 24. London: Wiley-Blackwell.

Muller, S. 1999. "Myths, media and politics: Implications for koala management decisions in Kangaroo Island, South Australia." Paper presented to the Institute of Australian Geographers conference, 27 September to 1 October, Sydney.

Mullings, B. 1999. "Insider or outsider, both or neither: Some dilemmas of interviewing in a cross-cultural setting." *Geoforum* 30: 337–50.

Mulvey, L. 1992. "Pandora: Topographies of the mask and curiosity." In B. Colomina, ed., *Sexuality and Space*. New York: Princeton Architectural Press.

Murdoch, J. 2006. *Post-Structuralist Geography*. London: Sage.

Murton, B. 2012. "Being in the place world: toward a Māori 'geographical self'." *Journal of Cultural Geography* 29 (1): 87–104.

Myers, G. 1998. "Displaying opinions: Topics and disagreement in focus groups." *Language in Society* 27: 85–111.

Nader, L. 1974. "Up the anthropologist—Perspective gained from studying up." In D. Hymes, ed., *Reinventing Anthropology*, 284–311. New York: Vintage Books.

Nagar, R. 2013. "Storytelling and co-authorship in feminist alliance work: reflections from a journey." *Gender, Place and Culture* 20 (1): 1–18.

Nansen, B., M. Arnold, M. Gibbs, and H. Davis. 2011. "Dwelling with media stuff: Latencies and logics of materiality in four Australian homes." *Environment and Planning D: Society and Space* 29 (4): 693–715.

Nast, H. 1994. "Opening remarks: Women in the field." *Professional Geographer* 46 (1): 54–66.

National Council on Public History. 2003. "NCPH ethics guidelines." http://ncph.org/ethics.html.

Nelson, S. 2003. "It's I mean like uh disrespectful." *Times Higher Educational Supplement* 28 March: 16.

Neutens, T, T. Schwanen, and F. Witlox. 2011. "The prism of everyday life: Towards a new research agenda for time geography." *Transport Reviews: A Transnational Transdisciplinary Journal* 31, 25–47.

NHMRC (National Health and Medical Research Council). 2003. *Values and Ethics—Guidelines for Ethical Conduct in Aboriginal and Torres Strait Islander Health Research*. Canberra: NHMRC.

Nicholls, R. 2010. "Research and indigenous participation: Critical reflexive methods." In A. Possamai-Inesedy and G. Gwyther, eds., *New Methods in Social Justice Research for the Twenty-First Century*, 18–27. London: Routledge.

Nicholson, B. 2000. "Something there is . . ." In K. Reed-Gilbert, ed., *The Strength of Us as Women: Black Women Speak*, 27–30. Canberra: Ginninderra Press.

Nicholson, H. 2002. "Telling travelers' tales: The world through home movies." In T. Creswell and D. Dixon, eds, *Engaging Film: Geographies of Mobility and Identity*. Lanham, MD: Rowman and Littlefield.

Nietschmann, B.Q. 1973. *Between Land and Water: The Subsistence Ecology of the Miskito Indians, Eastern Nicaragua*. New York: Seminar Press.

———. 1979. *Caribbean Edge: The Coming of Modern Times to Isolated People and Wildlife*. Indianapolis: Bobbs-Merrill.

———. 1987. "The third world war." *Cultural Survival Quarterly* 11 (3): 1–16.

———. 1995. "Defending the Miskito reefs with maps and GPS: Mapping with sail, scuba, and satellite." *Cultural Survival Quarterly* 18 (4): 34–7.

———. 1997. "Protecting indigenous coral reefs and sea territories, Miskito Coast, RAAN, Nicaragua." In S. Stevens, ed., *Conservation through Cultural Survival: Indigenous Peoples and Protected Areas*, 193–224. Washington: Island Press.

———. 2001. "The Nietschmann syllabus: A vision of the field." *Geographical Review* 91 (1–2): 175–84.

Nietzsche, F.W. 1969. *On the Genealogy of Morals*. tr. W. Kaufmann and R.J. Hollingdale. New York: Vintage.

Nolan, N. 2003. "The ins and outs of skateboarding and transgression in public space in Newcastle, Australia." *Australian Geographer* 34 (3): 311–27.

Noxolo, P., C. Madge, and P. Raghuram. 2012. "Unsettling responsibility: Postcolonial interventions." *Transactions of the Institute of British Geographers* 373, 418–29.

Nunavut Research Institute and Inuit Tapiriit Kanatami. n.d. "Negotiating research relationships with Inuit communities: A guide for researchers." http://www.nri.nu.ca/pdf/06-068%20ITK%20NRR%20booklet.pdf.

Oakley, A. 1981. *From Here to Maternity: Becoming a Mother.* Harmondsworth: Penguin.

Oberhauser, A., and M.A. Yeboah, 2011. "Heavy burdens: gendered livelihood strategies of porters in Accra, Ghana." *Singapore Journal of Tropical Geography* 32 (1): 22–37.

O'Brien, K. 1993. "Improving survey questionnaires through focus groups." In D.L. Morgan, ed., *Successful Focus Groups: Advancing the State of the Art.* Newbury Park: Sage.

O'Connell-Davidson, J., and D. Layder. 1994. *Methods, Sex and Madness.* London: Routledge.

O'Connor, H., and C. Madge. 2003. "'Focus groups in cyberspace': Using the Internet for qualitative research." *Qualitative Market Research: An International Journal* 6 (2): 133–43.

Ogborn, M. 2006. "Finding historical data." In N. Clifford and G. Valentine, eds, *Key Methods in Geography*, 101–15. London: Sage.

Oman-Reagan, M. 2012. "Occupying cyberspace: Indonesian cyberactivists and Occupy Wall Street." *Critical Quarterly* 54 (2): 39–45.

O'Neill, P.M. 2001. "Financial narratives of the modern corporation." *Journal of Economic Geography* 1: 181–99.

Opie, A. 1992. "Qualitative research, appropriation of the other and empowerment." *Feminist Review* 40: 52–69.

Ortner, S.B. 1978. *Sherpas through Their Rituals.* Cambridge: Cambridge University Press.

———. 1989. *High Religion: A Cultural and Political History of Sherpa Buddhism.* Princeton, NJ: Princeton University Press.

———. 1999. *Life and Death on Mt. Everest: Sherpas and Himalayan Mountaineering.* Princeton, NJ: Princeton University Press.

O'Steen, B., L. Perry, P. Cammock, S. Kingham, E. Pawson, R. Stowell, and D. Perry. 2011. "Engaging teachers and learners through service-learning," in *Good Practice Publication Grants.* Wellington: Ako Aotearoa, available at http://akoaotearoa.ac.nz/ako-hub/good-practice-publication-grants-e-book (accessed March 22, 2014).

Pain, R. 2003. "Social geography: On action-oriented research." *Progress in Human Geography* 27 (5): 677–85.

———. 2004 "Social geography: Participatory research." *Progress in Human Geography* 28, 652–63.

———. 2014. "Impact: Striking a blow or walking together?" *ACME: An International E-Journal for Critical Geographies*, 13 (1): 19–23.

Pain, R., and K. Askins. 2011. "Contact zones: participation, materiality and the messiness of interaction." *Environment and Planning D: Society and Space.* 29 (5): 803–21.

Pain, R., and P. Francis. 2003. "Reflections on Participatory research." *Area* 35 (1): 46–54.

Panelli, R. 2004. *Social Geographies.* London: Sage.

Parfitt, J. 2005. "Questionnaire design and sampling." In R. Flowerdew and D. Martin, eds, *Methods in Human Geography: A Guide for Students Doing a Research Project.* Harlow: Longman.

Park, P. 1993. "What is participatory research? A theoretical and methodological perspective." In P. Park, M. Brydon-Miller, B. Hall, and T. Jackson, eds, *Voices of Change: Participatory Research in the United States and Canada.* Westport, CT: Bergin and Garvey.

Parkes, M., and R. Panelli. 2001. "Integrating catchment ecosystems and community health: The value of participatory action research." *Ecosystem Health* 7 (2): 85–106.

Parnwell, M. 2003. "Consulting the poor in Thailand: Enlightenment or delusion?" *Progress in Development Studies* 3 (2): 99–112.

Parr, H. 1998. "Mental health, ethnography and the body." *Area* 30: 28–37.

———. 2001. "Feeling, reading, and making bodies in space." *Geographical Review* 9: 158–67.

Patton, M.Q. 1990. *Qualitative Evaluation and Research Methods.* 2nd edn. Beverly Hills: Sage.

———. 2002. *Qualitative Evaluation and Research Methods.* 3rd edn. Beverly Hills: Sage.

Pavlovskaya, M. 2004. "Other transitions: Multiple economies of Moscow households in the 1990s." *Annals of the Association of American Geographers* 94 (2): 329–51.

Pawson, E., and E. Teather. 2002. "Geographical expeditions: Assessing the benefits of a student-driven fieldwork method." *Journal of Geography in Higher Education* 26 (3): 275–89.

Peake, L. (on behalf of Red Thread Women's Development Programme). 2000. *Women Researching Women: Methodology Report and Research Projects on the Study of Domestic Violence and Women's Reproductive Health in Guyana*. Georgetown, Guyana: Interamerican Bank.

Pearce, M.W., E.M. Willis, B.A. Wadham, and B. Binks. 2010. "Attitudes to drought in outback communities in South Australia." *Geographical Research* 48 (4): 359–69.

Pearson, C. 2012. "Researching militarized landscapes: A literature review on war and the militarization of the environment." *Landscape Research* 37 (1): 115–33.

Pearson, L.J. 1996. "Place re-identification: The 'Leisure Coast' as a partial representation of Wollongong." (School of Geography, University of New South Wales, BSc Honours thesis.).

Pfaff, J. 2010. "A mobile phone: Mobility, materiality and everyday Swahili trading practices." *Cultural Geographies* 17 (3): 341–57.

Philip, L.J. 1998. "Combining quantitative and qualitative approaches to social research in human geography—An impossible mixture?" *Environment and Planning A* 30: 261–76.

Phillips, L., and Jørgensen, M.W. 2002. *Discourse Analysis as Theory and Method*. London: Sage.

Phillips, N., and C. Hardy. 2002. *Discourse Analysis: Investigating Processes of Social Construction*. Thousand Oaks, CA: Sage.

Pickerill, J. 2009. "Finding common ground? Spaces of dialogue and the negotiation of indigenous interests in environmental campaigns in Australia." *Geoforum* 40: 66–79.

Pickles, K. 2002. "Kiwi icons and the re-settlement of New Zealand as colonial space." *New Zealand Geographer* 58 (2): 5–16.

Pile, S. 2010. "Emotions and effect in recent human geography." *Transactions of the Institute of British Geographers* 35 (1): 5–20.

PLA notes. December 2003. London: International Institute for Environment and Development.

Platt, J. 1988. "What can case studies do?" *Studies in Qualitative Methodology* 1: 1–23.

———. 1992. "'Case study' in American methodological thought." *Current Sociology* 40: 17–48.

Popper, K. 1959. *The Logic of Scientific Discovery*. London: Hutchinson.

Porteous, D.J. 1985. "Smellscape." *Progress in Human Geography* 9: 356–78.

Potter, J. 1996. "Discourse analysis and constructionist approaches: Theoretical background." In J.T.E. Richardson, ed., *Handbook of Qualitative Methods for Psychology and the Social Sciences*. Leicester: British Psychological Society.

Powell, J.M. 1988. *An Historical Geography of Modern Australia: The Restive Fringe*. Cambridge: Cambridge University Press.

Powell, R.C. 2008. "Becoming a geographical scientist: Oral histories of Arctic fieldwork." *Transactions of the Institute of British Geographers* 33 (4): 548–65.

Power, M.J., P. Neville, E. Devereux, A. Haynes, and C. Barnes. 2013. "'Why bother seeing the world for real?': Google Street View and the representation of a stigmatised neighbourhood." *New Media & Society*, 15(7), 1022–1040. doi:10.1177/1461444812465138.

Pratt, G. 1994. "Poststructuralism," in R.J. Johnston, D. Gregory, and D.M. Smith, eds, *The Dictionary of Human Geography*, 3rd edn, 468–69. Blackwell: London.

———. 2000. "Participatory action research." In R. Johnston, D. Gregory, G. Pratt, and M. Watts, eds, *Dictionary of Human Geography*, 4th edn. Oxford: Blackwell.

———. 2002. "Studying immigrants in focus groups." In P. Moss, ed., *Feminist Geography in Practice: Research and Methods*. Oxford: Blackwell.

Pratt, G., in collaboration with the Philippine Women's Centre. 1999. "Is this Canada? Domestic workers' experiences in Vancouver, B.C." In J. Momson, ed., *Gender, Migration and Domestic Service*. London: Routledge.

Pratt, G., in collaboration with the Philippine Women's Centre of B.C. and Ugnayan Ng Kabataang Pilipino sa Canada/Filipino Canadian Youth Alliance. 2007. "Working with migrant communities: Collaborating with the Kalayaan Centre in Vancouver, Canada." In S. Kindon, R. Pain, and M. Kesby, eds, *Participatory Action Research Approaches and Methods: Connecting People, Participation and Place*, 95–103. London: Routledge.

Pretty, J., et al. 1995. *Participatory Learning*

and Action: A Trainer's Guide. London: International Institute for Environment and Development.

Pringle, T. 1988. "The privation of history: Landseer, Victoria and the Highland myth." In D. Cosgrove and S. Daniels, eds., *The Iconography of Landscape: Essays on the Symbolic Representation, Design and Use of Past Environments,* 142–60. Cambridge: Cambridge University Press.

Professional Geographer 1994. 46 (1: special issue on Women in the Field).

Professional Geographer 1995. 47 (4: special issue on Should Women Count?).

Pulvirenti, M. 1997. "Unwrapping the parcel: An examination of culture through Italian home ownership." *Australian Geographical Studies* 35: (1): 32–9.

Quanchi, M. 2006. "Photography and history in the Pacific Islands." *The Journal of Pacific History* 41 (2): 165–73.

Radicati Group. 2011. *Email Statistics Report, 2011–2015,* The Radicati Group, Palo Alto, see also: http://www.radicati.com. (accessed 29th November 2013).

Radice, H. 2013. "How we got here: UK higher education under neoliberalism." *ACME: An International E-Journal for Critical Geographies* 12 (3): 407–418.

Radley, A., D. Hodgetts, and A. Cullen. 2005. "Visualizing homelessness: A study in photography and estrangement." *Journal of Community and Applied Social Psychology* 15 (4): 273–95.

Raghuram, P., and C. Madge. 2006. "Towards a method for postcolonial development geography: Possibilities and challenges." *Singapore Journal of Tropical Geography* 27 (3): 270–88.

———. 2007. "Feminist theorising as practice." In A-H.K. Falconer and P. Moss, eds, *Feminisms in Geography: Space, Place and Environment,* 221–9. Lanham: Rowman and Littlefield.

Raju, S. 2005. "Gender and empowerment: Creating 'thus far and no further' supportive structures. A Case from India." In L. Nelson and J. Seager, eds, *A Companion to Feminist Geography,* 194–208. Oxford: Blackwell.

Reason, P., and J. Rowan, eds. 1981. *Human Inquiry: A Sourcebook of New Paradigm Research.* Chichester: John Wiley and Sons.

Reinharz, S. 1992. *Feminist Methods in Social Research.* New York: Oxford University Press.

Rekrut, A. 2011. "Connected constructions, constructing connections, materiality of archival records as historical evidence." In K. Gray and C. Verduyn, eds., *Archival Narratives for Canada Re-telling Stories in a Changing Landscape,* 135–157. Halifax and Winnipeg: Fernwood.

Reynolds, G. 2011. *presentationzen: Simple Ideas on Presentation, Design and Delivery.* 2nd edn. Berkeley, CA: New Riders.

Reynolds, H. 1998. *This Whispering in Our Hearts.* Sydney: Allen and Unwin.

Richardson, L., and E.A. St Pierre. 2005. "Writing: A method of inquiry." In N.K. Denzin and Y.S. Lincoln, eds, *Collecting and Interpreting Qualitative Materials Research,* 3rd edn, 473–99. Los Angeles: Sage.

Richardson, L., and E.A. St Pierre. 2008. "Writing: A method of inquiry." In N.K. Denzin and Y.S. Lincoln, eds, *Collecting and Interpreting Qualitative Materials Research,* 3rd edn, 473–99. Los Angeles: Sage.

Riley, M., and D. Harvey. 2007. "Editorial: Talking landscapes: On oral history and the practice of geography." *Social and Cultural Geography* 8 (4): 391–415.

Rinaldi, A.H. 1994. "The Net: User guidelines and netiquette." Academic/Institutional Support Services, Florida Atlantic University. http://www.wifak.uni-wuerzburg.de/wilan/sysgroup/texte/netiquet/netiquet.txt.

Rittel, H.W.J., and M.M. Webber. 1973. "Dilemmas in a general theory of planning." *Policy Sciences,* 4 (2): 155–69.

Rivera, M. 1997. "Various definitions of geography." http://www2.westga.edu/~geograph/define.html.

Robards, B. 2013. "Friending participants: managing the researcher-participant relationship on social network sites." *Young* 21 (3): 217–235.

Robertson, B.M. 1994. *Oral History Handbook.* Adelaide: Oral History Association of Australia.

———. 2006. *Oral History Handbook.* 5th edn. Adelaide: Oral History Association of Australia SA Branch Inc.

Robinson, G. 1998. *Methods and Techniques in Human Geography.* Chichester: John Wiley and Sons.

Robinson, W.D. 2007. "Formative Work and Preliminary Results Using RDS: New Orleans After Katrina." Paper presented at the American Public Health Association annual meetings, Washington, DC. Abstract, recording, and slides available at http://apha.confex.com/

apha/135am/techprogram/paper_167109.htm (accessed March 22, 2014).

Roche, M. 2011. "New Zealand geography, biography and autobiography." *New Zealand Geographer* 67 (2): 73–78.

Rofe, M.W. 2003. "'I Want to be global': Theorising the gentrifying class as an emergent elite global community." *Urban Studies* 40 (12): 2511–26.

———. 2007. "Urban revitalisation and masculine memories: Towards a critical awareness of gender in the postindustrial landscape." *Australian Planner* 44 (2): 26–33.

———. 2009. "Globalization, gentrification and spatial hierarchies in and beyond New South Wales: The local/global nexus." *Geographical Research* 47 (3): 292–305.

Rofe, M.W., and H.P.M. Winchester. 2003. "Masculine scripting and the mythology of motorcycling." *Journal of Interdisciplinary Gender Studies* 7 (1 and 2): 161–79.

———. 2007. "Lobethal the *Valley of Praise*: Inventing tradition for the purposes of place making in rural South Australia." In R. Jones and B.J. Shaw, eds, *Loving a Sunburned Country? Geographies of Australian Heritage*, 133–50. London: Ashgate.

Rogoff, I. 2000. *Terra Infirma: Geography's Visual Culture*. London and New York: Routledge.

Rose, C. 1988. "The concept of reach and the anglophone minority in Quebec." In J. Eyles and D. Smith, eds, *Qualitative Methods in Human Geography*. Cambridge: Polity Press.

Rose, D.B. 1996a. "Histories and rituals: Land claims in the Territory." In B. Attwood, ed., *In the Age of Mabo: History, Aborigines and Australia*, 35–52. Sydney: Allen and Unwin.

———. 1996b. *Nourishing Terrains: Australian Aboriginal Views of Landscape and Wilderness*. Canberra: Australian Heritage Commission.

———. 1999. "Indigenous ecologies and an ethic of connection." In N. Low, ed., *Global Ethics and Environment*, 175–87. London: Routledge.

Rose, G. 1993. *Feminism and Geography*. Minneapolis: University of Minnesota Press.

———. 1996. "Teaching visualised geographies: Towards a methodology for the interpretation of visual materials." *Journal of Geography in Higher Education* 20 (3): 281–94.

———. 1997. "Situating knowledges: Positionalities, reflexivities and other tactics." *Progress in Human Geography* 21 (3): 305–20.

———. 2001. *Visual Methodologies: An Introduction to the Interpretation of Visual Materials*. London: Sage.

———. 2003a. "Family photographs and domestic spacings: A case study." *Transactions of the Institute of British Geographers* 28: 5–18.

———. 2003b. "On the need to ask how, exactly, is geography 'visual'?" *Antipode*, 35(2): 212–221.

———. 2007. *Visual Methodologies: An Introduction to the Interpretation of Visual Materials*. 2nd edn. London: Sage.

———. 2008. "Using photographs as illustrations in human geography." *Journal of Geography in Higher Education* 32 (1): 151–60.

Routledge, P. 2002. "Travelling east as Walter Kurtz: Identity, performance and collaboration in Goa, India." *Environment and Planning D: Society and Space* 20 (4): 477–96.

Rowles, G.D. 1978. *Prisoners of Space: Exploring the Geographical Experience of Older People*. Boulder, CO: Westview Press.

Rowley, C.D. 1970. *The Destruction of Aboriginal Society: Aboriginal Policy and Practice, Volume I*. Canberra: Australian National University Press.

———. 1971a. *Outcasts in White Australia: Aboriginal Policy and Practice, Volume II*. Canberra: Australian National University Press.

———. 1971b. *The Remote Aborigines: Aboriginal Policy and Practice, Volume III*. Canberra: Australian National University Press.

Rowse, T. 2000. *Obliged to Be Difficult: Nugget Coombs' Legacy in Indigenous Affairs*. Cambridge: Cambridge University Press.

Ruddick, S. 2004. "Activist geographies: Building possible worlds." In P. Cloke, P. Crang, and M. Goodwin, eds, *Envisioning Human Geographies*. London: Arnold.

Rugendyke, B. 2005. "W(h)ither development geography in Australia?" *Geographical Research* 43 (3): 306–18.

Ruppert, E., J. Law, and M. Savage. 2013. "Reassembling Social Science Methods: The Challenge of Digital Devices." *Theory, Culture and Society* 30 (4): 22–46.

Sackett, H. 2005. "Nothing is true but change: Archaeology, time and landscape in the writing of Lewis Grassic Gibbon." *Scottish Archaeological Journal* 27 (1): 13–29.

Said, E. 1978. *Orientalism*. New York: Vintage Books.

———. 1993. *Culture and Imperialism*. New York: Vintage Books.

Saldaña, J. 2012. *The Coding Manual for Qualitative Researchers*. London: Sage.

Sanders, R. 2007. "Developing geographers through photography: Enlarging concepts." *Journal of Geography in Higher Education* 31 (1): 181–95.

Sanderson, E., with Holy Family Settlement Research Team, R. Newport, and *Umaki* Research Participants. 2007. "Participatory cartographies: Reflections from research performances in Fiji and Tanzania." In S. Kindon, R. Pain, and M. Kesby, eds, *Participatory Action Research Approaches and Methods: Connecting People, Participation and Place*, 122–31. London: Routledge.

Sanderson, E., and S. Kindon. 2004. "Progress in participatory development: Opening up the possibility of knowledge through progressive participation." *Progress in Development Studies* 4 (2): 114–26.

Sanjek, R., ed. 1990. *Fieldnotes: The Makings of Anthropology*. Ithaca, NY: Cornell University Press.

Sarantakos, S. 1993. *Social Research*. South Melbourne: Macmillan.

———. 2005. *Social Research*. 3rd edn. Melbourne: Palgrave.

———. 2012. *Social Research*. 4th edn. New York: Palgrave Macmillan.

Sauer, C.O. 1925. "The morphology of landscape." *University of California Publications in Geography* 2: 19–54.

———. 1941. "Foreword to historical geography." *Annals of the Association of American Geographers* 31 (1): 1–24.

Saunders, A. 2010. "Literary geography: Reforging the connections." *Progress in Human Geography* 34 (4): 436–452.

Saussure, F de. 1959. *Course in General Linguistics*. New York: Philosophical Library.

———. 1983. *Course in General Linguistics*. London: Duckworth.

Sayer, A. 1992. *Method in Social Science: A Realist Approach*. 2nd edn. London: Routledge.

———. 2000. *Realism and Social Science*. London & Newbury Park: Sage.

———. 2010. *Method in Social Science: A Realist Approach*. Revised 2nd edn. Milton Park and New York: Routledge .

Sayer, A., and K. Morgan. 1985. "A modern industry in a declining region: Links between method, theory and policy." In D. Massey and R. Meegan, eds, *Politics and Method: Contrasting Studies in Industrial Geography*. London: Methuen.

Schaffer, K. 1988. *Women and the Bush: Forces of Desire in the Australian Cultural Tradition*. Cambridge: Cambridge University Press.

Schein, R.H. 1997. "The place of landscape: A conceptual framework for interpreting an American scene." *Annals of the Association of American Geographers* 87 (4): 660–80.

Scheuermann, L., and G. Taylor. 1997. "Netiquette." *Internet Research: Electronic Networking Applications and Policy* 7 (4): 269–73.

Scheyvens, R. 2009. "Empowerment." In R. Kitchin and N. Thrift, eds, *International Encyclopedia of Human Geography*, 464–70. Elsevier: Oxford.

Schoenberger, E. 1991. "The corporate interview as a research method in economic geography." *Professional Geographer* 43 (2): 180–9.

———. 1992. "Self-criticism and self-awareness in research: A reply to Linda McDowell." *Professional Geographer* 44 (2): 215–18.

Schollmann, A., H.C. Perkins, and K. Moore. 2000. "Intersecting global and local influences in urban place promotion: The case of Christchurch, New Zealand." *Environment and Planning D: Society and Space* 32: 55–76.

Schwartz, J., and R. Ryan, eds. 2003. *Picturing Place, Photography and the Geographical Imagination*. London: I.B. Tauris.

Scott, K., J. Park, and C. Cocklin. 1997. "From 'sustainable rural communities' to 'social sustainability': Giving voice to diversity in Mangakahia Valley, New Zealand." *Journal of Rural Studies* 16 (4): 433–46.

Secor, A. 2004. "'There is an Istanbul that belongs to me': Citizenship, space, and identity in the city." *Annals of the Association of American Geographers* 94 (2): 352–68.

Seidman, I. 2013. *Interviewing as Qualitative Research: a Guide for Researchers in Education and the Social Sciences*. 4th edn. New York: Teachers College Press.

Selen, E. 2012. "The stage: A space for queer subjectification in contemporary Turkey." *Gender, Place and Culture* 19 (6): 730–49.

Sharp, J. 2005. "Geography and gender: Feminist methodologies in collaboration and in the field." *Progress in Human Geography* 29 (3): 304–9.

———. 2005. "On tricky ground: Researching the native in the age of uncertainity." In N.K. Denzin and Y.S. Lincoln, eds, *The Sage Handbook of Qualitative Research*, 85–107. Thousand Oaks, CA: Sage Publications.

———. 2009. "Geography and gender: What belongs to feminist geography? Emotion, power and change." *Progress in Human Geography* 33 (1): 74–80.

Sharp, J., K. Browne, and D. Thien. 2004. "Introduction." In *Women and Geography Study Group of the RGS-IBG, Geography and Gender Reconsidered* 1 (4). London: WGSG.

Sharpe, S., and A. Gorman-Murray. 2013. Special Issue: Bodies in Place, Bodies Displaced. *Geographical Research* 51 (2): 115–19.

Shaw, W.S. 2000. "Ways of whiteness: Harlemising Sydney's Aboriginal Redfern." *Australian Geographical Studies* 38 (3): 291–305.

———. 2007. *Cities of Whiteness*. Oxford: Blackwell.

———. 2013. "Redfern as the Heart(h): Living (Black) in Inner Sydney." *Geographical Research* 51 (3): 257–68.

Shelton, T., A. Poorthius, M. Graham, and M. Zook. 2014. "Mapping the Data Shadows of Hurricane Sandy: Uncovering the Sociospatial Dimensions of 'Big Data.'" *Geoforum* 52: 167–79.

Sheridan, G. 2001. "Dennis Norman Jeans: Historical geographer and landscape interpreter extraordinaire." *Australian Geographical Studies* 39 (1): 96–106.

Shurmer-Smith, P. 2002. *Doing Cultural Geography*. London: Sage.

Silverman, D. 1991. *Interpreting Qualitative Data: Methods for Analysing Talk, Text and Interaction*. Thousand Oaks, CA: Sage.

———. 1993. *Interpreting Qualitative Data: Methods for Analysing Talk, Text and Interaction*. 2nd edn. London: Sage.

———. 2001. *Interpreting Qualitative Data: Methods for Analysing Talk, Text, and Interaction*. 2nd edn. London: Sage.

———. 2005. *Doing Qualitative Research: A Practical Handbook*. London: Sage.

Simon, S. 2011. Inspiration in 140 characters, long before Twitter. *NPR Weekend Edition* (radio program). Available ://www.npr.org/2011/05/14/136302111/inspiration-in-140-characters-long-before-twitter (accessed Sept. 22, 2014).

Singer, A., and L. Woodhead. 1988. *Disappearing World: Television and Anthropology*. London: Boxtree and Granada Television.

Smith, D.M. 1977. *Human Geography: A Welfare Approach*. London: Edward Arnold.

Smith, L.T. 1999. *Decolonizing Methodologies: Research and Indigenous Peoples*. Dunedin and London: University of Otago Press and Zed Books.

———. 2005. "On tricky ground: Researching the native in the age of uncertainity." In N.K. Denzin and Y.S. Lincoln, eds, *The Sage Handbook of Qualitative Research*, 85–107. Thousand Oaks, CA: Sage Publications.

———. 2012. *Decolonizing Methodologies: Research and Indigenous Peoples*. 2nd edn. London: Zed Books.

Smith, S.J. 1981. "Humanistic method in contemporary social geography." *Area* 15: 355–8.

———. 1988. "Constructing local knowledge: The analysis of self in everyday life." In J. Eyles and D.M. Smith, eds, *Qualitative Methods in Human Geography*. Cambridge: Polity Press.

———. 1994. "Soundscape." *Area* 26: 232–40.

Solís, P. 2009. "Preparing competitive research grant proposals." In M. Solem, K. Foote, and J. Monk, eds, *Aspiring Academics: A Resource Book for Graduate Students and Early Career Faculty*. Upper Saddle River, NJ: Pearson/Prentice Hall.

Sparke, M. 1998. "A map that roared and an original atlas: Canada, cartography and the narration of a nation." *Annals of the Association of American Geographers* 88 (3): 463–95.

Spate, O.H.K., and A. Learmonth. 1967. *India and Pakistan: A General and Regional Geography*. London: Methuen.

Spinks, N., B. Baron Wells, and M. Meche. 1999. "Netiquette: A behavioral guide to electronic business communication." *Corporate Communications: An International Journal* 4 (3): 145–55.

Spradley, J.P. 1980. *Participant Observation*. New York: Holt, Rinehart and Wilson.

Stacey, J. 1988. "Can there be a feminist ethnography?" *Women's Studies International Forum* 11: 21–7.

Stake, R. 1995. *The Art of Case Study Research*. Thousand Oaks, CA: Sage.

Stanley, L., and S. Wise. 1993. *Breaking Out Again: Feminist Ontology and Epistemology*. 2nd edn. London: Routledge.

Stanton, N. 1996. *Mastering Communication*. 3rd edn. London: Macmillan.

Starkey, A. (video recording), and K. George (interviews). 2003. *Balfour's City Site, 1910–2003*. Adelaide: Corporation of the City of Adelaide.

Stevens, S. 1993. *Claiming the High Ground: Sherpas, Subsistence, and Environmental Change in the Highest Himalaya*. Berkeley: University of California Press.

———. 1997. "Consultation, co-management, and conflict in Sagarmatha (Mt Everest) National Park, Nepal." In S. Stevens, ed., *Conservation through Cultural Survival: Indigenous Peoples and Protected Areas*, 63–97. Washington: Island Press.

———. 2001. "Fieldwork as commitment." *Geographical Review* 91 (1–2): 66–73.

———. 2003. "Tourism and deforestation in the Mt Everest region of Nepal." *Geographical Journal* 169 (3): 255–77.

———. 2013. "National parks and ICCAs in the high Himalayan region of Nepal: Challenges and opportunities." *Conservation and Society*, 11 (1):29–45.

———. 2014. *Indigenous Peoples, National Parks, and Protected Areas: a New Paradigm Linking Conservation, Culture, and Rights*. Tucson, Arizona: University of Arizona Press.

Stevens, S., and M.N. Sherpa. 1993. "Indigenous peoples and protected areas: New approaches to conservation in highland Nepal." In L.S. Hamilton, D.P. Bauer, and H.F. Takeuchi, eds, *Parks, Peaks, and People*. Honolulu, HI: East-West Center, Program on Environment.

Stewart, D., P. Shamdasani, and D. Rook. 2007. *Focus Groups: Theory and Practice*. 2nd edn. Thousand Oaks, CA: Sage.

Stewart, E.J. 2009. "Comparing resident attitudes toward tourism: Community-based cases from Arctic Canada." (Department of Geography, University of Calgary, unpublished PhD thesis.)

Stewart, E.J., and D. Draper. 2009. "Reporting back research findings: A case study of community-based tourism research in northern Canada." *Journal of Ecotourism* 8 (2): 128–43.

Stockdale, A. 2002. "Tools for Digital Audio Recording in Qualitative Research." *Social Research Update* 38, see also http://sru.soc.surrey.ac.uk/SRU38.html (accessed 28th November, 2013).

Stratford, E. 1997. "Memory work in geography and environmental studies: Some suggestions for teaching and research." *Australian Geographical Studies* 35 (2): 208–21.

———. 1998. "Public spaces, urban youth and local government: The skateboard culture in Hobart's Franklin Square." In R. Freestone, ed., *20th Century Urban Planning Experience: Proceedings of the 8th International Planning History Conference*. Sydney: University of New South Wales.

———, ed. 1999. *Australian Cultural Geographies*. Melbourne: Oxford University Press.

———. 2001. "The Millennium Project on Australian geography and geographers: An introduction." *Australian Geographical Studies* 39 (1): 91–5.

———. 2002. "On the edge: A tale of skaters and urban governance." *Social and Cultural Geography* 3 (2): 193–206.

———. 2008. "Islandness and struggles over development: A Tasmanian case study." *Political Geography* 27 (2): 160–75.

———. 2012. "Vantage points: Observations on the emotional geographies of heritage." In G. Baldacchino, ed., *Extreme Heritage Management: Policies and Practices from Island Territories*, 1–20. New York: Berghahn Books.

———. 2015. *Geographies, Mobilities, and Rhythms Over the Life-Course: Adventures in the Interval*. New York and London: Routledge.

Stratford, E., and A. Harwood. 2001. "The regulation of skating in Australia: An overview and commentary on the Tasmanian case." *Urban Policy and Research* 19 (2): 61–76.

Stratford, E., and T. Henderson. In review. "Expert views on urban design, public spaces, and sense of place: Planning challenges and insights from Tasmania, Australia."

Stratford, E., and C. Langridge. 2012. "Critical artistic interventions into the geopolitical spaces of islands." *Social and Cultural Geography* 13 (7): 821–843.

Strauss, A., and J. Corbin. 1990. *Basics of Qualitative Research: Grounded Theory, Procedures and Techniques*. Newbury Park: Sage.

Sue, V., and L. Ritter. 2007. *Conducting Online Surveys*. London: Sage.

———. 2012. *Conducting Online Surveys*. 2nd edn. London: Sage.

Sui, D., and D. DeLyser. 2012. "Crossing the qualitative-quantitative chasm 1: Hybrid geographies, the spatial turn, and volunteered geographic information (VGI)." *Progress in Human Geography* 36 (1): 111–24.

Summerby-Murray, R. 2011. "Marshland memories, constructing narrative in an online archival exhibition." In K. Gray, and C. Verduyn, eds, *Archival Narratives for Canada: Re-telling Stories in a Changing Landscape*, 117–33. Halifax and Winnipeg: Fernwood.

Sundberg, J. 2003. "Masculinist epistemologies and the politics of fieldwork in Latin Americanist geography." *Professional Geographer* 55: 181–91.

———. 2014. "Decolonizing posthumanist geographies." *Cultural Geographies* 21: 33–47.

Swanson, K. 2007. "'Bad mothers' and 'delinquent children': Unravelling anti-begging rhetoric in the Ecuadorian Andes." *Gender, Place and Culture* 14 (6): 703–20.

———. 2008. "Witches, children and Kiva-the-research dog: Striking problems encountered in the field." *Area* 40 (1): 55–64.

Sweet, C. 2001. "Designing and conducting virtual focus groups." *Qualitative Market Research: An International Journal* 4 (3): 130–5.

Szili, G., and M.W. Rofe. 2007. "Greening port misery: The 'green face' of waterfront redevelopment in Port Adelaide, South Australia." *Urban Policy and Research* 25 (3): 363–84.

Tarrant, A. 2010. "Constructing a social geography of grandparenthood: A new focus for intergenerationality." *Area* 42 (2): 190–197.

Tashakkori, A., and C. Teddlie. 1998. *Mixed Methodology: Combining Qualitative and Quantitative Approaches*. Thousand Oaks, CA: Sage.

Teye, J.K. 2012. "Benefits, challenges, and dynamism of positionalities associated with mixed methods research in developing countries: Evidence from Ghana." *Journal of Mixed Method Research* 6 (4): 379–391.

Thatcher, J. 2013. "Avoiding the Ghetto through hope and fear: an analysis of immanent technology using ideal types." *GeoJournal* 78 (6): 967–80.

Thiesmeyer, L., ed. 2003. *Discourse and Silencing: Representation and the Language of Displacement*. Amsterdam and Philadelphia: John Benjamins Publishing.

Thomas, M. 2004. "Pleasure and propriety: Teen girls and the practice of straight space." *Environment and Planning D: Society and Space* 22 (5): 773–89.

Thomas, W., and F. Znaniecki. 1918. *The Polish Peasant in Europe and America*. New York: Dover Publications.

Thomas-Slayter, B. 1995. "A brief history of participatory methodologies." In R. Slocum, L. Wichart, D. Rocheleau, and B. Thomas-Slayter, eds, *Power, Process and Participation: Tools for Change*. London: Intermediate Technology Publications.

Thompson, P. 2000. *The Voice of the Past: Oral History*. 3rd edn. New York: Oxford University Press.

Thrift, N. 1996. *Spatial Formations*. Thousand Oaks, CA: Sage.

———. 2000. "Dead or alive?" In I. Cook, D. Crouch, S. Naylor, and J. Ryan, eds, *Cultural Turns/Geographical Turns*, 1–6. London: Sage.

Tomsen, S., and K. Markwell. 2007. *When the Glitter Settles: Safety and Hostility at and around Gay and Lesbian Public Events*. Survey Report, Centre for Cultural Industries and Practices, University of Newcastle, New South Wales.

Tonkiss, F. 1998. "Analysing discourse." In C. Seale, ed., *Researching Society and Culture*, 245–60. London: Sage.

Townsend, J., with U. Arrevillaga, J. Bain, S. Cancino, S. Frenk, S. Pacheco, and E. Perez. 1995. *Women's Voices from the Rainforest*. London: Routledge.

Tremblay, M.A. 1982. "The key informant technique: A non-ethnographic application." In R.G. Burgess, ed., *Field Research: A Sourcebook and Field Manual*. London: Allen and Unwin.

Tuan, Y.F. 1977. *Space and Place: The Perspective of Experience*. Minneapolis: University of Minnesota Press.

Tully, J. 1995. *Strange Multiplicity: Constitutionalism in an Age of Diversity*. Cambridge: Cambridge University Press.

Turner, F.J. 1920. *The Frontier in American History*. New York: Henry Holt.

United Nations Office of the High Commissioner for Human Rights. 2008. United Nations declaration on the rights of indigenous peoples: Adopted by the General Assembly on 13 September 2007. Geneva: OHCHR.

University of California, Berkeley. The Oral History Centre. http://bancroft.berkeley.edu/researchprograms/roho.html.

Urry, J. 2002. *The Tourist Gaze*. 2nd edn. London: Sage.

Valentine, G. 1993. "(Hetero)sexing space: Lesbian perceptions and experiences of everyday spaces." *Environment and Planning D: Society and Space* 11 (4): 395–413.

———. 1997. "Tell me about . . . : Using interviews as a research methodology." In R. Flowerdew and D. Martin, eds, *Methods in Human Geography: A Guide for Students Doing a Research Project*. Harlow: Longman.

———. 2002. "People like us: Negotiating sameness and difference in the research process." In P. Moss, ed., *Feminist Geography in Practice: Research and Methods*. Oxford: Blackwell.

———. 2003. "Geography and ethics: In pursuit of social justice—Ethics and emotions in geographies of health and disability research." *Progress in Human Geography* 27 (3): 375–80.

Valentine, G., and S. Holloway. 2001. "On-line dangers?: Geographies of parents' fears for children's safety in cyberspace." *Professional Geographer* 53 (1): 71–83.

Valentine, G., L. Jackson, and L. Mayblin. 2014. "Ways of Seeing: Sexism the Forgotten Prejudice?" *Gender, Place & Culture*, 21 (4), 401–14.

Valentine, G., T. Skelton, and R. Butler. 2003. "Coming out and outcomes: Negotiating lesbian and gay identities with, and in, the family." *Environment and Planning D: Society and Space* 21: 479–99.

Valentine, G., and L. Waite. 2012. "Negotiating difference through everyday encounters: The case of sexual orientation and religion and belief." *Antipode* 44 (2): 474–92.

Van der Plas, A. 2014. Creating a dialogue between people and place. (MA thesis [Geography] The University of Auckland, NZ).

van Hoven, B. 2010. "Computer assisted qualitative data analysis." In N. Clifford, S. French, and G. Valentine, eds, *Key Methods in Geography*, 2nd edn, 453–65. London: Sage.

Van Selm, M., and N. Jankowski. 2006. "Conducting online surveys." *Quality and Quantity* 40: 435–56.

Vehovar, V., and K.L. Manfreda. 2008. "Overview: Online surveys." In N. Fielding, R.M. Lee, and G. Blank, eds, *The Sage Handbook of Online Research Methods*, London: Sage.

Verd Pericas, J.M., and S. Porcel. 2012. "An application of qualitative geographic information systems (GIS) in the field of urban sociology using Atlas.ti: Uses and reflections." *Forum: Qualitative Social Science* 13 (2), Article 14.

Vincent, K.A. 2013. "The advantages of repeat interviews in a study with pregnant schoolgirls and schoolgirl mothers: Piecing together the jigsaw." *International Journal of Research & Method in Education* 36 (4): 341–54.

von Humboldt, A. 1811. *Political Essay on the Kingdom of New Spain Containing Researches Relative to the Military Defense of New Spain: With Physical Sections and Maps.* New York: I. Riley.

Wadsworth, Y. 1998. "What is participatory action research?" Action Research International Paper 2. http://www.scu.edu.au/schools/gecm/ar/ari/p-ywadsworth98.html.

Wainwright, J. 2008. *Decolonizing Development: Colonial Power and the Maya.* Oxford: Blackwell Press.

Waitt, G. 1997. "Selling paradise and adventure: Representations of landscape in the tourist advertising of Australia." *Australian Geographical Studies* 35 (1): 47–60.

———. 2013. "Bodies that sweat: the affective responses of young women in Wollongong, New South Wales, Australia." *Gender, Place and Culture* DOI: 10.1080/0966369X.2013.802668

Waitt, G., and L. Head. 2002. "Postcards and frontier mythologies: Sustaining views of the Kimberley as timeless." *Environment and Planning D: Society and Space* 20: 319–44.

Waitt, G., and P.M. McGuirk. 1996. "Marking time: Tourism and heritage representation at Millers Point, Sydney." *Australian Geographer* 27 (1): 11–29.

Waitt, G., and K. Markwell. 2006. *Gay Tourism: Culture and Context.* Binghamton, NY: Haworth Press.

Waitt, G., and C. Stapel. 2011. "Fornicating on floats: The Sydney Mardi Gras Parade beyond the metropolis." *Leisure Studies* 30 (20): 197–216.

Waitt, G., and A. Warren. 2008. "'Talking shit over a brew after a good session with your mates': Surfing, space and masculinity." *Australian Geographer* 39 (3): 53–365.

Wakefield, S. 2007. "Reflective action in the

academy: Exploring praxis in critical geography using a 'food movement' case study." *Antipode* 39 (2): 331–54.

Wang, C., and M.A. Burris. 1997. "Photovoice: Concept, methodology, and use for participatory needs assessment." *Health Education and Behaviour* 24: 369–87.

Ward, B. 1972. *What's Wrong with Economics?* London: Macmillan.

Ward, R. 1958. *The Australian Legend.* Melbourne: Oxford University Press.

Ward, V.M., J.T. Bertrand, and L.F. Brown. 1991. "The comparability of focus group and survey results: Three case studies." *Evaluation Review* 15 (2): 266–83.

Warf, B. 2011. "Geographies of global Internet censorship." *GeoJournal* 76 (1): 1–23.

Watson, A. 2012. "I crashed the boat and wept: Localizing the 'field' in critical geographic practice." *Emotion, Space and Society* 5 (3): 192–200.

Watson, A., and K. Till. 2010. "Ethnography and participant observation." In D. DeLyser, S. Herbert, S. Aitken, M. Crang, and L. McDowell, eds., *The Sage Handbook of Qualitative Geography*, 121–37. London: Sage.

Webb, B. 1982. "The art of note-taking." In R.G. Burgess, ed., *Field Research: A Sourcebook and Field Manual.* London: Allen and Unwin.

Weis, T. 2000. "Beyond peasant deforestation: Environment and development in rural Jamaica." *Global Environmental Change* 10: 299–305.

White, P., and P.A. Jackson. 1995. "(Re)theorising population geography." *International Journal of Population Geography* 1: 111–23.

Whyte, K.P. 2012. "Now this! Indigenous sovereignty, political obliviousness and governance models for SRM research." *Ethics, Policy & Environment* 15 (2): 172–187.

Whyte, W.F. 1982. "Interviewing in field research." In R.G. Burgess, ed., *Field Research: A Sourcebook and Field Manual.* London: Allen and Unwin.

Whyte, W.H. 1943. *Street Corner Society.* Chicago: University of Chicago Press.

Widdowfield, R. 2000. "The place of emotions in academic research." *Area* 32 (2): 199–208.

Wijnendaele, B.V. 2014. "The politics of emotion in participatory processes of empowerment and change." *Antipode* 46: 266–82.

Wilding, R. 2006. "'Virtual' Intimacies? Families Communicating Across Transnational Contexts." *Global Networks* 6 (2): 125–142.

Wiles, J. 2003. "Daily geographies of caregivers: Mobility, routine, scale." *Social Science and Medicine* 57: 1307–25.

Williams, C.C., and J. Round. 2007. "Re-thinking the nature of the informal economy: Some lessons from Ukraine." *International Journal of Urban and Regional Research* 31 (2): 425–41.

Williams, G., et al. 2003. "Enhancing pro-poor governance in eastern India: Participation, politics and action." *Progress in Development Studies* 3 (2): 159–78.

Williams, K., and C. Johnstone. 2000. "The politics of the selective gaze: Closed circuit television and the policing of public space." *Crime, Law and Society* 34: 183–210.

Williams, R. 1993. *Television: Technology and Cultural Form.* Middletown, CT.: Wesleyan University Press.

———. 1977. *Marxism and Literature.* New York: Oxford University Press.

Wilson, G.A. 2013. "Community resilience, social memory and the post-2010 Christchurch (New Zealand) earthquakes." *Area* 45 (2): 207–15.

Wilson, K., and E.J. Peters. 2005. "'You can make a place for it': Remapping urban First Nations spaces of identity." *Environment and Planning D: Society and Space* 23: 395–413.

Wilson, S. 2008. *Research Is Ceremony: Indigenous Research Methods.* Black Point, NS: Fernwood Pub.

Wilton, R., G. DeVerteuil, and J. Evans. 2014. "'No more of this macho bullshit': Drug treatment, place and the reworking of masculinity." *Transactions of the Institute of British Geographers*, 39(2): 291–303.

Winchester, H.P.M. 1996. "Ethical issues in interviewing as a research method in human geography." *Australian Geographer* 27 (1): 117–31.

———. 1999. "Lone fathers and the scales of justice: Renegotiating masculinity after divorce." *Journal of Interdisciplinary Gender Studies* 4 (2): 81–98.

Winchester, H.P.M., and L.N. Costello. 1995. "Living on the street: Social organisation and gender relations of Australian street kids." *Environment and Planning D: Society and Space* 13: 329–48.

Winchester, H.P.M., and K.M. Dunn. 1999. "Cultural geographies of film: Tales of urban reality." In K.J. Anderson and F. Gale, eds,

Inventing Places: Studies in Cultural Geography. Melbourne: Addison Wesley Longman.

Winchester, H.P.M., K.M. Dunn, and P.M. McGuirk. 1997. "Uncovering Carrington." In J. Moore, J. Ostwald, and A. Chawner, eds, *Hidden Newcastle: The Invisible City and the City of Memory*, 174–81. Arlington, VA: Gadfly Media.

Winchester, H.P.M., L. Kong, and K.M. Dunn. 2003. *Landscapes: Ways of Imagining the World*. Harlow: Pearsons.

Winchester, H.P.M., P.M. McGuirk, and K. Everett. 1999. "Celebration and control: Schoolies Week on the Gold Coast, Queensland." In E. Teather, ed., *Embodied Geographies: Spaces, Bodies and Rites of Passage*. London: Routledge.

Winchester, H.P.M., and M.W. Rofe. 2005. "Christmas in the 'Valley of Praise': Intersections of the rural idyll, heritage and community in Lobethal, South Australia." *Journal of Rural Studies* 21: 265–79.

Withers, C.W.J. 2010. *Geography and Science in Britain, 1831–1939*. Manchester: Manchester University Press.

Wolcott, H. 2001. *Writing up Qualitative Research*. 2nd edn. Newbury Park: Sage.

Wolf, D., ed. 1996. *Feminist Dilemmas in Fieldwork*. Boulder, CO: Westview Press.

Wolf, E. 1982. *Europe and the People without History*. Berkeley: University of California Press.

Women and Geography Study Group, eds. 1997. "Methods and methodologies in feminist research: Politics, practice and power." In *Feminist Geographies: Explorations in Diversity and Difference*. 86–111. London: Longmans.

Wood, L., and S. Williamson. 1996. *Consultants' Report on Franklin Square: Users, Activities and Conflicts*. Hobart: UNITAS Consulting.

Wood, L.A., and R.O. Kroger. 2000. *Doing Discourse Analysis: Methods for Studying Action in Talk and Text*. Thousand Oaks, CA: Sage.

Wood, N., M. Duffy, and S.J. Smith. 2007. "The art of doing (geographies of) music." *Environment and Planning D: Society and Space* 25 (5): 867–89.

Wood, N., and S. Smith. 2004. "Instrumental routes to emotional geographies." *Social and Cultural Geography* 5: 533–48.

Woodyer, T. 2012. "Ludic geographies: Not merely child's play." *Geography Compass* 6 (6): 313–26.

Wooldridge, S.W. 1955. "The status of geography and the role of fieldwork." *Geography* 40: 73–83.

Woon, C.Y. 2011. "'Protest is just a click away': Responses to the 2003 Iraq War on a bulletin board system in China." *Environment and Planning D: Society and Space* 29 (1): 131–49.

World Internet Project. 2013. *International Report*. Available: www.worldinternetproject.net. Accessed 3 April 2014.

Wright, S., K. Lloyd, S. Suchet-Pearson, L. Burarrwanga, M. Tofa, and Bawaka Country. 2012. "Telling stories in, through and with Country: engaging with Indigenous and more-than-human methodologies at Bawaka, NE Australia." *Journal of Cultural Geography* 29 (1): 39–60.

Wrigley, E.A. 1970. "Changes in the philosophy of geography." In R.J. Chorley and P. Haggett, eds, *Frontiers in Geographical Teaching*. London: Methuen.

Wylie, J.W. 2007. *Landscape*. New York: Routledge.

Yin, R. 2003. *Case Study Research: Design and Methods*. Los Angeles: Sage.

Young, L., and H. Barratt. 2001. "Adapting visual methods: Action research with Kampala street children." *Area* 33 (2): 141–52.

Zanotti, L. 2013. "Resistance and the politics of negotiation: Women, place and space among the Kayapó in Amazonia, Brazil." *Gender, Place & Culture: A Journal of Feminist Geography* 20 (3): 346–62.

Zeigler, D.J., S.D. Brunn, and J.H. Johnson. 1996. "Focusing on Hurricane Andrew through the eyes of the victims." *Area* 28 (2) 124–9.

Zelinsky, W. 1973. *Cultural Geography of the United States*. New York: Prentice Hall.

———. 2001. "The geographer as voyeur." *Geographical Review* 91 (1–2): 1–8.

Zimmerer, K.S., and T.J. Bassett, eds. 2003. *Political Ecology: An Integrative Approach to Geography and Environment–Development Studies*. New York: Guilford Press.

Index